Lecture Notes in Computer Science 10472

Commenced Publication in 1973
Founding and Former Series Editors:
Gerhard Goos, Juris Hartmanis, and Jan van Leeuwen

Advanced Research in Computing and Software Science
Subline of Lecture Notes in Computer Science

More information about this series at http://www.springer.com/series/7407

Ralf Klasing · Marc Zeitoun (Eds.)

Fundamentals of Computation Theory

21st International Symposium, FCT 2017
Bordeaux, France, September 11–13, 2017
Proceedings

 Springer

Editors
Ralf Klasing
CNRS
University of Bordeaux
Talence cedex
France

Marc Zeitoun
LaBRI
University of Bordeaux
Talence cedex
France

ISSN 0302-9743   ISSN 1611-3349   (electronic)
Lecture Notes in Computer Science
ISBN 978-3-662-55750-1   ISBN 978-3-662-55751-8   (eBook)
DOI 10.1007/978-3-662-55751-8

Library of Congress Control Number: 2017950046

LNCS Sublibrary: SL1 – Theoretical Computer Science and General Issues

Printed on acid-free paper

This Springer imprint is published by Springer Nature
The registered company is Springer-Verlag GmbH Germany
The registered company address is: Heidelberger Platz 3, 14197 Berlin, Germany

Preface

The 21st International Symposium on Fundamentals of Computation Theory (FCT 2017) took place during September 11–13, 2017 in Bordeaux, France.

The Symposium on Fundamentals of Computation Theory (FCT) was established in 1977 for researchers interested in all aspects of theoretical computer science, and in particular algorithms, complexity, and formal and logical methods. FCT is a biennial conference. Previous symposia have been held in Gdansk, Liverpool, Oslo, Wrocław, Budapest, and Lübeck.

The Program Committee (PC) of FCT 2017 received 99 submissions. Each submission was reviewed by at least three PC members and some trusted external referees, and evaluated on its quality, originality, and relevance to the symposium. The PC selected 29 papers, leading to an acceptance rate of 29%.

Four invited talks were given at FCT 2017, by Thomas Colcombet (CNRS, University of Paris-Diderot), Martin Dietzfelbinger (Technische Universität Ilmenau), Juraj Hromkovič (ETH Zürich), and Anca Muscholl (University of Bordeaux). There was also one invited talk in memoriam of Zoltán Ésik given by Jean-Éric Pin (CNRS, University of Paris-Diderot). This volume contains the papers of the five invited talks.

We thank the Steering Committee and its chair, Marek Karpinski, for giving us the opportunity to serve as the program chairs of FCT 2017, and for the responsibilities of selecting the Program Committee, the conference program, and publications.

The Program Committee selected two contributions for the best paper and the best student paper awards, sponsored by Springer and IDEX Bordeaux.

- The best paper award went to Albert Atserias, Phokion Kolaitis, and Simone Severini for their paper "Generalized Satisfiability Problems via Operator Assignments".
- The best student paper award was given to Matthias Bentert, Till Fluschnik, André Nichterlein, and Rolf Niedermeier for their paper "Parameterized Aspects of Triangle Enumeration".

We gratefully acknowledge additional financial support from the following institutions: University of Bordeaux, LaBRI, CNRS, Bordeaux INP, GIS Albatros, EATCS, Région Nouvelle-Aquitaine, and the French National Research Agency (ANR).

We would like to thank all the authors who responded to the call for papers, the invited speakers, the members of the Program Committee, the external referees, and—last but not least—the members of the Organizing Committee.

We would like to thank Springer for publishing the proceedings of FCT 2017 in their ARCoSS/LNCS series and for their support.

Finally, we acknowledge the help of the EasyChair system for handling the submission of papers, managing the review process, and generating these proceedings.

September 2017

Ralf Klasing
Marc Zeitoun

Organization

Steering Committee

Bogdan Chlebus	University of Colorado, USA
Marek Karpinski (Chair)	University of Bonn, Germany
Andrzej Lingas	Lund University, Sweden
Miklos Santha	CNRS and University Paris Diderot, France
Eli Upfal	Brown University, USA

Program Committee

Parosh Aziz Abdulla	Uppsala University, Sweden
Petra Berenbrink	University of Hamburg, Germany
Nathalie Bertrand	Inria, France
Benedikt Bollig	CNRS & ENS Cachan, Univ. Paris-Saclay, France
Patricia Bouyer-Decitre	CNRS & ENS Cachan, Univ. Paris-Saclay, France
Véronique Bruyère	University of Mons, Belgium
Arnaud Casteigts	University of Bordeaux, France
Hubie Chen	Univ. of the Basque Country and Ikerbasque, Spain
Colin Cooper	King's College London, UK
Kevin Costello	University of California at Riverside, USA
Jurek Czyzowicz	Université du Québec en Outaouais, Canada
Martin Dietzfelbinger	Ilmenau University of Technology, Germany
Robert Elsässer	University of Salzburg, Austria
Thomas Erlebach	University of Leicester, UK
Paola Flocchini	University of Ottawa, Canada
Pierre Fraigniaud	CNRS and University Paris Diderot, France
Luisa Gargano	University of Salerno, Italy
Sun-Yuan Hsieh	National Cheng Kung University, Taiwan
Stefan Kiefer	University of Oxford, UK
Ralf Klasing (Co-chair)	CNRS and University of Bordeaux, France
Dieter Kratsch	University of Lorraine, France
Manfred Kufleitner	University of Stuttgart, Germany
Fabian Kuhn	University of Freiburg, Germany
Thierry Lecroq	University of Rouen, France
Jérôme Leroux	CNRS and University of Bordeaux, France
Leo Liberti	CNRS and École Polytechnique, France
Markus Lohrey	University of Siegen, Germany
Frédéric Magniez	CNRS and University Paris Diderot, France
Wim Martens	University of Bayreuth, Germany
Pierre McKenzie	Université de Montréal, Canada
Madhavan Mukund	Chennai Mathematical Institute, India

Nicolas Nisse	Inria, Univ. Nice Sophia Antipolis, CNRS, France
Vangelis Th. Paschos	Paris Dauphine University, France
Joseph G. Peters	Simon Fraser University, Canada
Guido Proietti	Univ. of L'Aquila and IASI-CNR, Italy
Tomasz Radzik	King's College London, UK
R. Ramanujam	Institute of Mathematical Sciences, Chennai, India
Jean-François Raskin	Université Libre de Bruxelles, Belgium
José Rolim	University of Geneva, Switzerland
Sylvain Salvati	University of Lille, France
Maria Serna	Universitat Politècnica de Catalunya, Spain
Jean-Marc Talbot	Aix-Marseille University, France
Laurent Viennot	Inria, France
Gerhard Woeginger	RWTH Aachen, Germany
Marc Zeitoun (Co-chair)	University of Bordeaux, France

Organizing Committee

Arnaud Casteigts (Co-chair)	University of Bordeaux, France
Auriane Dantes	University of Bordeaux, France
Isabelle Garcia	University of Bordeaux, France
Ralf Klasing (Co-chair)	CNRS and University of Bordeaux, France
Sofian Maabout	University of Bordeaux, France
Yessin M. Neggaz	University Toulouse 1 Capitole, France
Antoine Rollet	Bordeaux INP, France
Marc Zeitoun	University of Bordeaux, France
Akka Zemmari	University of Bordeaux, France

External Reviewers

Adamatzky, Andrew
Adler, Isolde
Amano, Kazuyuki
Åman Pohjola, Johannes
Avrachenkov, Konstantin
Bampas, Evangelos
Bannach, Max
Bärtschi, Andreas
Bensmail, Julien
Bes, Alexis
Bienvenu, Laurent
Bilke, Andreas
Böckenhauer, Hans-Joachim
Bonamy, Marthe
Bournez, Olivier

Buchin, Kevin
Calderbank, Robert
Carton, Olivier
Cechlarova, Katarina
Chalopin, Jérémie
Chen, Li-Hsuan
Chen, Yu-Fang
Cheval, Vincent
Chistikov, Dmitry
Cordasco, Gennaro
Côté, Hugo
Cucker, Felipe
Dartois, Luc
Dvir, Zeev
Epstein, Leah

Fabre, Éric
Figueira, Diego
Fluschnik, Till
Friedetzky, Tom
Gajarský, Jakub
Gąsieniec, Leszek
Gawrychowski, Pawel
Georgatos, Konstantinos
Giannakos, Aristotelis
Gillis, Nicolas
Godard, Emmanuel
Godon, Maxime
Golovach, Petr
Grassl, Markus
Haase, Christoph
Hahn, Gena
Hansen, Kristoffer Arnsfelt
Hermo, Montserrat
Hertrampf, Ulrich
Hirvensalo, Mika
Hoffmann, Michael
Hojjat, Hossein
Hung, Ling-Ju
Jacobé De Naurois, Paulin
Jones, Mark
Kaaser, Dominik
Kachigar, Ghazal
Killick, Ryan
Klauck, Hartmut
Kling, Peter
Koiran, Pascal
Kong, Hui
Kopczynski, Eryk
Kosolobov, Dmitry
Kosowski, Adrian
Krebs, Andreas
Krnc, Matjaž
Lampis, Michael
Lavoie, Martin
Le Gall, Francois
Lenglet, Sergueï
Limaye, Nutan
Lin, Chuang-Chieh
Liśkiewicz, Maciej
Lodaya, Kamal
Lugiez, Denis

Manlove, David
Markham, Damian
Mayordomo, Elvira
Mazauric, Dorian
McDowell, Andrew P.
Michel, Pascal
Mittal, Rajat
Mnich, Matthias
Monmege, Benjamin
Morrill, Glyn
Mulzer, Wolfgang
Myoupo, Jean Frederic
Nederhof, Mark-Jan
Niewerth, Matthias
Ono, Hirotaka
Otterbach, Lena
Oum, Sang-Il
Pagourtzis, Aris
Pająk, Dominik
Patitz, Matthew
Paulusma, Daniel
Pighizzini, Giovanni
Pouly, Amaury
Renault, David
Rescigno, Adele
Reynier, Pierre-Alain
Rosamond, Frances
Salvail, Louis
Saurabh, Saket
Saxena, Nitin
Schaeffer, Luke
Schmidt, Johannes
Schneider, Jon
Schnoebelen, Philippe
Schwarzentruber, François
Scornavacca, Giacomo
Seidl, Helmut
Shen, Alexander
Siebertz, Sebastian
Sikora, Florian
Stamoulis, Georgios
Stephan, Frank
Suresh, S.P.
Tavenas, Sébastien
Tesson, Pascal
Toruńczyk, Szymon

Tzevelekos, Nikos
van Dijk, Tom
Variyam, Vinodchandran
Viglietta, Giovanni
Volk, Ben Lee
Walen, Tomasz

Wang, Hung-Lung
Watrigant, Rémi
Winslow, Andrew
Yedidia, Adam
Zémor, Gilles
Zhang, Shengyu

Sponsors

Abstracts of Invited Papers

Automata and Program Analysis

Thomas Colcombet[1], Laure Daviaud[2], and Florian Zuleger[3]

[1] IRIF, Case 7014, Université Paris Diderot, 75205 Paris Cedex 13, France
thomas.colcombet@irif.fr
[2] MIMUW, Banacha 2, 02-097 Warszawa, Poland
ldaviaud@mimuw.edu.pl
[3] Institut für Informationssysteme 184/4, Technische Universität Wien,
Favoritenstraße 9–11, 1040 Wien, Austria
zuleger@forsyte.at

Abstract. We show how recent results concerning quantitative forms of automata help providing refined understanding of the properties of a system (for instance, a program). In particular, combining the size-change abstraction together with results concerning the asymptotic behavior of tropical automata yields extremely fine complexity analysis of some pieces of code.

This abstract gives an informal, yet precise, explanation of why termination and complexity analysis are related to automata theory.

Optimal Dual-Pivot Quicksort: Exact Comparison Count

Martin Dietzfelbinger

Technische Universität Ilmenau, Fakultät für Informatik und Automatisierung,
Fachgebiet Komplexitätstheorie und Effiziente Algorithmen, P.O. Box 100565,
98684 Ilmenau, Germany
martin.dietzfelbinger@tu-ilmenau.de

Abstract. Quicksort, proposed by Hoare in 1961, is a venerable sorting algorithm - it has been thoroughly analyzed, it is taught in basic algorithms classes, and it is routinely used in practice. Can there be anything new about Quicksort today? Dual-pivot quicksort refers to variants of classical quicksort where in the partitioning step two pivots are used to split the input into three segments. Algorithms of this type had been studied by Sedgewick (1975) and by Hennequin (1991), with no further consequences. They received new attention starting from 2009, when a dual-pivot algorithm due to Yaroslavskiy, Bentley, and Bloch replaced the well-engineered quicksort algorithm in Oracle's Java 7 runtime library. An analysis of a variant of this algorithm by Nebel and Wild from 2012, where the two pivots are chosen randomly, showed there are about $1.9n \ln n$ comparisons on average for n input numbers. (Other works ensued. Standard quicksort has $2n \ln n$ expected comparisons. It should be noted that on modern computers parameters other than the comparison count will determine the running time.) In the center of the analysis is the partitioning procedure. Given two pivots, it splits the input keys in "small" (smaller than small pivot), "medium" (between the two pivots), "large" (larger than large pivot). We identify a partitioning strategy with the minimum average number of key comparisons in the case where the pivots are chosen from a random sample. The strategy keeps count of how many large and small elements were seen before and prefers the corresponding pivot. The comparison count is closely related to a "random walk" on the integers which keeps track of the difference of large and small elements seen so far. An alternative way of understanding what is going on is a Pólya urn with three colors. For the fine analysis it is essential to understand the expected number of times this random walk hits zero. The expected number of comparisons can be determined exactly and as a formula up to lower terms: It is $1.8n \ln n + 2.38..n + 1.675 \ln n + O(1)$. Extensions to larger numbers of pivots will be discussed.

Based on joint work with Martin Aumüller, Daniel Krenn, Clemens Heuberger, and Helmut Prodinger.

What One Has to Know When Attacking P vs. NP (Extended Abstract)

Juraj Hromkovič[1], and Peter Rossmanith[2]

[1] Department of Computer Science, ETH Zürich,
Universitätstrasse 6, 8092 Zürich, Switzerland
juraj.hromkovic@inf.ethz.ch

[2] Department of Computer Science, RWTH Aachen University. 52056 Aachen,
Germany
rossmani@cs.rwth-aachen.de

Abstract. Mathematics was developed as a strong research instrument with fully verifiable argumentations. We call any consistent and sufficiently powerful formal theory that enables to algorithmically verify for any given text whether it is a proof or not algorithmically verifiable mathematics (AV-mathematics for short). We say that a decision problem $L \subseteq \Sigma^*$ is almost everywhere solvable if for all but finitely many inputs $x \in \Sigma^*$ one can prove either "$x \in L$" or "$x \notin L$" in AV-mathematics.

First, we formalize Rice's theorem on unprovability, claiming that each nontrivial semantic problem about programs is not almost everywhere solvable in AV-mathematics. Using this, we show that there are infinitely many algorithms (programs that are provably algorithms) for which there do not exist proofs that they work in polynomial time or that they do not work in polynomial time. We can prove the same also for linear time or any time-constructible function.

Note that, if P $\neq$ NP is provable in AV-mathematics, then for each algorithm A it is provable that "A does not solve SATISFIABILITY or A does not work in polynomial time". Interestingly, there exist algorithms for which it is neither provable that they do not work in polynomial time, nor that they do not solve SATISFIABILITY. Moreover, there is an algorithm solving SATISFIABILITY for which one cannot prove in AV-mathematics that it does not work in polynomial time.

Furthermore, we show that P = NP implies the existence of algorithms X for which the true claim "X solves SATISFIABILITY in polynomial time" is not provable in AV-mathematics. Analogously, if the multiplication of two decimal numbers is solvable in linear time, one cannot decide in AV-mathematics for infinitely many algorithms X whether "X solves multiplication in linear time".

Finally, we prove that if P vs. NP is not solvable in AV-mathematics, then P is a proper subset of NP in the world of complexity classes based on algorithms whose behavior and complexity can be analyzed in AV-mathematics. On the other hand, if P = NP is provable, we can construct an algorithm that provably solves SATISFIABILITY almost everywhere in polynomial time.

A Tour of Recent Results on Word Transducers

Anca Muscholl

LaBRI, University of Bordeaux, Bordeaux, France
anca.muscholl@gmail.com

Abstract. Regular word transductions extend the robust notion of regular languages from acceptors to transformers. They were already considered in early papers of formal language theory, but turned out to be much more challenging. The last decade brought considerable research around various transducer models, aiming to achieve similar robustness as for automata and languages.

In this talk we survey some recent results on regular word transducers. We discuss how classical connections between automata, logic and algebra extend to transducers, as well as some genuine definability questions. For a recent, more detailed overview of the theory of regular word transductions the reader is referred to the excellent survey of E. Filiot and P.-A. Reynier (Siglog News 3, July 2016).

Based on joint work with Félix Baschenis, Olivier Gauwin and Gabriele Puppis. Work partially supported by the Institute of Advance Studies of the Technische Universität München and the project DeLTA (ANR-16-CE40-0007).

Some Results of Zoltán Ésik on Regular Languages

Jean-Éric Pin

IRIF, CNRS, Paris, France
`Jean-Eric.Pin@irif.fr`

Abstract. Zoltán Ésik published 2 books as an author, 32 books as editor and over 250 scientific papers in journals, chapters and conferences. It was of course impossible to survey such an impressive list of results and in this lecture, I will only focus on a very small portion of Zoltán's scientific work. The first topic will be a result from 1998, obtained by Zoltán jointly with Imre Simon, in which he solved a twenty year old conjecture on the shuffle operation. The second topic will be his algebraic study of various fragments of logic on words. Finally I will briefly describe some results on commutative languages obtained by Zoltán, Jorge Almeida and myself.

Contents

Invited Papers

Automata and Program Analysis 3
  Thomas Colcombet, Laure Daviaud, and Florian Zuleger

What One Has to Know When Attacking P vs. NP (Extended Abstract) 11
  Juraj Hromkovič and Peter Rossmanith

A Tour of Recent Results on Word Transducers 29
  Anca Muscholl

Some Results of Zoltán Ésik on Regular Languages 34
  Jean-Éric Pin

Contributed Papers

Contextuality in Multipartite Pseudo-Telepathy Graph Games 41
  Anurag Anshu, Peter Høyer, Mehdi Mhalla, and Simon Perdrix

Generalized Satisfiability Problems via Operator Assignments 56
  Albert Atserias, Phokion G. Kolaitis, and Simone Severini

New Results on Routing via Matchings on Graphs 69
  Indranil Banerjee and Dana Richards

Energy-Efficient Fast Delivery by Mobile Agents 82
  Andreas Bärtschi and Thomas Tschager

Parameterized Aspects of Triangle Enumeration 96
  Matthias Bentert, Till Fluschnik, André Nichterlein,
  and Rolf Niedermeier

Testing Polynomial Equivalence by Scaling Matrices 111
  Markus Bläser, B.V. Raghavendra Rao, and Jayalal Sarma

Strong Duality in Horn Minimization 123
  Endre Boros, Ondřej Čepek, and Kazuhisa Makino

Token Jumping in Minor-Closed Classes 136
  Nicolas Bousquet, Arnaud Mary, and Aline Parreau

Expressive Power of Evolving Neural Networks Working on Infinite
Input Streams   150
   Jérémie Cabessa and Olivier Finkel

Minimal Absent Words in a Sliding Window and Applications
to On-Line Pattern Matching   164
   Maxime Crochemore, Alice Héliou, Gregory Kucherov,
   Laurent Mouchard, Solon P. Pissis, and Yann Ramusat

Subquadratic Non-adaptive Threshold Group Testing.   177
   Gianluca De Marco, Tomasz Jurdziński, Michał Różański,
   and Grzegorz Stachowiak

The Snow Team Problem: (Clearing Directed Subgraphs
by Mobile Agents)   190
   Dariusz Dereniowski, Andrzej Lingas, Mia Persson, Dorota Urbańska,
   and Paweł Żyliński

FO Model Checking on Map Graphs.   204
   Kord Eickmeyer and Ken-ichi Kawarabayashi

Multiple Context-Free Tree Grammars and Multi-component
Tree Adjoining Grammars   217
   Joost Engelfriet and Andreas Maletti

On $\Sigma \wedge \Sigma \wedge \Sigma$ Circuits: The Role of Middle Σ Fan-In, Homogeneity
and Bottom Degree.   230
   Christian Engels, B.V. Raghavendra Rao, and Karteek Sreenivasaiah

Decidable Weighted Expressions with Presburger Combinators   243
   Emmanuel Filiot, Nicolas Mazzocchi, and Jean-François Raskin

The Complexity of Routing with Few Collisions.   257
   Till Fluschnik, Marco Morik, and Manuel Sorge

Parikh Image of Pushdown Automata   271
   Pierre Ganty and Elena Gutiérrez

Tropical Combinatorial Nullstellensatz and Fewnomials Testing   284
   Dima Grigoriev and Vladimir V. Podolskii

On Weak-Space Complexity over Complex Numbers   298
   Pushkar S. Joglekar, B.V. Raghavendra Rao, and Siddhartha Sivakumar

Deterministic Oblivious Local Broadcast in the SINR Model   312
   Tomasz Jurdziński and Michał Różański

Undecidability of the Lambek Calculus with Subexponential
and Bracket Modalities   326
  Max Kanovich, Stepan Kuznetsov, and Andre Scedrov

Decision Problems for Subclasses of Rational Relations over Finite
and Infinite Words   341
  Christof Löding and Christopher Spinrath

Listing All Fixed-Length Simple Cycles in Sparse Graphs
in Optimal Time   355
  George Manoussakis

Reliable Communication via Semilattice Properties of Partial Knowledge. . . .   367
  Aris Pagourtzis, Giorgos Panagiotakos, and Dimitris Sakavalas

Polynomial-Time Algorithms for the Subset Feedback Vertex
Set Problem on Interval Graphs and Permutation Graphs   381
  Charis Papadopoulos and Spyridon Tzimas

Determinism and Computational Power of Real Measurement-Based
Quantum Computation   395
  Simon Perdrix and Luc Sanselme

Busy Beaver Scores and Alphabet Size   409
  Holger Petersen

Automatic Kolmogorov Complexity and Normality Revisited   418
  Alexander Shen

Author Index   431

Invited Papers

Automata and Program Analysis

Thomas Colcombet[1](✉), Laure Daviaud[2], and Florian Zuleger[3]

[1] IRIF, Case 7014 Université Paris Diderot, 75205 Paris Cedex 13, France
`thomas.colcombet@irif.fr`
[2] MIMUW, Banacha 2, 02-097 Warszawa, Poland
`ldaviaud@mimuw.edu.pl`
[3] Institut für Informationssysteme 184/4, Technische Universität Wien,
Favoritenstraße 9–11, 1040 Wien, Austria
`zuleger@forsyte.at`

Abstract. We show how recent results concerning quantitative forms of automata help providing refined understanding of the properties of a system (for instance, a program). In particular, combining the size-change abstraction together with results concerning the asymptotic behavior of tropical automata yields extremely fine complexity analysis of some pieces of code.

This abstract gives an informal, yet precise, explanation of why termination and complexity analysis are related to automata theory.

1 Program Analysis and Termination

Program analysis is concerned with the automatic inference of properties of a chunk of code (or of a full program). Such analysis may serve many purposes, such as guaranteeing that some division by zero cannot occur in the execution of a program, or that types are properly used (when the language is not statically typed), or that there is no memory leakage, etc. Here, to start with, we are concerned with termination analysis, *i.e.*, proving that all executions of a program (or pieces of a program) eventually halt.

Consider the following *code C*:

```
void f() {
  uint x,y;
  x = read_input();
  y = read_input();
  while (x > 0) {
    if (y > 0) { // branch a
      y--;
    }
    else { // branch b
      x--;
      y = read_input();
    }
  }
}
```

© Springer-Verlag GmbH Germany 2017
R. Klasing and M. Zeitoun (Eds.): FCT 2017, LNCS 10472, pp. 3–10, 2017.
DOI: 10.1007/978-3-662-55751-8_1

It is clear for a human being that this piece of code eventually *terminates* whatever are the input values read during its execution. The question is how this can be automatically inferred? Of course, such a problem is in general undecidable unless restrictions are assumed (using variations around the theorem of Rice). The path we follow here consists in approximating the behavior of this code using size-change abstraction. We shall see that this abstraction transforms the code into a formalism, the size-change abstraction, for which termination is decidable.

2   Size-Change Abstraction

The *size-change abstraction* amounts to abstract a piece of code in the following manner:

- We identify some *size-change variables* that are considered of interest, that range over non-negative integers (in our example x and y). In general, size-change variables can represent any norm on the program state (a function which maps the state to the non-negative integers), such as the length of a list, the height of a tree, the sum of two non-negative variables, etc.
- We construct the *control-flow graph* (possibly simplified) of the code: vertices are positions in the code, and *edges* are steps of computations. We also identify some *entry* and *exit vertices* of the graph as one can expect.
- We abstract tests, *i.e.*, we replace all tests by non-determinism. This means that we consider possible executions independently of whether the tests in `if` statements or in `while` loops are true or not.
- Finally, each edge of the control-flow graph is labeled by *guards* expressing how the values of the size-change variables may evolve while taking the edge. The language for these relations is very restricted: it consists of a conjunction of properties of the form $x \geqslant y'$ or $x > y'$ where $x, y, \ldots$ represent the value of the variables before the edge is taken, while $x', y', \ldots$ represent the value of the variables after the edge is taken. We add guards *conservatively* in order to ensure the correctness of the abstraction, *i.e.*, we only add a $x \geqslant y'$ to some edge guard if we can guarantee that the value of y is not greater than the value of x before edge is taken, and similarly for $x > y'$.

For the code $\mathcal{C}$, we obtain the following size-change abstraction $\mathcal{S}$:

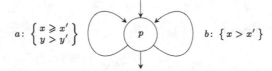

We comment on the size-change abstraction $\mathcal{S}$ of code $\mathcal{C}$. Here, we only consider one position in the code, which is the beginning of the `while` loop. This needs not be the case in general. The two edges a and b correspond respectively to executing the `if` branch and the `else` branch of code $\mathcal{C}$. In the first case (edge a), the value of y strictly decreases while the value of x does not change. In the second case (edge b), the value of x strictly decreases while we have no information about the value that y might take after the transition (because of the 'y = read_input()' in the code). We see that the two transitions a and b of S are an *abstraction* of the branches a and b of C: If the code C executes branch a resp. b and the variable values (x, y) change to some (x', y'), then transition a resp. b of S also allows the variable values (x, y) to change to (x', y'). This has the following important consequence for termination analysis: Every execution of C is also an execution of S. Thus, if we can show that S terminates, then we can deduce the termination of C.

In the following we explain how to reason about the termination of the size-change abstraction S.

In order to be precise, we have to define the semantics of the model. We call *execution path* of the size-change abstraction a sequence of edges that (1) starts in an entry vertex, and (2) is formed of compatible edges, meaning that for any two consecutive edges in the sequence, the target of the first one should coincide with the source of the second one. Such an execution path is *halting* if it is finite and ends in an exit vertex. Such a definition of an execution path does not yet capture the semantics of variables. For this, we shall consider the traces that realize an execution. Formally, a *trace* of the size-change abstraction is a sequence of *configurations* consisting of a vertex and a valuation of the variables by non-negative integers, that respect the transitions of the size-change abstraction. This is best seen in an example. Consider the execution path *aabaabaa*. One possible trace that realizes this execution is the following one (where the second component represents the value of the variable x and the third one the value of the variable y):

$$(p, 2, 2) \xrightarrow{a} (p, 2, 1) \xrightarrow{a} (p, 2, 0) \xrightarrow{b} (p, 1, 2) \xrightarrow{a} (p, 1, 1) \xrightarrow{a} (p, 1, 0) \xrightarrow{b} (p, 0, 2) \xrightarrow{a} (p, 0, 1) \xrightarrow{a} (p, 0, 0)$$

We say that the *size-change abstraction terminates* if every execution path that is realized by a trace is finite. As argued above, this implies that the program also terminates since all the executions of the program are captured by the abstraction. Of course, the converse does not hold, in particular because we threw away a good part of the original semantics, and thus many reasons for the program to terminate are not recovered.

3 Max-Plus Automata

The above definition of size-change abstraction does not make the termination property obviously decidable, yet. What we show now is how this question can be reduced to a problem of universality in automata theory. The key concept behind this last reduction is that the pattern that prevents an execution path to be realizable would be an infinite sequence of variables that are related by the guards via (non-necessarily stricts) inequalities, and infinitely many times by strict inequalities (finite sequences of arbitrarily large number of strict inequalities would also be a witness in some cases). Indeed, such a sequence would mean the existence in a trace of an infinite decreasing sequence of non-negative integers; a contradiction. For instance, the following infinite execution path of the size-change abstraction:

$$\overbrace{a \quad \cdots \quad a}^{n_0 \text{ times}} b \overbrace{a \quad \cdots \quad a}^{n_1 \text{ times}} b \overbrace{a \quad \cdots \quad a}^{n_2 \text{ times}} b \cdots$$

is impossible since the a edge contains the guard $x \leqslant x'$, and the b relation $x < x'$. Hence, the following infinite sequence of relations is impossible:

$$x_0 \overbrace{\leqslant x_1 \leqslant \cdots \leqslant x_{n_0-1}}^{n_0 \text{ inequalities}} < x_{n_0} \overbrace{\leqslant x_{n_0+1} \leqslant \cdots \leqslant x_{n_0+n_1}}^{n_1 \text{ inequalities}} < x_{n_0+n_1+1} \leqslant \quad \cdots \quad ,$$

in which x_i accounts for the value assumed by variable x at time i. No valuation of the x_i's by *non-negative integers* can fulfill these constraints. For similar reasons, an infinite execution path that would eventually consist only of the edge b would be impossible, this time because of the variable y.

We shall now define an automaton model that can measure these "bad sequences" of dependencies: max-plus automata[1].

We will define a *max-plus automaton* that is able to "count" the maximal length of such sequences of strict inequalities. The recipe is the following:

- The *input alphabet* is the set of edges of the size-change abstraction.
- The *states* of the automaton are the size-change variables, plus two extra states, called $\top$ and $\bot$.
- The *transitions* of the automaton are labeled by the edges of the size-change abstraction, and there is a transition from state x to state y labeled by δ if the guard of edge δ contains either '$x \geqslant y'$' or '$x > y'$'; Furthermore the state $\top$ is the origin of all possible transitions to every state (including itself), and $\bot$ is the target all possible transitions originating from any states (including itself).
- Some transitions are marked *special* (or *costly*): the ones that arise from the case '$x > y'$'.[2]
- All states are marked both *initial* and *final* (the feature of initial and final states, which is important in the theory of tropical automata, happens to be irrelevant for this application).

In our case, this yields the automaton below, in which the $*$ symbol means 'all possible labels', double arrows identify costly transitions, and we omitted to represent initial and final states:

---

[1] Though we take the principle, we do not use the standard notation of max-plus automata, which are traditionally defined as automata weighted over the max-plus semiring.

[2] In the standard terminology, non-costly transitions would be given *weight* 0, while costly ones would be attributed weight 1.

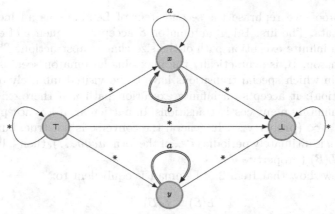

In this automaton, a sequence of inequalities over size-change variables can be witnessed by a path (formally, a run). Costly transitions correspond to strict inequalities in the guards.

The semantics of this max-plus automaton is to count the number of costly transitions and to maximize this cost among all runs. Formally, an *input* of the automaton is a sequence of edges of the size-change abstraction, such as the word *aaabaabaa* (it is not necessarily an execution path so far). A priori, we consider both finite and infinite sequences of edges. A *run* of the automaton over input u is a sequence of transitions which forms a path in the graph of the automaton, and it is accepting if either it is infinite and starts in an initial state or it is finite, starts in an initial state and ends in a final one. The max-plus automaton $\mathcal{A}$ can be used to compute a quantity given some input word u:

$$[\![\mathcal{A}]\!](u) = \sup\{\mathsf{cost}(\rho) \mid \rho \text{ accepting run over the input } u\} \in \mathbb{N} \cup \{\infty\},$$

where $\mathsf{cost}(\rho) =$ number of costly transitions in ρ.

For instance, over the input $u = aaabaabaa$, $[\![\mathcal{A}]\!](u) = 3$. It corresponds to a run (there are in fact several of them) that assumes state y during the first three letters of the word. This is the maximal one, since in our example the automaton computes the maximum of the number of b-edges with the maximum of the longest block of consecutive a-edges.

The following lemma formalizes the correction of this reduction to automata:

Lemma 1. *The following properties are equivalent:*

1. The size-change abstraction terminates.
2. All infinite execution paths u satisfy $[\![\mathcal{A}]\!](u) = \infty$.

It happens that Item 2 of Lemma 1 is decidable, from which we get:

Corollary 1 [9,10]**.** *The termination of size-change abstractions is decidable.*

Let us establish the decidability of Item 2. This requires some knowledge about Büchi automata theory. The reader may as well proceed to the next section.

In this proof, we rephrase the second item of Lemma 1 as an inclusion of Büchi automata. The first Büchi automaton $\mathcal{E}$ accepts a sequence of edges if it forms a valid infinite execution path of the size-change abstraction. The second Büchi automaton, $\mathcal{B}$, is syntactically the max-plus automaton seen as a Büchi automaton, in which special transitions have to be visited infinitely often (the Büchi condition): it accepts an infinite execution path u if there exists a run containing infinitely many costly transitions. In particular, if u is accepted by $\mathcal{B}$, then $[\![\mathcal{A}]\!](u) = \infty$ (property $\star$). In general the converse is not true, but one can check that if u is ultimately periodic (i.e., of the form $uvvv\dots$), then $[\![\mathcal{A}]\!](u) = \infty$ implies $u \in L(\mathcal{B})$ (property $\star\star$).

Let us now show that Item 2 of Lemma 1 is equivalent to:

$$L(\mathcal{E}) \subseteq L(\mathcal{B}).$$

Indeed, $L(\mathcal{E}) \subseteq L(\mathcal{B})$ means that all infinite execution paths u are accepted by $\mathcal{B}$, and thus $[\![\mathcal{A}]\!](u) = \infty$ by $\star$. For the converse direction, assume that $L(\mathcal{E}) \subseteq L(\mathcal{B})$ does not hold, i.e., there is an input that is accepted by $\mathcal{E}$ but not by $\mathcal{B}$. It is known from Büchi that in this case there exists such an input u which is ultimately periodic. By $\star\star$, this means that $[\![\mathcal{A}]\!](u)$ is finite, contradicting Item 2 of Lemma 1.

4   Complexity Analysis

Counting the number of costly transitions also gives an idea of the worst-case complexity of the program. Indeed, a possible execution of the program corresponds to an execution path of the size-change abstraction that is realized by a trace, and the time complexity of the execution is nothing but the length of the execution path.

To put this idea in action, let us consider the following slightly modified code:

```
void f(uint n) {
  uint x,y;
  x = read_input(n);
  y = read_input(n);
  while (x > 0) {
    if (y > 0) { // branch a
      y--;
    }
    else { // branch b
      x--;
      y = read_input(n);
    }
  }
}
```

This new code takes a non-negative integer as input, and the `read_input(n)` calls now guarantee that the value produced is in the interval $\{0, \dots, n\}$. A careful look at this code reveals that it terminates in $O(n^2)$ steps. We would like the analysis to reach this level of precision.

In fact, everything is similar to the termination case we have explained up to now. The only change is that the values of the variables are implicitly ranging over the interval $\{0, \ldots, n\}$. Under this assumption, the size-change abstraction also terminates within a quadratic bound.

The reduction to a max-plus automaton also remains valid, as shown by this variation around the ideas of Lemma 1:

Lemma 2. *The following properties are equivalent for all n and k:*

- *The size-change abstraction terminates within time bound k, assuming the variable values range in $\{0, \ldots, n\}$.*
- *All execution paths u such that $[\![\mathcal{A}]\!](u) \leqslant n$ have length at most k.*

However, we need now a much more delicate result of automata theory than the inclusion of Büchi automata. Here follows what we can do:

Theorem 1 [7]. *One can effectively compute, given as input a max-plus automaton $\mathcal{A}$, the value*

$$\liminf_{|u| \to \infty} \frac{\log([\![\mathcal{A}]\!](u) + 1)}{\log |u|}$$

which happens to be a rational in $[0, 1]$ or ∞.

Now, as a corollary, we get:

Theorem 2 [7]. *The length of the longest execution path realized by a trace in a size-change abstraction is of order $\Theta(n^\alpha)$ if the variables are restricted to take values in $[0, n]$, where $\alpha \geqslant 1$ is a rational number. Moreover, there is an algorithm that given a terminating size-change abstraction computes such an α.*

It can also be proved that all the rationals $\alpha \geq 1$ can be achieved by a given size-change abstraction.

5   Related Work

The goal of this paper was to illustrate the size-change abstraction (SCA), which is a popular technique for automated termination analysis. The last decade has seen considerable interest in automated techniques for proving the termination of programs. In this short paper we limit ourselves to describing the related work on SCA. SCA has been introduced by Lee et al. [10]. SCA is employed for the termination analysis of functional [10,11], logical [12] and imperative [1, 6] programs, term rewriting systems [5], and is implemented in the industrial-strength systems ACL2 [11] and Isabelle [8]. Recently, SCA has also been used for resource bound and complexity analysis of imperative programs [14], which motivated the results on complexity analysis presented in this paper. SCA is an attractive domain for an automated analysis because of several strong theoretical results on termination analysis [10], complexity analysis [7,13] and the extraction of ranking functions [3,13]. Further research has investigated the generalization of size-change constraints to richer classes of constraints, including difference constraints [2], gap-order constraints [4] and monotonicity constraints [3].

References

1. Anderson, H., Khoo, S.-C.: Affine-based size-change termination. In: Ohori, A. (ed.) APLAS 2003. LNCS, vol. 2895, pp. 122–140. Springer, Heidelberg (2003). doi:10.1007/978-3-540-40018-9_9

2. Ben-Amram, A.M.: Size-change termination with difference constraints. ACM Trans. Program. Lang. Syst. **30**(3), 16 (2008)

3. Ben-Amram, A.M.: Monotonicity constraints for termination in the integer domain. Logical Methods Comput. Sci. **7**(3), 1–43 (2011)

4. Bozzelli, L., Pinchinat, S.: Verification of gap-order constraint abstractions of counter systems. In: Kuncak, V., Rybalchenko, A. (eds.) VMCAI 2012. LNCS, vol. 7148, pp. 88–103. Springer, Heidelberg (2012). doi:10.1007/978-3-642-27940-9_7

5. Codish, M., Fuhs, C., Giesl, J., Schneider-Kamp, P.: Lazy abstraction for size-change termination. In: Fermüller, C.G., Voronkov, A. (eds.) LPAR 2010. LNCS, vol. 6397, pp. 217–232. Springer, Heidelberg (2010). doi:10.1007/978-3-642-16242-8_16

6. Codish, M., Gonopolskiy, I., Ben-Amram, A.M., Fuhs, C., Giesl, J.: Sat-based termination analysis using monotonicity constraints over the integers. TPLP **11**(4–5), 503–520 (2011)

7. Colcombet, T., Daviaud, L., Zuleger, F.: Size-change abstraction and max-plus automata. In: Csuhaj-Varjú, E., Dietzfelbinger, M., Ésik, Z. (eds.) MFCS 2014. LNCS, vol. 8634, pp. 208–219. Springer, Heidelberg (2014). doi:10.1007/978-3-662-44522-8_18

8. Krauss, A.: Certified size-change termination. In: Pfenning, F. (ed.) CADE 2007. LNCS (LNAI), vol. 4603, pp. 460–475. Springer, Heidelberg (2007). doi:10.1007/978-3-540-73595-3_34

9. Lee, C.S.: Ranking functions for size-change termination. ACM Trans. Program. Lang. Syst. **31**(3), 10:1–10:42 (2009)

10. Lee, C.S., Jones, N.D., Ben-Amram, A.M.: The size-change principle for program termination. In: POPL, pp. 81–92 (2001)

11. Manolios, P., Vroon, D.: Termination analysis with calling context graphs. In: Ball, T., Jones, R.B. (eds.) CAV 2006. LNCS, vol. 4144, pp. 401–414. Springer, Heidelberg (2006). doi:10.1007/11817963_36

12. Vidal, G.: Quasi-terminating logic programs for ensuring the termination of partial evaluation. In: PEPM, pp. 51–60 (2007)

13. Zuleger, F.: Asymptotically precise ranking functions for deterministic size-change systems. In: Beklemishev, L.D., Musatov, D.V. (eds.) CSR 2015. LNCS, vol. 9139, pp. 426–442. Springer, Cham (2015). doi:10.1007/978-3-319-20297-6_27

14. Zuleger, F., Gulwani, S., Sinn, M., Veith, H.: Bound analysis of imperative programs with the size-change abstraction. In: Yahav, E. (ed.) SAS 2011. LNCS, vol. 6887, pp. 280–297. Springer, Heidelberg (2011). doi:10.1007/978-3-642-23702-7_22

What One Has to Know When Attacking P vs. NP (Extended Abstract)

Juraj Hromkovič[1]($\boxtimes$) and Peter Rossmanith[2]($\boxtimes$)

[1] Department of Computer Science, ETH Zürich,
Universitätstrasse 6, 8092 Zürich, Switzerland
juraj.hromkovic@inf.ethz.ch
[2] Department of Computer Science, RWTH Aachen University, 52056 Aachen,
Germany
rossmani@cs.rwth-aachen.de

Abstract. Mathematics was developed as a strong research instrument with fully verifiable argumentations. We call any consistent and sufficiently powerful formal theory that enables to algorithmically verify for any given text whether it is a proof or not algorithmically verifiable mathematics (AV-mathematics for short). We say that a decision problem $L \subseteq \Sigma^*$ is almost everywhere solvable if for all but finitely many inputs $x \in \Sigma^*$ one can prove either "$x \in L$" or "$x \notin L$" in AV-mathematics.

First, we formalize Rice's theorem on unprovability, claiming that each nontrivial semantic problem about programs is not almost everywhere solvable in AV-mathematics. Using this, we show that there are infinitely many algorithms (programs that are provably algorithms) for which there do not exist proofs that they work in polynomial time or that they do not work in polynomial time. We can prove the same also for linear time or any time-constructible function.

Note that, if P $\neq$ NP is provable in AV-mathematics, then for each algorithm A it is provable that "A does not solve SATISFIABILITY or A does not work in polynomial time". Interestingly, there exist algorithms for which it is neither provable that they do not work in polynomial time, nor that they do not solve SATISFIABILITY. Moreover, there is an algorithm solving SATISFIABILITY for which one cannot prove in AV-mathematics that it does not work in polynomial time.

Furthermore, we show that P = NP implies the existence of algorithms X for which the true claim "X solves SATISFIABILITY in polynomial time" is not provable in AV-mathematics. Analogously, if the multiplication of two decimal numbers is solvable in linear time, one cannot decide in AV-mathematics for infinitely many algorithms X whether "X solves multiplication in linear time".

Finally, we prove that if P vs. NP is not solvable in AV-mathematics, then P is a proper subset of NP in the world of complexity classes based on algorithms whose behavior and complexity can be analyzed in AV-mathematics. On the other hand, if P = NP is provable, we can construct an algorithm that provably solves SATISFIABILITY almost everywhere in polynomial time.

© Springer-Verlag GmbH Germany 2017
R. Klasing and M. Zeitoun (Eds.): FCT 2017, LNCS 10472, pp. 11–28, 2017.
DOI: 10.1007/978-3-662-55751-8_2

1   Introduction

Mathematics was developed as a special language in which each word and thus each sentence has a clear, unambiguous meaning, at least for anybody who mastered this language. The goal was not only to communicate with unambiguous interpretations, but to create a powerful research instrument that enables everybody to verify any claim formulated in this language.

This way, experiments and mathematics became the main tools for discovering the world and for creating our technical world. The dream of Leibniz was to develop such a formal language, in which almost every problem can be formulated and successfully analyzed by a powerful calculus (thus his famous words "Let us calculate, without further ado, to see who is right."). After introducing logic as a calculus for verifying the validity of claims and proofs, there was hope to create mathematics as a perfect research instrument (see, for instance, Hilbert [5]). In 1930, Gödel [4] showed that mathematics will never be perfect, that is, that the process of increasing the power of mathematics as a research instrument is infinite. An important fact is that, in each nontrivial mathematics based on finitely many axioms, one can formulate claims in the language of mathematics whose validity cannot be verified inside the same mathematics.

Since the introduction of the concept of computational complexity, computer scientists have not been able to prove nontrivial lower bounds on the complexity of concrete problems. For instance, we are unable to prove that the multiplication of two decimal numbers cannot be computed in linear time, that matrix multiplication cannot be computed in $O(n^2)$ time, or that reachability cannot be solved in logarithmic space. In this paper, we strive to give an explanation for this trouble by showing that the concept of computational complexity may be too complex for being successfully mastered by mathematics. In particular, this means that open problems like P vs. NP or DLOG vs. NLOG could be too hard to be investigated inside of current mathematics. In fact, we strive to prove results about the unprovability of some mathematical claims like Gödel [4] did. However, the difference is that we do not focus on meta-statements about mathematics itself, but on concrete fundamental problems of complexity theory that are open for more than 40 years. Interestingly, fundamental contributions in this direction were made by Baker et al. [2], who showed that proof techniques that are sensible to relativization cannot help to solve the P vs. NP problem, and by Razborov and Rudich [8], who showed that natural proofs covering all proof techniques used in complexity theory cannot help to prove P $\neq$ NP. Aaronson gives an excellent survey on this topic [1]. Here, we first prove that there are infinitely many algorithms whose asymptotic time complexity or space complexity cannot be analyzed in mathematics. Our results pose the right questions. How hard could it be to prove a superlinear lower bound on multiplication of two decimal numbers (that is, the nonexistence of linear time algorithms for multiplication) if there exist algorithms for which mathematics cannot find out whether they work in linear time or not, or even recognize what they really do? Similarly, we can discuss SATISFIABILITY and polynomial time, or REACHABILITY and logarithmic space.

In what follows, we focus on the unprovability of theorems in AV-mathematics. First, we reformulate Rice's theorem [9] about the undecidability of nontrivial semantic problems about programs (Turing machines) to unprovability. More precisely, we say that a decision problem $L \subseteq \Sigma^*$ is *almost everywhere solvable* in AV-mathematics if for all but finitely many inputs $x \in \Sigma^*$, one can prove either "$x \in L$" or "$x \notin L$" in AV-mathematics. Here, we prove that each nontrivial semantic problem about programs is not almost everywhere solvable. This means, for instance, that there exist infinitely many programs for which one cannot prove whether they compute a constant function or not. Note that this has a deep consequence for our judgement about undecidability. Originally one was allowed to view the existence of an algorithm as a reduction of the infinity of the problem (given by its infinite set of problem instances) to finiteness (given by the finite description of the algorithm). In this sense, one can view the undecidability of a problem as the impossibility to reduce the infinite variety of a problem to a finite size. Here, we see another true reason. If, for particular problem instances, one cannot discover in AV-mathematics what the correct output is, then, for sure, there does not exist any provably correct algorithm for the problem.

We use Rice's theorem on unprovability as the first step for establishing the hardness of the analysis of computational complexity in AV-mathematics. We will succeed to switch from programs to algorithms as inputs,[1] and ask which questions about algorithms are not solvable almost everywhere. We prove results such as

(i) for each time-constructible function $f(n) \geq n$, the problem whether a given algorithm works in time $O(f(n))$ is not almost everywhere solvable in AV-mathematics, and

(ii) the problem whether a given algorithm solves SATISFIABILITY (REACHABILITY, multiplication of decimal numbers, etc.) is not almost everywhere solvable in AV-mathematics.

Particularly, this also means that there are infinitely many algorithms for which one cannot distinguish in AV-mathematics whether they work in polynomial time or not. Note that this is essential because, if P $\neq$ NP is provable in AV-mathematics, then, for each algorithm A, the statement

"A does not work in polynomial time
or A does not solve SATISFIABILITY"   (*)

would be provable in AV-mathematics.

In this paper, we show that there exist algorithms, for which it is neither provable that they do not work in polynomial time nor that they do not recognize SATISFIABILITY. Moreover, we show that if P = NP, then there would exist algorithms for which (*) is not provable.

---

[1] This means that one has a guarantee that a given program is an algorithm, or even a proof that the given program is an algorithm may be part of the input.

Finally, we show that if P = NP is not provable in AV-mathematics, then $P_{ver} \neq NP_{ver}$ where P_{ver} and NP_{ver} are counterparts of P and NP in the world of algorithms that can be analyzed in AV-mathematics. On the other hand, if P = NP is provable, then there exists a constructive proof of this fact: There is a concrete algorithm that provably solves an NP-complete problem in polynomial time.

We do not present the shortest way of proving our results. We present here the genesis of our ideas to reach this goal. This is not only better for a deeper understanding of the results and proofs, but also for deriving several interesting byproducts of interest. Note that combining some ideas from the following chapters with some fundamental theorems from ZFC, one can get shorter proofs for some of our results in ZFC if ZFC is consistent.

2   Rice's Theorem on Unprovability

The starting point for our unprovability results is the famous theorem of Chaitin [3], stating that one can discover the Kolmogorov complexity[2] for at most finitely many binary strings. Let, for each binary string $w \in \{0,1\}^*$, $K(w)$ denote the Kolmogorov complexity of w. As the technique is repeatedly used in this paper, we prefer to present our version of this theorem as well as a specific proof. In what follows, let Σ_{math} be an alphabet in which any mathematical proof can be written. Furthermore, let λ denote the empty word.

Theorem 1 (Chaitin [3]). *There exists $d \in \mathbb{N}$ such that, for all $n \geq d$ and all $x \in \{0,1\}^*$, there does not exist any proof in AV-mathematics of the fact "$K(x) \geq n$".*

Proof. Let us prove Theorem 1 by contradiction. Suppose there exists an infinite sequence of natural numbers $\{n_i\}_{i=1}^{\infty}$ with $n_i < n_{i+1}$ for $i = 1, 2, \ldots$ such that for each n_i, there exists a proof in AV-mathematics of the claim

$$\text{"}K(w_i) \geq n_i\text{"}$$

for some $w_i \in \{0,1\}^*$. If, for some i, there exist several such proofs for different w_i's, let w_i be the word with the property that the proof of "$K(w_i) \geq n_i$" is the first one with respect to the canonical order of the proofs. Then we design an infinite sequence $\{A_i\}_{i=1}^{\infty}$ of algorithms as follows:

---

[2] Recall that the Kolmogorov complexity of a binary string w is the length of the binary code of the shortest program (in some fixed programming language with a compiler) generating w [6,7].

```
A_i:    Input: n_i
        Output: w_i
        begin
            x := λ;
            repeat
                verify algorithmically whether x is a proof of "K(w) ≥ n_i"
                    for some w ∈ {0,1}*;
                if x is a proof of "K(w) ≥ n_i" then
                    output(w); exit;
                else
                    x := successor of x in Σ*_math in the canonical order;
                end
            end
        end
```

Obviously, A_i generates w_i. All the algorithms A_i are identical except for the number n_i. Hence, there exists a constant c such that each A_i can be described by
$$c + \lceil \log_2(n_i + 1) \rceil$$
bits. This way we get, for all $i \in \mathbb{N}$, that

$$n_i \leq K(w_i) \leq c + \lceil \log_2(n_i + 1) \rceil. \tag{1}$$

However, (1) can clearly hold for at most finitely many different $i \in \mathbb{N}$, and so we got a contradiction. □

In what follows, we use the terms "program" and "Turing machine" (TM) as synonyms. Likewise, we use the terms "algorithm" and "Turing machine that always halts" as synonyms. The language of a TM M is denoted by $L(M)$. Let $c(M)$ denote the string representation of a TM M for a fixed coding of TMs. Obviously,
$$\text{code-TM} = \{c(M) \mid M \text{ is a TM}\}$$
is a recursive set, and this remains true if we exchange TMs by programs in any programming language possessing a compiler.

Now consider

$$\text{HALT}_\lambda = \{c(M) \mid M \text{ is a TM (a program) and } M \text{ halts on } \lambda\}.$$

If a program (a TM) M halts on λ, there is always a proof of this fact. To generate a proof one can simply let run M on λ, and the finite computation of M on λ is a proof that M halts on λ.

Theorem 2. *There exists a program P that does not halt on λ, and there is no proof in AV-mathematics of this fact.*

Proof. Assume the opposite, that is, for each program there exists a proof that the program halts or does not halt on λ. Then one can compute $K(w)$ for each $w \in \{0,1\}^*$ as follows.

```
A:    Input: w
      Output: K(w)
      begin
          generate in the canonical order all programs P_1, P_2, ...;
          for each P_i do
              search for a proof of "P_i halts on λ" or of
                  "P_i does not halt on λ" in the canonical order of proofs
              if P_i halts on λ then
                  simulate P_i on λ;
                  if P_i generates w then output |P_i|; exit;
                  else continue with P_{i+1}; end
              else
                  continue with P_{i+1};
              end
          end
      end
```

Following Theorem 1, one can estimate $K(w)$ for at most finitely many w's, and so we have a contradiction.   □

Theorem 3. *There exist infinitely many TMs (programs) A that do not halt on λ, and such that there is no proof in AV-mathematics for any of them that A does not halt on λ.*

Proof. Following Theorem 2, there exists a program P such that "P does not halt on λ" and there is no proof of this fact in AV-mathematics. There are several ways how to construct infinitely many programs P' such that there is a proof that "P does not halt on λ" iff there is a proof that "P' does not halt on λ".

We present the following two ways:

(i) Take an arbitrary program P_0 that halts on λ. Modify P to P' by taking the simulation of P at the beginning and if the simulation finishes, P' continues with the proper computation of P_0.

(ii) For each line of P containing **end**, insert some finite sequence of dummy operations before **end**.   □

Following Rice [9], a set $\mathcal{A} \subseteq$ code-TM is a semantically nontrivial decision problem on TMs if

(i) $\mathcal{A} \neq \emptyset$,
(ii) $\mathcal{A} \neq$ code-TM, and
(iii) if $c(M) \in \mathcal{A}$ for some TM M, then $c(M') \in \mathcal{A}$ for each TM M' with $L(M') = L(M)$.

Let $\overline{\mathcal{A}} =$ code-TM $- \mathcal{A}$ for any $\mathcal{A} \subseteq$ code-TM.

Observation 1. *The following is true for any $\mathcal{A} \subseteq$ code-TM. If, for each TM M, there exists a proof in AV-mathematics of either "$c(M) \in \mathcal{A}$" or "$c(M) \notin \mathcal{A}$", then, for each TM M', there exists a proof in AV-mathematics of either "$c(M') \notin \overline{\mathcal{A}}$" or "$c(M') \in \overline{\mathcal{A}}$".*

Proof. A proof of "$c(M) \in \mathcal{A}$" is simultaneously a proof of "$c(M) \notin \overline{\mathcal{A}}$". A proof of "$c(M) \notin \mathcal{A}$" is simultaneously a proof of "$c(M) \in \overline{\mathcal{A}}$". □

Theorem 4 (Rice's Theorem on Unprovability). *For each semantically nontrivial decision problem $\mathcal{A}$, there exist infinitely many TMs M' such that there is no proof of "$c(M') \in \mathcal{A}$" and no proof of "$c(M') \notin \mathcal{A}$", that is, one cannot investigate in AV-mathematics whether $c(M')$ is in $\mathcal{A}$ or not.*

Proof. Let $\mathcal{A}$ be a semantically nontrivial decision problem. The scheme of the proof is depicted in Fig. 2 in the appendix. According to property (iii), either for all D with $L(D) = \emptyset$ we have $c(D) \in \mathcal{A}$, or for all such D we have $c(D) \notin \mathcal{A}$. Following Observation 1, we assume without loss of generality that $c(D) \in \mathcal{A}$ for all D with $L(D) = \emptyset$. Let $M_\emptyset$ be a fixed, simple TM with the property $L(M_\emptyset) = \emptyset$, and thus $c(M_\emptyset) \in \mathcal{A}$. Let $\overline{M}$ be a TM such that $c(\overline{M}) \notin \mathcal{A}$. In particular, $L(\overline{M}) \neq \emptyset$.

We prove Theorem 4 by contradiction. For all but finitely many TMs M' let there exist a proof of either "$c(M') \in \mathcal{A}$" or "$c(M') \notin \mathcal{A}$". Then we prove that, for all but finitely many TMs M there exists a proof of either "M halts on λ" or "M does not halt on λ", which contradicts Theorem 3.

Let M be an arbitrary TM. We describe an algorithm that produces either the proof of "M does not halt on λ" if M does not halt on λ or the proof of "M halts on λ" if M halts on λ.

First we apply the procedure A (Fig. 2) that transforms M into a TM M'_A with the following properties:

(1.1) $L(M'_A) = \emptyset$ (and thus $c(M'_A) \in \mathcal{A}$) $\iff$ M does not halt on λ,
(1.2) $L(M'_A) = L(\overline{M})$ (and thus $c(M'_A) \notin \mathcal{A}$) $\iff$ M halts on λ.

This is achieved by constructing M'_A in such a way that M'_A starts to simulate the work of M on λ without reading its proper input. If the simulation finishes, M'_A continues to simulate the work of $\overline{M}$ on its proper input. This way, if M does not halt on λ, M'_A simulates the work of M on λ infinitely long and does not accept any input. If M halts on λ, then $L(M'_A) = L(\overline{M})$, because M'_A simulates the work of $\overline{M}$ on each of its inputs.

After that, one algorithmically searches for a proof of "$c(M'_A) \in \mathcal{A}$" or a proof of "$c(M'_A) \notin \mathcal{A}$" by constructing all words over Σ_{math} in the canonical order and algorithmically checking for each word whether it is a proof of "$c(M'_A) \in \mathcal{A}$" or a proof of "$c(M'_A) \notin \mathcal{A}$". If such a proof exists, one will find it in finite time. Due to (1.1) and (1.2), this proof can be viewed as (or modified to) a proof of "M does not halt on λ" or a proof of "M halts on λ".

The construction of M'_A from M done by A is an injective mapping. As a consequence, if there exists a proof of "$c(B) \in \mathcal{A}$" or a proof of "$c(B) \notin \mathcal{A}$" for all but finitely many TMs B, then there exist proofs of "M halts on λ" or "M does not halt on λ" for all but finitely many TMs M. This contradicts Theorem 3. □

Using concrete choices for $\mathcal{A}$, one can obtain a number of corollaries, such as the following ones.

Corollary 1. *For infinitely many TMs M, one cannot prove in AV-mathematics whether $L(M)$ is in P or not.*

Proof. Choose
$$\mathcal{A} = \{\, c(M) \mid M \text{ is a TM and } L(M) \in \mathsf{P} \,\}$$
in Theorem 4. □

Corollary 2. *For infinitely many TMs, one cannot prove in AV-mathematics whether they accept SATISFIABILITY or not.*

Proof. Choose
$$\mathcal{A} = \{\, c(M) \mid M \text{ is a TM and } L(M) = \text{SATISFIABILITY} \,\}$$
in Theorem 4. □

Corollary 3. *For infinitely many TMs M, one cannot prove in AV-mathematics whether M is an algorithm working in polynomial time or not.*

Proof. Choose
$$\mathcal{A} = \{\, c(M) \mid M \text{ is an algorithm working in polynomial time}\}$$
in Theorem 4. □

Still, we are not satisfied with the results formulated above. One can argue that the specification of languages (decision problems) by TMs can be so crazy that, as a consequence, one cannot recognize what they really do. Therefore we strive to prove the unprovability of claims about algorithms, preferably for algorithms for we which we even have a proof that they indeed are algorithms. This is much closer to the common specifications of NP-hard problems that can be usually expressed by algorithms solving them.

3 Hardness of Complexity Analysis of Concrete Algorithms

Among others, we prove here that, for each time-constructible function f, there exist infinitely many algorithms working in time $f(n)+\mathcal{O}(1)$ for which there is no proof in AV-mathematics that they do. To this end, we construct an algorithm $X_{A,B,f}(M)$ for given

(i) algorithm A working in $\mathrm{Time}_A(n)$ and $\mathrm{Space}_A(n)$,
(ii) algorithm B working in $\mathrm{Time}_B(n)$ and $\mathrm{Space}_B(n)$,
(iii) time-constructible function f with $f(n) \geq n$ (or some other "nice" unbounded, nondecreasing function f), and
(iv) TM M.

Here, A, B, and f are considered to be fixed by an appropriate choice, and $X_{A,B,f}(M)$ is examined for all possible TMs M. The algorithm $X_{A,B,f}(M)$ works as follows.

$X_{A,B,f}(M)$: Input: w
     begin
          simulate at most $f(|w|)$ steps of M on λ;
          if M halts on λ during this simulation **then**
               simulate A on w;
          else
               simulate B on w;
          end
     end

We say that two languages L_1 and L_2 are *almost everywhere* equal, $L_1 =_\infty L_2$ for short, if the symmetric difference of L_1 and L_2 is finite. We say that M *almost everywhere* accepts L if $L(M) =_\infty L$.

Claim. If M halts on λ, then $L(X_{A,B,f}(M)) =_\infty L(A)$ and $X_{A,B,f}(M)$ works in time $\mathrm{Time}_A(n) + \mathcal{O}(1)$ and space $\mathrm{Space}_A(n) + \mathcal{O}(1)$.

Claim. If M does not halt on λ, then $L(X_{A,B,f}(M)) = L(B)$ and $X_{A,B,f}(M)$ works in time $\mathrm{Time}_B(n) + f(n)$ and space $\mathrm{Space}_B(n) + f(n)$.

If $L(A)$ and $L(B)$ are not almost everywhere equal, then one can replace the implications in the above two claims by equivalences. Moreover, $X_{A,B,f}(M)$ is an algorithm for each TM M, and if it is provable that A and B are algorithms, it is also provable that $X_{A,B,f}(M)$ is an algorithm.

Let us now present a few applications of the construction of $X_{A,B,f}(M)$. Choose A and B in such a way that $L(A) = L(B)$ and that $\mathrm{Time}_B(n)$ grows asymptotically slower or faster than $\mathrm{Time}_A(n)$. Let $f(n) = n$. Then $L(X_{A,B,f}(M)) = L(A)$ and

- M halts on $\lambda \iff X_{A,B,f}(M)$ works in $\text{Time}_A(n) + \mathcal{O}(1)$,
- M does not halt on $\lambda \iff X_{A,B,f}(M)$ works in $\text{Time}_B(n) + n$.

Corollary 4. *Suppose $\text{Time}_A(n) \in o(\text{Time}_B(n))$ and $\text{Time}_B(n) \in \Omega(n)$. Then there are infinitely many algorithms for which one cannot distinguish in AV-mathematics whether they run in $\Theta(\text{Time}_A(n))$ or in $\Theta(\text{Time}_B(n))$.*

Proof. If one can prove "$X_{A,B,f}(M)$ works in $\Theta(\text{Time}_A(n))$", then one can also prove that "M halts on λ".

If one can prove "$X_{A,B,f}(M)$ works in $\Theta(\text{Time}_B(n))$", then there exists a proof that "M does not halt on λ". $\qquad\square$

Choosing $\text{Time}_A(n)$ as a polynomial function and $\text{Time}_B(n)$ as an exponential function, and vice versa, implies the following statement.

Theorem 5. *There exist infinitely many algorithms which do not work in polynomial time, but for which this fact is not provable in AV-mathematics. Similarly, there exist infinitely many algorithms which work in polynomial time, but for which this fact is not provable in AV-mathematics.*

Proof. Let $f(n) = n$. Note that, for a TM M that halts on λ, the claim "M halts on λ" is always provable, but for a TM M that does not halt on λ, the claim "M does not halt on λ" is not provable for infinitely many TMs M (as shown in Theorem 3). Let M_1 be a TM that does not halt on λ, but for which this fact is not provable in AV-mathematics. Taking A as a polynomial time algorithm and B as an algorithm running in superpolynomial time, the algorithm $X_{A,B,f}(M_1)$ does not run in polynomial time, but the claim "$X_{A,B,f}(M_1)$ does not run in polynomial time" is not provable in AV-mathematics.

Now, if one takes A as a superpolynomial time algorithm and B as a polynomial time algorithm, then "M_1 does not halt on λ" iff "$X_{A,B,f}(M_1)$ runs in polynomial time", but this fact is not provable in AV-mathematics, because otherwise "M_1 does not halt on λ" would be provable as well. $\qquad\square$

Theorem 5 shows how complex it may be to prove that some problem is not solvable in polynomial time since there are algorithms for which it is not provable whether they work in polynomial time or not. But if one takes $\text{Time}_A(n) \in \mathcal{O}(n)$ and $\text{Time}_B(n) \in \Omega(n^2)$, then we even realize that there are algorithms for which it is not provable whether they work in linear time or not. This could indicate why proving superlinear lower bounds on any problem in NP is hard. We are not able to analyze the complexity of some concrete algorithms for any problem, and the complexity of a problem should be something like the complexity of the "best" algorithm for that problem.

Similarly, one can look at the semantics of algorithms. Assume B solves SATISFIABILITY, A solves something else, and both A and B work in time smaller than $f(n) = 1000 \cdot n^n$ (or any sufficiently large time-constructible function f). In that case, $X_{A,B,f}(M)$ works in time $\mathcal{O}(n^n)$, and it solves SATISFIABILITY iff M does not halt on λ. One can also exchange the role of A and B in order to get that $X_{A,B,f}(M)$ solves SATISFIABILITY almost everywhere iff M halts on λ.

Theorem 6. *There are infinitely many algorithms for which it is not provable in AV-mathematics that they do not solve SATISFIABILITY.*

What is clear from Theorems 5 and 6 is that one cannot start proving P ≠ NP with the set of all polynomial time algorithms and try to show that none of them solves SATISIFIABILITY, because one cannot decide in AV-mathematics for all algorithms whether they are in the set of polynomial time algorithms or not. Analogously, one cannot start with the set of all algorithms solving SATISFIABILITY and then to try to show that their complexity is superpolynomial, because the set of all algorithms solving SATISFIABILITY is also not exactly determinable in AV-mathematics.

In our considerations, one can exchange SATISFIABILITY for any other NP-hard problem or for FACTORIZATION in order to see that proving that these problems are not in P may be very hard.

Let us look at the problem from another point of view. If P ≠ NP is provable in AV-mathematics, then, as already stated, for each algorithm A, the following statement is provable in AV-mathematics:

> "A does not work in polynomial time   (∗)
> or A does not solve SATISFIABILITY".

On the other hand, if P = NP, then one can take $f(n) = n$, and B as a polynomial time algorithm solving SATISFIABILITY and A as a superpolynomial time algorithm computing something else, and consequently get the following theorem.

Theorem 7. *If P = NP, then there exist infinitely many algorithms X for which one cannot prove or disprove in AV-mathematics the statement[3] "X solves SATISFIABILITY in polynomial time".*

One can play the same game for investigating the computational complexity of the multiplication of two decimal numbers.

Theorem 8. *If multiplication of two decimal numbers is feasible in linear time, then there exist infinitely many algorithms X, for which one cannot decide in AV-mathematics whether "X solves multiplication in linear time", or "X does not solve multiplication or does not work in linear time".*

Proof. Take A as an algorithm for multiplication with $\text{Time}_A(n) = \Theta(n^2)$, and B as a linear time algorithm for multiplication. Let $f(n) = n$. Then $X_{A,B,f}(M)$ solves multiplication in linear time iff "M does not halt on λ". Hence, if "$X_{A,B,f}(M)$ solves multiplication in linear time" is provable in AV-mathematics, then "M does not halt on λ" is provable as well.   □

Similarly, one can consider space complexity, look at DLOG vs. NLOG with respect to REACHABILITY, and prove similar versions of Theorems 5 to 8.

---

[3] That is, statement (∗).

One only needs to modify our scheme by taking a reasonable, unbounded, non-decreasing function g that bounds the space complexity of the simulation of M on λ.

The previous results look promising, but we are still far from proving the unprovability of "P $\neq$ NP" in AV-mathematics. This is because we only proved for some algorithms that it is not provable that they "do not work in polynomial time", and maybe for some other ones that it is not provable that "they do not solve SATISFIABILITY". We now prove that the intersection of these two sets of algorithms is not empty, i.e., that there exists an algorithm X for which it is neither provable that "X does not solve SATISFIABILITY", nor provable that "X does not work in polynomial time". To do that, we use the following construction.

Construction of the Algorithm X_1

Let M_1 be a TM that does not halt on λ, and for which this fact is not provable in AV-mathematics (such TMs exist due to Theorem 3). Let C be an algorithm that provably solves SATISFIABILITY in exponential time, and works in exponential time on every input. We define, for each TM M, an algorithm $X_C(M)$ as follows.

$X_C(M)$: Input: w
   begin
      simulate at most $|w|$ steps of M on λ;
      if M halts on λ within $|w|$ steps **then**
         reject w;
      else
         simulate the work of C on w;
      end
   end

The following statements are true:

- M halts on λ $\iff$ $X_C(M)$ accepts almost everywhere the empty set $\iff$ $X_C(M)$ works in polynomial time (even in linear time).
- M does not halt on λ $\iff$ $X_C(M)$ solves SATISFIABILITY $\iff$ $X_C(M)$ works in exponential time (and does not work in polynomial time).

If, for any TM M, there exists a proof of either "$X_C(M)$ works in polynomial time", or "$X_C(M)$ does not work in polynomial time", then correspondingly "M halts on λ" or "M does not halt on λ" would be provable in AV-mathematics. Analogously, if, for any TM M, there exists a proof of "$X_C(M)$ recognizes SATISFIABILITY" or "$X_C(M)$ does not recognize SATISFIABILITY", it is also provable whether M halts on λ or not.

Since M_1 does not halt on λ, and this fact is not provable in AV-mathematics, we have

$$X_1 := X_C(M_1) \text{ solves SATISFIABILITY},$$

but it is neither provable that

"X_1 does not work in polynomial time"

nor that

"X_1 solves SATISFIABILITY"

Hence, we have the following theorem.

Theorem 9. *There exists an algorithm X_1, for which it is neither provable whether X_1 recognizes SATISFIABILITY nor provable whether X_1 works in polynomial time.*

Unfortunately, this is not a proof of the fact that (*) is not provable for X_1, i. e., that

"X_1 does not work in polynomial time or X_1 does not solve
SATISFIABILITY"

is not provable in AV-mathematics (i.e., we did not prove this way that "P $\neq$ NP" is not provable in AV-mathematics). Even the opposite is true. From the construction of $X_C(M)$, we see that, for each TM M, the statement (*) is provable for $X_C(M)$. Hence, we have something like an uncertainty principle about properties of algorithms. There is a proof of the statement "$\alpha(X_1) \vee \beta(X_1)$" for the algorithm X_1, but there does neither exist a proof of "$\alpha(X_1)$" nor a proof of "$\beta(X_1)$".

Again, note that we can do the same as in Theorem 9, due to the construction of $X_C(M)$, for

1. any NP-hard problem or for FACTORIZATION by exchanging SATISFIABILITY by one of these in the construction of X_1,
2. the multiplication of two decimal numbers by taking C as an algorithm that computes multiplication in superlinear time.

4   P vs. NP in AV-Mathematics and the Existence of Constructive Proofs

In this chapter, we outline and discuss some important consequences of our work. As we showed, the computational complexity of some algorithms cannot be analyzed in AV-mathematics. Let us consider classes based only on algorithms that can be analyzed in AV-mathematics. Let Φ be a logic (formal system) that is powerful enough to specify any language in NP. Let $L(\alpha)$ denote a language determined by a specification α from Φ. We define the following classes

$\mathsf{P}_{ver} = \{L(M) \mid \alpha$ is a specification from Φ and M is an algorithm (a TM that always halts) and there exists a proof in AV-mathematics that M works in polynomial time and recognizes $L(\alpha)\}$,

$\mathsf{NP}_{ver} = \{L(M) \mid \alpha$ is a specification from Φ and M is a nondeterministic TM that provably in AV-mathematics works in polynomial time and accepts $L(\alpha)\}$,

Analogously, one can define the class A_{ver} for each complexity class A.

Now we show that the unprovability of P = NP in AV-mathematics immediately implies $P_{ver} \neq NP_{ver}$.

Theorem 10. *If the claim "P = NP" ("NLOG = DLOG") is not provable in AV-mathematics, then*

$$P_{ver} \neq NP_{ver}$$
$$(DLOG_{ver} \neq NLOG_{ver})$$

Proof. There is no doubt about the fact that each known NP-complete problem L is in NP_{ver}, because for each such L, we have a polynomial time nondeterministic TM M with a proof of $L = L(M)$.

If $P_{ver} = NP_{ver}$ would hold, then SATISFIABILITY $\in P_{ver}$. Since $P_{ver} \subseteq P$, one obtains "SATISFIABILITY $\in$ P" is provable and consequently "P = NP" is provable.   □

We can get similar results for the comparisons of other classes for which the upper classes contain complete problems, with respect to the lower ones, e.g., DLOG vs. NLOG, P vs. PSPACE, etc.

5   Making Nonconstructive Proofs Constructive

It has been noted on several occasions that there is the theoretical possibility that we could be able to prove P = NP in a nonconstructive way and still had no concrete algorithm for any NP-complete problem. Scott Aaronson writes in a recent survey [10]:

> **Objection:** Even if P = NP, the proof could be nonconstructive—in which case it wouldn't have any of the amazing implications discussed in Sect. 1.1, because we wouldn't know the algorithm.
>
> **Response:** A nonconstructive proof that an algorithm exists is indeed a theoretical possibility, though one that's reared its head only a few times in the history of computer science. [...] Even then, however, once we knew that an algorithm existed, we'd have a massive inducement to try to find it. [...]

In a poll about what theoretists think about the future of the P vs. NP-question, including whether they believe them to be equal and when the question will be settled, two persons (out of a total of 100) commented they fear that P = NP will be proved in a non-constructive way and call it a "worst case scenario" [11].

On the other hand, it is well known that a *completely nonconstructive* proof cannot exist – an exercise in the textbook *Computational Complexity* asks the student to show that (under the assumption that P = NP) there is a fixed algorithm that "solves" SATISFIABILITY in the following way: it provides satisfying assignments for all yes-instances in polynomial time, but is allowed arbitrary behavior on no-instances [12, p. 350].

While we cannot completely answer the question whether a nonconstructive proof can be converted into a constructive one, we can improve upon the answer to this exercise: if P = NP, we can present a concrete algorithm that solves SATISFIABILITY on all but finitely many instances in polynomial time (and not only on yes-instances).

While the possibility of having only a nonconstructive proof might appear to be "disturbing" [12], having a concrete algorithm that solves SATISFIABILITY in polynomial time iff P = NP is also strange: in principle, we could just use such an algorithm to solve NP-complete problems in polynomial time even if we cannot prove that P = NP. It is sufficient that P = NP holds. While we can arrive at this strange situation almost only for the P = NP-question, there are other similar questions, where such concrete algorithms indeed exist: we will show that there is a randomized algorithm that solves QSAT in expected polynomial time iff PSPACE = BPP and a deterministic one that solves graph reachability with logarithmic space iff LOGSPACE = NLOGSPACE. We could implement these algorithm today and let them run. Owing to large constants, we probably could not observe the asymptotic behavior, but it is still a strange situation.

In the following, we will use an arbitrary, but fixed enumeration $M_1, M_2, \ldots$, of all Turing machines with the input alphabet $\{0, 1\}$ such that a description of M_i can be computed in time polynomial in i and that M_i can be simulated with only polynomial overhead. In the following, *SlowSAT* denotes an exact algorithm that solves SATISFIABILITY in $O^*(2^n)$ steps, where n is the number of variables. It returns a satisfying assignment on yes-instances and "no" on no-instances. *Simulate*(M_i, w, t) denotes the result of running TM M_i on input $w \in \{0, 1\}^*$ for up to t steps. Figure 1 contains an algorithm that attempts to solve SATISFIABILITY in polynomial time. It succeeds to do so for all but finitely many inputs iff P = NP.

Theorem 11. *If P = NP, then Algorithm S solves SATISFIABILITY, runs in polynomial time, returns "no" on all no-instances, and returns a satisfying assignment on all but finitely many yes-instances.*

Proof. Assume that P = NP. Then there is a number d and infinitely many TMs M_i that solve SATISFIABILITY in time n^d. Among those, we choose one with $i \geq d$ and conclude that there exists a Turing machine M_k that solves SATISFIABILITY in time n^k for all input instances of size n.

If $i < k$, then there is at least one instance $w_i \in \{0, 1\}^{\leq m}$ such that M_i does not solve w_i correctly or runs for more than $|w_i|^i$ steps on w_i. Let $N = \min\{2^{2^{|w_1|}}, \ldots, 2^{2^{|w_{k-1}|}}\}$. If Algorithm S is run on an input instance I with $|I| > N$, then $m \geq w_i$ for all $i = 1, \ldots, k-1$. Hence, M_i is simulated on w_i for at least $|w_i|^i$ steps. This shows that M_i fails to solve w_i as a SATISFIABILITY instance or exceeds its running time bound. Hence, i is added to F. On the other hand, k is *not* added to F, because M_k solves every instance w correctly within $|w|^k$ steps. In the end, Algorithm S returns the result of this simulation.

The running time is polynomial: for inputs of length n, there is only a polynomial number of TMs that are simulated for at most $(\log \log^n)^{O(\log \log n)}$ steps

Input: A SATISFIABILITY instance $I \in \{0,1\}^*$
Output: A satisfying assignment if I is a well-formed yes-instance,
                no otherwise
$m := \lceil \log \log |I| \rceil$
$F := \emptyset$
for $i = 1$ **to** m **do**
  for $w \in \{0,1\}^{\leq m}$ **do**
    $r_1 := SlowSAT(w)$
    $r_2 := Simulate(M_i, w, |w|^i)$
    if $r_1 \not\equiv_w r_2$ **then** $F := F \cup \{i\}$ **fi**
  od
od
$k := \min\{ i \in \mathbf{N} \mid i \notin F \}$
if $k \leq m$ **then**
  $r := Simulate(M_k, I, |I|^k)$
  if r is a satisfying assignment to I **then return** r
  else return "no" **fi**
fi
return $SlowSAT(I)$

Fig. 1. Algorithm S

(which is polynomial in n). In the end, one more simulation is carried out for at most n^m steps if $n \geq N$. For inputs of length smaller than N, we cannot be sure what the running time is, but it is bounded by a (large) constant. □

Using similar ideas and concepts from [11–14], we can establish the following results.

Theorem 12. *There is a concrete deterministic logspace-bounded algorithm that solves graph reachability iff LOGSPACE = NLOGSPACE.*

Theorem 13. *If graph isomorphism is in P, then for every $\epsilon > 0$, there exists an algorithm that solves graph isomorphism and has the following properties:*

1. *It is a randomized algorithm.*
2. *It runs in expected polynomial time.*
3. *For all but finitely many yes-instances, it always answers correctly.*
4. *For the remaining yes-instances, the answer is correct with probability at least $1 - 2^{n^2}$.*
5. *For all no-instances, it always answers correctly.*

Theorem 14. *There exists a concrete randomized algorithm that solves QBF in expected polynomial time with error probability at most $1/3$ iff BPP = PSPACE.*

Acknowledgment. We would like to thank Hans-Joachim Böckenhauer, Dennis Komm, Rastislav Královič, Richard Královič, and Georg Schnitger for interesting discussions related to the first verification of the proofs presented here. Essential progress was made during the 40th Mountain Workshop on Algorithms organized by Xavier Muñoz from UPC Barcelona that offered optimal conditions for research work.

A   Concept of the Proof of Theorem 4

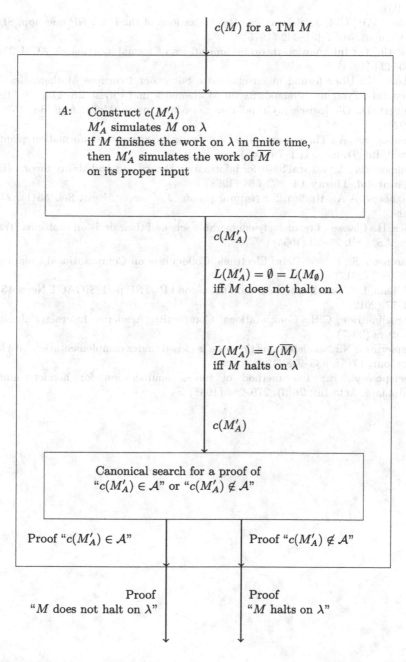

$c(M)$ for a TM M

A: Construct $c(M'_A)$
M'_A simulates M on λ
if M finishes the work on λ in finite time,
then M'_A simulates the work of $\overline{M}$
on its proper input

$c(M'_A)$

$L(M'_A) = \emptyset = L(M_\emptyset)$
iff M does not halt on λ

$L(M'_A) = L(\overline{M})$
iff M halts on λ

$c(M'_A)$

Canonical search for a proof of
"$c(M'_A) \in \mathcal{A}$" or "$c(M'_A) \notin \mathcal{A}$"

Proof "$c(M'_A) \in \mathcal{A}$"   Proof "$c(M'_A) \notin \mathcal{A}$"

Proof   Proof
"M does not halt on λ"   "M halts on λ"

Fig. 2. The schema of the reduction for the existence of proofs in Theorem 4.

References

1. Aaronson, S.: Is P versus NP formally independent? Bull. EATCS **81**, 109–136 (2003)
2. Baker, T.P., Gill, J., Solovay, R.: Relativizations of the P =? NP question. SIAM J. Comput. **4**(4), 431–442 (1975)
3. Chaitin, G.: Information-theoretic limitations of formal systems. J. ACM **21**(3), 403–424 (1974)
4. Gödel, K.: Über formal unentscheidbare Sätze der Principia Mathematica und verwandte Systeme. Monatshefte für Mathematik und Physik **28**, 173–198 (1931)
5. Hilbert, D.: Die logischen Grundlagen der Mathematik. Math. Ann. **88**, 151–165 (1923)
6. Kolmogorov, A.: Three approaches for defining the concept of information quantity. Probl. Inf. Transm. **1**, 1–7 (1965)
7. Kolmogorov, A.: Logical basis for information theory and probability theory. IEEE Transit. Inf. Theory **14**, 662–664 (1968)
8. Razborov, A.A., Rudich, S.: Natural proofs. J. Comput. Syst. Sci. **55**(1), 24–35 (1997)
9. Rice, H.: Classes of recursively enumerable sets and their decision problems. Transact. ASM **89**, 25–59 (1953)
10. Aaronson, S.: P $\overset{?}{=}$ NP. In: Electronic Colloquium on Computational Complexity (ECCC) (2017)
11. William, I.: Gasarch: guest column: the second P=?NP poll. SIGACT News **43**(2), 53–77 (2012)
12. Papadimitriou, C.H.: Computational Complexity. Academic Internet Publishers, Ventura (2007)
13. Immerman, N.: Nondeterministic space is closed under complementation. SIAM J. Comput. **17**(5), 935–938 (1988)
14. Szelepcsényi, R.: The method of forced enumeration for nondeterministic automata. Acta Inf. **26**(3), 279–284 (1988)

A Tour of Recent Results on Word Transducers

Anca Muscholl[✉]

LaBRI, University of Bordeaux, Bordeaux, France
anca.muscholl@gmail.com

Abstract. Regular word transductions extend the robust notion of regular languages from acceptors to transformers. They were already considered in early papers of formal language theory, but turned out to be much more challenging. The last decade brought considerable research around various transducer models, aiming to achieve similar robustness as for automata and languages.

In this talk we survey some recent results on regular word transducers. We discuss how classical connections between automata, logic and algebra extend to transducers, as well as some genuine definability questions. For a recent, more detailed overview of the theory of regular word transductions the reader is referred to the excellent survey [22].

Since the early times of computer science, the notion of transduction has played a fundamental role, since computers typically process data and transform it between different formats. Numerous fields of computer science are ultimately concerned with transformations, ranging from databases to image processing, and an important issue is to perform transformations with low cost, whenever possible.

The most basic form of transformers are devices that process inputs and thereby produce outputs using finite memory. Such devices are called finite-state transducers. Word-to-word finite-state transducers were considered in very early work in formal language theory [1,8,11,16,24,32], and it was soon clear that achieving a good understanding of transducers will be much more challenging than for the classical finite-state automata. One essential difference between transducers and automata over words is that the capability to process the input in both directions strictly increases the expressive power in the case of transducers, whereas this is not the case for automata [29,34]. In other words, two-way word transducers are strictly more expressive than one-way word transducers.

We consider in this overview functional word transducers, so non-deterministic finite-state transducers that compute functions[1]. It turns out that two-way functional transducers capture very nicely the notion of regularity for word-to-word transductions. Engelfriet and Hoogeboom showed in [17] that two-way finite-state transducers have the same expressive power as Courcelle's

---

Based on joint work with Félix Baschenis, Olivier Gauwin and Gabriele Puppis. Work partially supported by the Institute of Advance Studies of the Technische Universität München and the project DeLTA (ANR-16-CE40-0007).

[1] Also called "single-valued" in the literature, as a special instance of "k-valued".

R. Klasing and M. Zeitoun (Eds.): FCT 2017, LNCS 10472, pp. 29–33, 2017.
DOI: 10.1007/978-3-662-55751-8_3

monadic second-order logic definable graph transductions, restricted to words. This equivalence supports thus the notion of "regular" word functions, in the spirit of classical results on regular languages from automata theory and logic (Büchi, Elgot, Trakhtenbrot, Rabin, ...). Recently, Alur and Cerný [2] fostered new research around this topic by introducing streaming transducers and showing that they are equi-expressive to the two previous models. A streaming transducer processes the input word from left to right, and stores (partial) output words in finitely many, write-only registers.

Regular word functions inherit pleasant algorithmic properties of the robust class of regular word languages. It was long known that the equivalence problem for various types of transducers is decidable [13,14,26,27]. For functional streaming transducers [3] and functional two-way transducers the problem is actually PSPACE-complete, so not harder than for NFAs. In contrast, the equivalence problem for unrestricted (i.e., fully relational) transducers, even one-way, is undecidable [25]. Interestingly, the decidability frontier lies beyond functional transducers: using Ehrenfeucht's conjecture [13] showed that k-valued, one-way transducers have a decidable equivalence problem, and stated the result also for the two-way case. It is interesting to note that for k-valued streaming transducers the status of the equivalence problem is open. It is conjectured that for k-valued transductions, streaming transducers and two-way transducers are equivalent, and this would settle the decidability of the equivalence problem. A more algorithmic proof would go through a decomposition theorem of k-valued transducers into k functional transducers [31,35]. The paper [23] provides such a decomposition in the case of streaming transducers with only one register.

Two-way and streaming transducers raise new and challenging questions about resource requirements. A crucial resource for streaming transducers is the number of registers. For two-way transducers it is the number of times the transducer needs to re-process the input word. In particular, the case where the input can be processed (deterministically) in a single pass, from left to right, is very attractive as it corresponds to the setting of *streaming*, where the – potentially very large – inputs do not need to be stored in order to be processed. It was shown in [20] that it is decidable whether the transduction defined by a functional two-way transducer is one-way-definable, i.e., if it can be implemented by a non-deterministic, one-way transducer. However, the decision procedure of [20] has non-elementary complexity, and it was natural to ask whether one can do better. In [7] we provided a decision procedure for one-way definability of double exponential space complexity, that allows to construct equivalent one-way transducers of triple exponential size (if possible). The result on one-way definability, together with [11], answers to the streaming question. However, it would be interesting to have a direct construction from two-way (or streaming) transducers to deterministic, one-way transducers, and to settle the precise complexities. For the one-way definability problem there is a double exponential lower bound on the size of the transducer, and this lower bound is tight in the case of sweeping transducers [5]. It can be also noted that the technique developed in [7] allows to characterize sweeping transducers within the class of two-way transducers.

The minimization problem of the number of registers for streaming transducers is still an open problem, in spite of some progress. In [4] this problem was shown to be solvable in polynomial space for deterministic, concatenation-free streaming transducers over unary output alphabet. Extending this result, [15] showed how to compute the minimal number of registers over any commutative group, provided that group operations are computable. In addition, the results of [15] allow to minimize the number of registers for deterministic streaming transducers with only right updates. On the other hand, in the non-deterministic case, the minimal number of passes of sweeping transducers is computable in exponential space [6]. By exploiting a tight connection between the number of passes of sweeping transducers and the number of registers of concatenation-free streaming transducers, the last result also allows to minimize the number of registers of concatenation-free streaming transducers.

Another line of research that appears to be a significant challenge, concerns algebraic characterizations for functional word transducers. This problem is tightly related to the existence of canonical objects, like minimal automata in the case of languages. For word languages, the existence of minimal automata is crucial for determining whether a language belongs to some subclass. A renowned example are first-order definable languages, that coincide with aperiodic and star-free languages [28,33]. It is well-known that minimal transducers exist in the deterministic, one-way case [11,12]. For functional one-way transducers, [30] studied bimachines, that roughly correspond to one-way transducers with regular look-ahead, and showed how to obtain canonical bimachines. This result led to a PSPACE decision procedure to determine if the transduction realized by a functional one-way transducer is definable by an order-preserving, first-order transduction [18,19]. It is interesting to note that adding origin information to transductions greatly simplifies the setting: there is an algebraic characterization that provides a decision procedure for knowing whether a transduction with origin information is equivalent to a first-order transduction [9]. However, for the general case of regular word functions all we know is that first-order definable transductions are equivalent to transductions defined by aperiodic streaming transducers [21] and to aperiodic two-way transducers [10]. Coming up with a decision procedure for knowing whether a regular word function is first-order definable is a challenging endeavour.

Acknowledgments. I thank Félix Baschenis, Emmanuel Filiot, Olivier Gauwin, Nathan Lhote, Gabriele Puppis and Sylvain Salvati for numerous discussions and feedback, as well as Mikolaj Bojanczyk for the origins of this work.

References

1. Aho, A., Hopcroft, J., Ullman, J.: A general theory of translation. Math. Syst. Theory **3**(3), 193–221 (1969)
2. Alur, R., Cerný, P.: Expressiveness of streaming string transducer. In: Proceedings of FSTTCS 2010. LIPIcs, vol. 8, pp. 1–12. Schloss Dagstuhl - Leibniz-Zentrum für Informatik (2010)

3. Alur, R., Deshmukh, J.V.: Nondeterministic streaming string transducers. In: Aceto, L., Henzinger, M., Sgall, J. (eds.) ICALP 2011. LNCS, vol. 6756, pp. 1–20. Springer, Heidelberg (2011). doi:10.1007/978-3-642-22012-8_1

4. Alur, R., Raghothaman, M.: Decision problems for additive regular functions. In: Fomin, F.V., Freivalds, R., Kwiatkowska, M., Peleg, D. (eds.) ICALP 2013. LNCS, vol. 7966, pp. 37–48. Springer, Heidelberg (2013). doi:10.1007/978-3-642-39212-2_7

5. Baschenis, F., Gauwin, O., Muscholl, A., Puppis, G.: One-way definability of sweeping transducers. In: IARCS Annual Conference on Foundation of Software Technology and Theoretical Computer Science (FSTTCS 2015). LIPIcs, vol. 45, pp. 178–191. Schloss Dagstuhl - Leibniz-Zentrum für Informatik (2015)

6. Baschenis, F., Gauwin, O., Muscholl, A., Puppis, G.: Minimizing resources of sweeping, streaming string transducers. In: International Colloquium on Automata, Languages, Programming (ICALP 2016). LIPIcs, vol. 55, pp. 114:1–114:14. Schloss Dagstuhl - Leibniz-Zentrum für Informatik (2016)

7. Baschenis, F., Gauwin, O., Muscholl, A., Puppis, G.: Untwisting two-way transducers in elementary time. In: ACM/IEEE Symposium on Logic in Computer Science (LICS 2017). IEEE Computer Society (2017)

8. Berstel, J.: Transductions and context-free languages. Teubner Studienbücher Stuttgart (1979)

9. Bojańczyk, M.: Transducers with origin information. In: Esparza, J., Fraigniaud, P., Husfeldt, T., Koutsoupias, E. (eds.) ICALP 2014. LNCS, vol. 8573, pp. 26–37. Springer, Heidelberg (2014). doi:10.1007/978-3-662-43951-7_3

10. Carton, O., Dartois, L.: Aperiodic two-way transducers and FO-transductions. In: Proceedings of CSL 2015. LIPIcs, pp. 160–174. Schloss Dagstuhl - Leibniz-Zentrum für Informatik (2015)

11. Choffrut, C.: Une caractérisation des fonctions séquentielles et des fonctions sous-séquentielles en tant que relations rationnelles. Theor. Comput. Sci. **5**, 325–338 (1977)

12. Choffrut, C.: Minimizing subsequential transducers: a survey. Theor. Comput. Sci. **292**, 131–143 (2003)

13. Culik, K., Karhumäki, J.: The equivalence of finite valued transducers (on HDT0L languages) is decidable. Theor. Comput. Sci. **47**, 71–84 (1986)

14. Culik II, K., Karhumäki, J.: The equivalence problem for single-valued two-way transducers (on NPDT0L languages) is decidable. SIAM J. Comput. **16**(2), 221–230 (1987)

15. Daviaud, L., Reynier, P.-A., Talbot, J.-M.: A generalised twinning property for minimisation of cost register automata. In: Annual ACM/IEEE Symposium on Logic in Computer Science (LICS 2016), pp. 857–866. ACM (2016)

16. Eilenberg, S.: Automata, Languages and Machines, vol. B. Academic Press, New York (1976)

17. Engelfriet, J., Hoogeboom, H.J.: MSO definable string transductions and two-way finite-state transducers. ACM Trans. Comput. Log. **2**(2), 216–254 (2001)

18. Filiot, E., Gauwin, O., Lhote, N.: Aperiodicity of rational functions is PSPACE-complete. In: IARCS Annual Conference on Foundations of Software Technology and Theoretical Computer Science (FSTTCS 2016). LIPIcs, vol. 65, pp. 13:1–13:15. Schloss Dagstuhl - Leibniz-Zentrum für Informatik (2016)

19. Filiot, E., Gauwin, O., Lhote, N.: First-order definability of rational transductions: an algebraic approach. In: Annual ACM/IEEE Symposium on Logic in Computer Science (LICS 2016), pp. 387–396. ACM (2016)

20. Filiot, E., Gauwin, O., Reynier, P.-A., Servais, F.: From two-way to one-way finite state transducers. In: ACM/IEEE Symposium on Logic in Computer Science (LICS 2013), pp. 468–477 (2013)
21. Filiot, E., Krishna, S.N., Trivedi, A.: First-order definable string transformations. In: IARCS Annual Conference on Foundations of Software Technology and Theoretical Computer Science (FSTTCS 2014). LIPIcs, pp. 147–159. Schloss Dagstuhl - Leibniz-Zentrum für Informatik (2014)
22. Filiot, E., Reynier, P.-A.: Transducers, logic and algebra for functions of finite words. ACM SIGLOG News **3**(3), 4–19 (2016)
23. Gallot, P., Muscholl, A., Puppis, G., Salvati, S.: On the decomposition of finite-valued streaming string transducers. In: STACS 2017. LIPIcs, vol. 66, pp. 34:1–34:14. Schloss Dagstuhl - Leibniz-Zentrum für Informatik (2017)
24. Ginsburg, S., Rose, G.: A characterization of machine mappings. Canad. J. Math. **18**, 381–388 (1966)
25. Griffiths, T.V.: The unsolvability of the equivalence problem for lambda-free nondeterministic generalized machines. J. ACM **15**(3), 409–413 (1968)
26. Gurari, E.: The equivalence problem for deterministic two-way sequential transducers is decidable. SIAM J. Comput. **11**(3), 448–452 (1982)
27. Gurari, E., Ibarra, O.H.: A note on finite-valued and finitely ambiguous transducers. Math. Syst. Theory **16**(1), 61–66 (1983)
28. McNaughton, R., Papert, S.: Counter-Free Automata. MIT Press, Cambridge (1971)
29. Rabin, M.O., Scott, D.: Finite automata and their decision problems. IBM J. Res. Dev. **3**(2), 114–125 (1959)
30. Reutenauer, C., Schützenberger, M.-P.: Minimization of rational word functions. SIAM J. Comput. **20**(4), 669–685 (1991)
31. Sakarovitch, J., de Souza, R.: On the decomposition of k-valued rational relations. In: Annual Symposium on Theoretical Aspects of Computer Science (STACS 2008). LIPIcs, vol. 1, pp. 621–632. Schloss Dagstuhl - Leibniz-Zentrum für Informatik (2008)
32. Schützenberger, M.-P.: A remark on finite transducers. Inf. Control **4**(2–3), 185–196 (1961)
33. Schützenberger, M.P.: On finite monoids having only trivial subgroups. Inf. Control **8**, 190–194 (1965)
34. Shepherdson, J.: The reduction of two-way automata to one-way automata. IBM J. Res. Dev. **3**(2), 198–200 (1959)
35. Weber, A.: Decomposing a k-valued transducer into k unambiguous ones. RAIRO-ITA **30**(5), 379–413 (1996)

Some Results of Zoltán Ésik on Regular Languages

Jean-Éric Pin[✉]

IRIF, CNRS, Paris, France
Jean-Eric.Pin@irif.fr

Abstract. Zoltán Ésik published 2 books as an author, 32 books as editor and over 250 scientific papers in journals, chapters and conferences. It was of course impossible to survey such an impressive list of results and in this lecture, I will only focus on a very small portion of Zoltán's scientific work. The first topic will be a result from 1998, obtained by Zoltán jointly with Imre Simon, in which he solved a twenty year old conjecture on the shuffle operation. The second topic will be his algebraic study of various fragments of logic on words. Finally I will briefly describe some results on commutative languages obtained by Zoltán, Jorge Almeida and myself.

1 Regular Languages and Varieties

Let A be a finite alphabet. Let L be a language of A^* and let x and y be words of A^*. The *quotient* $x^{-1}Ly^{-1}$ of L by x and y is defined by the formula

$$x^{-1}Ly^{-1} = \{u \in A^* \mid xuy \in L\}.$$

A *lattice of languages* of A^* is a set $\mathcal{L}$ of languages of A^* containing $\emptyset$ and A^* and closed under finite union and finite intersection. It is *closed under quotients* if every quotient of a member of $\mathcal{L}$ is also in $\mathcal{L}$. A *Boolean algebra of languages* is a lattice of languages closed under complement.

A *class of languages* is a correspondence $\mathcal{C}$ which associates with each alphabet A a set $\mathcal{C}(A^*)$ of regular languages of A^*. A *variety of languages* is a class of languages $\mathcal{V}$ closed under Boolean operations, quotients and inverses of morphisms. This means that, for each alphabet A, $\mathcal{V}(A^*)$ is a Boolean algebra of regular languages closed under quotients and that if $\varphi : A^* \to B^*$ is a monoid morphism, then $L \in \mathcal{V}(B^*)$ implies $\varphi^{-1}(L) \in \mathcal{V}(A^*)$.

A *variety of finite monoids* is a class of finite monoids which is closed under taking submonoids, homomorphic images and finite direct products. Eilenberg's variety theorem [6] gives a bijective correspondence between varieties of monoids and varieties of languages. Let **V** be a variety of finite monoids and, for each alphabet A, let $\mathcal{V}(A^*)$ be the set of all languages of A^* whose syntactic monoid is in **V**. Then $\mathcal{V}$ is a variety of languages. Furthermore, the correspondence $\mathbf{V} \to \mathcal{V}$ is a bijection between varieties of monoids and varieties of languages.

R. Klasing and M. Zeitoun (Eds.): FCT 2017, LNCS 10472, pp. 34–37, 2017.
DOI: 10.1007/978-3-662-55751-8_4

A language L is *commutative* if any word obtained by permuting the letters of a word of L also belongs to L. A variety of languages is *commutative* if all of its languages are commutative, or equivalently, if all the monoids of the corresponding variety of finite monoids are commutative.

A *renaming* or *length-preserving morphism* is a morphism φ from A^* into B^*, such that, for each word u, the words u and $\varphi(u)$ have the same length. It is equivalent to require that, for each letter a, $\varphi(a)$ is also a letter, that is, $\varphi(A) \subseteq B$. Similarly, a morphism is *length-decreasing* if the image of each letter is either a letter or the empty word. Finally, a morphism is *length-multiplying* if all letters have images of the same length, that is, if $\varphi(A) \subseteq B^k$ for some integer k.

2   The Shuffle Operation

Recall that the *shuffle product* (or simply *shuffle*) of two languages L_1 and L_2 over A is the language

$$L_1 \sqcup\!\sqcup L_2 = \{w \in A^* \mid w = u_1 v_1 \cdots u_n v_n \text{ for some words } u_1, \ldots, u_n,$$
$$v_1, \ldots, v_n \text{ of } A^* \text{ such that } u_1 \cdots u_n \in L_1 \text{ and } v_1 \cdots v_n \in L_2\}.$$

The shuffle product defines a commutative and associative operation over the set of languages over A.

Zoltán Ésik has long been interested in the shuffle operation. In a series of papers with Bertol or with Bloom [3–5,8,9], he studied the free shuffle algebra and proved that the equational theory of shuffle has no finite axiomatization. This was probably the reason why he got interested in the conjecture proposed by Perrot [12] in 1978.

Perrot wanted to characterize the varieties of languages closed under shuffle. He was able to characterize all commutative varieties closed under shuffle, but failed to characterize the noncommutative ones. However, he conjectured that the only noncommutative variety of languages closed under shuffle is the variety of all regular languages.

This conjecture remained open for twenty years, until Zoltán Ésik, in collaboration with another famous Hungarian-born computer scientist, Imre Simon, managed to solve the conjecture positively [11]. Their proof is very ingenious and was the starting point of further development.

3   Logic on Words

In December 2001, Ésik and Ito published a BRICS report entitled *Temporal logic with cyclic counting and the degree of aperiodicity of finite automata*. In this paper, which was published in 2003 [10], the authors studied an enrichment of temporal logic involving cyclic counting and they provided an algebraic characterization of the corresponding class of regular languages. An instance of Eilenberg's variety theorem? Not quite, because this class is closed under inverses of length-preserving morphisms, but is not closed under inverses of arbitrary morphisms.

The next year (2002), Ésik published another BRICS report, entitled *Extended temporal logic on finite words and wreath product of monoids with distinguished generators*, which became a DLT paper in 2003 [7]: in this paper, he further developed his idea of enriching temporal logic, in the spirit of Wolper [15]. He managed to give an algebraic characterization of several fragments by using monoids with distinguished generators. This led to a series of new results as well as a unified and elegant proof of known results.

It turns out that a similar idea was developed independently and at the same time by Straubing [14]. This gave rise to the theory of $\mathcal{C}$-varieties, which is an extension of Eilenberg's variety theory. The letter $\mathcal{C}$ refers to a class of morphisms (called $\mathcal{C}$-*morphisms*) between free monoids (for instance *length-preserving* morphisms, *length-decreasing* or *length-multiplying* morphisms). Now a $\mathcal{C}$-variety of languages is defined as a *variety of languages* except that it is only closed under inverses of $\mathcal{C}$-morphisms.

The corresponding algebraic objects are no longer monoids, but *stamps*, that are surjective monoid morphisms $\varphi : A^* \to M$ from a finitely generated free monoid A^* onto a finite monoid M.

4   Back to the Shuffle Operation

In 1995, I proposed another extension of Eilenberg's variety theorem [13]. A positive variety of languages is defined exactly like a variety of languages, except that it is not closed under complement. In other words, for each alphabet A, $\mathcal{V}(A^*)$ is not required to be a Boolean algebra of languages, but only a lattice of languages. For the algebraic counterpart, one needs to consider ordered monoids instead of monoids.

The question now arises to describe all positive varieties closed under shuffle. Some progress in this direction can be found in [1], but the first step, namely the commutative case, was clarified only recently, in a paper of Jorge Almeida, Zoltán Ésik and myself [2].

I am sure that Zoltán would have liked to further investigate this type of questions among the numerous topics he was interested in. I deeply miss him, as a scientist and as a personal friend.

References

1. Almeida, J., Cano, A., Klíma, O., Pin, J.-É.: Fixed points of the lower set operator. Int. J. Algebra Comput. **25**(1–2), 259–292 (2015)
2. Almeida, J., Ésik, Z., Pin, J.-É.: Commutative positive varieties of languages. Acta Cybern. **23**, 91–111 (2017)
3. Bloom, S.L., Ésik, Z.: Free shuffle algebras in language varieties extended abstract. In: Baeza-Yates, R., Goles, E., Poblete, P.V. (eds.) LATIN 1995. LNCS, vol. 911, pp. 99–111. Springer, Heidelberg (1995). doi:10.1007/3-540-59175-3_84

4. Bloom, S.L., Ésik, Z.: Nonfinite axiomatizability of shuffle inequalities. In: Mosses, P.D., Nielsen, M., Schwartzbach, M.I. (eds.) CAAP 1995. LNCS, vol. 915, pp. 318–333. Springer, Heidelberg (1995). doi:10.1007/3-540-59293-8_204
5. Bloom, S.L., Ésik, Z.: Free shuffle algebras in language varieties. Theoret. Comput. Sci. **163**(1–2), 55–98 (1996)
6. Eilenberg, S.: Automata, Languages and Machines, vol. B. Academic Press, New York (1976)
7. Ésik, Z.: Extended temporal logic on finite words and wreath product of monoids with distinguished generators. In: Ito, M., Toyama, M. (eds.) DLT 2002. LNCS, vol. 2450, pp. 43–58. Springer, Heidelberg (2003). doi:10.1007/3-540-45005-X_4
8. Ésik, Z., Bertol, M.: Nonfinite axiomatizability of the equational theory of shuffle. In: Fülöp, Z., Gécseg, F. (eds.) ICALP 1995. LNCS, vol. 944, pp. 27–38. Springer, Heidelberg (1995). doi:10.1007/3-540-60084-1_60
9. Ésik, Z., Bertol, M.: Nonfinite axiomatizability of the equational theory of shuffle. Acta Inform. **35**(6), 505–539 (1998)
10. Ésik, Z., Ito, M.: Temporal logic with cyclic counting and the degree of aperiodicity of finite automata. Acta Cybern. **16**, 1–28 (2003)
11. Ésik, Z., Simon, I.: Modeling literal morphisms by shuffle. Semigroup Forum **56**, 225–227 (1998)
12. Perrot, J.-F.: Variétés de langages et operations. Theoret. Comput. Sci. **7**, 197–210 (1978)
13. Pin, J.-E.: A variety theorem without complementation. Russ. Math. (Iz. VUZ) **39**, 80–90 (1995)
14. Straubing, H.: On logical descriptions of regular languages. In: Rajsbaum, S. (ed.) LATIN 2002. LNCS, vol. 2286, pp. 528–538. Springer, Heidelberg (2002). doi:10. 1007/3-540-45995-2_46
15. Wolper, P.: Temporal logic can be more expressive. Inform. Control **56**(1–2), 72–99 (1983)

Contributed Papers

Contextuality in Multipartite Pseudo-Telepathy Graph Games

Anurag Anshu[1], Peter Høyer[2], Mehdi Mhalla[3], and Simon Perdrix[4(✉)]

[1] Centre for Quantum Technologies, National University of Singapore,
Singapore, Singapore
[2] University of Calgary, Calgary, Canada
[3] University of Grenoble Alpes, CNRS, Grenoble INP, LIG, 38000 Grenoble, France
[4] CNRS, LORIA, Université de Lorraine, Inria Carte, Nancy, France
simon.perdrix@loria.fr

Abstract. Analyzing pseudo-telepathy graph games, we propose a way to build contextuality scenarios exhibiting the quantum supremacy using graph states. We consider the combinatorial structures generating equivalent scenarios. We introduce a new tool called multipartiteness width to investigate which scenarios are harder to decompose and show that there exist graphs generating scenarios with a linear multipartiteness width.

1 Introduction

Contextuality is an active area of research that describes models of correlations and interpretations, and links to some fundamental questions about the natural world. It also provides a framework where one can utilize the understanding of quantum mechanics (and quantum information) in order to better analyze, understand, and interpret macroscopic phenomena [7,16,19,29,42].

The theoretical and experimental study of quantum world has proven that a scenario involving many parties (each having access to a local information) can contain correlations that do not possess any classical interpretation that relies on decomposition of these correlations using local functions. Contextuality can be viewed as a tool to describe the combinatorial structures present in these correlations.

Recent works on the mathematical structures of contextuality [3,4,13] are based on a model introduced by Abramsky and Brandenburger [1] which uses sheaf theory to naturally translate the consistency of interpretation by the pre-sheaf structure obtained by a distribution functor on the sheaf of events. The authors introduce three levels of contextuality: (*i*) Probabilistic contextuality, which corresponds to the possibility of simulating locally and classically a probability distribution. It extends the celebrated Bell's theorem [9] which shows that quantum probabilities are inconsistent with the predictions of any local realistic theory; (*ii*) Logical contextuality or possibilistic contextuality, which extends Hardy's construction [27] and considers only the support of a probability distribution; (*iii*) Strong contextuality, which extends the properties of the

© Springer-Verlag GmbH Germany 2017
R. Klasing and M. Zeitoun (Eds.): FCT 2017, LNCS 10472, pp. 41–55, 2017.
DOI: 10.1007/978-3-662-55751-8_5

GHZ state [24] and relies on the existence of a global assignment consistent with the support.

More recently Acín et al. [5] have presented contextuality scenarios defined as hypergraphs, in which vertices are called outcomes and hyperedges are called measurements. A general interpretation model is an assignment of non negative reals to the vertices that can be interpreted as a probability distribution for any hyperedge (weights of the vertices of each hyperedge sum to 1). Each hypergraph H admits a set $\mathcal{C}(H)$ (resp. $\mathcal{Q}(H)$, $\mathcal{G}(H)$) of classical (resp. quantum, general probabilistic) models with $\mathcal{C}(H) \subseteq \mathcal{Q}(H) \subseteq \mathcal{G}(H)$.

They have shown that the Foulis Randall product of hypergraphs [20] allows one to describe the set of no-signaling models in product scenarios $\mathcal{G}(H_1 \otimes H_2)$. They have also investigated the multipartite case, showing that the different products for composition produce models that are observationally equivalent.

A particular case of contextuality scenarios is the *pseudo-telepathy games* [10], which are games that can be won by non-communicating players that share quantum resources, but cannot be won classically without communication. A family of pseudo-telepathy games based on graph states have been introduced in [6]. The pseudo-telepathy game associated with a graph G of order n (on n vertices), is a collaborative n-player game where each player receives a binary input (question) and is asked to provide, without communication, a binary output (answer). Some global pairs of (answers—questions) are forbidden and correspond to losing positions. Given such a scenario, to quantify its multipartiteness, we define the **multipartiteness width**: a model on n parties has a multipartiteness width less than k if it has an interpretation (assignment of real positive numbers to the vertices) that can be obtained using as ressources interpretations of contextual scenarios on less than k parties.

It has been shown in [13] that even though GHZ type scenarios are maximally non local (strongly contextual), they can be won with 2 partite nonlocal boxes. So the multipartiteness width is different from the usual measures of contextuality [2,23]. However, it has potential application for producing device independent witnesses for entanglement depth [32].

In Sect. 2, we define the graph pseudo-telepathy games, investigate in detail the quantum strategy and link them to contextuality scenarios. The quantum strategy consists in sharing a particular quantum state called graph state [28]. Graph states have multiple applications in quantum information processing, e.g. secret sharing [22,34,35], interactive proofs [12,33,39], and measurement-based quantum computing [14,17,18,37,38,40]. We show in Sect. 3 that provided that the players share multipartite randomness, it is enough to surely win the associated pseudo-telepathy game, in order to simulate the associated quantum probability distribution. In Sect. 4, we prove that graphs obtained by a combinatorial graph transformation called pivoting correspond to equivalent games. Finally, we prove that there exist graphs for which the multipartiteness width is linear in the number of players, improving upon the previous logarithmic bound given in [6].

Note that even though the rules of these graph games appear non-trivial, they naturally correspond to the correlations present in outcomes of a quantum process that performs X and Z measurements on a graph state. Thus, they might be easy to produce empirically. Furthermore even if the space of events is quite large, the scenarios have the advantage of possessing concise descriptions, quite similar to the separating scenarios using Johnson graphs in [21]. Requiring such large structures to achieve possibilistic contextuality for quantum scenarios seems to be unavoidable. Indeed, it has been shown that multiparty XOR type inequalities involving two-body correlation functions cannot achieve pseudo-telepathy [25].

2   Pseudo-Telepathy Graph Games, Multipartiteness and Contextuality Scenarios

Graph Notations. We consider finite simple undirected graphs. Let $G = (V, E)$ be a graph. For any vertex $u \in V$, $N_G(u) = \{v \in V \mid (u, v) \in E\}$ is the neighborhood of u. For any $D \subseteq V$, the odd neighborhood of D is the set of all vertices which are oddly connected to D in G: $\mathrm{Odd}(D) = \{v \in V : |D \cap N(v)| = 1 \bmod 2\}$. $\mathrm{Even}(D) = V \setminus \mathrm{Odd}(D)$ is the even neighborhood of D, and $\mathrm{loc}(D) = D \cup \mathrm{Odd}(D)$ is the local set of D which consists of the vertices in D and those oddly connected to D. For any $D \subseteq V$, $G[D] = (D, E \cap D \times D)$ is the subgraph induced by D, and $|G[D]|$ its size, i.e. the number of edges of $G[D]$. Note that Odd can be realized as linear map (where we consider subsets as binary vectors), which implies that for any two subset of vertices A, B, $\mathrm{Odd}(A \oplus B) = \mathrm{Odd}(A) \oplus \mathrm{Odd}(B)$ where $\oplus$ denotes the symmetric difference.

We introduce the notion of *involvement*:

Definition 1 (Involvement). *Given a graph $G = (V, E)$, a set $D \subseteq V$ of vertices is* involved *in a binary labelling $x \in \{0, 1\}^V$ of the vertices if $D \subseteq \mathrm{supp}(x) \subseteq \mathrm{Even}(D)$, where $\mathrm{supp}(x) = \{u \in V, x_u = 1\}$.*

In other words, D is involved in the binary labelling x, if all the vertices in D are labelled with 1 and all the vertices in $\mathrm{Odd}(D)$ are labelled with 0. Notice that when $G[D]$ is not a union of Eulerian graphs[1], there is no binary labelling in which D is involved. On the other hand, if $G[D]$ is a union of Eulerian graphs, there are $2^{|\mathrm{Even}(D)| - |D|}$ binary labellings in which D is involved.

Collaborative Games. A multipartite collaborative game $\mathcal{G}$ for a set V of players is a scenario characterised by a set $\mathcal{L} \subseteq \{0, 1\}^V \times \{0, 1\}^V$ of losing pairs: each player u is asked a binary question x_u and has to produce a binary answer a_u. The collaborative game is won by the players if for a given question $x \in \{0, 1\}^V$ they produce an answer $a \in \{0, 1\}^V$ such that the pair formed by a and x, denoted $(a|x)$, is not a losing pair, i.e. $(a|x) \notin \mathcal{L}$.

---

[1] The following three properties are equivalent: (i) $D \subseteq \mathrm{Even}(D)$; (ii) every vertex of $G[D]$ has an even degree; (iii) $G[D]$ is a union of Eulerian graphs. Notice that $D \subseteq \mathrm{Even}(D)$ does not imply that $G[D]$ is Eulerian as it may not be connected.

A game is pseudo-telepathic if classical players using classical resources cannot perfectly win the game (unless they cheat by exchanging messages after receiving the questions) whereas using entangled states as quantum resources the players can perfectly win the game, giving the impression to a quantum non believer that they are telepathic (as the only classical explanation to a perfect winning strategy is that they are communicating).

Example 1: The losing set associated with the Mermin parity game [36] is $\mathcal{L}_{\mathbf{Mermin}} = \{(a|x) : \sum x_i = 0 \bmod 2 \text{ and } \sum a_i + (\sum x_i)/2 = 1 \bmod 2\}$. Notice that the losing set admits the following simpler description: $\mathcal{L}_{\mathbf{Mermin}} = \{(a|x) : 2|a| = |x| + 2 \bmod 4\}$, where $|x| = |\mathrm{supp}(x)|$ is the Hamming weight of x.

Collaborative Graph Games MCG(G): A multipartite collaborative game $\mathbf{MCG}(G)$ associated with a graph $G = (V, E)$, where V is a set of players, is the collaborative game where the set of losing pairs is $\mathcal{L}_G := \{(a|x) : \exists D \text{ involved in } x \text{ s.t. } \sum_{u \in \mathrm{loc}(D)} a_u = |G[D]| + 1 \bmod 2\}$. In other words, the collaborative game is won by the players if for a given question $x \in \{0,1\}^V$ they produce an answer $a \in \{0,1\}^V$ such that for any non-empty D involved in x, $\sum_{u \in \mathrm{loc}(D)} a_u = |G[D]| \bmod 2$.

Example 2: Consider $\mathbf{MCG}(K_n)$ the collaborative game associated with the complete graph K_n of order n. When a question x contains an even number of 1s the players trivially win since there is no non-empty subset of vertices involved in such a question. When x has an odd number of 1s, the set of players (vertices) involved in this question is $D = \mathrm{supp}(x)$. In this case, all the players are either in D or $\mathrm{Odd}(D)$ thus the sum of all the answers has to be equal to $|G[D]| = \frac{|D|(|D|-1)}{2} = \frac{|D|-1}{2} \bmod 2$. Thus for the complete graph K_n, $\mathcal{L}_{K_n} = \{(a|x) : |x| = 1 \bmod 2 \text{ and } |a| = \frac{|x|-1}{2} + 1 \bmod 2\} = \{(a|x) : 2|a| = |x| + 1 \bmod 4\}$. Note that for this particular graph, the constraints are global in the sense that the sum of the answers of all the players is used for all the questions. Notice also that the set of losing pairs $\mathcal{L}_{K_n} = \{(a|x) : 2|a| = |x| + 1 \bmod 4\}$ is similar to the one of the Mermin parity game, $\mathcal{L}_{\mathbf{Mermin}} = \{(a|x) : 2|a| = |x| + 2 \bmod 4\}$. In Sect. 4, we actually show the two games simulate each other.

Quantum Strategy (Qstrat): In the following we show that for any graph G, the corresponding multipartite collaborative game can be won by the players if they share a particular quantum state. More precisely the state they share is the so-called graph state $|G\rangle = \frac{1}{\sqrt{2^{|V|}}} \sum_{y \in \{0,1\}^V} (-1)^{|G[\mathrm{supp}(y)]|} |y\rangle$, and they apply the following strategy: every player u measures his qubit according to X if $x_u = 1$ or according to Z if $x_u = 0$. Every player answers the outcome $a_u \in \{0,1\}$ of this measurement.

This quantum strategy **QStrat**, not only produces correct answers, but provides all the good answers uniformly:

Lemma 1. *Given a graph $G = (V, E)$ and question $x \in \{0,1\}^V$, the probability $p(a|x)$ to observe the outcome $a \in \{0,1\}^V$ when each qubit u of a graph state*

$|G\rangle$ is measured according to Z if $x_u = 0$ or according to X if $x_u = 1$ satisfies:

$$p(a|x) = \begin{cases} 0 & \text{if } (a|x) \in \mathcal{L} \\ \frac{|\{D \text{ involved in } x\}|}{2^{|V|}} & \text{otherwise.} \end{cases}$$

Proof. According to the Born rule, the probability to get the answer $a \in \{0,1\}^V$ to a given question $x \in \{0,1\}^V$ is:

$$p(a|x) = \langle G| \left(\bigotimes_{v \in V \setminus \mathrm{supp}(x)} \frac{I + (-1)^{a_v} Z_v}{2} \right) \otimes \left(\bigotimes_{u \in \mathrm{supp}(x)} \frac{I + (-1)^{a_u} X_u}{2} \right) |G\rangle$$

$$= \frac{1}{2^n} \sum_{D \subseteq V} (-1)^{\sum_{u \in D} a_u} \langle G| Z_{D \setminus \mathrm{supp}(x)} X_{D \cap \mathrm{supp}(x)} |G\rangle$$

The basic property which makes this strategy work is that for any $u \in V$, $X_u |G\rangle = Z_{N(u)} |G\rangle$. As a consequence, since X and Z anti-commute and $X^2 = Z^2 = I$, for any $D \subseteq V$, $X_D |G\rangle = (-1)^{|G[D]|} Z_{\mathrm{Odd}(D)} |G\rangle$. Thus,

$$p(a|x) = \frac{1}{2^n} \sum_{D \subseteq V} (-1)^{|G[D \cap \mathrm{supp}(x)]| + \sum_{u \in D} a_u} \langle G| Z_{(\mathrm{Odd}(D \cap \mathrm{supp}(x))) \oplus (D \setminus \mathrm{supp}(x))} |G\rangle$$

where $\oplus$ denotes the symmetric difference. Since $\langle G| Z_C |G\rangle = \begin{cases} 1 & \text{if } C = \emptyset \\ 0 & \text{otherwise} \end{cases}$,

$$p(a|x) = \frac{1}{2^n} \sum_{D \subseteq V, D \setminus \mathrm{supp}(x) = \mathrm{Odd}(D \cap \mathrm{supp}(x))} (-1)^{|G[D \cap \mathrm{supp}(x)]| + \sum_{u \in D} a_u}$$

$$= \frac{1}{2^n} \sum_{D_1 \subseteq \mathrm{supp}(x)} \sum_{D_0 \subseteq V \setminus \mathrm{supp}(x), D_0 = \mathrm{Odd}(D_1)} (-1)^{|G[D_1]| + \sum_{u \in D_0 \cup D_1} a_u}$$

$$= \frac{1}{2^n} \sum_{D_1 \subseteq \mathrm{supp}(x), \mathrm{Odd}(D_1) \cap \mathrm{supp}(x) = \emptyset} (-1)^{|G[D_1]| + \sum_{u \in \mathrm{loc}(D_1)} a_u}$$

$$= \frac{1}{2^n} \sum_{D_1 \text{ involved in } x} (-1)^{|G[D_1]| + \sum_{u \in \mathrm{loc}(D_1)} a_u} = \frac{|R_0^{(x,a)}| - |R_1^{(x,a)}|}{2^n}$$

where $R_d^{(x,a)} = \{D \text{ involved in } x : |G[D]| + \sum_{u \in \mathrm{loc}(D)} a_u = d \bmod 2\}$. If $(a|x) \notin \mathcal{L}$, then $R_1^{(x,a)} = \emptyset$, so $p(a|x) = \frac{|\{D \text{ involved in } x\}|}{2^n} > 0$ since $\emptyset$ is involved in x. Otherwise, there exists $D' \in R_1^{(x,a)}$. Notice that $R_0^{(x,a)}$ is a vector space ($\forall D_1, D_2 \in R_0^{(x,a)}, D_1 \oplus D_2 \in R_0^{(x,a)}$) and $R_1^{(x,a)}$ an affine space $R_1^{(x,a)} = \{D' \oplus D \mid D \in R_0^{(x,a)}\}$. Thus $|R_0^{(x,a)}| = |R_1^{(x,a)}|$ which implies $p(a|x) = 0$. $\square$

The probability distribution produced by **QStrat** depends on the number of sets D involved in a given question x. Notice that a set $D \subseteq \mathrm{supp}(x)$ is involved in x if and only if $D \in \mathrm{Ker}(L_x)$, where L_x linearly[2] maps $A \subseteq \mathrm{supp}(x)$

---

[2] L_x is linear for the symmetric difference: $L_x(D_1 \oplus D_2) = L_x(D_1) \oplus L_x(D_2)$.

to $\mathrm{Odd}(A) \cap \mathrm{supp}(x)$. Thus $|\{D \text{ involved in } x\}| = 2^{|x|-rk_G(x)}$, where $rk_G(x) = \log_2(|\{L_x(A) : A \subseteq \mathrm{supp}(x))\}|)$ is the rank of $L_x = A \mapsto \mathrm{Odd}(A) \cap \mathrm{supp}(x)$.

Contextuality Scenario. Following the hypergraph model of [5], we associate with every graph G a contextuality scenario, where each vertex is a pair $(a|x)$ and each hyperedge corresponds, roughly speaking, to a constraint. There are two kinds of hyperedges, those (H_{Nsig_V}) which guarantee no-signaling and those (H_G), depending on the graph G, which avoid the losing pairs:

- H_{Nsig_V} is the hypergraph representing the no-signaling polytope. It corresponds [5] to the Bell scenario $B_{V,2,2}$ where $|V|$ parties have access to 2 local measurements each, each of which has 2 possible outcomes (see Fig. 1), which is obtained as a product[3] of the elementary scenario $B_{1,2,2}$.
- The hypergraph H_G defined on the same vertex set, corresponds to the game constraints: for each question[4] $x \in \{0,1\}^V$ we associate an hyperedge e_x containing all the answers which make the players win on x i.e., $e_x = \{(a|x) \in \{0,1\}^V \times \{0,1\}^V, (a|x) \notin \mathcal{L}\}$.

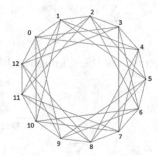

Fig. 1. H_{Nsig_2}: hyperedges of the Bell scenario $B_{2,2,2}$ from [21]

Fig. 2. Paley Graph of order 13

Given a graph $G = (V, E)$, **MCG**(G) is a *pseudo-telepathy* game if it admits a quantum model $(\mathcal{Q}(H_G \cup H_{\mathrm{Nsig}_V}) \neq \emptyset)$ but no classical model $(\mathcal{C}(H_G \cup H_{\mathrm{Nsig}_V}) =$

---

[3] The Foulis Randall product of scenarios [5] is the scenario $H_A \otimes H_B$ with vertices $V(H_A \otimes H_B) = V(H_A) \times V(H_B)$ and edges $E(H_A \otimes H_B) = E_{A \to B} \cup E_{A \leftarrow B}$ where $E_{A \to B} := \{\cup_{a \in e_A}\{a\} \times f(a) : e_A \in E_A, f : e_A \to E_B\}$ and $E_{A \leftarrow B} := \{\cup_{b \in e_A} f(b) \times \{b\} : e_b \in E_b, f : E_B \to E_A\}$. In the multipartite case there are several ways to define products, however they all correspond to the same non-locality constraints [5]. Therefore one can just consider the minimal product $^{\mathrm{min}}\otimes_{i=1}^n H_i$ which has vertices in the cartesian product $V = \Pi V_i$ and edges $\cup_{k \in [1,n]} E_k$ where $E_k = \{(v_1 \ldots, v_n), v_i \in e_i \, \forall i \neq k, v_k \in f(\overrightarrow{v})\}$ for some edge $e_i \in E(H_i)$ for every party $i \neq k$ and a function $\overrightarrow{v} \mapsto f(\overrightarrow{v})$ which assigns to every joint outcome $\overrightarrow{v} = (v_1 \ldots v_{k-1}, v_{k+1}, \ldots v_n)$ an edge $f(\overrightarrow{v}) \in E(H_k)$ (the kth vertex is replaced by a function of the others).

[4] Note that for the questions x for which there exists no D involved in x, all the answers are allowed thus the constraints represented by the associated edge is a hyperedge of no-signaling scenario H_{Nsig}.

$\emptyset$). It has been proven in [6] that $\mathbf{MCG}(G)$ is pseudo-telepathic if and only if G is not bipartite.

Example 3: In a complete graph K_n of order n, there exists a non-empty set D involved in a question $x \in \{0,1\}^V$ if and only if $|x| = 1 \bmod 2$. With each such question x, the associated hyperedge is $e_x = \{(a|x) \in \{0,1\}^V \times \{0,1\}^V \, s.t. \, 2|a| \neq |x| + 1 \bmod 4\}$.

Example 4: In the graph Paley 13 (see Fig. 2), $\mathrm{Odd}(\{0,1,4\}) = \{2,7,8,9,11,12\}$ thus if $\{0,1,4\}$ is involved in x i.e. $x_i = 1$ for $i \in \{0,1,4\}$ and $x_i = 0$ for $i \in \{2,7,8,9,11,12\}$ then the associated pseudo-telepathy game requires that the sum of the outputs of these nine players $\sum_{i \notin \{3,5,6,10\}} a_i$ has to be odd. This corresponds to 8 hyperedges e_{jkl} for $j,k,l \in \{0,1\}$ in the contextuality scenario where $e_{jkl} = \{(a|x), \sum_{i \notin \{5,6,10\}} a_i = 1 \bmod 2, x_i = 1$ for $i \in \{0,1,4\}, x_i = 0$ for $i \in \{2,7,8,9,11,12\}, x_5 = j, x_6 = k, x_{10} = l\}$.

The probabilistic contextuality is what was considered in [6] as it corresponds to investigating the possibility of simulating a probability distribution of a quantum strategy playing with graph states. The two other levels of contextuality gain some new perspectives when iewed as games: indeed the possibilistic contextuality coincides with the fact that the players cannot give all the good answers with non zero probability using classical local strategies, and strong contextuality just means that classical players cannot win the game (even by giving a strict subset of the good answers).

Definition 2. *An interpretation* $p : \{0,1\}^V \times \{0,1\}^V \to [0,1]$ *is* k-*multipartite if it can be obtained by a strategy without communication using nonlocal boxes that are at most* k-*partite: for any set* $I \subset V$ *with* $|I| \leq k$, *each player has access to one bit of a variable* $\lambda_I(a_I|x_I)$ *that has a no-signaling probability distribution.*

In other words, a k-multipartite interpretation can be obtained with no-signaling correlations involving at most k players. For example the strategy to win the Mermin game proposed in [13] where each pair among n players share a (2-partite) non localbox and each player outputs the sum of his boxes' ouputs is a 2-multipartite interpretation. Similarly, the result in [8] where they prove that a probability distribution that can be obtained by 5 players measuring a quantum state cannot be simulated without communication using any number of bi-partite non local boxes shows that it is not a 2-multipartite interpretation.[5]

Definition 3 (multipartiteness width). *A scenario has a multipartiteness width* k *if it admits a* k-*multipartite interpretation but no* $(k-1)$-*multipartite interpretation.*

In a contextual scenario, the more hyperedges one adds the less possible interpretations exist. A scenario has a multipartiteness width k if its hyperedges already forbids all the interpretations of a product of Bell scenarios on less

---

[5] The probability distribution described in [8] corresponds to the quantum winning strategy on the graph state obtained from a cycle with 5 vertices.

than k parties. For a scenario, having a classical interpretation means being decomposable: one can think of the probability distribution as local actors acting each on his bit and that's a classical interpretation. The multipartiteness width measure how non-decomposable a scenario is: it can not be decomposed with interpretations where each subspace has a small width.

It implies that the players cannot perfectly win the game if they have only quantum systems on less than k qubits, this corresponds to using k separable states as ressources as defined in [26].

Note that from the observations in [6] the multipartiteness width of the scenario generated by the Paley graph on 13 (see Fig. 2) is strictly larger than 4.

In the next section, we will show how for the scenarios we describe, being able to give only good answers allows for simulation of the quantum distribution with random variables. Thus, the contextuality lies in the combinatorial structure of the graph and the three levels collapse for these games.

3   Simulating a Probability Distribution is the Same as Winning the Pseudo-Telepathy Graph Game

In [6] it was proven that for some graphs, the probability distributions of the quantum strategy using the graph states cannot be simulated using non local boxes on less than k parties, we show here that any strategy that allows to win the game can be extended using random variables shared between neighbors (in the graph) to simulate the uniform probability distribution arising from the quantum strategy.

We start by describing a classical strategy **CStrat** based on shared random variables rather than quantum states. We show that **CStrat** is a winning strategy if and only if the graph is bi-partite. We also show that **CStrat** can be used to make any winning strategy a uniform winning strategy, i.e. each valid answer to a given question are equiprobable. We show that **CStrat** can be locally adapted to collaborative games on graphs that can be obtained by a sequence of local complementations.

Classical Strategy (Cstrat): Given a graph $G = (V, E)$, pick uniformly at random $\lambda \in \{0, 1\}^V$. Each player $u \in V$ receives a pair of bits (λ_u, μ_u), where $\mu_u = \sum_{v \in N_G(u)} \lambda_u \bmod 2$. Given a question $x \in \{0, 1\}^V$, each player $u \in V$ locally computes and answers $a_u = (1 - x_u).\lambda_u + x_u.\mu_u \bmod 2$.

Lemma 2. *Given a graph $G = (V, E)$ and a question $x \in \{0, 1\}^V$, CStrat produces an answer uniformly at random in $\{a \in \{0, 1\}^V \mid \exists D \subseteq S, (A \oplus Odd(A \oplus D)) \cap S = \emptyset$ where $A = supp(a)$ and $S = supp(x)\}$.*

Proof. Given a graph $G = (V, E)$, a question $x \in \{0, 1\}^V$ and $a \in \{0, 1\}^V$, the probability that **CStrat** outputs a is

$$p(a|x) = p\left(\forall u \in V \backslash S, a_u = \lambda_u\right) p(\forall u \in S, a_u = \sum_{v \in N(u)} \lambda_v \bmod 2 \mid \forall u \in V \backslash S, a_u = \lambda_u)$$

$$= p\left(A \backslash S = \Lambda \backslash S\right) p(A \cap S = Odd(\Lambda) \cap S \mid A \backslash S = \Lambda \backslash S)$$

where $S = \text{supp}(x)$, $A = \text{supp}(a)$ and $\Lambda = \text{supp}(\lambda)$. Since $p(A \setminus S = \Lambda \setminus S) = \frac{1}{2^{n-|x|}}$, $p(a|x) = \frac{1}{2^{n-|x|}} p(A \cap S = \text{Odd}(\Lambda \cap S \oplus \Lambda \setminus S) \cap S | A \setminus S = \Lambda \setminus S) = \frac{1}{2^{n-|x|}} p(A \cap S = \text{Odd}(D \oplus (A \setminus S)) \cap S | A \setminus S = \Lambda \setminus S)$, where $D = \Lambda \cap S$. If $A \cap S \neq \text{Odd}(D \oplus (A \setminus S)) \cap S$ for all $D \subseteq S$, then $p(a|x) = 0$. Otherwise, the set of subsets D of S which satisfy the condition is the affine space $\{D_0 \oplus D \mid D \subseteq S \wedge \text{Odd}(D) \cap S = \emptyset\}$, where D_0 is a fixed set which satisfies $A \cap S = \text{Odd}(D_0 \oplus A \setminus S) \cap S$. Thus the $p(a|x) = \frac{1}{2^{n-|x|}} \cdot \frac{|\{D \subseteq S | \text{Odd}(D) \cap S = \emptyset\}|}{2^{|x|}} = 2^{|x|-rk_G(x)-n}$, which is independent of a, proving the uniformity of the answer. Finally notice $\exists D_0 \subseteq S, A \cap S = \text{Odd}(D_0 \oplus (A \setminus S)) \cap S$ if and only if $\exists D_1 \subseteq S, (A \oplus \text{Odd}(A \oplus D_1)) \cap S = \emptyset$, by taking $D_1 = D_0 \oplus (A \cap S)$. □

We consider some standard graph transformations: Given a graph $G = (V, E)$ the local complementation [11] on a vertex $u \in V$ produces the graph $G * u = (V, E \oplus K_{N(v)})$ where the sum is taken modulo 2 (it is the symmetric difference) and K_U is the complete graph on $U \subset V$. $G * u$ is obtained from G by exchanging the edges by non edges and vice versa in the neighborhood of the vertex u. Pivoting using an edge (u, v), is a sequence of three local complementations $G \wedge uv = G * u * v * u$. We denote by $\delta_{loc}(G)$ [15,30,31] (resp. $\delta_{piv}(G)$) the minimum degree taken over all graphs that can be obtained from G through some sequence of local complementations (edge pivots).

Given the shared randomness $(\lambda_v, \mu_v)_{v \in V}$ associated with G, if player u replaces its first bit by the XOR of its two bits, and each of his neighbors replaces his second bit by the XOR of his two bits, one gets the shared randomness associated with $G * u$. (proof given in Appendix)

Lemma 3. *Given the probability distribution $(\lambda_v, \mu_v)_{v \in V}$ associated with G, if player u replaces its first bit by the XOR of its two bits, and each of its neighbors replaces their second bit by the XOR of their two bits, one gets the probability distribution associated with $G * u$.*

Thus the probability distribution corresponding to the classical strategy for G can be locally transformed into the probability distribution associated with the $G * u$, thus one can use local complementation to optimise the cost of preparing the shared randomness. For instance the classical strategy **CStrat** for a graph G requires shared random bits on at most $\Delta_{loc}(G) + 1$ players, where $\Delta_{loc}(G) = \min(\Delta(G'), \text{s.t. } \exists u_1, \ldots, u_k, G' = G * u_1 * \ldots * u_k)$ and $\Delta(G)$ is its maximum degree. If there is no pre-shared random bits, the probability distribution can be prepared using at most $2|G|_{loc}$ communications in-between the players, where $|G|_{loc} = \min(|G'|, \text{s.t. } \exists u_1, \ldots, u_k, G' = G * u_1 * \ldots * u_k)$ is the minimum number of edges by local complementation.

Now we show how, using the classical strategy **CStrat**, one can simulate the quantum strategy **QStrat** given an oracle that provides only good answers.

Lemma 4. *For any collaborative game on a graph G, for any strategy Q that never loses, there exists a strategy Q' using the outputs of Q and shared random variables that simulate **QStrat**.*

Proof. Given a collaborative graph game on a graph G, let Q be a strategy that always outputs permissible outputs for any set of inputs x, so we have pairs $(a|x) \notin \mathcal{L}$. We consider the strategy which combines Q and **CStrat** for this graph: For a given question x, Q' outputs the XOR of the Q answer and **CStrat** answer for x. First we prove that such an answer is a valid answer and then the uniform probability among the possible answer to a given question. Given a question $x \in \{0,1\}^V$, suppose Q' outputs $a' \in \{0,1\}^V$: $\forall u \in V$, $a'_u = a_u + (1-x_u)\lambda_u + x_u\mu_u$ where a_u is the answer produced by Q and λ and μ are as defined in the classical strategy. By contradiction, assume $(a'|x) \in \mathcal{L}$, so there exists D involved in x such that $\sum_{u \in \text{loc}(D)} a'_u = |G[D]| + 1 \bmod 2$.

$\sum_{u \in \text{loc}(D)} a'_u = \sum_{u \in \text{loc}(D)} (a_u + (1-x_u)\lambda_u + x_u\mu_u) \bmod 2 = \sum_{u \in \text{loc}(D)} a_u + \sum_{u \in \text{loc}(D) \backslash \text{supp}(x)} \lambda_u + \sum_{u \in \text{loc}(D) \cap \text{supp}(x)} \mu_u \bmod 2 = \sum_{u \in \text{loc}(D)} a_u + \sum_{u \in \text{Odd}(D)} \lambda_u + \sum_{u \in D} \sum_{v \in N(u)} \lambda_v \bmod 2 = \sum_{u \in \text{loc}(D)} a_u + \sum_{u \in \text{Odd}(D)} \lambda_u + \sum_{v \in \text{Odd}(D)} \lambda_v \bmod 2 = \sum_{u \in \text{loc}(D)} a_u \bmod 2$. Thus $(a|x) \in \mathcal{L}$ which is a contradiction thus $p(a'|x) = 0$ if $(a'|x) \in \mathcal{L}$. Now we prove that $p(a'|x) = 2^{|x|-n-rk_G(x)}$. First assume Q is determinist, thus $p(a'|x)$ is the probability that the classical strategy outputs $a + a' := (a_u + a'_u \bmod 2)_{u \in V}$. Since this probability is non zero it must be $2^{|x|-n-rk_G(x)}$. If Q is probabilistic, $p(a'|x) = \sum_{a \in \{0,1\}^V} p(Q \text{ outputs } a \text{ on } x) p(\text{classical strategy outputs } a + a' \text{ on } x) \leq 2^{|x|-n-rk_G(x)} \sum_{a \in \{0,1\}^V} p(Q \text{ outputs } a \text{ on } x) \leq 2^{|x|-n-rk_G(x)}$. Thus each answer a produced by the strategy on a given question x is s.t. $(a|x) \notin \mathcal{L}$ and occurs with probability at most $2^{|x|-n-rk_G(x)}$. Since $|\{a \in \{0,1\}^V \mid (a|x) \notin \mathcal{L}\}| = 2^{|x|-n-rk_G(x)}$, each of the possible answers is produced by the strategy and occurs with probability $2^{|x|-n-rk_G(x)}$. □

4   Locally Equivalent Games

A pseudo telepathy game $\mathcal{G}$ locally simulates another pseudo telepathy game $\mathcal{G}'$ if any winning strategy for $\mathcal{G}$ can be locally turned into a winning strategy for $\mathcal{G}'$:

Definition 4 (Local Simulation). *Given two pseudo telepathy games $\mathcal{G}$ and $\mathcal{G}'$ on a set V of players which sets of losing pairs are respectively $\mathcal{L}_{\mathcal{G}}$ and $\mathcal{L}_{\mathcal{G}'}$, $\mathcal{G}$ locally simulates $\mathcal{G}'$ if for all $u \in V$, there exist $f_1, \ldots, f_n : \{0,1\} \to \{0,1\}$ and $g_1, \ldots, g_n : \{0,1\} \times \{0,1\} \to \{0,1\}$ s.t. $\forall x, a \in \{0,1\}^V$ $(g(a,x), x) \in \mathcal{L}_{\mathcal{G}'} \Rightarrow (a|f(x)) \in \mathcal{L}_{\mathcal{G}}$ where $f(x) = (f_u(x_u))_{u \in V}$ and $g(a,x) = (g_u(a_u, x_u))_{u \in V}$.*

Assuming $\mathcal{G}$ locally simulates $\mathcal{G}'$ and that the players have a strategy to win $\mathcal{G}$, the strategy for $\mathcal{G}'$ is as follows: given an input x of $\mathcal{G}'$, each player u applies the preprocessing f_u turning her input x_u into $f_u(x_u)$, then they collectively play the game $\mathcal{G}$ with this input $f(x)$ getting an output a s.t. $(a|f(x)) \notin \mathcal{L}_{\mathcal{G}}$. Finally each player u applies a postprocessing g_u which depends on her output a_u and her initial input x_u to produce the output $g_u(a_u, x_u)$ to the game $\mathcal{G}'$. This output is valid since, by contradiction, $(g(a,x), x) \in \mathcal{L}_{\mathcal{G}'}$ would imply $(a|f(x)) \in \mathcal{L}_{\mathcal{G}}$.

Definition 5 (Local Equivalence). $\mathcal{G}$ and $\mathcal{G}'$ are locally equivalent games if $\mathcal{G}$ locally simulates $\mathcal{G}'$ and $\mathcal{G}'$ locally simulates $\mathcal{G}$.

In the following we give two examples of locally equivalent games (the proofs of equivalence are given in Appendix): first we show that the games associated with the complete graphs are locally equivalent to Mermin parity games, and then that pivoting, a graph theoretical transformation, produces a graph game locally equivalent to the original one:

Lemma 5. For any n, the game associated with the complete graph K_n is locally equivalent to the Mermin parity game on n players.

Lemma 6. Given a graph $G = (V, E)$ and $(u, v) \in E$, the games associated with G and $G \wedge uv$ are locally equivalent.

Therefore, the important quantity for the pre-shared randomness for the strategies defined with a graph is $\Delta_{piv}(G) = \min\{\Delta(G'), G' \text{ pivot}$ equivalent to $G\}$.

5 Scenarios with Linear Multipartiteness Width

We prove that there exist contextuality scenarios with linear multipartiteness width. We use a graph property called k-odd domination which is related [6] to the classical simulation of the quantum probability distribution obtained by playing the associated graph game. Since bipartite graphs correspond to graph games that can be won classically [6], we focus on the non-bipartite case by showing that there exist non-bipartite $0.11n$-odd dominated graphs of order n.

Definition 6 (k-odd domination [6]). A graph $G = (V, E)$ is k-odd dominated (k-o.d.) iff for any $S \in \binom{V}{k}$, there exists a labelling of the vertices in $S = \{v_1, \ldots, v_k\}$ and $C_1, \ldots C_k$, s.t. $\forall i$, $C_i \subseteq V \setminus S$ and $Odd(C_i) \cap \{v_i, \ldots v_k\} = \{v_i\}$ and $C_i \subseteq Even(C_i)$.

Lemma 7. For any $k \geq 0$, $r \geq 0$ and any graph $G = (V, E)$ a graph of order n having two distinct independent sets V_0 and V_1 of order $|V_0| = |V_1| = \lfloor \frac{n-r}{2} \rfloor$, G is k-odd dominated if for any $i \in \{0, 1\}$, and any non-empty $D \subseteq V \setminus V_i$, $|Odd_G(D) \cap V_i| > k - |D|$

Proof. Given $S_0 \subseteq V_0$, $S_1 \subseteq V_1$, and $S_2 \subseteq V_2 = V \setminus (V_0 \cup V_1)$ s.t. $|S_0| + |S_1| + |S_2| = k$, we show that for any $u \in S = S_0 \cup S_1 \cup S_2$, there exists $C_u \subseteq V \setminus S$ s.t. $Odd(C_u) \cap S = \{u\}$ and $C_u \subseteq Even(C_u)$. For any $u \in S$, there exists $i \in \{0, 1\}$ s.t. $u \in S_i \cup S_2$. Let $L_i : 2^{S_i \cup S_2} \to 2^{V_{1-i} \setminus S_{1-i}}$ be the function which maps $D \subseteq S_i \cup S_2$ to $L_i(D) = Odd_G(D) \cap (V_{1-i} \setminus S_{1-i})$. L_i is linear according to the symmetric difference. L_i is injective: for any $D \subseteq S_i \cup S_2$, $Odd(D) \cap (V_{1-i} \setminus S_{1-i}) = \emptyset$ implies $Odd(D) \cap V_{1-i} \subseteq S_{1-i}$, thus $|Odd(D) \cap V_{1-i}| \leq |S_{1-i}|$. notice that $|D| \leq |S_i| + |S_2|$, so $|Odd(D) \cap V_{1-i}| \leq |S_{1-i}| \leq |S_0| + |S_1| + |S_2| - |D| = k - |D|$, so $D = \emptyset$. The matrix representing L_i is nothing but the submatrix $\Gamma_{[S_i \cup S_2, V_{1-i} \setminus S_{1-i}]}$ of the adjacency matrix Γ of G. So its transpose $\Gamma_{[V_{1-i} \setminus S_{1-i}, S_i \cup S_2]}$ is surjective which means that the corresponding linear map $L_i^T : 2^{V_{1-i} \setminus S_{1-i}} \rightarrow 2^{S_i \cup S_2} = C \mapsto Odd_G(C) \cap (V_{1-i} \setminus S_{1-i})$ is surjective, so $\exists C_u \subseteq V_{1-i} \setminus S_{1-i}$ s.t. $Odd_G(C_u) \cap (S_i \cup S_2) = \{u\}$, which implies, since V_{1-i} is an independent set, that $Odd_G(C_u) \cap S = \{u\}$ and $C_u \subseteq Even(C_u)$.   □

Theorem 1. *For any even $n > n_0$, there exists a non-bipartite $\lfloor 0.110n \rfloor$-odd dominated graph of order n.*

Proof. Given n, $r \leq n$ s.t. $r = n \mod 2$, and $k \geq 0$. Let $p = (n-r)/2$, and let $G = (V_0 \cup V_1 \cup V_2, E)$ s.t. $|V_0| = |V_1| = p$, $|V_2| = r$ be a random graph on n vertices s.t. for any $u \in V_i$, $v \in V_j$ there is an edge between u and v with probability 0 if $i = j$ and with probability $1/2$ otherwise. For any $i \in \{0, 1\}$, and any non empty $D \subseteq V \setminus V_i$ s.t. $|D| \leq k$, let $A_D^{(i)}$ be the bad event $|Odd_G(D) \cap V_i| \leq k - |D|$. Since each vertex of V_i is in $Odd_G(D)$ with probability $1/2$, $Pr(A_D^{(i)}) = \sum_{j=0}^{k-|D|} \binom{p}{j} 2^{-p} \leq 2^{p[H(\frac{k-|D|}{p})-1]}$. Another bad event is that G is bipartite which occurs with probability less than $(\frac{7}{8})^{pr}$. Indeed, the probability that given $u \in V_0, v \in V_1, w \in V_2$, (u, v, w) do not form a triangle is $\frac{7}{8}$, so given a bijection $f : V_0 \rightarrow V_1$, the probability that $\forall u \in V_0, \forall w \in V_2$, $(u, f(u), w)$ do not form a triangle is $(\frac{7}{8})^{pr}$. Let X be the number of bad events. $E[X] = 2 \sum_{d=1}^{k} \binom{p+r}{d} \sum_{j=0}^{k-d} \binom{p}{j} 2^{-p} + (\frac{7}{8})^{pr} \leq 2 \sum_{d=1}^{k} 2^{(p+r)H(\frac{d}{p+r})+pH(\frac{k-d}{p})-p} + (\frac{7}{8})^{pr} \leq 2 \sum_{d=1}^{k} 2^{pH(\frac{d}{p+r})+pH(\frac{k-d}{p})-p+r} + (\frac{7}{8})^{pr}$. The function $d \mapsto pH(\frac{d}{p+r})+pH(\frac{k-d}{p}) - p+r$ is maximal for $d = \frac{k(p+r)}{2p+r}$. Thus, $E[X] \leq 2k 2^{2pH(\frac{k}{2p+r})-p+r} + (\frac{7}{8})^{pr}$. By taking $r = 1$, and $k = 0.11n = 0.11(2p+1)$, $E[X] < 1$ when p large enough, thus G has no bad event with a non zero probability.   □

Corollary 1. *There exist contextuality scenarios with linear multipartiteness width: for any even $n > n_0$, there exist graph games on n players producing contextuality scenarios of multipartiteness width at least $\lfloor 0.11n \rfloor$.*

Proof. Using the result from [6], for any non bipartite graph of order n being $0.11n$-o.d ensures that the probability distribution obtained by using the quantum strategy cannot be simulated using non local boxes involving at most $0.11n$ parties. Thus lemma 4 allows to conclude that the associated pseudo-telepathy game cannot be won classically. Therefore there is no interpretation that is k-multipartite with $k < 0.11n$ which means that the contextuality scenario has linear width.   □

6   Conclusion

We have shown that there exist graphs with linear multipartiteness width, however the proof is non constructive and the best known bound for explicit families is logarithmic. A natural future direction of research would be to find explicit families with linear multipartiteness width or to improve the bounds proven for the Paley graph states. An other important question is to consider lower bounds for the scenarios associated with the graph games. A promising area of investigation for multipartite scenarios is: what happens if we limit the width of shared randomness? Indeed, for the proof of how winning the game allows to simulate the quantum probability distributions, one needs only shared random variables that are correlated in local neighborhoods in the graph. One can also consider the link with building entanglement witnesses for graph states, generalizing the construction of [28]. It would be also very interesting to link the multipartiteness width with the structures of the groups of the associated binary linear system defining the two-player bipartite non-local games [41]. Finally, one can expect that the multipartiteness width of the Paley graph states might have cryptographic applications to ensure security against cheating for some protocols for example.

Acknowledgements. We would like to thank an anonymous reviewer for noticing a mistake in an earlier version and helpful comments.

References

1. Abramsky, S., Brandenburger, A.: The sheaf theoretic structure of non locality and contextuality. New J. Phys. **13**, 113036 (2011)
2. Abramsky, S., Barbosa, R.S., Mansfield, S.: Quantifying contextuality via linear programming. In: Informal Proceedings of Quantum Physics and Logic (2016)
3. Abramsky, S., Mansfield, S., Barbosa, R.S.: The cohomology of non-locality and contextuality. Electron. Proc. Theor. Comput. Sci. **95**, 1–14 (2012)
4. Abramsky, S., Barbosa, R.S., Carù, G., Perdrix, S.: A complete characterisation of all-versus-nothing arguments for stabiliser states. arXiv preprint arXiv:1705.08459 (2017)
5. Acín, A., Fritz, T., Leverrier, A., Sainz, B., et al.: A combinatorial approach to nonlocality and contextuality. Comm. Math. Phys. **334**(2), 533–628 (2015)
6. Anshu, A., Mhalla, M.: Pseudo-telepathy games and genuine NS k-way nonlocality using graph states. Quantum Inf. Comput. **13**(9–10), 0833–0845 (2013). Rinton Press
7. Badanidiyuru, A., Langford, J., Slivkins, A.: Resourceful contextual bandits. In: COLT (2014). http://arxiv.org/abs/1402.6779
8. Barrett, J., Pironio, S.: Popescu-Rohrlich correlations as a unit of nonlocality. Phys. Rev. Lett. **95**, 140401 (2005)
9. Bell, J.S.: On the Einstein-Podolsky-Rosen paradox. Physics **1**, 195–200 (1964)
10. Brassard, G., Broadbent, A., Tapp, A.: Multi-party pseudo-telepathy. In: Dehne, F., Sack, J.-R., Smid, M. (eds.) WADS 2003. LNCS, vol. 2748, pp. 1–11. Springer, Heidelberg (2003). doi:10.1007/978-3-540-45078-8_1
11. Bouchet, A.: Connectivity of isotropic systems. In: Proceedings of the Third International Conference on Combinatorial Mathematics, pp. 81–93. New York Academy of Sciences (1989)
12. Broadbent, A., Fitzsimons, J., Kashefi, E.: Universal blind quantum computation. In: 50th Annual IEEE Symposium on Foundations of Computer Science, FOCS 2009 (2009)

13. Broadbent, A., Methot, A.A.: On the power of non-local boxes. Theor. Comput. Sci. C **358**, 3–14 (2006)
14. Browne, D.E., Kashefi, E., Mhalla, M., Perdrix, S.: Generalized flow and determinism in measurement-based quantum computation. New J. Phys. (NJP) **9**(8), 250 (2007). http://iopscience.iop.org/1367-2630/9/8/250/fulltext/
15. Cattanéo, D., Perdrix, S.: Minimum degree up to local complementation: bounds, parameterized complexity, and exact algorithms. In: Elbassioni, K., Makino, K. (eds.) ISAAC 2015. LNCS, vol. 9472, pp. 259–270. Springer, Heidelberg (2015). doi:10.1007/978-3-662-48971-0_23
16. Coecke, B.: From quantum foundations via natural language meaning to a theory of everything. arXiv:1602.07618v1 (2016)
17. Danos, V., Kashefi, E., Panangaden, P., Perdrix, S.: Extended Measurement Calculus. Cambridge University Press, Cambridge (2010)
18. Danos, V., Kashefi, E.: Determinism in the one-way model. Phys. Rev. A **74**(5), 052310 (2006)
19. Dzhafarov, E., Jordan, S., Zhang, R., Cervantes, V. (eds.): Contextuality from Quantum Physics to Psychology. Advanced Series on Mathematical Psychology, vol. 6. World Scientific Press, New Jersey (2015)
20. Foulis, D.J., Randall, C.H.: Empirical logic and tensor products. J. Math. Phys. **5**, 9–20 (1981). MR683888
21. Fritz, T., Sainz, A.B., Augusiak, R., Bohr Brask, J., Chaves, R., Leverrier, A., Acín, A.: Local orthogonality as a multipartite principle for quantum correlations. Nat. Commun. **4**, 2263 (2013)
22. Gravier, S., Javelle, J., Mhalla, M., Perdrix, S.: Quantum secret sharing with graph states. In: Kučera, A., Henzinger, T.A., Nešetřil, J., Vojnar, T., Antoš, D. (eds.) MEMICS 2012. LNCS, vol. 7721, pp. 15–31. Springer, Heidelberg (2013). doi:10. 1007/978-3-642-36046-6_3
23. Grudka, A., Horodecki, K., Horodecki, M., Horodecki, P., Horodecki, R., Joshi, P., Klobus, W., Wójcik, A.: Quantifying contextuality. Phys. Rev. Lett. **112**, 120401 (2014)
24. Greenberger, D.M., Horne, M.A., Shimony, A., Zeilinger, A.: Bell's theorem without inequalities. Am. J. Phys. **58**, 1131 (1990)
25. Gnaciński, P., Rosicka, M., Ramanathan, R., Horodecki, K., Horodecki, M., Horodecki, P., Severini, S.: Linear game non-contextuality and Bell inequalities - a graph-theoretic approach. e-print arXiv:1511.05415, November 2015
26. Gühne, O., Tóth, G., Briegel, H.J.: Multipartite entanglement in spin chains. New J. Phys. **7**(1), 229 (2005)
27. Hardy, L.: Nonlocality for two particles without inequalities for almost all entangled states. Phys. Rev. Lett. **71**, 1665–1668 (1993)
28. Hein, M., Dür, W., Eisert, J., Raussendorf, R., Nest, M., Briegel, H.J.: Entanglement in graph states and its applications. arXiv preprint quant-ph/0602096 (2006)
29. Howard, M., Wallman, J.J., Veitch, V., Emerson, J.: Contextuality supplies the "magic" for quantum computation. Nature **510**, 351–355 (2014)
30. Høyer, P., Mhalla, M., Perdrix, S.: Resources required for preparing graph states. In: Asano, T. (ed.) ISAAC 2006. LNCS, vol. 4288, pp. 638–649. Springer, Heidelberg (2006). doi:10.1007/11940128_64
31. Javelle, J., Mhalla, M., Perdrix, S.: On the minimum degree up to local complementation: bounds and complexity. In: Golumbic, M.C., Stern, M., Levy, A., Morgenstern, G. (eds.) WG 2012. LNCS, vol. 7551, pp. 138–147. Springer, Heidelberg (2012). doi:10.1007/978-3-642-34611-8_16

32. Liang, Y.-C., Rosset, D., Bancal, J.-D., Pütz, G., Barnea, T.J., Gisin, N.: Family of bell-like inequalities as device-independent witnesses for entanglement depth. Phys. Rev. Lett. **114**(19), 190401 (2015)
33. McKague, M.: Interactive proofs for BQP via self-tested graph states. Theory Comput. **12**(3), 1–42 (2016)
34. Marin, A., Markham, D., Perdrix, S.: Access structure in graphs in high dimension and application to secret sharing. In: 8th Conference on the Theory of Quantum Computation, Communication and Cryptography, p. 308 (2013)
35. Markham, D., Sanders, B.C.: Graph states for quantum secret sharing. Phys. Rev. A **78**, 042309 (2008)
36. Mermin, N.D.: Extreme quantum entanglement in a superposition of macroscopically distinct states. Phys. Rev. Lett. **65**(15), 1838–1849 (1990)
37. Mhalla, M., Murao, M., Perdrix, S., Someya, M., Turner, P.S.: Which graph states are useful for quantum information processing? In: Bacon, D., Martin-Delgado, M., Roetteler, M. (eds.) TQC 2011. LNCS, vol. 6745, pp. 174–187. Springer, Heidelberg (2014). doi:10.1007/978-3-642-54429-3_12
38. Mhalla, M., Perdrix, S.: Graph states, pivot minor, and universality of (X-Z) measurements. IJUC **9**(1–2), 153–171 (2013)
39. Perdrix, S., Sanselme, L.: Determinism and computational power of real measurement-based quantum computation. In: 21st International Symposium on Fundamentals of Computation Theory (FCT 2017) (2017)
40. Raussendorf, R., Briegel, H.J.: A one-way quantum computer. Phys. Rev. Lett. **86**, 5188–5191 (2001)
41. Slofstra, W.: Tsirelson's problem and an embedding theorem for groups arising from non-local games. e-print arXiv:1606.03140 (2016)
42. Zeng, W., Zahn, P.: Contextuality and the weak axiom in the theory of choice. In: Atmanspacher, H., Filk, T., Pothos, E. (eds.) QI 2015. LNCS, vol. 9535, pp. 24–35. Springer, Cham (2016). doi:10.1007/978-3-319-28675-4_3

Generalized Satisfiability Problems via Operator Assignments

Albert Atserias[1]([⊠]), Phokion G. Kolaitis[2,3]([⊠]), and Simone Severini[4,5]([⊠])

[1] Universitat Politècnica de Catalunya, Barcelona, Spain
atserias@cs.upc.edu
[2] University of California Santa Cruz, Santa Cruz, USA
kolaitis@cs.ucsc.edu
[3] IBM Research - Almaden, San Jose, USA
[4] University College London, London, UK
s.severini@ucl.ac.uk
[5] Shanghai Jiao Tong University, Shanghai, China

Abstract. Schaefer introduced a framework for generalized satisfiability problems on the Boolean domain and characterized the computational complexity of such problems. We investigate an algebraization of Schaefer's framework in which the Fourier transform is used to represent constraints by multilinear polynomials. The polynomial representation of constraints gives rise to a relaxation of the notion of satisfiability in which the values to variables are linear operators on some Hilbert space. For constraints given by a system of linear equations over the two-element field, this relaxation has received considerable attention in the foundations of quantum mechanics, where such constructions as the Mermin-Peres magic square show that there are systems that have no solutions in the Boolean domain, but have solutions via operator assignments on some finite-dimensional Hilbert space. We completely characterize the classes of Boolean relations for which there is a gap between satisfiability in the Boolean domain and the relaxation of satisfiability via operator assignments. To establish our main result, we adapt the notion of primitive-positive definability (pp-definability) to our setting, a notion that has been used extensively in the study of constraint satisfaction. Here, we show that pp-definability gives rise to gadget reductions that preserve satisfiability gaps, and also give several additional applications.

1 Introduction and Summary of Results

In 1978, Schaefer [17] classified the computational complexity of generalized satisfiability problems. Each class A of Boolean relations gives rise to the generalized satisfiability problem SAT(A). An instance of SAT(A) is a conjunction of relations from A such that each conjunct has a tuple of variables as arguments; the question is whether or not there is an assignment of Boolean values to the variables, so that, for each conjunct, the resulting tuple of Boolean values belongs to the underlying relation. Schaefer's main result is a dichotomy theorem for the computational complexity of SAT(A), namely, depending on A, either SAT(A) is NP-complete or SAT(A) is solvable in polynomial time. Schaefer's

© Springer-Verlag GmbH Germany 2017
R. Klasing and M. Zeitoun (Eds.): FCT 2017, LNCS 10472, pp. 56–68, 2017.
DOI: 10.1007/978-3-662-55751-8_6

dichotomy theorem provided a unifying explanation for the NP-completeness of many well-known variants of Boolean satisfiability, such as POSITIVE 1-IN-3 SAT and MONOTONE 3SAT, and became the catalyst for numerous subsequent investigations in computational complexity and constraint satisfaction.

Every Boolean relation can be identified with its characteristic function, which, via the Fourier transform, can be represented as a multilinear polynomial (i.e., a polynomial in which each variable has degree at most one) in a unique way. In carrying out this transformation, the truth values *false* and *true* are represented by $+1$ and -1, instead of 0 and 1. Thus, the multilinear polynomial representing the conjunction $x \wedge y$ is $\frac{1}{2}(1 + x + y - xy)$. The multilinear polynomial representation of Boolean relations makes it possible to consider relaxations of satisfiability in which the variables take values in some suitable space, instead of the two-element Boolean algebra. Such relaxations have been considered in the foundations of physics several decades ago, where they have played a role in singling out the differences between classical theory and quantum theory. In particular, it has been shown that there is a system of linear equations over the two-element field that has no solutions over $\{+1, -1\}$, but the system of the associated multilinear polynomials has a solution in which the variables are assigned linear operators on a Hilbert space of dimension four. The Mermin-Peres magic square [12,13,15] is the most well known example of such a system. These constructions give *small proofs* of the celebrated Kochen-Specker Theorem [8] on the impossibility to explain quantum mechanics via hidden-variables [2]. More recently, systems of linear equations with this relaxed notion of solvability have been studied under the name of *binary constraint systems*, and tight connections have been established between solvability and the existence of perfect strategies in non-local games that make use of entanglement [4,5].

A Boolean relation is *affine* if it is the set of solutions of a system of linear equations over the two-element field. The collection LIN of all affine relations is prominent in Schaefer's dichotomy theorem, as it is one of the main classes A of Boolean relations for which SAT(A) is solvable in polynomial time. The discussion in the preceding paragraph shows that SAT(LIN) has instances that are unsatisfiable in the Boolean domain, but are satisfiable when linear operators on a Hilbert space are assigned to variables (for simplicity, from now on we will use the term "operator assignments" for such assignments). Which other classes of Boolean relations exhibit such a gap between satisfiability in the Boolean domain and the relaxation of satisfiability via operator assignments? As a matter of fact, this question bifurcates into two separate questions, depending on whether the relaxation allows linear operators on Hilbert spaces of arbitrary (finite or infinite) dimension or only on Hilbert spaces of finite dimension. In a recent breakthrough paper, Slofstra [18] showed that these two questions are different for LIN by establishing the existence of systems of linear equations that are satisfiable by operator assignments on some infinite-dimensional Hilbert space, but are not satisfiable by operator assignments on any finite-dimensional Hilbert space. In a related vein, Ji [11] showed that a 2CNF-formula is satisfiable in the Boolean domain if and only if it is satisfiable by an operator assignment in some finite-dimensional Hilbert space. Moreover, Ji showed that the same holds true for Horn formulas. Note that 2SAT, HORN SAT, and DUAL HORN SAT also feature prominently in Schaefer's dichotomy theorem as, together with SAT(LIN) which from now on we will denote by LIN SAT, they constitute the main tractable cases of generalized satisfiability problems (the other tractable cases are the trivial cases of SAT(A), where A is a class of 0-valid relations or a class of 1-valid relations, i.e., Boolean relations that contain the tuple consisting entirely of 0's or, respectively, the tuple consisting entirely of 1's).

In this paper, we completely characterize the classes A of Boolean relations for which SAT(A) exhibits a gap between satisfiability in the Boolean domain and satisfiability via operator assignments. Clearly, if every relation in A is 0-valid or every relation in A is 1-valid, then there is no gap, as every constraint is satisfied by assigning to every variable the identity operator or its negation. Beyond this, we first generalize and extend Ji's results [11] by showing that if A is a *bijunctive* class of Boolean relations (i.e., every relation in A is the set of satisfying assignments of a 2-CNF formula), or A is *Horn* (i.e., every relation in A is the set of satisfying assignments of a Horn formula), or A is *dual Horn* (i.e., every relation in A is the set of satisfying assignments of a dual Horn formula), then there is no gap whatsoever between satisfiability in the Boolean domain and satisfiability via operators of Hilbert spaces of any dimension. In contrast, we show that for all other classes A of Boolean relations, SAT(A) exhibits a two-level gap: there are instances of SAT(A) that are not satisfiable in the Boolean domain, but are satisfiable by an operator assignment on some finite-dimensional Hilbert space, and there are instances of SAT(A) that are not satisfiable by an operator assignment on any finite-dimensional Hilbert space, but are satisfiable by an operator assignment on some (infinite-dimensional) Hilbert space.

The proof of this result uses several different ingredients. First, we use the substitution method [5] to show that there is no satisfiability gap for classes of relations that are bijunctive, Horn, and dual Horn. This gives a different proof of Ji's results [11], which were for finite-dimensional Hilbert spaces, but also shows that, for such classes of relations, there is no difference between satisfiability by linear operators on finite-dimensional Hilbert spaces and satisfiability by linear operators on arbitrary Hilbert spaces. The main tool for proving the existence of a two-level gap for the remaining classes of Boolean relations is the notion of *pp-definability*, that is, definability via primitive-positive formulas, which are existential first-order formulas having a conjunction of (positive) atoms as their quantifier-free part. In the past, primitive-positive formulas have been used to design polynomial-time reductions between decision problems; in fact, this is one of the main techniques in the proof of Schaefer's dichotomy theorem. Here, we show that primitive-positive formulas can also be used to design *gap-preserving* reductions, that is, reductions that preserve the gap between satisfiability on the Boolean domain and satisfiability by operator assignments. To prove the existence of a two-level gap for classes of Boolean relations we combine gap-preserving reductions with the two-level gap for LIN discussed earlier (i.e., the results of Mermin [12, 13], Peres [15], and Slofstra [18]) and with results about Post's lattice of clones on the Boolean domain [16].

We also give two additional applications of pp-definability. First, we consider an extension of pp-definability in which the existential quantifiers may range over linear operators on some finite-dimensional Hilbert space. By analyzing closure operations on sets of linear operators, we show that, perhaps surprisingly, this extension of pp-definability is not more powerful than standard pp-definability, i.e., if a Boolean relation is pp-definable in the extended sense from other Boolean relations, then it is also pp-definable from the same relations. Second, we apply pp-definability to the problem of quantum realizability of contextuality scenarios. Recently, Fritz [8] used Slofstra's results [18] to resolve two problems raised by Acín et al. in [1]. Using pp-definability and Slofstra's results, we obtain new proofs of Fritz's results that have the additional feature that the parameters involved are optimal. Complete proofs of all results are found in the full version of the paper at https://arxiv.org/abs/1704.01736.

2 Definitions and Technical Background

For an integer n, we write $[n]$ for the set $\{1, \ldots, n\}$. We use mainly the $1, -1$ representation of the Boolean domain (1 for "false" and -1 for "true"). We write $\{\pm 1\}$ for the set $\{+1, -1\}$. Every $f : \{\pm 1\}^n \to \mathbb{C}$ has a unique representation as a multilinear polynomial P_f in $\mathbb{C}[X_1, \ldots, X_n]$ given by the Fourier or Hadamard-Welsh transform [14]. The polynomial represents f in the sense that $P_f(a) = f(a)$, for every $a \in \{\pm 1\}^n$. All Hilbert spaces of finite dimension d are isomorphic to $\mathbb{C}^d$ with the standard complex inner product, so we identify its linear operators with the matrices in $\mathbb{C}^{d \times d}$. A matrix A is *Hermitian* if it is equal to its conjugate transpose A^*. A matrix A in *unitary* if $A^* A = A A^* = I$. Two matrices A and B *commute* if $AB = BA$; a collection of matrices $A_1, \ldots, A_r$ *pairwise commute* if $A_i A_j = A_j A_i$, for all $i, j \in [r]$. See [10], for the basics of Hilbert spaces, including the concepts of bounded linear operator and of adjoint A^* of a densely defined linear operator A. See [7], for the definitions of L^2-spaces and L^∞-spaces.

A *Boolean constraint language* A is a collection of relations over the Boolean domain $\{\pm 1\}$. Let $V = \{X_1, \ldots, X_n\}$ be a set of variables. An *instance* $\mathcal{I}$ on the variables V over the constraint language A is a finite collection of pairs

$$\mathcal{I} = ((Z_1, R_1), \ldots, (Z_m, R_m)) \tag{1}$$

where each R_i is a relation from A and $Z_i = (Z_{i,1}, \ldots, Z_{i,r_i})$ is a tuple of variables from V or constants from $\{\pm 1\}$, where r_i is the arity of R_i. Each pair (Z_i, R_i) is a *constraint*, and each Z_i is its *constraint-scope*. A *Boolean assignment* is an assignment $f : X_1, \ldots, X_n \mapsto a_1, \ldots, a_n$ of a Boolean value in $\{\pm 1\}$ to each variable. The assignment *satisfies* the i-th constraint if the tuple $f(Z_i) = (f(Z_{i,1}), \ldots, f(Z_{i,r_i}))$ belongs to R_i. The *value of f on $\mathcal{I}$* is the fraction of constraints that are satisfied by f. The *value* of $\mathcal{I}$, denoted by $\nu(\mathcal{I})$, is the maximum value over all Boolean assignments; $\mathcal{I}$ is *satisfiable* if $\nu(\mathcal{I}) = 1$.

Let $\mathcal{H}$ be a Hilbert space. An *operator assignment* for $X_1, \ldots, X_n$ over $\mathcal{H}$ is an assignment $f : X_1, \ldots, X_n \mapsto A_1, \ldots, A_n$ of a bounded linear operator A_j on $\mathcal{H}$ to each variable X_j, so that each A_j is self-adjoint and squares to the identity, i.e., $A_j^* = A_j$ and $A_j^2 = I$, for all $j \in [n]$. If S is a subset of $\{X_1, \ldots, X_n\}$, the operator assignment $A_1, \ldots, A_n$ *pairwise commutes* on S if, in addition, $A_j A_k = A_k A_j$ holds for all X_j and X_k in the set S. If it pairwise commutes on the whole set $\{X_1, \ldots, X_n\}$, then the assignment *fully commutes*.

Let A be a Boolean constraint language, let $\mathcal{I}$ be an instance over A, with n variables $X_1, \ldots, X_n$ as in (1), and let $\mathcal{H}$ be a Hilbert space. An *operator assignment for $\mathcal{I}$ over $\mathcal{H}$* is an operator assignment $f : X_1, \ldots, X_n \mapsto A_1, \ldots, A_n$ that pairwise commutes on the set of variables of each constraint scope Z_i in $\mathcal{I}$; explicitly, $A_j A_k = A_k A_j$ holds, for all X_j and X_k in Z_i and all $i \in [m]$. We also require that f maps -1 and $+1$ to $-I$ and I, respectively. We say that the assignment f *satisfies* the i-th constraint if $P_{R_i}(f(Z_i)) = P_{R_i}(f(Z_{i,1}), \ldots, f(Z_{i,r_i})) = -I$, where P_{R_i} denotes the unique multilinear polynomial representation of the characteristic function of the relation R_i that takes value -1 on tuples in R_i and 1 on tuples not in R. Since $f(Z_{i,1}), \ldots, f(Z_{i,r_i})$ are required to commute by definition, this notation is unambiguous, despite the fact that P_{R_i} is defined as a polynomial in commuting variables. We abbreviate the terms "operator assignment" over a finite-dimensional Hilbert space by *fd-operator assignment*, and "operator assignment over a finite-dimensional or infinite-dimensional Hilbert" space by *operator assignment*. We write $\nu^*(\mathcal{I})$ to denote the maximum fraction of constraints of $\mathcal{I}$ that can be satisfied by an fd-operator assignment. We write $\nu^{**}(\mathcal{I})$ to denote the maximum fraction of constraints of $\mathcal{I}$ that can be satisfied by an operator assignment.

3   The Strong Spectral Theorem

The Spectral Theorem plays an important role in linear algebra and functional analysis. It has also been used in the foundations of quantum mechanics (for some recent uses, see [5,11]). We will make a similar use of it, but we will also need the version of this theorem for infinite-dimensional Hilbert spaces.

The basic form of the Spectral Theorem for complex matrices states that if A is a $d \times d$ Hermitian matrix, then there exist a unitary matrix U and a diagonal matrix E such that $A = U^{-1}EU$. The Strong Spectral Theorem (SST) applies to sets of pairwise commuting Hermitian matrices and is stated as follows.

Theorem 1 (Strong Spectral Theorem (SST): finite-dimensional case). *Let $A_1, \ldots, A_r$ be $d \times d$ Hermitian matrices, where d is a positive integer. If $A_1, \ldots, A_r$ pairwise commute, then there exist a unitary matrix U and diagonal matrices $E_1, \ldots, E_r$ such that $A_i = U^{-1}E_iU$, for every $i \in [r]$.*

This form of the SST will be enough to discuss satisfiability via fd-operators. For operator assignments over arbitrary Hilbert spaces, we need to appeal to the most general form of the SST in which the role of diagonal matrices is played by the *multiplication operators* on an $L^2(\Omega, \mu)$-space. For each a in $L^\infty(\Omega, \mu)$, the multiplication operator T_a acts on $L^2(\Omega, \mu)$ and is defined by $(T_a(f))(x) = a(x)f(x)$, for all $x \in \Omega$. The result we need [6, Theorem 1.47] is as follows: If $A_1, \ldots, A_r$ are pairwise commuting normal bounded linear operators on a Hilbert space $\mathcal{H}$, then there is a measure space $(\Omega, \mathcal{M}, \mu)$, a unitary map $U : \mathcal{H} \to L^2(\Omega, \mu)$, and $a_1, \ldots, a_r$ in $L^\infty(\Omega, \mu)$ such that $A_i = U^{-1}T_{a_i}U$, for all $i \in [r]$.

The following lemma encapsulates a frequently used application of the SST.

Lemma 1. *Let $X_1, \ldots, X_r$ be variables, let $Q_1, \ldots, Q_m, Q$ be polynomials in $\mathbb{C}[X_1, \ldots, X_r]$, and let $\mathcal{H}$ be a Hilbert space. If every Boolean assignment that satisfies the equations $Q_1 = \cdots = Q_m = 0$ also satisfies the equation $Q = 0$, then every fully commuting operator assignment over $\mathcal{H}$ that satisfies the equations $Q_1 = \cdots = Q_m = 0$ also satisfies the equation $Q = 0$.*

4   Reductions via Primitive Positive Formulas

A *primitive-positive* (pp-)formula is one of the form

$$\phi(x_1, \ldots, x_r) = \exists y_1 \cdots \exists y_t \, (S_1(w_1) \wedge \cdots \wedge S_m(w_m)), \qquad (2)$$

where each S_i is a relation symbol of arity r_i and each w_i is an r_i-tuple of variables or constants from $\{x_1, \ldots, x_r\} \cup \{y_1, \ldots, y_s\} \cup \{\pm 1\}$. Let A be a Boolean constraint language. A relation $R \subseteq \{\pm 1\}^r$ is *pp-definable from* A if there exists a pp-formula as in (2) with symbols for the relations of A such that R is the set of all tuples $(a_1, \ldots, a_r) \in \{\pm 1\}^r$ such that $\phi(x_1/a_1, \ldots, x_r/a_r)$ is true in A. A Boolean constraint language A is *pp-definable from* a Boolean constraint language B if every relation in A is pp-definable from B.

Let A and B be two Boolean constraint languages such that A is pp-definable from B. For R in A, let ϕ_R be a formula as in (2) that defines R from B, where now $S_1, \ldots, S_m$ are relations from B. For every instance $\mathcal{I}$ of A, we construct an instance $\mathcal{J}$ of B as follows.

Consider a constraint (Z, R) in $\mathcal{I}$, where $Z = (Z_1, \ldots, Z_r)$ is a tuple of variables of $\mathcal{I}$ or constants in $\{\pm 1\}$. In addition to the variables in Z, we augment $\mathcal{J}$ with fresh variables $Y_1, \ldots, Y_t$ for the quantified variables $y_1, \ldots, y_t$ in ϕ_R. We also add one constraint (W_j, S_j) for each $j \in [m]$, where W_j is the tuple of variables and constants obtained from w_j by replacing the variables in $x_1, \ldots, x_r$ by the corresponding components $Z_1, \ldots, Z_r$ of Z, replacing every variable y_i occurring in w_j by the corresponding Y_i, and leaving all constants untouched. We do this for each constraint in $\mathcal{I}$, one by one. The collection of variables $Z_1, \ldots, Z_r, Y_1, \ldots, Y_t$ is referred to as the *block* of (Z, R) in $\mathcal{J}$.

This construction is referred to as a *gadget reduction* in the literature. Its main property for satisfiability in the Boolean domain is that $\mathcal{I}$ is satisfiable in the Boolean domain if and only if $\mathcal{J}$ is. We omit the very easy proof of this fact. Our goal in the rest of this section is to show that one direction of this basic property of gadget reductions is also true for satisfiability via operators, for both finite-dimensional and infinite-dimensional Hilbert spaces, and that the other direction is *almost true* in a sense that we will make precise in due time.

Lemma 2. *Let $\mathcal{I}$ and $\mathcal{J}$ be as above and let $\mathcal{H}$ be a Hilbert space. For every satisfying operator assignment f for $\mathcal{I}$ over $\mathcal{H}$, there exists a satisfying operator assignment g for $\mathcal{J}$ over $\mathcal{H}$ that extends f. Moreover, g is pairwise commuting on each block of $\mathcal{J}$.*

So far we discussed that satisfying operator assignments for $\mathcal{I}$ lift to satisfying operator assignments for $\mathcal{J}$. We do not know if the converse is true. To achieve a version of the converse, we modify slightly the instance $\mathcal{J}$ into a new instance $\hat{\mathcal{J}}$. This new construction will be useful later on.

In the sequel, let T denote the full binary Boolean relation, i.e., $T = \{\pm 1\}^2$. Observe that the indicator polynomial $P_T(X_1, X_2)$ of the relation T is just the constant function -1; the letter T stands for *true*.

Let A and B be two constraint languages such that A is pp-definable from B. Let $\mathcal{I}$ and $\mathcal{J}$ be the instances over A and B as defined above. The modified version of $\mathcal{J}$ will be an instance over the expanded constraint language $B \cup \{T\}$. This instance is denoted by $\hat{\mathcal{J}}$ and it is defined as follows: the variables and the constraints of $\hat{\mathcal{J}}$ are defined as in $\mathcal{J}$, but we also add all binary constraints of the form $((X_i, X_j), T)$, $((X_i, Y_k), T)$ or $((Y_k, Y_\ell), T)$, for every four different variables X_i, X_j, Y_k and Y_ℓ that come from the same block in $\mathcal{J}$.

We show that, in this new construction, satisfying assignments not only lift from $\mathcal{I}$ to $\hat{\mathcal{J}}$, but also project from $\hat{\mathcal{J}}$ to $\mathcal{I}$.

Lemma 3. *Let $\mathcal{I}$ and $\hat{\mathcal{J}}$ be as above and let $\mathcal{H}$ be a Hilbert space. Then the following statements are true.*

1. *For every satisfying operator assignment f for $\mathcal{I}$ over $\mathcal{H}$, there exists a satisfying operator assignment g for $\hat{\mathcal{J}}$ over $\mathcal{H}$ that extends f.*
2. *For every satisfying operator assignment g for $\hat{\mathcal{J}}$ over $\mathcal{H}$, the restriction f of g to the variables of $\mathcal{I}$ is a satisfying operator assignment for $\mathcal{I}$ over $\mathcal{H}$.*

5   Satisfiability Gaps via Operator Assignments

Let A be a Boolean constraint language and let $\mathcal{I}$ be an instance over A. It is easy to see that the following inequalities hold:

$$\nu(\mathcal{I}) \leq \nu^*(\mathcal{I}) \leq \nu^{**}(\mathcal{I}). \tag{3}$$

Indeed, the first inequality holds because if we interpret the field of complex numbers $\mathbb{C}$ as a 1-dimensional Hilbert space, then the only solutions to the equation $X^2 = 1$ are $X = -1$ and $X = +1$. The second inequality is a direct consequence of the definitions. For the same reason, if $\mathcal{I}$ is satisfiable in the Boolean domain, then it is satisfiable via fd-operators, and if it is satisfiable via fd-operators, then it is satisfiable via operators. The converses are, in general, not true; however, finding counterexamples is a non-trivial task. For the Boolean constraint language LIN of affine relations, counterexamples are given by Mermin's magic square [12,13] for the first case, and by Slofstra's recent construction [18] for the second case. These will be discussed at some length in due time. In the rest of this section, we characterize the constraint languages that exhibit such gaps.

We distinguish three types of gaps. We say that an instance $\mathcal{I}$ is

1. a *satisfiability gap of the first kind* if $\nu(\mathcal{I}) < 1$ and $\nu^*(\mathcal{I}) = 1$;
2. a *satisfiability gap of the second kind* if $\nu(\mathcal{I}) < 1$ and $\nu^{**}(\mathcal{I}) = 1$;
3. a *satisfiability gap of the third kind* if $\nu^*(\mathcal{I}) < 1$ and $\nu^{**}(\mathcal{I}) = 1$.

As a mnemonic rule, count the number of stars * that appear in the defining inequalities in 1, 2 or 3 to recall what kind the gap is.

We say that a Boolean constraint language A has a *satisfiability gap of the i-th kind*, $i = 1, 2, 3$, if there is at least one instance $\mathcal{I}$ over A that witnesses such a gap. Clearly, a gap of the first kind or a gap of the third kind implies a gap of the second kind. In other words, if A has no gap of the second kind, then A has no gap of the first kind and no gap of the third kind. A priori no other relationships seem to hold. We show that, in a precise sense, either A has no gaps of any kind or A has a gap of every kind. Recall from Sect. 4 that T denotes the full binary Boolean relation; i.e. $T = \{\pm 1\}^2$. We are now ready to state and prove the main result of this section.

Theorem 2. *Let A be a Boolean constraint language. Then the following statements are equivalent.*

1. *A does not have a satisfiability gap of the first kind.*
2. *A does not have a satisfiability gap of the second kind.*
3. *$A \cup \{T\}$ does not have a satisfiability gap of the third kind,*
4. *A is 0-valid, or 1-valid, or bijunctive, or Horn, or dual Horn.*

The proof of Theorem 2 has two main parts. In the first part, we show that if A satisfies at least one of the conditions in the fourth statement, then A has no satisfiability gaps of the first kind or the second kind, and $A \cup \{T\}$ has no satisfiability gaps of the third kind. In the second part, we show that, in all other cases, A has satisfiability gaps of the first kind and the second kind, and $A \cup \{T\}$ has satisfiability gaps of the third kind. The ingredients in the proof of the second part are the existence of gaps of all three kinds for LIN, results about Post's lattice [16], and *gap-preserving* reductions that use the results about pp-definability established in Sect. 4.

5.1 No Gaps of Any Kind

Assume that A satisfies at least one of the conditions in the fourth statement in Theorem 2. First, we observe that the full relation T is 0-valid, 1-valid, bijunctive, Horn, and dual Horn. Indeed, T is obviously 0-valid and 1-valid. Moreover, it is bijunctive, Horn, and dual Horn because it is equal to the set of satisfying assignments of the Boolean formula $(x \vee \neg x) \wedge (y \vee \neg y)$, which is bijunctive, Horn, and dual Horn. Therefore, to prove that the fourth statement in Theorem 2 implies the other three statement, it suffices to prove that if A satisfies at least one of the conditions in the fourth statement, then A has no gaps of any kind.

The cases in which A is 0-valid or A is 1-valid are trivial. Next, we have to show that if A is bijunctive or Horn or dual Horn, then A has no gaps of any kind. It suffices to show that A does not have a gap of the second kind.

Ji [11] proved that if $\mathcal{I}$ is a 2SAT instance or a HORN SAT instance that is satisfiable via fd-operators, then $\mathcal{I}$ is also satisfiable in the Boolean domain.

We give an alternative proof that does not rely on the existence of eigenvalues and thus applies to Hilbert spaces of arbitrary dimension. Our proof is based on the manipulation of non-commutative polynomial identities, a method that has been called *the substitution method* (see, e.g., [5]).

Lemma 4. *Let $\mathcal{I}$ be an instance of* 2SAT, HORN SAT, *or* DUAL HORN SAT. *Then $\mathcal{I}$ is satisfiable in the Boolean domain if and only if it is satisfiable via fd-operators, if and only if it is satisfiable via operators.*

Note that Lemma 4 does not immediately yield that *4 $\Longrightarrow$ 1* in Theorem 2: for bijunctive constraint-languages, for example, the relations are defined by conjunctions of 2-clauses, but need not be defined by individual 2-clauses. To show that *4 $\Longrightarrow$ 1*, we apply Lemma 1.

5.2   Gaps of Every Kind

Assume that A satisfies none of the conditions in the fourth statement in Theorem 2, i.e., A is not 0-valid, A is not 1-valid, A is not bijunctive, A is not Horn, and A is not dual Horn. We will show that A has a satisfiability gap of the first kind (hence, A also has a satisfiability gap of the second kind) and $A \cup \{T\}$ has a satisfiability gap of the third kind. As a stepping stone, we will use the known fact that LIN has gaps of every kind. We now discuss the proof of this fact and give the appropriate references to the literature.

Recall that LIN is the class of all affine relations, i.e., Boolean relations that are the set of solutions of a system of linear equations over the two-element field. In the ± 1-representation, every such equation is a *parity* equation of the form $\prod_{i=1}^{r} x_i = y$, where $y \in \{\pm 1\}$. Mermin [12,13] considered the following system:

$$
\begin{array}{ll}
X_{11}X_{12}X_{13} = 1 & X_{11}X_{21}X_{31} = 1 \\
X_{21}X_{22}X_{23} = 1 & X_{12}X_{22}X_{32} = 1 \\
X_{31}X_{32}X_{33} = 1 & X_{13}X_{23}X_{33} = -1.
\end{array}
\tag{4}
$$

We denote this system of parity questions by $\mathcal{M}$. It is easy to see that $\mathcal{M}$ has no solutions in the Boolean domain. Mermin [12,13] showed it has a solution consisting of linear operators on a Hilbert space of dimension four. Thus, in our terminology, Mermin established the following result.

Theorem 3 [12,13]. *$\mathcal{M}$ is a satisfiability gap of the first kind for* LIN.

Cleve and Mittal [5, Theorem 1] have shown that a system of parity equations has a solution consisting of linear operators on a finite-dimensional Hilbert space if and only if there is a perfect strategy in a certain non-local game in the tensor-product model. Cleve et al. [4, Theorem 4] have shown that a system of parity equations has a solution consisting of linear operators on a (finite-dimensional or infinite-dimensional) Hilbert space if and only if there is a perfect strategy in a certain non-local game in the commuting-operator model. Slofstra [18] obtained a breakthrough result that has numerous consequences about these models. In particular, Corollary 3.2 in Slofstra's paper [18] asserts that there is a system S of parity equations whose associated non-local game has a perfect strategy in the commuting-operator model, but not in the tensor-product model. Thus, by combining Theorem 1 in [5], Theorem 4 in [4], and Corollary 3.2 in [18], we obtain the following result.

Theorem 4 [4,5,18]. S *is a satisfiability gap of the third kind for* LIN.

LIN has a rather special place among all classes of Boolean relations that are not 0-valid, not 1-valid, not bijunctive, not Horn, and not dual Horn. This role is captured by the next lemma, which follows from Post's analysis of the lattice of clones of Boolean functions (Post's lattice) [16] and the Galois connection between clones of Boolean functions and co-clones of Boolean relations discovered by Geiger [9] and, independently, by Bodnarchuk et al. [3].

Lemma 5. *If A is a Boolean constraint language that is not 0-valid, not 1-valid, not bijunctive, not Horn, and not dual Horn, then* LIN *is pp-definable from A.*

The final lemma in this section asserts that reductions based on pp-definitions preserve satisfiability gaps upwards. Its proof makes use of Lemmas 2 and 3.

Lemma 6. *Let B and C be Boolean constraint languages such that B is pp-definable from C. If B has a satisfiability gap of the first kind, then so does C, and if B has a satisfiability gap of the third kind, then so does $C \cup \{T\}$.*

We now have all the machinery needed to put everything together.

Let A be a Boolean constraint language that is not 0-valid, not 1-valid, not bijunctive, not Horn, and not dual Horn. By Lemma 5, we have that LIN is pp-definable from A. Since, by Theorem 3, LIN has a satisfiability gap of the first kind, the first part of Lemma 6 implies that A has a satisfiability gap of the first kind. Since, by Theorem 4, LIN has a satisfiability gap of the third kind, the second part of Lemma 6 implies that $A \cup \{T\}$ has a satisfiability gap of the third kind. The proof of Theorem 2 is now complete.

If A is a Boolean constraint language, then SAT(A) is the following decision problem: Given an instance $\mathcal{I}$ over A, is $\mathcal{I}$ satisfiable in the Boolean domain? Theorem 2 and Schaefer's dichotomy theorem [17] imply that if A is a Boolean constraint language such that SAT(A) is NP-complete, then A has satisfiability gaps of the first kind and the second kind, and $A \cup \{T\}$ has a satisfiability gap of the third kind. In particular, this holds for the languages expressing the most widely used variants of Boolean satisfiability in Schaefer's framework, including 3SAT, MONOTONE 3SAT, NOT-ALL-EQUAL 3SAT, and 1-IN-3 SAT.

6 Further Applications

In this section, we discuss two applications of the results obtained thus far. For a Boolean constraint language A, let SAT*(A) and SAT**(A) be the versions of SAT(A) in which the question is whether a given instance $\mathcal{I}$ is satisfiable via fd-operators, or, respectively, via operators. Slofstra's Corollary 3.3 in [18], combined with Theorem 4 in [4], implies the undecidability of LIN SAT**, where we write LIN SAT** to denote SAT**(LIN).

Theorem 5 [4,18]. *LIN SAT** is undecidable.*

Theorem 5 and Lemmas 3, 4, and 5 yield the following dichotomy theorem.

Theorem 6. *Let A be a Boolean constraint language and let $A' = A \cup \{T\}$. Then, exactly one of the following two cases holds.*

1. *SAT**(A') is decidable in polynomial time.*
2. *SAT**(A') is undecidable.*

Moreover, the first case holds if and only if A is 0-valid, or A is 1-valid, or A is bijunctive, or A is Horn, or A is dual Horn.

Note that the second case of Theorem 6 states that SAT**(A') is undecidable, but it says nothing about SAT**(A). Luckily, in most cases, it is possible to infer the undecidability of SAT**(A) from the undecidability of SAT**(A'). This holds, for example, for 3LIN SAT** and 3SAT**, (i.e., for SAT**(A), where $A = \{\{(a_1, a_2, a_3) \in \{\pm 1\}^3 : a_1 a_2 a_3 = b\} : b \in \{\pm 1\}\}$ and $A = \{\{\pm 1\}^3 \setminus \{(a_1, a_2, a_3)\}\} : a_1, a_2, a_3 \in \{\pm 1\}\}$, respectively).

Theorem 7. *3LIN SAT** and 3SAT** are undecidable. Moreover, there is an instance of 3SAT that witnesses a satisfiability gap of the third kind.*

Now we turn to the second application. We follow the terminology in [1]. A *contextuality scenario* is a hypergraph H with a set $V(H)$ of vertices and a set $E(H) \subseteq 2^{V(H)}$ of edges such that $\bigcup_{e \in E(H)} e = V(H)$. A contextuality scenario H is *quantum realizable* if there exists a Hilbert space $\mathcal{H}$ and an assignment of bounded linear operators P_v on $\mathcal{H}$ to each vertex v in $V(H)$ in such a way that P_v is self-adjoint, $P_v^2 = P_v$, and $\sum_{v \in e} P_v = I$, for each $e \in E(H)$. In [1], the question was raised whether there exist contextuality scenarios that are quantum realizable but only over infinite-dimensional Hilbert spaces (see Problem 8.3.2 in [1]). A related computational question was also raised: Is it decidable whether a contextuality scenario given as input is quantum realizable? (see Conjecture 8.3.3 in [1]). This problem is called ALLOWS-QUANTUM. See [1] for a discussion on why these problems are important, and their relationship to Connes Embedding Conjecture in functional analysis. Fritz [8] answered these questions and, in particular, showed that ALLOWS-QUANTUM is undecidable. Here, we give an alternative proof that, as we shall see, yields optimal parameters.

Let k-ALLOWS-QUANTUM be the restriction of ALLOWS-QUANTUM where the input hypergraph has edges of cardinality at most k. Using Theorem 7 and, crucially, a device invented by Ji [11, Lemma 5], we show that 3-ALLOWS-QUANTUM is undecidable. We show that Ji's device extends to the realm of infinite-dimensional Hilbert spaces and use it for the following result.

Corollary 1. 3-ALLOWS-QUANTUM *is undecidable. Moreover, there exists a hypergraph with edges of size at most three that is quantum realizable on some Hilbert space, but not on a finite-dimensional Hilbert space.*

The parameters are optimal since 2-ALLOWS-QUANTUM reduces to 2SAT**.

7   Closure Operations

We say that R is *pp*-definable from A* if there is a pp-formula $\phi(\overline{x}) = \exists \overline{y}(\psi(\overline{x}, \overline{y}))$ over A such that, for every $\overline{a} \in \{\pm 1\}^{|\overline{x}|}$, the tuple $\overline{a}$ is in R if and only if the instance $\psi(\overline{x}/\overline{a}, \overline{y}/\overline{Y})$ is satisfiable via fd-operator assignments. One of the goals of this section is to prove the following conservation result.

Theorem 8. *Let A be a Boolean constraint language and let R be a Boolean relation. If R is pp*-definable from A, then R is also pp-definable from A.*

To prove this theorem, we need to adapt the concept of *closure* operation. For a Boolean r-ary relation R, let R^* denote the set of fully commuting r-variable fd-operator assignments that satisfy the equation $P_R(X_1, \ldots, X_r) = -I$. If A is a set of Boolean relations, let $A^* = \{R^* : R \in A\}$. Let $\mathcal{H}_1, \ldots, \mathcal{H}_m$ and $\mathcal{H}$ be Hilbert spaces, and let f be a function that takes as inputs m linear operators, one on each $\mathcal{H}_i$, and produces as output a linear operator on $\mathcal{H}$. We say that f is an *operation* if it maps the 1-variable operator assignments $A_1, \ldots, A_m$ to a 1-variable operator assignment $f(A_1, \ldots, A_m)$, and it maps the commuting 2-variable operator assignments $(A_1, B_1), \ldots, (A_m, B_m)$ to a commuting operator assignment $(f(A_1, \ldots, A_m), f(B_1, \ldots, B_m))$. Let R be a set of fully commuting operator assignments of arity r and let F be a collection of operations. We say that R is *invariant* under F if each $f \in F$ maps tuples $(A_{1,1}, \ldots, A_{1,r}), \ldots, (A_{m,1}, \ldots, A_{m,r})$ in R to $(f(A_{1,1}, \ldots, A_{m,1}), \ldots, f(A_{1,r}, \ldots, A_{m,r}))$ in R. If A is a collection of sets of operator assignments, we say that A is *invariant* under F if every relation in A is invariant under F. We also say that F is a *closure operation* of A.

We show that every Boolean closure operation (i.e., a 1-dimensional operation) gives a closure operation over finite-dimensional Hilbert spaces. If S is a set, we put $S(i) = 1$ if $i \in S$, and $S(i) = 0$ if $i \notin S$.

Theorem 9. *Let A be a Boolean constraint language and let $f : \{\pm 1\}^m \to \{\pm 1\}$ be a Boolean closure operation of A. Let F be the function on linear operators of Hilbert spaces defined by $F(X_1, \ldots, X_m) = \sum_{S \subseteq [m]} \hat{f}(S) \bigotimes_{i \in [m]} X_i^{S(i)}$, where $\hat{f}$ is the Fourier transform of f. Then F is a closure operation of A^*. Moreover, $F(a_1 I, \ldots, a_m I) = f(a_1, \ldots, a_m) I$ holds for every $(a_1, \ldots, a_m) \in \{\pm 1\}^m$.*

Theorem 8 follows from combining Theorem 9 with Geiger's Theorem [3, 9].

Acknowledgments. This work was initiated when all three authors were in residence at the Simons Institute for the Theory of Computing. Albert Atserias partly funded by

European Research Council (ERC) under the European Union's Horizon 2020 research and innovation programme, grant agreement ERC-2014-CoG 648276 (AUTAR), and by MINECO through TIN2013-48031-C4-1-P (TASSAT2); Simone Severini partly funded by The Royal Society, Engineering and Physical Sciences Research Council (EPSRC), National Natural Science Foundation of China (NSFC).

References

1. Acín, A., Fritz, T., Leverrier, A., Sainz, A.B.: A combinatorial approach to nonlocality and contextuality. Commun. Math. Phys. **2**(334), 533–628 (2015)
2. Bell, J.S.: On the problem of hidden variables in quantum mechanics. Rev. Mod. Phys. **38**(3), 447 (1966)
3. Bodnarchuk, V.G., Kaluzhnin, L.A., Kotov, V.N., Romov, B.A.: Galois theory for post algebras. I. Cybernetics **5**(3), 243–252 (1969)
4. Cleve, R., Liu, L., Slofstra, W.: Perfect commuting-operator strategies for linear system games (2016). arXiv preprint arXiv:1606.02278
5. Cleve, R., Mittal, R.: Characterization of binary constraint system games. In: Esparza, J., Fraigniaud, P., Husfeldt, T., Koutsoupias, E. (eds.) ICALP 2014. LNCS, vol. 8572, pp. 320–331. Springer, Heidelberg (2014). doi:10.1007/978-3-662-43948-7_27
6. Folland, G.B.: A Course in Abstract Harmonic Analysis. Studies in Advanced Mathematics. Taylor & Francis, London (1994)
7. Folland, G.B.: Real Analysis: Modern Techniques and Their Applications. Pure and Applied Mathematics: A Wiley Series of Texts, Monographs and Tracts. Wiley, New York (2013)
8. Fritz, T.: Quantum logic is undecidable (2016). arXiv preprint arXiv:1607.05870
9. Geiger, D.: Closed systems of functions and predicates. Pac. J. Math. **27**(1), 95–100 (1968)
10. Halmos, P.R.: Introduction to Hilbert Space and the Theory of Spectral Multiplicity. Benediction Classics, Oxford (2016)
11. Ji, Z.: Binary constraint system games and locally commutative reductions (2013). arXiv preprint arXiv:1310.3794
12. Mermin, N.D.: Simple unified form for the major no-hidden-variables theorems. Phys. Rev. Lett **65**(27), 3373 (1990)
13. Mermin, N.D.: Hidden variables and the two theorems of John Bell. Rev. Mod. Phys. **65**(3), 803 (1993)
14. O'Donnell, R.: Analysis of Boolean Functions. Cambridge University Press, Cambridge (2014)
15. Peres, A.: Incompatible results of quantum measurements. Phys. Lett. A **151**(3–4), 107–108 (1990)
16. Post, E.L.: The Two-Valued Iterative Systems of Mathematical Logic. Annals of Mathematical Studies, vol. 5. Princeton University Press, Princeton (1941)
17. Schaefer, T.J.: The complexity of satisfiability problems. In: Proceedings of the 10th Annual ACM Symposium on Theory of Computing, San Diego, California, USA, 1–3 May 1978, pp. 216–226 (1978)
18. Slofstra, W.: Tsirelson's problem and an embedding theorem for groups arising from non-local games (2016). arXiv preprint arXiv:1606.03140

New Results on Routing via Matchings on Graphs

Indranil Banerjee$^{(\boxtimes)}$ and Dana Richards

Department of Computer Science, George Mason University, Fairfax, VA 22030, USA
{ibanerje,richards}@gmu.edu

Abstract. In this paper we present some new complexity results on the routing time of a graph under the *routing via matching* model. This is a parallel routing model which was introduced by Alon et al. [1]. The model can be viewed as a communication scheme on a distributed network. The nodes in the network can communicate via matchings (a step), where a node exchanges data (pebbles) with its matched partner. Let G be a connected graph with vertices labeled from $\{1, ..., n\}$ and the destination vertices of the pebbles are given by a permutation π. The problem is to find a minimum step routing scheme for the input permutation π. This is denoted as the routing time $rt(G, \pi)$ of G given π. In this paper we characterize the complexity of some known problems under the routing via matching model and discuss their relationship to graph connectivity and clique number. We also introduce some new problems in this domain, which may be of independent interest.

1 Introduction

Originally introduced by Alon et al. [1] the routing via matching model explores a parallel routing problem on connected undirected graphs. Consider a undirected labeled graph G. Each vertex of G contains a pebble with the same label as the vertex. Pebbles move along edges through a sequence of swaps. A set of swaps (necessarily disjoint) that occurs concurrently is called a step. This is determined by a matching. A permutation π gives the destination of each pebble. That is, the pebble p_v on vertex v is destined for the vertex $\pi(v)$. The task is to route each pebble to their destination via a sequence of matchings. The *routing time* $rt(G, \pi)$ is defined as the minimum number of steps necessary to route all the pebbles for a given permutation. The *routing number* of G, $rt(G)$, is defined as the maximum routing time over all permutation. Let $G = (V, E)$, $m = |E|$ and $|G| = n = |V|$.

Determining the routing time is a special case of the *minimum generator sequence* problem for groups. In this problem instead of a graph we are given a permutation group $\mathcal{G}$ and a set of generators S. Given a permutation $\pi \in \mathcal{G}$ the task is to determine if there exists a generator sequence of length $\leq k$ that generates π from the identity permutation. It was first shown to be NP-hard by Evan and Goldreich [11]. Later Jerrum [15] showed that it is in fact PSPACE-complete, even when the generating set is restricted to only two generators.

© Springer-Verlag GmbH Germany 2017
R. Klasing and M. Zeitoun (Eds.): FCT 2017, LNCS 10472, pp. 69–81, 2017.
DOI: 10.1007/978-3-662-55751-8_7

The serial version, where swaps takes place one at a time, is also of interest. This has recently garnered interest after its introduction by Yamanaka et al. [8]. They have termed it the *token swapping problem*. This problem is also NP-complete as shown by Miltzow et al. [9] in a recent paper, where the authors prove token swapping problem is hard to approximate within $(1+\delta)$ factor. They also provide a simple 4-approximation scheme for the problem. A generalization of the token swapping problem (and also the permutation routing problem) is the colored token swapping problem [4,8]. In this model the vertices and the tokens are partitioned into equivalence classes (using colors) and the goal is to route all pebbles in such a way that each pebble ends up in some vertex with the same class as the pebble. If each pebble (and vertex) belong to a unique class then this problem reduces to the original token swapping problem. This problem is also proven to be NP-complete by Yamanaka et al. [8] when the number of colors is at least 3. The problem is polynomial time solvable for the bi-color case.

1.1   Prior Results

Almost all previous literature on this problem focused on determining the routing number of typical graphs. In the introductory paper, Alon et al. [1] show that for any connected graph G, $rt(G) \leq 3n$. This was shown by considering a spanning tree of G and using only the edges of the tree to route permutations in G. Note that, one can always route a permutation on a tree, by iteratively moving a pebble that belongs to some leaf node and ignoring the node afterward. The routing scheme is recursive and uses an well known property of trees: a tree has a centroid (vertex) whose removal results in a forest of trees with size at most $n/2$. Later Zhang et al. [5] improve this upper bound to $3n/2 + O(\log n)$. This was done using a new decomposition called the caterpillar decomposition. This bound is essentially tight as it takes $\lfloor 3(n-1)/2 \rfloor$ steps to route some permutations on a star $K_{1,n-1}$. There are also some known results for routing numbers of graphs besides trees. We know that for the complete graph and the complete bipartite graph the routing number is 2 and 4 respectively [1], where the latter result is attributed to W. Goddard. Li et al. [6] extend these results to show $rt(K_{s,t}) = \lfloor 3s/2t \rfloor + O(1)$ $(s \geq t)$. For the n-cube Q_n we know that $n + 1 \leq rt(Q_n) \leq 2n - 2$. The lower bound is quite straightforward. The upper bound was discovered by determining the routing number of the Cartesian product of two graphs [1]. If $G = G_1 \square G_2$ be the Cartesian product of G_1 and G_2 then:

$$rt(G) \leq 2\min(rt(G_1), rt(G_2)) + \max(rt(G_1), rt(G_2))$$

Since $Q_n = K_2 \square Q_{n-1}$ the result follow.[1].

1.2   Our Results

In this paper we present several complexity results for the routing time problem and some variants of it. We summarize these results below.

---

[1] The base case, which computes $rt(Q_3)$ was determined to be 4 via a computer search [6].

Complexity results on routing time:

1. If G is bi-connected then determining whether $rt(G,\pi) = k$ for any arbitrary permutation and $k > 2$ is NP-complete.[2]
2. For any graph, determining $rt(G,\pi) \leq 2$ can be done in polynomial time, for which we give a $O(n^{2.5})$ time algorithm.
3. As a consequence of our NP-completeness proof of the routing time we show that the problem of determining a minimum sized partitioning scheme of a colored graph such that each partition induces a connected subgraph is NP-complete.
4. We introduce a notion of approximate routing called *maximum routability* for a graph and give an approximation algorithm for it.

Structural results on routing number:

5. If G is h-connected then G has a routing number of $O(nr_G)$. Here $r_G = \min rt(G_h)/|G_h|$, over all induced connected subgraphs G_h, with $|G_h| \leq h$.
6. A connected graph with a clique number of κ has a routing number of $O(n-\kappa)$.

Routing on general graphs is a natural question and the swapping model is a natural model in synchronous networks. Our results are some of the first to address these models when the graph has certain topological properties. Connectivity properties are basic, especially for network algorithms. While the hope is to have matching upper and lower bounds for, say, h-connected graphs, we give new algorithms and techniques towards that end.

2 Computational Results

2.1 An $O(n^{2.5})$ Time Algorithm for Determining if $rt(G,\pi) \leq 2$

In this section we present a polynomial time deterministic algorithm to compute a two step routing scheme if one exists. It is trivial to determine whether $rt(G,\pi) = 1$. Hence, we only consider the case if $rt(G,\pi) > 1$. The basic idea centers around whether we can route the individual cycles of the permutation within 2 steps. Let $\pi = \pi_1\pi_2\ldots\pi_k$ consists of k cycles and $\pi_i = (\pi_{i,1}\ldots\pi_{i,a_i})$, where a_i is the number of elements in π_i. A cycle π_i is identified with the vertex set $V_i \subset V$ whose pebbles need to be routed around that cycle. We say a cycle π_i is *self-routable* if it can be routed on the induced subgraph $G[V_i]$ in 2 steps.

If all cycles were self-routable we would be done, so suppose that there is a cycle π_i that needs to match across an edge between it and another cycle π_j. Let $G[V_i, V_j]$ be the induced bipartite subgraph corresponding to the two sets V_i and V_j.

Lemma 1. *If π_i is not self-routable and it is routed with an edge from V_i to V_j then π_i and π_j are both routable in 2 steps when only the edges used are from $G[V_i, V_j]$ and when $|V_i| = |V_j|$.*

---

[2] After publication of our results to arXiv [16] a similar result was independently discovered in the context of parallel token swapping by Kawahara et al. [4].

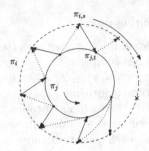

Fig. 1. The two cycles are shown as concentric circles. The direction of rotation for the outer circle is clockwise and the inner circle is counter-clockwise. Once, we choose $(\pi_{i,s}, \pi_{j,t})$ as the first matched pair, the rest of the matching is forced. Solid arrows indicate matched vertices during the first round. Note that the cycles are unequal and the crossed vertices in the figure will not be routed.

Proof. We prove this assuming G is a complete graph. Since for any other case the induced subgraph $G[V_i, V_j]$ would have fewer edges, hence this is a stronger claim. Let the cycle $\pi_i = (\pi_{i,1}, \ldots, \pi_{i,s}, \ldots, \pi_{i,a_i})$. If there is an edge used between the cycles then there must be such an edge in the first step, since pebbles need to cross from one cycle to another and back. Assume $\pi_{i,s}$ is matched with $\pi_{j,t}$ in the first step. From Fig. 1 we see that the crossing pattern is forced, and unless $|V_i| = |V_j|$, the pattern will fail. □

A pair of cycles π_i, π_j is *mutually-routable* in the case described by Lemma 1. Naively verifying whether a cycle π_i is self-routable, or a pair (π_i, π_j) is mutually-routable takes $O(|V_i|^2)$ and $O((|V_i| + |V_j|)^2)$ time respectively. However, with additional bookkeeping we can compute this in linear time on the size of the induced graphs. This can be done by considering the fact that no edge can belong to more than one routing scheme on $G[V_i]$ or on $G[V_i, V_j]$. Hence the set of edges are partitioned by the collection of 2 step routing schemes. Self-routable schemes, if they exist, are forced by the choice of any edge to be in the first step; no edge is forced by more than four initial choices, leading to a test that runs in time proportional in $|G[V_i]|$. Mutually-routable schemes, if they exist, are one of $|V_i|$ ($= |V_j|$) possible schemes; each edge votes for a scheme and it is routable if a scheme gets enough votes, leading to a test that runs in time proportional in $|G[V_i, V_j]|$. All the tests can be done in $O(m)$ time.

We define a graph $G_{cycle} = (V_{cycle}, E_{cycle})$ whose vertices are the cycles ($V_{cycle} = \{\pi_i\}$) and two cycle are adjacent iff they are mutually-routable in 2 steps. Additionally, G_{cycle} has loops corresponding to vertices which are self-routable cycles. We can modify any existing maximum matching algorithm to check whether G_{cycle} has a perfect matching (assuming self loops) with only a linear overhead. We omit the details. Then the next lemma follows immediately:

Lemma 2. $rt(G, \pi) = 2$ iff there is a perfect matching in G_{cycle}.

The graph G_{cycle} can be constructed in $O(m)$ time by determining self and mutual routability of cycles and pair of cycles respectively. Since we have at most k cycles, G_{cycle} has k vertices and thus $O(k^2)$ edges. Hence we can determine a maximum matching in G_{cycle} in $O(k^{2.5})$ time [7]. This gives a total runtime of $O(n+m+k^{2.5})$ for our algorithm to find a 2-step routing scheme of a connected graph if one exists.

Corollary 1. $rt(G) = 2$ *iff G is a clique.*

Proof. ($\Rightarrow$) A two step routing scheme for K_n was given in [1].
($\Leftarrow$) If G is not a clique then there is at least a pair of non-adjacent vertices. Let (i,j) be a non-edge. Then by Lemma 1 the permutation $(ij)(1)(2)\ldots(n)$ cannot be routed in two steps. $\square$

2.2   Determining $rt(G,\pi) \le k$ is Hard for any $k \ge 3$

Theorem 1. *For $k \ge 3$ computing $rt(G,\pi)$ is NP-complete.*

Proof. Proving it is in NP is trivial, we can use a set of matchings as a witness. We give a reduction from 3-SAT. We first define three *atomic* gadgets (see Fig. 2) which will be use to construct the variable and clause gadgets. Vertices whose pebbles are fixed (1-cycles) are represented as red circles. Otherwise they are represented as black dots. So in the first three sub-figures ((a)–(c)) the input permutation is $(a,b)^3$. In all our constructions we shall use permutations consisting of only 1 or 2 cycles. Each cycle labeled i will be represented as the pair (a_i, b_i). If the correspondence between a pair is clear from the figure then we shall omit the subscript. It is an easy observation that $rt(P_3, ((a,b))) = rt(P_4, ((a,b))) = rt(H, ((a,b))) = 3$. In the case of the hexagon H we see that in order to route the pebbles within 3 steps we have to use the left or the right path, but we cannot use both paths simultaneously (i.e., a goes along the left path but b goes along the right and vice-versa). Figure 2(e) shows a chain of diamonds connecting u to v. Where each diamond has a 2-cycle, top and bottom. If vertex u is used to route any pebble other than the two pebbles to its right then the chain construction forces v to be used in routing the two pebbles to its left. This chain is called a *diamond-chain*. In our construction we only use chains of constant length to simplify the presentation of our construction.

Clause Gadget: Say we have a clause $C = x \vee y \vee \neg z$. In Fig. 2(d) we show how to create a clause gadget. This is referred to as the *clause graph G_C* for the clause C. The graph in Fig. 2(d) can route $\pi_C = (a_C, b_C)$ in three steps by using one of the three paths between a_C and b_C. Say, a_C is routed to b_C via x. Then it must be the case that vertex x is not used to route any other pebbles. We say the vertex x is *owned* by the clause. Otherwise, it would be not possible to route a_C to b_C in three steps via x. We can interpret this as follows, a clause has a satisfying assignment iff its clause graph has a owned vertex.

----
[3] We do not write the 1 cycles explicitly as is common.

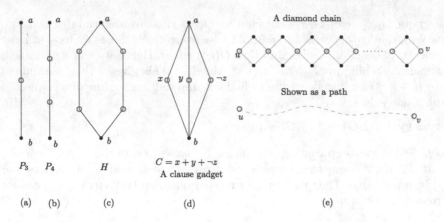

Fig. 2. Atomic gadgets, pairs (a, b) need to swap their pebbles. The unmarked red circles have pebbles that are fixed. (Color figure online)

Variable Gadget: Construction of the variable gadgets is done in a similar manner. The variable gadget G_X corresponding to X is shown in Fig. 3(b). Figure 3(a) is essentially a smaller version of 4(b) and is easier to understand. If we choose to route a_1 and b_1 via the top-left path passing through x_1 and u_1 then (a_2, b_2) must be routed via x_2 and u_2. This follows from the fact that since u_1 is occupied the pebbles in the diamond chain C (the dashed line connecting u_1 with u_3) must use u_3 to route the right most pair. By symmetry, if we choose to route (a_1, b_1) using the bottom right path (via $\neg x_1$, u_2) then we also have to choose the bottom right path for (a_2, b_2). These two (and only two) possible (optimal) routing scheme can be interpreted as variable assignment. Let G_X be the graph corresponding to the variable X (Fig. 3(b)). The top-left routing scheme leaves the vertices $\neg x_1, \neg x_2, \ldots$ free to be used for other purposes since they will not be able take part in routing pebbles in G_X. This can be interpreted as setting the variable X to false. This "free" vertex can be used by a clause (if the clause has that literal) to route its own pebble pair. That is they can become owned vertices of some clause. Similarly, the bottom right routing scheme can be interpreted as setting X to true. For each variable we shall have a separate graph and a corresponding permutation on its vertices. The permutation we will route on G_X is $\pi_X = (a_1 b_1)(a_2 b_2) \ldots (a_{m_X}, b_{m_X}) \pi_{f_X}$. The permutation π_{f_X} corresponds to the diamond chain connecting u_1 with u_{m_X+1}. The size of the graph G_X is determined by m_X, the number of clauses the variable X appears in.

Reduction: For each clause C, if the literal $x \in C$ then we connect $x_i \in G_X$ (for some i) to the vertex labeled $x \in G_C$ via a diamond chain. If $\neg x \in C$ then we connect it with $\neg x_i$ via a diamond chain. This is our final graph G_ϕ corresponding to an instance of a 3-SAT formula. The input permutation is $\pi = \pi_X \ldots \pi_C \ldots \pi_f \ldots$, which is the concatenation of all the individual permutations on the variable graphs, clause graphs and the diamond chains. This completes

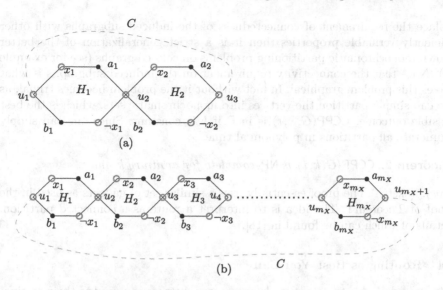

Fig. 3. Variable graph of X. (a) is a special case for $m_X = 2$, (b) is the general case.

our construction. We need to show, $rt(G_\phi, \pi) = 3$ iff ϕ is satisfiable. Suppose ϕ is satisfiable. Then for each variable X, if the literal x is true then we use bottom-right routing on G_X, otherwise we use top-left routing. This ensures in each clause graph there will be at least one owned vertex. Now suppose $(G_\phi, \pi) = 3$. Then each clause graph has at least one owned vertex. If x is a free vertex in some clause graph then $\neg x$ is not a free vertex in any of the other clause graphs, otherwise variable graph G_X will not be able route its own permutation in 3 steps. Hence the set of free vertices will be a satisfying assignment for ϕ. It is an easy observation that the number of vertices in G_ϕ is polynomially bounded in n, m; the number of variables and clauses in ϕ respectively and that G_ϕ can be explicitly constructed in polynomial time. □

Corollary 2. *Computing $rt(G, \pi)$ remains hard even when G is restricted to being 2-connected.*

2.3   Connected Colored Partition Problem (CCPP)

Our proof technique for Theorem 1 can be used to prove that the following problem is also NP-hard. Let G be a graph whose vertices are colored with k colors. We say a partition $\mathcal{S} = \{S_1, \ldots, S_r\}$ of the vertex set V respects the coloring C (where $C : V \rightarrow \{1, \ldots, k\}$) if each partition either contains all vertices of some color or none of the vertices of that color (necessarily $r \leq k$). Further, we require the induced subgraph $G[S_i]$ be connected, for every i. Given a graph G, a coloring C (with k colors) and a integer $t \leq n$ the decision version of the problem asks, whether there exists a valid partitioning whose largest block has a size of at most t. We denote this problem by CCPP(G, k, t). If we replace the requirement of connectedness of the induced subgraphs with other efficiently verifiable properties then it is a strict generalization of the better known monochromatic partitioning problems on colored graphs (see for example [2]). Note that the connectivity requirement on the induced subgraphs is what makes this problem graphical. In fact without it the problem becomes trivial, as one can simply partition the vertices into monochromatic sets, which is the best possible outcome. $CCPP(G, k, t)$ is in P if k is constant. Since one can simply enumerate all partitions in polynomial time.

Theorem 2. $CCPP(G, k, t)$ *is* NP-*complete for arbitrary k and t.*

Proof. (sketch) The proof essentially uses a similar set of gadgets as used in the proof of Theorem 1. The idea is to interpret a route as a connected partition. Details of which can be found in [18]. $\qquad\square$

2.4   Routing as Best You Can

It is often desirable to determine how many packets we can send to their destination within a certain number of steps. Such as propagating information in social media. In the context of permutation routing this leads to a notion of *maximum routability*. Given two permutation π and σ let $|\pi - \sigma|$ be the number of fixed points in τ such that $\tau\pi = \sigma$. Let us define *maximum routability* $mr(G, \pi, k)$ as follows:

$$mr(G, \pi, k) = \max_{\sigma \in S_n,\ rt(G,\sigma) \leq k} |\pi - \sigma|$$

We denote by MaxRoute the problem of computing maximum routability. Essentially, σ is a permutation out of all permutations that can be routed in $\leq k$ steps and that has the maximum number of elements in their correct position as given by π. The permutation σ may not be unique. It can be easily shown (as a corollary to Theorem 1) that the decision version of this problem is NP-hard, since we can determine $rt(G, \pi)$ by asking whether $mr(G, \pi, k) = n$. (Of course $rt(G, \pi) = O(n)$ for any graph, hence $O(\log n)$ number of different choices of k is sufficient to compute $rt(G, \pi)$ exactly.)

In this section we give an approximation algorithm for computing the maximum routability when the input graph G satisfies the following restriction. If the maximum degree of G is Δ such that $(\Delta + 1)^k = O(\log^2 n)$ then $mr(G, \pi, k)$ can be approximated within a factor of $O(n \log \log n / \log n)$ from the optimal. Unfortunately a good approximation for $rt(G, \pi)$ does not lead to a good approximation ratio when computing $mr(G, \pi, k)$ for any $k > 2$. The reason being that in an optimal algorithm for routing π on G it is conceivable that all pebbles are displaced at the penultimate stage and the last matching fixes all the displaced pebbles.

Our approximation algorithm is based on a reduction to the MaxClique problem. The MaxClique problem has been extensively studied. In fact it is one of the defining problems for PCP-type systems of probabilistic verifiers [3]. It has been shown that MaxClique can not be approximated within a $n^{1-o(n)}$ factor of the optimal [14]. The best known upper bound for the approximation ratio is by Feige [12] of $O(n(\log \log n / \log^3 n))$ which improves upon Boppana and Halldorsson's [13] result of $O(n / \log^2 n)$. Note that if there is a $f(n)$-approximation for MaxClique then whenever the clique number of the graph is $\omega(f(n))$, the approximation algorithm returns a non-trivial clique (not a singleton vertex).

Theorem 3. *Given a graph G whose maximum degree is Δ, in polynomial time we can construct another graph G_{clique}, with $|G_{clique}| = O(n(\Delta+1)^k)$, such that if the clique number of G_{clique} is κ then $mr(G, \pi, k) = \kappa$.*

In the above theorem the graph G_{clique} will be an n-partite graph. Hence $\kappa \leq n = O(|G_{clique}|/(\Delta+1)^k)$. As long as we have $(\Delta+1)^k = O(\log^2 n)$ we can use the approximation algorithm for MaxClique to get a non-trivial approximation ratio of $O(n \log \log n / \log n)$.

Proof. Here we give the reduction from MaxRoute to MaxClique. First we augment G by adding self-loops. Let this new graph be G'. Hence we can make every matching in G' perfect by assuming each unmatched vertex is matched to itself. Observe that any routing scheme on G' induces a collection of walks for each pebble. This collection of walks are constrained as follows. Let W_i and W_j corresponds to walks of pebbles starting at vertices i and j respectively. Let $W_i[t]$ be the position of the pebble at time step t. They must satisfy the following two conditions: (1) $W_i[t] \neq W_j[t]$ for all $t \geq 0$. (2) $W_i[t+1] = W_j[t]$ iff $W_i[t] = W_j[t+1]$. Now consider two arbitrary walks in G'. We call them compatible iff they satisfy the above two conditions. We can check if two walks are compatible in linear time.

Let W_i be the collection of all possible length k walks starting from i and ending at $\pi(i)$. Note that $|W_i| = O((\Delta+1)^k)$. For each $w \in W_i$ we create a vertex in G_{clique}. Two vertices u, v in G_{clique} are adjacent if they do not come from the same collection ($u \in W_i$ then $v \notin W_i$) and u and v are compatible walks in G'. Clearly, G_{clique} is n-partite, where each collection of vertices from W_i forming a block. Furthermore, if G_{clique} has a clique of size κ then it must be the case that there are κ mutually compatible walks in G'. These walk determines a routing scheme (since they are compatible) that routes κ pebbles to their destination. Now if G_{clique} has a clique number $< \kappa$ then the largest collection of mutually compatible length k walks must be $< \kappa$. Hence number of pebbles that can be routed to their destination in at most k steps will be $< \kappa$.

In order to get a non-trivial approximation ratio we require that $(\Delta+1)^k = O(\log^2 n)$ which implies that the above reduction is polynomial in n. This completes the proof. $\square$

3 Structural Results on the Routing Number

3.1 An Upper Bound for h-Connected Graphs

It was shown in [1] that for some h-connected graph G, its routing number has a lower bound of $\Omega(n/h)$. This is easy to see since there exists h-connected graphs which have a balanced bipartition with respect to some cut-set of size h. For such a graph the permutation that routes every pebble from one partition to the other and vice-versa takes at least $\Omega(n/h)$ matchings. In this section we give an upper bound. Let G_h be a induced connected subgraph of G having h vertices, we will show $rt(G) = O(n\ rt(G_h)/h)$. Hence if G has a h-clique then $rt(G) = O(n/h)$. In fact the result is more general. If G_h is an induced subgraph with $\leq h$ vertices such that $r = rt(G_h)/|G_h|$ is minimized then $rt(G) = O(nr)$.

We use the classical Lovasz-Gyori partition theorem for h-connected graphs for this purpose:

Theorem 4 (Lovasz-Gyori). *If G is a h-connected graph then for any choice of positive numbers $n_1, \ldots, n_h$ with $n_1 + \ldots + n_h = n$ and any set of vertices $v_1, \ldots, v_h$ there is a partition of the vertices $V_1, \ldots, V_h$ with $v_i \in V_i$ and $|V_i| = n_i$ such that the induced subgraph $G[V_i]$ is connected for all $1 \leq i \leq h$.*

We prove a combinatorial result. We have a lists L_i, $1 \leq i \leq a$, each of length b. Each element of a list is a number c, $1 \leq c \leq a$. Further, across all lists, each number c occurs exactly b times.

Lemma 3. *Given lists as described, there exists an $a \times b$ array A such that the ith row is a permutation of L_i and each column is a permutation of $\{1, 2, 3, \ldots, a\}$.*

Proof. By Hall's Theorem for systems of distinct representatives [10], we know that we can choose a representative from each L_i to form the first column of A. The criterion of Hall's Theorem is that, for any k, any set of k lists have at least k distinct numbers; but there are only b of each number so $k - 1$ numbers can not fill up k lists. Now remove the representative from each list, and iterate on the collection of lists of length $b - 1$.   □

To prove our upper bound we need an additional lemma.

Lemma 4. *Given a set S of k pebbles and tree T with k pebbles on its k vertices. Suppose we are allowed an operation that replaces the pebble at the root of T by a pebble from S. We can replace all the pebbles in T with the pebbles from S in $\Theta(k)$ steps, each a replace or a matching step.*

Proof. Briefly, as each pebble comes from S it is assigned a destination vertex in T, in reverse level order (the root is at level 0). After a replace-root operation, there are two matching steps; these three will repeat. The first matching step uses disjoint edges to move elements of S down to an odd level and the second matching step moves elements of S down to an even level. Each matching moves every pebble from S, that has not reached its destination, towards its destination. The new pebbles move without delay down their paths in this pipelined scheme. (The invariant is that each pebble from S is either at its destination, or at an even level before the next replace-root operation.)   □

Theorem 5. *If G is h-connected and G_h is an induced connected subgraph of order h then $rt(G) = O(n\ rt(G_h)/h)$.*

Proof. Let $V_h = \{u_1, \dots, u_h\}$ be the vertices in G_h. We take these vertices as the set of k vertices in Theorem 4. We call them ports as they will be used to route pebbles between different components. Without loss of generality we can assume $p = n/h$ is an integer. Let $n_1 = n_2 = \dots = n_h = p$ and V_i be the block of the partition such that $u_i \in V_i$. Let $H_i = G[V_i]$. Then for any permutation π on G:

1. Route the pebbles in H_i according to some permutation π_i. Since H_i has n/h vertices and is connected it takes $O(n/h)$ matchings.
2. Next use G_h, n/h times, to route pebbles between different partitions. We show that this can be done in $O(n \, rt(G_h)/h)$ matchings. (The "replace-root" step of Lemma 4, is actually the root replacements done by routing on G_h.)
3. Finally, route the pebbles in each H_i in parallel. Like step 1, this also can be accomplished in $O(n/h)$ matchings.

Clearly the two most important thing to attend to in the above procedure are the permutations in step 1 and the routing scheme of step 2. We can assume that each H_i is a tree rooted at u_i (since each H_i has a spanning tree). Thus the decomposition looks like the one shown in Fig. 4.

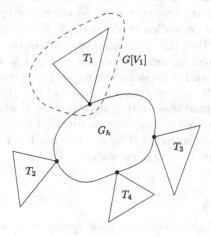

Fig. 4. G is decomposed into 4 connected blocks, which are connected to each other via G_h.

The permutation π on G indicates for each element of H_i, which H_j it wants to be routed to, where j could be i. So each H_i can build a list L_i of indices of the ports of G_h that it wants to route its elements to (again, possibly to its own port). The lists satisfy the conditions of Lemma 3, with $a = h$ and $b = n/h$, We will use the columns of the array A to specify the permutations routed using G_h in step 2. Note that step 1 will need to preprocess each H_i so that the algorithm of Lemma 4 will automatically deliver the elements of H_i up to u_i in the order specified by the ith row of A.

Once the pebbles are rearranged in step 1, we use the graph G_h to route them to their destination components. Each such routing takes $rt(G_h)$ steps. Between these routings on G_h the incoming pebble at any of the port vertices is replaced by the next pebble to be ported; this requires 2 matching steps as seen in Lemma 4. Hence, after $rt(G_h) + 2$ steps a set of h pebbles are routed to their destination components. This immediately gives the bound of the theorem.   □

3.2   Relation Between Clique Number and Routing Number

Theorem 6. *For a connected graph G with clique number κ its routing number is bounded by $O(n - \kappa)$.*

Proof. Let H be a clique in G of size κ. Let $G_{\backslash H}$ be the minor of G after the contraction of the subgraph H. Let the vertex that H has been contracted to be v. Further, let T be a spanning tree of $G_{\backslash H}$. When routing on $G_{\backslash H}$ we can treat v as any other vertex of $G_{\backslash H}$. Taking into account the fact that v can store more than one pebble internally. When v participates in a matching with some other vertex u in $G_{\backslash H}$ we assume that exchanging pebbles takes 3 steps. This accounts for the fact that the pebble thats need to be swapped with the pebbles at u was not on a vertex adjacent to u in the un-contracted graph G. The basic idea is to break the routing into two steps. In the first step we simply move all pebbles in v whose final detination is not in v (i.e. not in un-contracted H) out. For a tree, it is known that [17] we can route a subset of p pebbles where each pebble needs to be moved at most l distance in $\leq p + 2l$ steps. Since T has a diameter at most $n - \kappa$ and at most $min(\kappa, n - \kappa)$ pebbles need to be moved out of v the first step can be accomplished $\leq 3(n - \kappa) + O(1)$ steps. At this point we can employ any optimal tree routing algorithm on T where we charge 3 time units whenever v is part of the matching to route all the pebbles in $G_{\backslash H}$. If we use the algorithm presented in [5] then we see that the routing takes at most $15/2(n - \kappa) + o(n)$ steps for any permutation.   □

References

1. Alon, N., Chung, F.R., Graham, R.L.: Routing permutations on graphs via matchings. SIAM J. Discrete Math. **7**(3), 513–530 (1994)
2. Gyárfás, A., Ruszinkó, M., Sárközy, G., Szemerédi, E.: Partitioning 3-colored complete graphs into three monochromatic cycles. Electronic J. Comb. **18**, 1–16 (2011)
3. Feige, U., Goldwasser, S., Lovász, L., Safra, S., Szegedy, M.: Interactive proofs and the hardness of approximating cliques. J. ACM (JACM) **43**(2), 268–292 (1996)
4. Kawahara, J., Saitoh, T., Yoshinaka, R.: The time complexity of the token swapping problem and its parallel variants. In: Poon, S.-H., Rahman, Md, Yen, H.-C. (eds.) WALCOM 2017. LNCS, vol. 10167. Springer, Cham (2017). doi:10.1007/978-3-319-53925-6_35
5. Zhang, L.: Optimal bounds for matching routing on trees. SIAM J. Discrete Math. **12**(1), 64–77 (1999)
6. Li, W.T., Lu, L., Yang, Y.: Routing numbers of cycles, complete bipartite graphs, and hypercubes. SIAM J. Discrete Math. **24**(4), 1482–1494 (2010)

7. Micali, S., Vazirani, V.V.: An $O(\sqrt{|V|}|E|)$ algoithm for finding maximum matching in general graphs. In: 1980 21st Annual Symposium on Foundations of Computer Science, pp. 17–27. IEEE, October 1980

8. Yamanaka, K., Demaine, E.D., Ito, T., Kawahara, J., Kiyomi, M., Okamoto, Y., Toshiki, S., Akira, S., Kei, U., Uno, T.: Swapping labeled tokens on graphs. In: Theoretical Computer Science, vol. 586, pp. 81–94 (2015). Computer programming: sorting and searching (Vol. 3). Pearson Education

9. Miltzow, T., Narins, L., Okamoto, Y., Rote, G., Thomas, A., Uno, T.: Approximation and hardness for token swapping. arXiv preprint arXiv:1602.05150 (2016)

10. Hall, M.: Combinatorial Theory, vol. 71. Wiley, Hoboken (1998)

11. Even, S., Goldreich, O.: The minimum-length generator sequence problem is NP-hard. J. Algorithms **2**(3), 311–313 (1981)

12. Feige, U.: Approximating maximum clique by removing subgraphs. SIAM J. Discrete Math. **18**(2), 219–225 (2004)

13. Boppana, R., Halldrsson, M.M.: Approximating maximum independent sets by excluding subgraphs. BIT Numer. Math. **32**(2), 180–196 (1992)

14. Engebretsen, L., Holmerin, J.: Clique is hard to approximate within $n^{1-o(1)}$. In: Montanari, U., Rolim, J.D.P., Welzl, E. (eds.) ICALP 2000. LNCS, vol. 1853, pp. 2–12. Springer, Heidelberg (2000). doi:10.1007/3-540-45022-X_2

15. Jerrum, M.R.: The complexity of finding minimum-length generator sequences. Theor. Comput. Sci. **36**, 265–289 (1985)

16. Banerjee, I., Richards, D.: Routing and sorting via matchings on graphs. arXiv preprint arXiv:1604.04978 (2016)

17. Benjamini, I., Shinkar, I., Tsur, G.: Acquaintance time of a graph. SIAM J. Discrete Math. **28**(2), 767–785 (2014)

18. Banerjee, I., Richards, D.: New results on routing via matchings on graphs. arXiv preprint arXiv:1706.09355 (2017)

Energy-Efficient Fast Delivery by Mobile Agents

Andreas Bärtschi[(⊠)] and Thomas Tschager

Department of Computer Science, ETH Zürich, Zürich, Switzerland
{baertschi,tschager}@inf.ethz.ch

Abstract. We consider the problem of collaboratively delivering a package from a specified source node s to a designated target node t in an undirected graph $G = (V, E)$, using k mobile agents. Each agent i starts at time 0 at a node p_i and can move along edges subject to two parameters: Its *weight* w_i, which denotes the rate of energy consumption while travelling, and its *velocity* v_i, which defines the speed with which agent i can travel.

We are interested in operating the agents such that we minimize the *total energy consumption* $\mathcal{E}$ and the *delivery time* $\mathcal{T}$ (time when the package arrives at t). Specifically, we are after a schedule of the agents that lexicographically minimizes the tuple $(\mathcal{E}, \mathcal{T})$. We show that this problem can be solved in polynomial time $\mathcal{O}(k|V|^2 + \text{APSP})$, where $\mathcal{O}(\text{APSP})$ denotes the running time of an all-pair shortest-paths algorithm. This completes previous research which shows that minimizing only $\mathcal{E}$ or only $\mathcal{T}$ is polynomial-time solvable [6,7], while minimizing a convex combination of $\mathcal{E}$ and $\mathcal{T}$, or lexicographically minimizing the tuple $(\mathcal{T}, \mathcal{E})$ are both NP-hard [7].

1 Introduction

Recently, the production of simple autonomous mobile robots such as battery-powered vehicles and drones is getting increasingly cheaper, which allows for the deployment of such *agents* for tasks such as the delivery of *packages* [18]. This leads to new research questions focusing on the distance that the agents need to travel – quite often this is the main source of energy consumption and thus also a large fraction of the operational costs. Thus, for delivery over longer distances, there are many reasons to consider the use of several collaborative agents, for example because they differ in terms of energy-efficiency or maximum speed. Therefore an *energy-efficient* and *fast* operation becomes an integral part of algorithm design to operate a swarm of these agents. Here we consider the delivery problem of moving a single package with a team of k agents from a given source s to a specified destination t. Our goal is to coordinate the agents to deliver the package in an energy-efficient way and as fast as possible.

Our Model. We are given an undirected graph $G = (V, E)$ with specified edge lengths. There are k mobile agents, initially placed on arbitrary nodes $p_1, \ldots, p_k$ of G, which are heterogeneous in the following sense: They differ in their rate

R. Klasing and M. Zeitoun (Eds.): FCT 2017, LNCS 10472, pp. 82–95, 2017.
DOI: 10.1007/978-3-662-55751-8_8

of energy consumption (energy used by an agent per unit distance travelled), denoted by *weights* $0 < w_1, \ldots, w_k < \infty$, and in their *velocities* $0 < v_1, \ldots, v_k < \infty$. Together with the *length* l_e of an edge $e = \{u, v\} \in E$, the weight and the speed of an agent i describe the energy cost and time loss incurred: Every time agent i traverses e (in either direction), it consumes an energy amount of $w_i \cdot l_e$ and needs time $v_i^{-1} \cdot l_e$ to do so. We also allow agents to stop or pause; this can happen on a vertex or at a point p inside an edge e – in that case we think of p subdividing e into two edges $\{u, p\}, \{p, v\}$ with lengths proportional to the position of p in e. Thus we can define the distance $d(p, q)$ between two points p, q (nodes or points inside edges) of the graph as the length of a shortest path from p to q in G.

Furthermore we are given a single package, originally placed on a source node s, which has to be delivered to a target node t. Each agent can move to the current location of the package, pick it up, carry it to another location (a node or a point inside an edge), and drop it there. In that sense it is also possible to *hand over* the package between two agents. While pick-up and drop-off are assumed to occur instantaneously, the time between the package being dropped by an agent until being picked up by another agent is taken into account.

A *schedule* S describes the pick-up and drop-off actions of all agents such that the package is delivered from its origin s to its destination t. Without loss of generality, we only consider schedules in which each agent picks up the package at most once and travels from its starting position p_i to its pick-up position q_i^+ and from there to its drop-off location q_i^- along two shortest paths of respective length $d(p_i, q_i^+)$ and $d(q_i^+, q_i^-)$. In case an agent is not needed, we set $q_i^+ := p_i =: q_i^-$. Thus we can write the total energy consumption of the schedule as $\mathcal{E} = \sum_{i=1}^k w_i \cdot (d(p_i, q_i^+) + d(q_i^+, q_i^-))$ and the time needed to deliver the package as $\mathcal{T} = \sum_{i=1}^k v_i^{-1} \cdot d(q_i^+, q_i^-) +$ (overall time the package is not carried).

EFFICIENTFASTDELIVERY is the optimization problem of finding the fastest schedule among all energy-optimal ones, i.e. we ask for a schedule that lexicographically minimizes the tuple $(\mathcal{E}, \mathcal{T})$ (see e.g. Fig. 1).

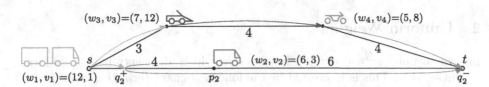

Fig. 1. Using agents $1, 3$ and 4 on the upper path results in an optimal schedule with energy cost and delivery time $(\mathcal{E}, \mathcal{T}) = (12 \cdot 3 + 7 \cdot 4 + 5 \cdot 4, \ 3/1 + 4/12 + 4/8) = (84, 3\frac{5}{6})$. A schedule using agents 1 and 2 on the (shorter) lower path has the same energy cost but higher delivery time $(\mathcal{E}, \mathcal{T}) = (12 \cdot 1 + 6 \cdot 12, \ \max\{1/1, 3/3\} + (3 + 6)/3) = (84, 4)$.

Our Results. In Sect. 2 we give a characterization of optimal schedules, motivated by which we first show that EFFICIENTFASTDELIVERY can be solved in time $\mathcal{O}(k^2 + \text{APSP})$ if we have uniform weights $(\forall i, j \colon w_i = w_j)$. Here, $\mathcal{O}(\text{APSP})$ denotes the running time of an all-pair shortest-paths algorithm.

In Sect. 3 we present a dynamic program for EFFICIENTFASTDELIVERY that runs in time $\mathcal{O}(k|V|^2 + \text{APSP})$, based on the assumption that no two weights are exactly a factor 2 apart $(\forall i, j\colon w_i \neq 2w_j)$ – in this case, we show that an optimal solution does not have any drop-off points inside edges.

Finally, in Sect. 4, we extend the techniques of the previous two sections to allow for arbitrary weights and thus in-edge-handovers as well, without increasing the running time.

Related Work. Energy-efficient DELIVERY has recently been introduced [6], showing that minimizing the overall energy consumption $\mathcal{E}$ by mobile agents of heterogeneous weights is solvable in time $\mathcal{O}(k + |V|^3)$. Similarly, *fast* DELIVERY (minimizing $\mathcal{T}$ for agents of heterogeneous speeds) was shown to have a dynamic programming algorithm running in time $\mathcal{O}(k^2|V|^2 + \text{APSP})$ [7]. The same paper also considers two natural combinations of energy- and time-efficiency which differ from the setting studied in this paper: lexicographically minimizing $(\mathcal{T}, \mathcal{E})$ and minimizing $\epsilon \cdot \mathcal{T} + (1 - \epsilon)\mathcal{E}$, for some input value $\epsilon \in (0, 1)$. Both problems turn out to be NP-hard [7], in contrast to our result, which shows that minimizing $(\mathcal{E}, \mathcal{T})$ is polynomial-time solvable. Furthermore, energy-efficient DELIVERY was also studied for multiple packages [6,8].

An earlier model [2] considered agents of uniform weights but *individually limited energy budget*, which restricts how far each agent can move. In that model, neither the delivery time nor the overall energy consumption are taken into account. Instead, one considers the decision problem of whether a single package can be delivered, given the restrictions on the agents. This is strongly NP-hard on planar graphs [4,5] and weakly NP-hard on paths [11]. DELIVERY with energy budgets was also studied in terms of resource augmentation [4,10] and approximation algorithms [10,17].

Mobile agents with distinct maximal speeds have been getting attention in areas such as searching [3], walking [12] and patrolling [13]. Furthermore, minimizing the average or maximum distance travelled per agent has been considered for tasks such as pattern formation [9,15] or graph exploration [1,14,16].

2   Uniform Weights

In this section, we will analyze instances in which all agents have a uniform weight $w_i = w$. This is motivated by the following characterization of agents in an optimum schedule:

Theorem 1 (Characterization of optimum schedules). *For every instance of* EFFICIENTFASTDELIVERY *there is an optimum schedule in which: (i) the tuples (w_i, v_i^{-1}) of the involved agents are strictly lexicographically decreasing, and (ii) for each pair of consecutive agents i, j with $w_i = w_j \wedge v_i^{-1} > v_j^{-1}$, we have $d(p_j, q_j^+) = 0$.*

Proof. We show both properties by an exchange argument. Starting from an arbitrary optimum schedule, we can transform it into an optimum satisfying properties (i) and (ii) in the following way:

(i) Relabel all involved agents i (agents with $d(q_i^+, q_i^-) > 0$) by $1, 2, \ldots, \ell$ according to the order in which they carry the package: $s = q_1^+, q_1^- = q_2^+, \ldots, q_\ell^- = t$. Assume for the sake of contradiction that there are agents $i, i+1$ such that $w_i < w_{i+1}$. In this case we can replace agent $i+1$ completely by handing its work to agent i. By our assumptions and by the triangle inequality we have $w_i \cdot d(q_i^+, q_{i+1}^-) \leq w_i(d(q_i^+, q_i^-) + d(q_{i+1}^+, q_{i+1}^-)) < w_i \cdot d(q_i^+, q_i^-) + w_{i+1} \cdot d(q_{i+1}^+, q_{i+1}^-)$, thus the replacement would results in a decrease of the energy consumption, contradicting the optimality of the original schedule.

Now consider the first pair of agents $i, i+1$ such that $w_i = w_{i+1}$ and $v_i^{-1} \geq v_{i+1}^{-1}$. As before, we can replace the agent $i+1$ by agent i without increasing the energy consumption. Furthermore, since $v_i \geq v_{i+1}$, also the delivery time either decreases or stays the same (if the package has to wait somewhere later on). Repeating this procedure results in an optimum schedule that adheres to property (i).

(ii) Assume that after the transformation above we remain with a schedule such that there are agents $i, i+1$ with $w_i = w_{i+1}$ and $d(p_{i+1}, q_{i+1}^+) > 0$. This leads to a contradiction to the optimality of the original schedule as well: Replacing agent $i+1$ with agent i strictly decreases $\mathcal{E}$ by at least $w_{i+1} \cdot d(p_{i+1}, q_{i+1}^+) > 0$. □

Furthermore, using the same replacements arguments, one can see that whenever multiple agents have the same starting position v, then only the agent with minimal (w_i, v_i^{-1}) among all agents on v is needed in an optimal solution:

Corollary 1. *After a preprocessing step of time $\mathcal{O}(k + |V|)$ – in which we remove in each vertex all but the agent with lexicographically smallest (w_i, v_i^{-1}) – we may assume that $k \leq |V|$.*

Weight Classes. Property (ii) of the characterization in Theorem 1 contains an important observation: among all agents of the same weight which contribute to an optimal schedule, only the first agent might actually walk *towards* the package – the other agents merely transport the package after it is dropped off at their respective starting position. Motivated by this, we partition the set $[k]$ of all agents into h *weight classes*, disjoint sets $W_1, \ldots, W_h$ of agents of the same decreasing weight and denote their sizes by $x_c = |W_c|$.[1] The idea then is to first solve EFFICIENTFASTDELIVERY for each weight class W_c on its own (albeit not only for *s-t*). This leads to the following result:

Theorem 2 (EFFICIENTFASTDELIVERY for uniform weights). *An optimum schedule for* EFFICIENTFASTDELIVERY *from s to t can be found in time $\mathcal{O}(k^2 + \text{APSP})$, assuming all agents have the same weight w.*

Proof (Analysis). Denote by $1, \ldots, \ell$ the agents involved in an optimum schedule that satisfies the characterization properties of Theorem 1. Since all agents

---

[1] Formally, we have $\bigcup_c W_c = [k]$ and $\sum_c x_c = k$ such that $\forall c, \forall i, j \in W_c : w_i = w_j$ and $\forall c_1 < c_2, \forall i \in W_{c_1}, \forall j \in W_{c_2} : w_i > w_j$.

have the same weight w, we know by (ii) that the total energy consumption can be written as $\mathcal{E} = w \cdot (d(p_1, s) + \sum_{i=1}^{\ell} d(q_i^+, q_i^-))$, where the sum $\sum_{i=1}^{\ell} d(q_i^+, q_i^-)) \geq d(s, t)$ achieves its minimum if the package is delivered along a shortest path (which is always possible). Thus we have $\mathcal{E} = w \cdot (d(p_1, s) + d(s, t))$. By minimality of the schedule, agent 1 must be closest to s among all k agents, $1 \in \arg\min_{1 \leq i \leq k}\{d(p_i, s)\}$.

Furthermore, we get $\mathcal{T} = v_1^{-1} \cdot (d(p_1, s) + d(s, q_1^-)) + \sum_{i=2}^{\ell} v_i^{-1} \cdot d(q_i^+, q_i^-)$. Thus we may assume without loss of generality, that among all agents of minimum distance to s, agent 1 has the highest velocity.

Proof (Preprocessing). Given a uniform-weight instance of EfficientFastDe-livery, we can find an optimum schedule as follows: We first run an all-pair shortest-path algorithm in time $\mathcal{O}(\mathrm{APSP})$. Then we find the first agent i_1 to pick up the package at s by searching for an agent of minimum $(d(p_i, s), v_i^{-1})$ in time $\mathcal{O}(k)$. To reach the package's origin s, this agent needs energy $\mathcal{E}_s := w \cdot d(p_{i_1}, s)$ and time $\mathcal{T}_s := v_{i_1}^{-1} \cdot d(p_{i_1}, s)$. Taking this into account, we transform our instance into an equivalent instance in which we have $p_{i_1} = s$ and where all other starting positions remain the same. Note that we have $p_i = q_i^+$ for all agents, hence lexico-graphically minimizing $(\mathcal{E}, \mathcal{T})$ amounts to finding a number of agents $i_1, i_2 \dots, i_\ell$ that minimize $(\mathcal{E}, \mathcal{T}) =$

$$\left(w \sum_{j=1}^{\ell} d(q_{i_j}^+, q_{i_j}^-), \ \sum_{j=1}^{\ell} v_{i_j}^{-1} d(q_{i_j}^+, q_{i_j}^-) \right) =$$

$$\left(w \sum_{j=1}^{\ell-1} d(p_{i_j}, p_{i_{j+1}}) + w \cdot d(p_{i_\ell}, t), \ \sum_{j=1}^{\ell-1} v_{i_j}^{-1} d(p_{i_j}, p_{i_{j+1}}) + v_{i_\ell}^{-1} d(p_{i_\ell}, t) \right).$$

By property (i) of Theorem 1, we can restrict ourselves to look at sequences of agents which have strictly increasing velocities. Since by property (ii) all pick-ups q_i^+ and drop-offs q_i^- of our transformed instance occur at the starting positions p_i (except for $q_{i_\ell}^- = t$), we model the delivery of the package up to the last involved agent as an auxiliary *directed acyclic graph* DAG with node set $\{p_1, \dots, p_k\}$ and directed edges $\{(p_i, p_j) \mid v_i < v_j\}$ of length $l_{(p_i, p_j)} = d(p_i, p_j)$, see Fig. 2.

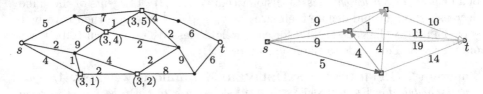

Fig. 2. (Left) original instance with 4 agents (□) of uniform weight 3 and speeds $1, 2, 4, 5$. The (unique) optimal schedule needs $(\mathcal{E}, \mathcal{T}) = (3(3 + 19), (3 + 5)/1 + 4/2 + 10/5) = (66, 12)$. **(Right)** transformed instance, represented by a DAG, with edges from p_i to p_j iff $v_i < v_j$.

Proof (Dynamic program). The DAG representing the transformed instance suggests an inductive approach from slow to fast agents. Therefore we relabel the agents i according to the topological ordering of their starting positions p_i, which can be done in time linear in the size ($\mathcal{O}(k^2)$) of the auxiliary DAG. Now we define the following subproblems:

$(\mathcal{E}, \mathcal{T})[p_i]$ = the energy consumption and delivery time of an optimum schedule
delivering the package from s to p_i, using only agents $1, \ldots, i - 1$.

The subproblems can be computed in increasing order of the agents: If an agent i is involved in an optimum schedule, the package must have been previously transported by agents with parameters (w, v_j^{-1}) lexicographically strictly smaller than the tuple (w, v_i^{-1}). All of these agents precede agent i in the topological ordering. Hence we initialize $(\mathcal{E}, \mathcal{T})[p_0]$ with the energy consumption and time needed in the original instance for the first agent to reach the package source s, $(\mathcal{E}, \mathcal{T})[p_1] := (\mathcal{E}_s, \mathcal{T}_s)$. Now we compute in topological order (as indicated by colors in Fig. 2 (right)):

$$(\mathcal{E}, \mathcal{T})[p_i] := \min_{j<i} \left\{ (\mathcal{E}, \mathcal{T})[p_j] + \left(w \cdot l_{(p_j, p_i)}, \; v_j^{-1} \cdot l_{(p_j, p_i)} \right) \right\},$$

where the operator $+$ denotes the element-wise addition of the tuple entries. Note that we already know that the respective energy consumption must amount to $w \cdot (d(p_{i_1}, s) + l_{(p_1, p_i)})$, it is really the optimal delivery time that we are after. Each of the k tuples $(\mathcal{E}, \mathcal{T})[p_i]$ can be computed in time $\mathcal{O}(i) \subseteq \mathcal{O}(k)$, needing time $\mathcal{O}(k^2)$ overall. Finally, we get the optimal $(\mathcal{E}, \mathcal{T})$ for delivery from s to t by taking the minimum over all candidates for the last agent i_ℓ in time $\mathcal{O}(k)$:

$$(\mathcal{E}, \mathcal{T}) := \min_{1 \leq j \leq k} \left\{ (\mathcal{E}, \mathcal{T})[p_j] + \left(w \cdot d_{(p_j, t)}, \; v_j^{-1} \cdot d_{(p_j, t)} \right) \right\},$$

We remark that a complete optimum schedule describing all pick-ups/drop-offs can be computed by backtracking, using the obtained values for $(\mathcal{E}, \mathcal{T})[p_i]$.   □

Application to Weight Classes. In the next section, we are going to apply the techniques described in Theorem 2 to each weight class W_c. However, if a subset of agents of the same weight class is involved in an optimum schedule, they will not necessarily deliver the package on their own, but work hand in hand with agents of preceding and following weight classes. In other words, the agents of W_c do not transport the package from s to t, but rather between two other points of the graph. For now, we assume these points to be any combination of two vertices $(u, v) \in V^2$. How can we – given the position of all agents of uniform weights – solve EFFICIENTFASTDELIVERY for all possible source/target tuples $(u, v) \in V^2$? Trivially, this can be done by running the dynamic program of Theorem 2 for all $\binom{|V|}{2}$ potential source-target pairs in overall time $\mathcal{O}(k^2 |V|^2 + \text{APSP})$.

However, we can do better than this. Note that we computed the dynamic program for the auxiliary DAG on $s = p_1, p_2, \ldots, p_k$ independent of $(\mathcal{E}, \mathcal{T})$, i.e. before we looked at $(\mathcal{E}, \mathcal{T})$ at the destination t. Using this independence, we can decrease the running time of the naïve solution by a linear factor: First we compute for each candidate $u \in V$ a dynamic program on a DAG with starting node u, in overall time $\mathcal{O}(k^2|V| + \text{APSP}) \subseteq \mathcal{O}(k|V|^2 + \text{APSP})$, assuming $k \leq |V|$ by Corollary 1. Then, for each of the $\mathcal{O}(|V|^2)$ many tuples (u, v) we compute $(\mathcal{E}, \mathcal{T})$ for an optimum delivery to v by looking at the $\mathcal{O}(k)$ many agents in the DAG with origin u, again in overall time $\mathcal{O}(k|V|^2 + \text{APSP})$. In total, we get:

Corollary 2. *An optimum schedule for* EFFICIENTFASTDELIVERY *from u to v for all possible tuples $(u, v) \in V^2$ can be found in time $\mathcal{O}(k|V|^2 + \text{APSP})$, assuming all agents have the same weight w.*

3   Vertex Handovers

Recall that in the last section we found that there is always an optimum schedule in which the involved agents have lexicographically strictly decreasing parameter tuples (w_i, v_i^{-1}) by Theorem 1 (i). These agents belong to weight classes W_c of decreasing weights, and we may assume that the involved agents of the same weight class have strictly increasing velocities. By property (ii), all handovers between agents of the same weight occur at nodes of the graph. What about agents of different weights? We show that in an optimum schedule, these handovers can only take place inside an edge if the weights of consecutive agents differ by a multiplicative factor of exactly 2:

Lemma 1 (In-edge-handovers I). *In any optimum schedule satisfying the two properties of Theorem 1, for any pair of consecutively involved agents i, j with handover point $q_i^- = q_j^+$ and weights $w_i \neq 2 \cdot w_j$, we have $q_j^+ \in V$.*

Proof. Assume for the sake of contradiction that $q_i^- = q_j^+$ lies strictly inside the edge $e = \{u, v\}$. Without loss of generality, the schedule delivers the package in direction from u to v. Denote by $\varepsilon > 0$ the minimum distance of q_i^- to u, v, or any other handover point in the interior of e. We distinguish three cases (Fig. 3):

1. Both agent i and agent j contain node u in their respective trajectory. Moving the handover point $q_i^- = q_j^+$ by ε in direction of u, we maintain a feasible schedule by minimality of ε. Furthermore, we strictly decrease agent i's travel distance while the total travel distance $d(p_j, q_j^+) + d(q_j^+, q_j^-)$ of agent j remains the same. Hence the overall energy consumption strictly decreases, contradicting the optimality of the original schedule.

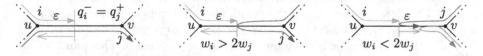

Fig. 3. (Left) there is no handover inside $\{u, v\}$ in which both agents come from u. **(Middle)** if $w_i > 2w_j$, reducing i's travel distance by $\varepsilon > 0$ and increasing j's travel distance by 2ε strictly decreases the overall energy consumption $\mathcal{E}$. **(Right)** vice versa.

2. Agent i approaches q_i^- in direction from u, agent j approaches q_j^+ coming from v, and $w_i > 2w_j$. As before, we move the handover point by ε in direction of u, thus changing the energy consumption $\mathcal{E}$ by $-w_i \cdot \varepsilon + w_j \cdot 2\varepsilon < 0$, contradicting the optimality of the original schedule.

3. Agent i approaches q_i^- in direction from u, agent j approaches q_j^+ coming from v, and $w_i < 2w_j$. We move the handover point by ε in direction of v, contradicting the optimality of $\mathcal{E}$ since $+w_i \cdot \varepsilon - w_j \cdot 2\varepsilon < 0$. $\square$

For the remainder of this section we will restrict ourselves to instances in which there are *no* two agents i, j with weights $w_i = 2w_j$. Thus, by Lemma 1 we may assume that there is an optimum schedule in which there are no in-edge-handovers.

Combining Weight Classes. Recall that we split the agents into h weight classes $W_1, W_2, \ldots, W_h$ of strictly decreasing weights, where each class W_c contains $x_c = |W_c|$ agents of the same weight. In an optimum schedule, the involved agents form subsets of a subset of the weight classes – our goal is hence to decide "in which part of the graph" agents of which weight class "can help the most". In order to do this, we will invoke Corollary 2 for each weight class (in decreasing order of their weights). For each class W_c, this allows us to find for each source/destination-pair $(u, v) \in V^2$ the contribution of its x_c many agents towards the total energy consumption $\mathcal{E}$ and delivery time T in time $\mathcal{O}(x_c|V|^2)$.

Dynamic Program. Again, we are interested in the distance between any two nodes in the graph, which we find by an all-pair shortest-path. Given the weight classes in decreasing order of the agents' weights, we define for each prefix $W_1, \ldots, W_j$ of the h weight classes and for each node v the following subproblem:

$(\mathcal{E}, T)[j, v] = $ the energy consumption $\mathcal{E}[j, v]$ and delivery time $T[j, v]$ of an

optimum schedule delivering the package from s up to v,

using only agents from the first j weight classes $W_1, \ldots, W_j$.

We are going to show how to compute $(\mathcal{E}, T)[j, v]$ from all smaller subproblems $(\mathcal{E}, T)[j - 1, u]$ using the dynamic program for uniform weights as a subroutine. The total energy consumption and delivery time of an optimum schedule can then be found in $(\mathcal{E}, T)[h, t]$. The optimum schedule itself can be found either by backtracking or by additionally storing in each node where the package came from and which agent brought it there. Our dynamic program is based on the assumption that there are no handover points inside any edges; we show in the next section how to adapt it to cover in-edge-handovers as well.

Theorem 3 (EFFICIENTFASTDELIVERY, vertex handovers only). *An optimum schedule for* EFFICIENTFASTDELIVERY *from s to t can be found in time* $\mathcal{O}(k|V|^2 + \text{APSP})$, *assuming that for all agents i, j we have weights $w_i \neq 2w_j$.*

Proof (Initialization). Set $(\mathcal{E}, T)[1, s] := (0, 0)$. To find all other values $(\mathcal{E}, T)[1, v]$, we use the dynamic program described in Theorem 2 as a subroutine. Specifically, we find the agent i_1 closest to s, $i_1 \in$ arg $\min_{i \in W_1} \{(w_i \cdot d(p_i, s), \ v_i^{-1} \cdot d(p_i, s))\}$ and store the energy and time it needs to reach s as $\mathcal{E}_s := (w_{i_1} \cdot d(p_{i_1}, s)$ and $\mathcal{T}_s := v_{i_1}^{-1} \cdot d(p_{i_1}, s))$.

To underline that the DAG on which our subroutine runs is rooted at s, we denote the subroutine subproblems by $(\mathcal{E}, \mathcal{T})_s[p_i]$. First we move agent i_1's starting position p_{i_1} to s and set $(\mathcal{E}, \mathcal{T})_s[p_{i_1}] := (\mathcal{E}_s, \mathcal{T}_s)$; then we compute all other values $\{(\mathcal{E}, \mathcal{T})_s[p_i]\}_{i \in W_1}$.

Hence for all $v \in V$, we complete our initialization phase by setting

$$(\mathcal{E}, \mathcal{T})[1, v] := \min_{i \in W_1} \left\{ (\mathcal{E}, \mathcal{T})_s[p_i] + \left(w_i \cdot d_{(p_i, v)}, \ v_i^{-1} \cdot d_{(p_i, v)} \right) \right\}.$$

Proof (Induction). After the first $j - 1$ weight classes have been considered, we compute a possible contribution by weight class W_j. Clearly, we have $(\mathcal{E}, \mathcal{T})[j, v] \leq (\mathcal{E}, \mathcal{T})[j - 1, v]$, since not using any agent of W_j is always an option. Therefore we start by setting $(\mathcal{E}, \mathcal{T})[j, v] := (\mathcal{E}, \mathcal{T})[j - 1, v]$ for all nodes $v \in V$.

Now, any node u is a potential starting position for the agents of weight class W_j to first pick up the package. Hence for *each* node $u \in V$, we will build a DAG and run a dynamic program, as described in Corollary 2. We denote the values of the corresponding subproblems by $(\mathcal{E}, \mathcal{T})_u[p_i]$. Finding all closest agents among all agents $i \in W_j$ to u and taking the fastest agent i_1 among them, we set $\mathcal{E}_u := w_{i_1} \cdot d(p_{i_1}, u)$ and $\mathcal{T}_u := v_{i_1}^{-1} \cdot d(p_{i_1}, u)$ before moving i_1's starting position to u. Note that for the package to be present at u, an energy amount of $\mathcal{E}[j - 1, u]$ was spent by the agents of the preceding weight classes. Furthermore these agents needed time $\mathcal{T}[j - 1, u]$ to bring the package to u; the package might be at u before or after agent i_1. We take this into account by defining

$$(\mathcal{E}, \mathcal{T})_u[p_{i_1}] := (\mathcal{E}[j - 1, u] + \mathcal{E}_u, \ \max\{\mathcal{T}[j - 1, u], \mathcal{T}_u\}).$$

For all $|V|$ many subroutines (dynamic programs on directed acyclic graphs rooted at one node $u \in V$ each), we compute all values $\{(\mathcal{E}, \mathcal{T})_u[p_i]\}_{i \in W_j}$.

Similarly, any node v is a potential destination position for the agents of W_j to deliver the package to. If indeed the agents of the current weight class can improve the delivery to v (i.e. if $(\mathcal{E}, \mathcal{T})[j, v] < (\mathcal{E}, \mathcal{T})[j - 1, v]$), then the package is transported to v by some agent in a DAG rooted at some u. Hence for each v we get $(\mathcal{E}, \mathcal{T})[j, v] :=$

$$\min \left\{ (\mathcal{E}, \mathcal{T})[j, v], \ \min_{u \in V} \left\{ \min_{i \in W_j} \left\{ (\mathcal{E}, \mathcal{T})_u[p_i] + \left(w_i \cdot d_{(p_i, v)}, \ v_i^{-1} \cdot d_{(p_i, v)} \right) \right\} \right\} \right\}. \quad (1)$$

Proof (Running time). After the initial all-pair shortest-paths computation in time $\mathcal{O}(\text{APSP})$, each of the h phases of the dynamic program (including the initialization) can be computed in time $\mathcal{O}(x_j|V|^2)$ by Corollary 2. Overall we get a running time of $\mathcal{O}(\text{APSP} + \sum_{j=1}^{h} x_j|V|^2) = \mathcal{O}(\text{APSP} + k|V|^2)$. $\qquad \square$

4   Full Solution

In this section, we consider EFFICIENTFASTDELIVERY in its full generality, that is we no longer restrict the possible weights of the agents. Instead we also allow weights to differ by a multiplicative factor of 2, in which case handovers inside an edge are possible and sometimes necessary in an optimum schedule (consider for example the delivery in Fig. 1 withstanding the presence of agents 3 and 4).

We first show that in an optimum schedule there can be at most one handover point inside any given edge. To account for this, we have to adapt the dynamic program of Sect. 3 as well as the subroutines given by the dynamic programs of Sect. 2. It turns out that this is possible without increasing the overall asymptotic running time, compared to Theorem 3.

Lemma 2 (In-edge-handovers II). *In any optimum schedule satisfying the two properties of Theorem 1, there is at most one handover point $q_{i_1}^- = q_{i_2}^+$ in the interior of any given edge $e = \{u, v\}$, where $i_1 \in W_{c_1}, i_2 \in W_{c_2}$. Furthermore, if the package is transported from u towards v, we may assume that: (i) $w_{i_1} = 2w_{i_2}$, (ii) $v_{i_1} < v_{i_2}$ and (iii) i_2 is the fastest among all agents i of the same weight w_{i_2} that have minimum distance to v.*

Proof. By Lemma 1 we know that for every edge-interior handover point $q_{i_1}^- = q_{i_2}^+$ we must have $w_{i_1} = 2w_{i_2}$, satisfying property (i). Now assume that we have two handover points in the interior of e: $q_{i_1}^- = q_{i_2}^+$ and $q_{i_2}^- = q_{i_3}^+$. By the preceding remark we must have $w_{i_1} = 2w_{i_2} = 4w_{i_3}$. Analogously to the first case in the proof of Lemma 1, we can assume that both i_2 and i_3 come from v towards their respective pick-up locations $q_{i_2}^+, q_{i_3}^+$, thereupon turning and moving towards v again, see Fig. 4 (left). But then we can replace agent i_2 by delegating its workload to agent i_3: This saves at least an energy amount of $2d(q_{i_2}^+, q_{i_2}^-) \cdot (w_{i_2} - w_{i_3})$, contradicting the optimality of the schedule.

Henceforth assume that there is exactly one handover point $q_{i_1}^- = q_{i_2}^+$ in e and that $w_{i_1} = 2w_{i_2}$. This means that agent i_1 must carry the package from u into e (otherwise we could reuse the exchange argument from before) and that i_2 enters the edge e from v. The distances travelled by the two agents inside e are thus $d(u, q_{i_1}^-)$ and $2d(v, q_{i_2}^+)$ – from an energy consumption point of view, the exact position of the handover point inside e is irrelevant. If $v_{i_1} \geq v_{i_2}$, we could move the handover to v without increasing the delivery time. Thus we may assume (ii).

Finally, assumption (iii) follows from the same arguments as given in the beginning of Theorem 2.   □

Definitions. How can we capture in-edge-handovers in our dynamic program from Sect. 3? Recall that by $(\mathcal{E}, \mathcal{T})[j, v]$ we denoted the energy consumption and delivery time of an optimum schedule delivering the package from s up to v, using only agents from the first j weight classes $W_1, \ldots, W_j$. We write $\mathcal{E}[j, v]$ and $\mathcal{T}[j, v]$ for the entries of the tuple $(\mathcal{E}, \mathcal{T})[j, v]$. Additionally, we now define

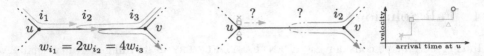

Fig. 4. (Left) if an optimum schedule had more than 1 handover point in the same edge, replacing the second agent by the third would reduce the total energy consumption. **(Right)** an agent of minimum weight in a minimum-energy subproblem schedule is a candidate to carry the package into $e = \{u, v\}$ if it is on the Pareto Frontier of all such agents (with respect to arrival time at u and their respective velocities).

$w[j-1, u]$ = the smallest weight $w_i, i \in W_1 \cup \ldots \cup W_{j-1}$, such that there is
a schedule delivering the package from s to u using:

- overall energy $\mathcal{E}[j-1, u]$,

- only agents from $W_1, \ldots, W_{j-1}$, and

- using agent i with $q_i^- = u$ in this schedule.

Intuitively, this means that there is an energy-optimal schedule from s to u for which there is some agent i_1 which we might consider in the next induction step j to move the package into the edge $e = \{u, v\}$. Note that we only require such a schedule to be optimal in the energy consumption, not with respect to the delivery time! Why is this the case? First of all, for an agent i_1 stationed at u to be considered to carry the package into the edge e, there are several necessary conditions on the fastest agent i_2 among all closest agents in W_j (Lemma 2 (iii)):

– For the weight w_{i_2} of the new agent i_2 we must have $2w_{i_2} = w_{i_1}$.
– For the velocity v_{i_2} we must have $v_{i_2} \geq v_{i_1}$.
– The package arrival time at v must be later than the arrival time of i_2:
  $\mathcal{T}[j-1, v] > v_{i_2}^{-1} \cdot d(p_{i_2}, v)$. Otherwise, agent i_2 has no incentive to enter e.
– The package arrival time at u must be earlier than the arrival time of i_2:
  $\mathcal{T}[j-1, u] < v_{i_2}^{-1} \cdot d(p_{i_2}, u)$. Otherwise, agent i_2 does not need the help of i_1.

Pareto Frontier. Assume that all these criteria are met and that agent i_2 moved to node v. Which agent(s) at u are in line for the tributary agent i_1? Only one (e.g. the fastest) or every agent of weight $w[j-1, u]$? Here we resume the reasoning why we consider arbitrary energy-optimal schedules up to u, instead of $(\mathcal{E}, \mathcal{T})$-optimal schedules as for example in the previous section: Compare a fast agent i_f that (due to preceding slow agents) brings the package to u quite late, with a slow but (because of fast enough preceding agents) early agent i_e: A priori it is not clear which of the two will win the race inside e, handing over the package to i_2: While i_e has a time advantage in the beginning, i_f might catch up due to its higher velocity.

Hence among all agents i of weight $w[j-1, u]$ for which there is a schedule with optimum energy consumption $\mathcal{E}[j-1, u]$ and in which $q_i^+ = u$, we consider those agents which are on the Pareto frontier with respect to their arrival time in the schedule at u and their respective velocity (Fig. 4 (right)). It remains to:

1. Incorporate the Pareto frontier into the subroutine (see Theorem 2).
2. Incorporate in-edge-handovers into the main dynamic program (Theorem 3).

Adapting the Subroutine. In the first part of the subroutine, we compute the dynamic programs $(\mathcal{E}, \mathcal{T})_u[p_i]$ on each DAG rooted at some node $u \in V$. In this dynamic program, we accelerate the package whenever possible. This is also the case in schedules which are Pareto-optimal rather than Delivery-time-optimal. The Pareto frontier comes into play when the DAG contains two different shortest u-v-paths or when v can be reached with the same optimum energy consumption from two different starting vertices u_1, u_2. Hence when computing the result of the subroutine as in Eq. (1), we do not only compute $(\mathcal{E}, \mathcal{T})[j, v]$ but we also update $w[j, v]$ and the agents in the Pareto front (if necessary). The asymptotic running time of the subroutine remains the same.

Adapting the Main Dynamic Program. Recall that our main dynamic program called as subroutines for each node the dynamic program for agents of uniform weights, based on the DAG rooted at a node. This was fine in Sect. 3, where we only had vertex handovers. Now we first have to check whether we already need the first agent i_2 of the new weight class W_j to pick up the package inside an edge $e = \{u, v\}$ and bring it to v. Observe that for a fixed v and a given candidate agent in the Pareto frontier at u, we can compute the energy consumption and arrival time for the delivery to v via a handover inside e in constant time. We iterate over all vertices v and check whether in any of the adjacent edges there is a possible handover that decreases the energy consumption and delivery time needed to transport the package to v. However, we need to consider the agents in the current Pareto frontier at u only if $w[j - 1, u] = 2w_{i_2}$. Therefore each agent appears during at most a single phase in a relevant Pareto frontier, since an agent i is only relevant when we consider the weight class W_j that contains agents of weight $w_i/2$. Thus all necessary precomputations before invoking the subroutines can be done in time $\mathcal{O}\left(\left(\sum_{v \in V} \deg(v)\right) \cdot \left(\sum_{j=1}^{h} x_j\right)\right) \subseteq \mathcal{O}(|E| \cdot k) \subseteq \mathcal{O}(k|V|^2)$. We conclude:

Theorem 4. *An optimum schedule for* EFFICIENTFASTDELIVERY *from s to t can be found in time* $\mathcal{O}(k|V|^2 + \text{APSP})$.

5 Conclusion

Two recent papers introduced DELIVERY by heterogenous energy-efficient agents [6] (where the rate of energy consumption is denoted by the agents' weights) and by agents of different velocities [7]. In these two models, minimizing the overall energy consumption $\mathcal{E}$, respectively the delivery time $\mathcal{T}$, are polynomial-time solvable. There are at least three natural ways to combine the models, with the goal of finding a schedule that simultaneously minimizes $\mathcal{E}$ and $\mathcal{T}$ in the sense of:

1. Lexicographically minimizing the tuple $(\mathcal{T}, \mathcal{E})$, or
2. Minimizing $\epsilon\mathcal{T} + (1 - \epsilon)\mathcal{E}$ for some input value $\epsilon \in (0, 1)$, or
3. Lexicographically minimizing the tuple $(\mathcal{E}, \mathcal{T})$: EFFICIENTFASTDELIVERY.

The first two variants were shown to be NP-hard [7]. In contrast, this paper shows that EFFICIENTFASTDELIVERY is polynomial-time solvable. We presented a dynamic program recursing over agents of *weight classes* of decreasing weight, in which for each weight class we call as a subroutine another dynamic program recursing over agents of increasing velocities. Overall, a schedule for EFFICIENT-FASTDELIVERY can be found in time $\mathcal{O}(k|V|^2 + \text{APSP})$, where k denotes the number of agents, $|V|$ the number of nodes in the graph and $\mathcal{O}(\text{APSP})$ the running time of an all-pair shortest-paths algorithm.

Acknowledgments. This work was partially supported by the SNF (project 200021L_156620, Algorithm Design for Microrobots with Energy Constraints).

References

1. Albers, S., Henzinger, M.R.: Exploring unknown environments. SIAM J. Comput. **29**(4), 1164–1188 (2000)
2. Anaya, J., Chalopin, J., Czyzowicz, J., Labourel, A., Pelc, A., Vaxès, Y.: Converge-cast and broadcast by power-aware mobile agents. Algorithmica **74**(1), 117–155 (2016)
3. Bampas, E., Czyzowicz, J., Gąsieniec, L., Ilcinkas, D., Klasing, R., Kociumaka, T., Pająk, D.: Linear search by a pair of distinct-speed robots. In: Suomela, J. (ed.) SIROCCO 2016. LNCS, vol. 9988, pp. 195–211. Springer, Cham (2016). doi:10. 1007/978-3-319-48314-6_13
4. Bärtschi, A., Chalopin, J., Das, S., Disser, Y., Geissmann, B., Graf, D., Labourel, A., Mihalák, M.: Collaborative delivery with energy-constrained mobile robots. In: Suomela, J. (ed.) SIROCCO 2016. LNCS, vol. 9988, pp. 258–274. Springer, Cham (2016). doi:10.1007/978-3-319-48314-6_17
5. Bärtschi, A., Chalopin, J., Das, S., Disser, Y., Geissmann, B., Graf, D., Labourel, A., Mihalák, M.: Collaborative delivery with energy-constrained mobile robots. In: Theoretical Computer Science (2017, to appear)
6. Bärtschi, A., Chalopin, J., Das, S., Disser, Y., Graf, D., Hackfeld, J., Penna, P.: Energy-efficient delivery by heterogeneous mobile agents. In: 34th Symposium on Theoretical Aspects of Computer Science STACS 2017, pp. 10:1–10:14 (2017)
7. Bärtschi, A., Graf, D., Mihalák, M.: Collective fast delivery by energy-efficient agents (2017, unpublished manuscript)
8. Bärtschi, A., Graf, D., Penna, P.: Truthful mechanisms for delivery with mobile agents. CoRR arXiv:1702.07665 (2017)
9. Bilò, D., Disser, Y., Gualà, L., Mihalák, M., Proietti, G., Widmayer, P.: Polygon-constrained motion planning problems. In: Flocchini, P., Gao, J., Kranakis, E., Meyer auf der Heide, F. (eds.) Algorithms for Sensor Systems. LNCS, vol. 8243, pp. 67–82. Springer, Heidelberg (2014). doi:10.1007/978-3-642-45346-5_6
10. Chalopin, J., Das, S., Mihalák, M., Penna, P., Widmayer, P.: Data delivery by energy-constrained mobile agents. In: Flocchini, P., Gao, J., Kranakis, E., Meyer auf der Heide, F. (eds.) ALGOSENSORS 2013. LNCS, vol. 8243, pp. 111–122. Springer, Heidelberg (2014). doi:10.1007/978-3-642-45346-5_9

11. Chalopin, J., Jacob, R., Mihalák, M., Widmayer, P.: Data delivery by energy-constrained mobile agents on a line. In: Esparza, J., Fraigniaud, P., Husfeldt, T., Koutsoupias, E. (eds.) ICALP 2014. LNCS, vol. 8573, pp. 423–434. Springer, Heidelberg (2014). doi:10.1007/978-3-662-43951-7_36
12. Czyzowicz, J., Gasieniec, L., Georgiou, K., Kranakis, E., MacQuarrie, F.: The beachcombers' problem: walking and searching with mobile robots. Theor. Comput. Sci. **608**, 201–218 (2015)
13. Czyzowicz, J., Gąsieniec, L., Kosowski, A., Kranakis, E.: Boundary patrolling by mobile agents with distinct maximal speeds. In: Demetrescu, C., Halldórsson, M.M. (eds.) ESA 2011. LNCS, vol. 6942, pp. 701–712. Springer, Heidelberg (2011). doi:10.1007/978-3-642-23719-5_59
14. Das, S., Dereniowski, D., Karousatou, C.: Collaborative exploration by energy-constrained mobile robots. In: Scheideler, C. (ed.) Structural Information and Communication Complexity. LNCS, vol. 9439, pp. 357–369. Springer, Cham (2015). doi:10.1007/978-3-319-25258-2_25
15. Demaine, E.D., Hajiaghayi, M., Mahini, H., Sayedi-Roshkhar, A.S., Oveisgharan, S., Zadimoghaddam, M.: Minimizing movement. ACM Trans. Algorithms **5**(3), 1–30 (2009)
16. Fraigniaud, P., Gąsieniec, L., Kowalski, D.R., Pelc, A.: Collective tree exploration. Networks **48**(3), 166–177 (2006)
17. Giannakos, A., Hifi, M., Karagiorgos, G.: Data Delivery by Mobile Agents with Energy Constraints over a fixed path. CoRR arXiv:1703.05496 (2017)
18. Weise, E.: Amazon delivered its first customer package by drone. USA Today, 14 December 2016. http://usat.ly/2hNgf0y

Parameterized Aspects of Triangle Enumeration

Matthias Bentert[✉], Till Fluschnik, André Nichterlein, and Rolf Niedermeier

Institut für Softwaretechnik und Theoretische Informatik, TU Berlin,
Berlin, Germany
{matthias.bentert,till.fluschnik,andre.nichterlein,
rolf.niedermeier}@tu-berlin.de

Abstract. Listing all triangles in an undirected graph is a fundamental graph primitive with numerous applications. It is trivially solvable in time cubic in the number of vertices. It has seen a significant body of work contributing to both theoretical aspects (e.g., lower and upper bounds on running time, adaption to new computational models) as well as practical aspects (e.g. algorithms tuned for large graphs). Motivated by the fact that the worst-case running time is cubic, we perform a systematic parameterized complexity study of triangle enumeration, providing both positive results (new enumerative kernelizations, "subcubic" parameterized solving algorithms) as well as negative results (uselessness in terms of possibility of "faster" parameterized algorithms of certain parameters such as diameter).

1 Introduction

Detecting, counting, and enumerating triangles in undirected graphs is a basic graph primitive. In an n-vertex graph, there can be up to $\binom{n}{3}$ different triangles and an algorithm checking for each three-vertex subset if it forms a triangle can list all triangles in $O(\binom{n}{3})$ time. As to counting the number of triangles in a graph, the best known algorithm takes $O(n^\omega) \subset O(n^{2.373})$ time [23] and is based on fast matrix multiplication.[1] Finally, detecting a triangle in a graph is doable in $O(n^\omega)$ time [23] and it is conjectured that every algorithm for detecting a triangle in a graph takes at least $\Theta(n^{\omega-o(1)})$ time [1]. We mention that for sparse m-edge graphs there is also an $O(m^{1.5})$-time algorithm [18]. This paper is motivated by trying to break such (relative or conjectured) lower bounds and improve on best known upper bounds—the twist is to introduce a secondary measurement beyond mere input size. This is also known as problem parameterization. While parameterizing problems with the goal to achieve fixed-parameter tractability results is a well-established line of research for NP-hard problems,

---

A full version is available at https://arxiv.org/abs/1702.06548.

T. Fluschnik—Supported by the DFG, project DAMM (NI 369/13-2).

A. Nichterlein—Supported by a postdoc fellowship of the DAAD while at Durham University, UK.

[1] ω is a placeholder for the best known $n \times n$-matrix multiplication exponent.

© Springer-Verlag GmbH Germany 2017
R. Klasing and M. Zeitoun (Eds.): FCT 2017, LNCS 10472, pp. 96–110, 2017.
DOI: 10.1007/978-3-662-55751-8_9

systematically applying and extending tools and concepts from parameterized algorithmics to polynomial-time solvable problems is still in its infancy [2,12–14,26]. Performing a closer study of mostly triangle enumeration, we contribute to this line of research, also referred to as "FPT-in-P" for short [14]. Our central leitmotif herein is the quest for parameterized subcubic triangle enumeration algorithms.

Related Work. Triangle enumeration, together with its relatives counting and detection, has many applications, ranging from spam detection [4] over complex network analysis [15,27] and database applications [19] to applications in bioinformatics [32]. Hence there has been substantial theoretical and practical work. The theoretically fastest algorithms are based on matrix multiplication and run in $O(n^\omega + n^{3(\omega-1)/(5-\omega)} \cdot (\#T)^{2(3-\omega)/(5-\omega)})$ time, where $\#T$ denotes the number of listed triangles [5]. Furthermore, there is (heuristic and experimental) work on listing triangles in large graphs [22,30], on triangle enumeration in the context of map reduce [28], and even on quantum algorithms for triangle detection [24].

As to parameterized results, early work by Chiba and Nishizeki [7] showed that all triangles in a graph can be counted in $O(m \cdot d)$ time, where d is the degeneracy of the graph.[2] This running time can be improved by saving polylogarithmic factors [20], but the 3SUM-conjecture[3] rules out more substantial improvements [21]. Green and Bader [16] described an algorithm for triangle counting running in $O(T_K + |K| \cdot \Delta_K^2)$ time, where K is a vertex cover of the input graph, Δ_K is the maximum degree of vertices in K, and T_K is the time needed to compute K. They also described several experimental results.

Our Contributions. We systematically explore the parameter space for triangle enumeration and classify the usefulness of the parameters for FPT-in-P algorithms. In doing so, we present an extended concept of enumerative kernelization and a novel hardness concept, as well as algorithmic results. Our concrete results are surveyed in Table 1. We defer to the respective sections for a formal definition of the various parameters. In particular, we provide *enumerative problem kernels* with respect to the parameters "feedback edge number" and "vertex deletion distance to d-degeneracy". Partially based on data reduction algorithms, we provide fast algorithms for several parameters such as feedback edge number, vertex deletion distance to cographs and to d-degeneracy (also with additional parameter maximum vertex degree), distance to cographs, and clique-width. On the negative side, using a concept we call *"General-Problem-hardness"*, we show that using the parameters domination number, chromatic number, and diameter do not help to get FPT-in-P algorithms for detecting triangles, that is, even for constant parameter values the problem remains as "hard" as the general version

---

[2] Degeneracy measures graph sparseness. A graph G has degeneracy d if every subgraph contains a vertex of degree at most d; thus G contains at most $n \cdot d$ edges.

[3] The 3SUM problem asks whether a given set S of n integers contains three integers $a, b, c \in S$ summing up to 0. The 3SUM-conjecture states that for any constant $\varepsilon > 0$ there is no $O(n^{2-\varepsilon})$-time algorithm solving 3SUM. The connection between 3SUM and listing/detecting triangles is well studied [24,29].

Table 1. Overview of our results. (n: number of vertices; m: number of edges; $\#\,T$: number of triangles; k: respective parameter; Δ: maximum degree)

	Parameter k	Result	Reference
Enum-kernel	Feedback edge number	Size at most $9k$ in $O(n+m)$ time	Proposition 9
	Distance to d-degenerate	at most $k+2^k+3$ vertices in $O(n \cdot d \cdot (k+2^k))$ time	Theorem 11
Solving	Feedback edge number	In $O(k^2+n+m)$ time	Theorem 10
	Distance to d-degenerate	In $O(n \cdot d \cdot (k+d)+2^{3k}+\#T)$ time	Corollary 12
	Distance to cographs	In $O(\#\,T+n+m \cdot k)$ time	Proposition 14
	Distance to d-degenerate + maximum degree Δ	In $O(k \cdot \Delta^2+n \cdot d^2)$ time	Proposition 7
	Clique-width	In $O(n^2+n \cdot k^2+\#T)$ time	Theorem 15
Hard	Domination number, chromatic number, and diameter	For $k \geq 3$ as hard as the general case	Proposition 4

with unbounded parameter. Due to space constraints, some proofs and details are omitted (marked with ($\star$)).

2 Preliminaries

Notation. For an integer $\ell \geq 1$, let $[\ell] = \{1,\ldots,\ell\}$. Let $G = (V,E)$ be an undirected simple graph. We set $n := |V|$, $m := |E|$, and $|G| = n+m$. We denote by $N(v)$ the (open) neighborhood of a vertex $v \in V$ and by $\deg(v) := |N(v)|$ the degree of v. By $G[U]$ we denote the subgraph of G induced by the vertex subset $U \subseteq V$ and $G - U := G[V \setminus U]$. If $\{x,y,z\} \subseteq V$ induces a triangle in a graph, we refer to $T = \{x,y,z\}$ as the triangle. We denote the number of triangles in the graph by $\#\,T$. Our central problem is as follows.

> TRIANGLE ENUMERATION ($\triangle$-ENUM)
> **Input:** An undirected graph G.
> **Task:** List all triangles contained in G.

Parameterized Complexity. A language $L \subseteq \Sigma^* \times \mathbb{N}$ is a *parameterized problem* over some finite alphabet Σ, where $(x,k) \in \Sigma^* \times \mathbb{N}$ denotes an instance of L and k is the parameter. Then L is called *fixed-parameter tractable* (equivalently, L is in the class FPT) if there is an algorithm that on input (x,k) decides whether $(x, k) \in L$ in $f(k) \cdot |x|^{O(1)}$ time, where f is some computable function only depending on k and $|x|$ denotes the size of x. We call an algorithm with a running time of the form $f(k) \cdot |x|$ a linear-time FPT algorithm. Creignou et al. [9, Definition 3.2] introduced the concept of FPT-delay algorithms for enumeration problems. An algorithm $\mathcal{A}$ is an FPT-delay algorithm if there exist a computable function f and a polynomial p such that $\mathcal{A}$ outputs for every input x all solutions of x with at most $f(k) \cdot p(|x|)$ time between two successive solutions. If the delay can be upper-bounded in $p(|x|)$, then the algorithm is called a p-delay algorithm. A *kernelization* for L is an algorithm that on input (x, k) computes in time polynomial in $|x| + k$ an output (x', k') (the *kernel*) such that (i) $(x, k) \in L \iff (x', k') \in L$, and (ii) $|x'| + k' \leq g(k)$ for some computable function g only depending on k. The value $g(k)$ denotes the *size of the kernel*.

This work focuses on enumeration, while the great majority of parameterized complexity works study decision (or search and optimization) problems.

Definition 1 [9, Definition 1]. *A parameterized enumeration problem is a pair (P, Sol) such that*

- *$P \subseteq \Sigma^* \times \mathbb{N}$ is a parameterized problem over a finite alphabet Σ and*
- *$\mathsf{Sol}\colon \Sigma^* \times \mathbb{N} \to \mathcal{P}(\Sigma^*)$ is a function such that for all $(x, k) \in \Sigma^* \times \mathbb{N}$, $\mathsf{Sol}(x, k)$ is a finite set and $\mathsf{Sol}(x, k) \neq \emptyset \iff (x, k) \in P$.*

Intuitively, the function Sol contains for each instance (x, k) of P the set of all solutions. Given an instance (x, k), the task is then to compute $\mathsf{Sol}(x, k)$.

3 New Notions of Hardness and Kernelization

In this section we introduce two notions and give simple proofs of concept for both of them. The first notion is a many-one reduction that relates parameterized problems to its unparameterized counterpart. We call it *"General-Problem-hardness"* as it proves the parameterized version to be as hard as the unparameterized (general) problem. We show hardness for the TRIANGLE DETECTION ($\triangle$-DETECT) problem: given an undirected graph G, decide whether G contains a triangle. Since $\triangle$-DETECT is a special case of $\triangle$-ENUM, it follows that any lower bound for $\triangle$-DETECT implies the same lower bound for $\triangle$-ENUM. Thus, if a certain parameter does not admit a solving algorithm for $\triangle$-DETECT in some (parameterized) time X, then $\triangle$-ENUM does not either.

3.1 Computational Hardness

Before giving a formal definition, consider as introductory example the parameter minimum degree. Adding an isolated vertex to any graph in constant time leaves the set of triangles unchanged and the resulting graph has minimum degree zero. Hence, one can not use the parameter minimum degree to design faster algorithms for $\triangle$-ENUM. Upon this trivial example, we study which parameters for $\triangle$-ENUM cannot be used to design linear-time FPT algorithms under the conjecture that $\triangle$-ENUM is not linear-time solvable [1]. To this end we reduce in linear time an arbitrary instance of $\triangle$-DETECT to a new equivalent (and not too large) instance of the problem with the parameter upper-bounded by a constant. The corresponding notion of a many-one reduction is as follows.

Definition 2. *Let* $P \subseteq \Sigma^* \times \mathbb{N}$ *be a parameterized problem, let* $Q \subseteq \Sigma^*$ *be the unparameterized decision problem associated to* P, *and let* $f \colon \mathbb{N} \to \mathbb{N}$ *be a function. We call* P ℓ-*General-Problem-hard*(f) *(ℓ-GP-hard*(f)*) if there exists an algorithm* $\mathcal{A}$ *transforming any input instance* I *of* Q *into a new instance* (I', k') *of* P *such that*

(G1) $\mathcal{A}$ *runs in* $O(f(|I|))$ *time,*   *(G3)* $k' \leq \ell$, *and*
(G2) $I \in Q \iff (I', k') \in P$,   *(G4)* $|I'| \in O(|I|)$.

We call P *General-Problem-hard*(f) *(GP-hard*(f)*) if there exists an integer* k *such that* P *is* k-*GP-hard*(f). *We omit the running time and call* P k-*General-Problem-hard* (k-GP-hard) *if* f *is a linear function.*

If one can exclude an algorithm solving Q in $O(f(|I|))$ time and can further prove that P is GP-hard, then one can (under the same assumptions that excluded an $O(f(|I|))$-time algorithm for Q) exclude an algorithm solving P in $O(g(k) \cdot f(|I|))$ time for any computable function g. This yields the following.

Lemma 3 ($\star$). *Let* $f \colon \mathbb{N} \to \mathbb{N}$ *be a function, let* $P \subseteq \Sigma^* \times \mathbb{N}$ *be a parameterized problem that is* ℓ-*GP-hard*(f), *and let* $Q \subseteq \Sigma^*$ *be the unparameterized decision problem associated to* P. *If there is an algorithm solving each instance* (I, k) *of* P *in* $O(g(k) \cdot f(|I|))$ *time, then there is an algorithm solving each instance* I' *of* Q *in* $O(f(|I'|))$ *time.*

It is folklore that $\triangle$-DETECT in tripartite graphs belongs to its hardest cases. Based on this, we show that $\triangle$-DETECT with respect to the combined parameters domination number, chromatic number, and diameter is 9-GP-hard. Indeed, $\triangle$-DETECT is 3-GP-hard for each of the (single) parameters. The *domination number* of a graph is the size of a minimum cardinality set S with $\bigcup_{v \in S} N(v) \cup S = V$. The *chromatic number* of a graph is the minimum number of colors needed to color the vertices such that no edge contains vertices of the same color. The *diameter* of a graph is the length of the longest shortest path between two vertices.

Proposition 4 ($\star$). $\triangle$-DETECT *is 9-GP-hard with respect to the sum of domination number, chromatic number, and diameter.*

3.2 Enum-Advice Kernelization

The second new notion we introduce in this paper is an adaption of an enumerative kernelization concept due to Creignou et al. [9].

The aim of kernelization is to efficiently reduce a large instance of a computationally hard, say NP-hard, problem to an "equivalent" smaller instance

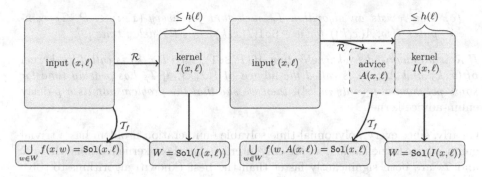

Fig. 1. A schematic picture of enum- (left) and enum-advice (right) kernelization. The kernelization $\mathcal{R}$ produces the kernel and, for enum-advice kernelization also the advice. Then, there is a polynomial-delay algorithm T_f that lists all solutions $\mathtt{Sol}(x)$ of the input x from the solutions of the kernel and either the input instance (in the enum-kernelization) or the advice (enum-advice kernel).

(called "kernel"). Then, solving the kernel by a trivial brute-force algorithm often significantly reduces the overall running time. This technique is by no means restricted to computationally hard problems even though it was invented to tackle problems for which no polynomial-time algorithms are known.

Note that kernelization is usually defined for decision problems only. Creignou et al. [9] developed a concept to address enumeration problems. Roughly speaking, their concept requires that *all* solutions of the input instance can be recovered from the input instance and the solutions of the kernel (see left side of Fig. 1). We modify the concept by adding a generic object which we call the *advice* of the kernelization. The intention of this change is that in order to compute all solutions of the *input* instance, one only needs the kernel and the advice (which might be much smaller than the input instance), see Fig. 1 for an illustration. In the examples we provide in this paper, we store in the advice information about all triangles that are destroyed by data reduction rules.

We will now give a formal definition of our new enumerative kernelization concept and then discuss the advantages compared to the concept by Creignou et al.

Definition 5. *Let $(P, \mathtt{Sol})$ be a parameterized enumeration problem. Let $\mathcal{R}$ be an algorithm which for every input (x, k) computes in time polynomial in $|x| + k$ a pair $(I(x, k), A(x, k))$. We call $\mathcal{R}$ an enum-advice kernelization of $(P, \mathtt{Sol})$ if*

(K1) there is a function h such that for all (x, k) it holds that $|I(x, k)| \leq h(k)$,
(K2) for all (x, k) it holds $(x, k) \in P \iff I(x, k) \in P$, and
(K3) there exists a function f such that for all $(x, k) \in P$
    (a) $\forall p, q \in \mathtt{Sol}(I(x, k)): p \neq q \implies f(p, A(x, k)) \cap f(q, A(x, k)) = \emptyset$,
    (b) $\bigcup_{w \in \mathtt{Sol}(I(x,k))} f(w, A(x, k)) = \mathtt{Sol}(x, k)$, and

(c) *there exists an algorithm $\mathcal{T}_f$ such that for every $(x, k) \in P$, $\mathcal{T}_f$ com-*
    putes $f(w, A(x, k))$ for $w \in \text{Sol}(I(x, k))$ in FPT-delay time [9].

*If $\mathcal{R}$ is an enum-advice kernelization of (P, Sol), then $I(x, k)$ is called the kernel
of (x, k) and $A(x, k)$ is called the advice of $I(x, k)$. If $\mathcal{T}_f$ has p-delay time for
some polynomial p (only in $|x|$), then we say that the problem admits a p-delay
enum-advice kernel.*

Clearly, since every polynomial-time solvable enumeration problem has a trivial
enum-advice kernelization, we are only interested in those kernelizations where $\mathcal{R}$
and $\mathcal{T}_f$ are both significantly faster than the best (known) algorithms to solve
the enumeration problem.

We will now discuss the advantages of our new definition compared to enum-
kernelization. First, note that one can set $A(x, \ell) = (x, \ell)$, and thus enum-advice
kernelization is a generalization of enum-kernelization. Second, the advice can
be used to design faster algorithms since the advice might be much smaller
than the input instance. In general, enumeration algorithms can be derived from
enum-advice kernels as stated in the next lemma.

Lemma 6 ($\star$). *Let $\mathcal{R}$ be an enum-advice kernelization of a parameterized enumeration problem (P, Sol) such that for every instance (x, k) of P:*

- *$\mathcal{R}$ runs in $O((|x| + k)^c)$ time for some constant c;*
- *the unparameterized version of P can be solved in $g(|x|)$ time on x;*
- *the kernelization computes the pair (I, A) where $|I| \leq h(k)$, and $\mathcal{T}_f$ takes at
  most $O(|I|^d)$ time between generating two solutions for some constant d;*
- *#s denotes the number of solutions in I and #S denotes the number of
  solutions in x.*

Then, (P, Sol) can be solved in $O((|x| + k)^c + g(h(k)) + (\#s + \#S) \cdot h(k)^d)$ time.

Note that in general we cannot give any meaningful upper bound on the
delay of the constructed algorithm as the kernel instance might be packed with
solutions p such that $f(w, A) = \emptyset$. If no such solutions exist, then the delay
of the described algorithm is $O((|x| + k)^c + f(h(k)) + h(k)^d)$. The delay of all
algorithms presented in this paper are only upper-bounded by the respective
running times of the algorithms.

4   Algorithms

In this section, we show FPT algorithms solving $\triangle$-ENUM exploiting several
parameters. We systematically explore the parameter landscape along a hierarchy of graph parameters (see [31]) in the following way (Fig. 2 surveys our
outline). We start from the fact that $\triangle$-ENUM allows for an $O(m \cdot d)$-time algorithm when parameterized by degeneracy d [7], and go into the following two
directions: First, we study in Sect. 4.1 whether parameters k lying above degeneracy in the parameter hierarchy admit algorithms running in $f(k) + O(n + m)$

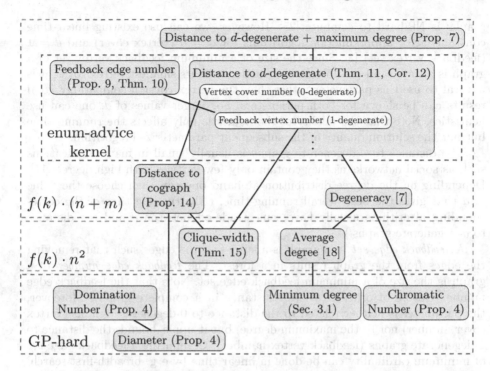

Fig. 2. Layering of considered parameters with respect to our and known results. Herein, the parameters are hierarchically arranged in the sense that if two parameters are connected by a line, then the lower one can be upper-bounded by the higher one. Thus, hardness results transfer downwards and tractability results upwards. For the parameter average degree σ an $O((\sigma n)^{1.5})$-time algorithm follows from an existing $O(m^{1.5})$-time algorithm [18] and the simple observation that any graph has $O(\sigma n)$ edges.

time. Kernelization is one way to achieve such *additive* ($f(k)+O(n+m)$) instead of *multiplicative* ($f(k) \cdot O(n+m)$) running times. Indeed, for the two parameters feedback edge number and distance to d-degenerate graphs (see definitions below) we show enum advice-kernels. Second, we study in Sect. 4.2 parameters that are incomparable with degeneracy and so far unclassified.

4.1 Parameters Lower-Bounded by Degeneracy

In this section we show results on parameters that are hierarchically above the degeneracy. We first describe the parameters and then turn to the results.

We first discuss distance to d-degenerate graphs. The *distance to d-degenerate graphs* of a graph G is the size of a minimum cardinality vertex set D such that $G - D$ is d-degenerate. This parameter generalizes several well-known parameters like vertex cover (distance to 0-degenerate graphs) and feedback vertex set (distance to 1-degenerate graphs). For any fixed d the distance to d-degenerate graphs is NP-hard to compute [25]. However, we can use existing linear-time constant-factor approximation algorithms for $d = 0$ (vertex cover) and $d = 1$ (feedback vertex set) [3]. Since the size of a minimum feedback vertex set of a graph is always (and possibly much) smaller than its smallest vertex cover, it is natural to use this parameter rather than the vertex cover if comparably good results can be shown for both parameters. For larger values of d, one can use heuristics. Note that the quality of the heuristic only affects the running time but not the solution quality of the subsequent parameterized algorithm.

The distance to d-degenerate graphs is usually small in many applications such as social networks as they contain only few vertices with high degree [11]. Depending on the degree distribution at hand one can then choose the value of d that gives the best overall running-time. (The running time of the corresponding algorithms usually have some trade-off between d and the distance to d-degenerate graphs.)

A *feedback edge set* in a graph is a subset of the edges such that removing the edges from the graph results in a forest. The *feedback edge number* of a graph is the size of a minimum feedback edge set. Note that the feedback edge number can be of order $O(m)$ (for instance in a complete graph). Moreover, the parameter is neither related to the distance to 0-degenerate graphs (vertex cover number) nor to the maximum degree, but it upper-bounds the distance to 1-degenerate graphs (feedback vertex number). Computing a feedback edge set of minimum cardinality can be done in linear time by e.g. breadth-first-search. We hence assume that a feedback edge set is given.

Distance to d-Degenerate Graphs plus Maximum Degree. Green and Bader [16] stated that TRIANGLE COUNTING parameterized by the size of a vertex cover V' and the maximum degree $d_{max} = \max(\{\deg(v) \mid v \in V'\})$ in this vertex cover can be solved in $O(|V'| \cdot d_{max}^2 + n)$ time. We will construct an algorithm which solves $\triangle$-ENUM parameterized by the size of D and the maximum degree Δ_D in D, where D is a set of vertices such that $G - D$ is d-degenerate, provided that D is given. The algorithm takes $O(|D| \cdot \Delta_D^2 + n \cdot d^2)$ time. Bear in mind that for each vertex cover V' it holds that $G - V'$ is 0-degenerate and hence $O(|D| \cdot \Delta_D^2 + n \cdot d^2) = O(|D| \cdot \Delta_D^2)$. Consequently, our result generalizes the result by Green and Bader.

Proposition 7 ($\star$). $\triangle$-ENUM *parameterized by distance to d-degenerate graphs and the maximum degree Δ_D in a set D such that $G - D$ is d-degenerate is solvable in $O(|D| \cdot \Delta_D^2 + n \cdot d^2)$ time provided that the set D is given.*

Feedback Edge Number. We provide a key lemma and then state a linear-size enum-advice kernel for $\triangle$-ENUM parameterized by feedback edge number.

Lemma 8 ($\star$). *Let $G = (V, E)$ be an undirected graph and let F be a feedback edge set in G. All triangles $\{u, v, w\}$ where at least one of the edges between the three vertices is not in F can be enumerated in $O(n + m)$ time. There are at most $2|F|$ such triangles.*

Proposition 9 ($\star$). $\triangle$-ENUM *parameterized by feedback edge number k admits a constant-delay enum-advice kernel with at most $2k + 3$ vertices and $k + 3$ edges which can be computed in $O(n + m)$ time.*

A straight-forward application of Lemma 6 given Proposition 9 yields the following.

Theorem 10 ($\star$). $\triangle$-ENUM *parameterized by feedback edge number k can be solved in $O(k^{1.5} + n + m)$ time.*

Distance to d-Degenerate Graphs. We next present an enum-advice kernel for $\triangle$-ENUM parameterized by distance to d-degenerate graphs.

Theorem 11. $\triangle$-ENUM *parameterized by distance to d-degenerate graphs admits a constant-delay enum-advice kernel provided that the distance deletion set D to d-degenerate graphs is given. The kernel contains at most $|D| + 2^{|D|} + 3$ vertices and can be computed in $O(n \cdot (d + 1) \cdot (|D| + d))$ time.*

Proof. Let G be an instance of $\triangle$-ENUM and let $k = |D|$ be the parameter. Construct the enum-advice kernel $(I(G, k), A(G, k))$ as follows. To this end, we abbreviate $(G_I, k') := I(G, k)$ and $A := A(G, k)$.

First, compute the d-degenerate graph $G' = G - D$. The graph G' contains exactly the triangles in G that do not contain any vertices in D. Compute the set of triangles in G' in $O(m \cdot d)$ time [7]. Next, compute all triangles with exactly one vertex in D. To this end, compute the degeneracy order in linear time, iterate over all $v \in D$, $u \in N(v) \setminus D$, and the at most d neighbors of u that are ordered after u in the degeneracy order, and list all triangles found. By this, all triangles in G containing exactly one vertex in D are found in $O(k \cdot n \cdot d)$ time. Altogether, we can compute the set T_1 of all triangles in G with at most one vertex in D in $O(n \cdot d \cdot (k + d))$ time.

Delete all edges which have no endpoint in D as they cannot be part of any further triangles. Next, compute all twin-classes in the current graph, that is, a partition $\mathcal{P}$ of the vertices according to their neighbors, using partition refinement in $O(n + m)$ time [17].

For each non-empty part $P \in \mathcal{P}$ pick one vertex $v_P \in P$ and store a function M such that $M(v_P) = P \setminus D$. Put all vertices in D, all of the chosen vertices, and all edges induced by these vertices into G_I. Add three new vertices a, b, c to G_I and if $T_1 \neq \emptyset$, then add three new edges $\{a, b\}, \{a, c\}, \{b, c\}$. Note that all edges have an endpoint in $D' = D \cup \{a, b\}$ and thus D' is a deletion set to d-degenerate graphs for every d. Complete the construction by setting $k' = |D'|$ and $A = (T_1, M, \{a, b, c\})$. Note that G_I contains at most $k + 2^k + 3$ vertices (K1). Observe that since $m \in O(n \cdot (k + d))$, the kernel can be constructed in $O(n \cdot d \cdot (k + d))$ time. For $x_1, x_2, x_3 \in V(G_I)$, define the function f as $f(\{x_1, x_2, x_3\}, A) = T_1$ if $\{x_1, x_2, x_3\} = \{a, b, c\}$ and $f(\{x_1, x_2, x_3\}, A) = \{\{v_1, v_2, v_3\} \mid v_1 \in M(x_1) \wedge v_2 \in M(x_2) \wedge v_3 \in M(x_3)\}$ otherwise. Next, we prove that the algorithm fulfills all conditions of Definition 5.

Observe that G_I is isomorphic to a subgraph of G and, hence, if there is a triangle G_I, then there is a triangle in G. Assume that there is a triangle X with vertices $\{x_1, x_2, x_3\}$ in G. If X contains at most one vertex in D, then $T_1 \neq \emptyset$ and thus there is the triangle formed by $\{a, b, c\}$ in G_I. Otherwise, X contains at least two vertices in D. Assume without loss of generality that $x_2, x_3 \in D$. If x_1 is in D, then X is also contained in G_I. Otherwise, there is a vertex v in G_I such that $x_1 \in M(v)$. Since $\{x_1, x_2, x_3\}$ forms a triangle in G, it follows that $\{v, x_2, x_3\}$ forms a triangle in G and G_I. Hence, condition (K2) (of Definition 5) is fulfilled.

Next we discuss the condition (K3). We will prove that for every triangle $X = \{x_1, x_2, x_3\}$ in G there is a unique solution $w \in \mathrm{Sol}(G_I, k')$ such that $X \in f(w, A)$ (K3b). If X contains at most one vertex in D, then by construction $X \in f(\{a, b, c\}, A)$. Since G_I contains only edges with an endpoint in D, no triangle $\{v_1, v_2, v_3\}$ where $v_1 \in M(x_1)$, $v_2 \in M(x_2)$, and $v_3 \in M(x_3)$ is contained in G_I. Thus $\{a, b, c\}$ is the only triangle in G_I such that $X \in f(\{a, b, c\}, A)$. If X contains at least two vertices $x_2, x_3 \in D$, then there exists a vertex v in G_I such that $x_1 \in M(v)$ and the triangle $\{v, x_2, x_3\}$ is contained in G. By construction, the triangle $\{v, x_2, x_3\}$ is also contained in G_I and $X \in f(\{v, x_2, x_3\}, A)$. Since $X \notin T_1$, it follows $X \notin f(\{a, b, c\}, A)$.

Next we show that for any two triangles $p = \{u_1, u_2, u_3\}$ and $q = \{v_1, v_2, v_3\}$ in G_I, it holds that $f(p, A) \cap f(q, A) = \emptyset$ (K3a). If either p or q is $\{a, b, c\}$ (let us assume without loss of generality p), then by definition $f(p, A)$ only contains triangles with at most one vertex in D and $f(q, A)$ only contains triangles with at least two vertices in D and thus $f(p, A) \cap f(q, A) = \emptyset$.

If neither p nor q is $\{a, b, c\}$, then both of them only contain vertices from the original graph G. As $p \neq q$, assume without loss of generality that $u_1 \notin q$ and $v_1 \notin p$. By construction all triangles in $f(p, A)$ contain one vertex in $M(u_1)$ and all triangles in $f(q, A)$ contain one vertex in $M(v_1)$. As shown above, $M(u_1)$ ($M(v_1)$, respectively) only contains u_1 (v_1) and vertices that have the same neighbors as u_1 (v_1) in D. Hence, no triangle in $f(p, A)$ ($f(q, A)$ respectively) contains a vertex in $M(v_1)$ ($M(u_1)$) and thus $f(p, A) \cap f(q, A) = \emptyset$.

Each triangle in $\{\{v_1, v_2, v_3\} \mid v_1 \in M(x_1) \wedge v_2 \in M(x_2) \wedge v_3 \in M(x_3)\}$ and in T_1 can be returned with constant delay between generating two successive solutions (K3c). □

The time needed to compute the kernel $(I(G, |D|), A(G, |D|))$ in Theorem 11 is upper-bounded by $O(n \cdot d \cdot (|D| + d) + |D| + m) = O(n \cdot (d + 1) \cdot (|D| + d))$. The equality holds since $m \in O(n \cdot (|D| + d))$.

To the best of our knowledge, there is no algorithm that solves $\triangle$-ENUM parameterized by distance to d-degenerate graphs within this time. All solutions can be reconstructed in constant-delay time and there is no known algorithm that solves $\triangle$-ENUM parameterized by distance to d-degenerate graphs in constant-delay time (and it seems unlikely that such an algorithm exists).

Using Lemma 6 we get the following result.

Corollary 12 ($\star$). $\triangle$-ENUM *parameterized by distance to d-degenerate graphs is solvable in $O(n \cdot (d + 1) \cdot (|D| + d) + 2^{3|D|} + \# \mathrm{T})$ time provided that the vertex deletion set D to d-degenerate is given.*

4.2 Parameters Incomparable with Degeneracy

In this section we present results on parameters that are unrelated to the degeneracy. Again, we first describe the parameters and then turn to our results.

A graph is called a cograph if it contains no induced path with four vertices (P_4). We study the (vertex deletion) distance to cographs since the vertex cover number allows for tractability results (see Sect. 4.1) but diameter does not (see Sect. 3.1), and distance to cographs is sandwiched between those two parameters. Furthermore, distance to cographs lower-bounds the cluster vertex number—a parameter advocated by Doucha and Kratochvíl [10]. Moreover, given a graph G we can determine in linear time whether it is a cograph and return an induced P_4 if this is not the case [6,8]. This implies that in $O(k \cdot (m + n))$ time we can compute a set $K \subseteq V$, $k = |K|$, of size at most $4k$ such that $G - K$ is a cograph.

Since treewidth is lower-bounded by the degeneracy, we know that there is an $O(\omega \cdot m)$-time algorithm for $\triangle$-ENUM. A parameter below treewidth ω in the parameter hierarchy is the clique-width k (it holds that $k \leq 2^\omega$ and k can be arbitrarily small compared to ω). Moreover, clique-width is also below distance to cograph. Thus, we study clique-width as it lies on the "border to tractability" of $\triangle$-ENUM.

Distance to Cograph. We give a linear-time FPT algorithm for $\triangle$-ENUM with respect to the distance to cographs. Before we do this, we provide a general lemma which can be used to enumerate all triangles in a graph parameterized by (vertex) deletion set to some graph class Π if all triangles in Π can be enumerated efficiently.

Lemma 13 ($\star$). $\triangle$-ENUM *parametrized by deletion set* K *to* Π *is solvable in* $O(m \cdot |K| + n + x)$ *time if* $\triangle$-ENUM *on a graph in* Π *is solvable in* $O(x)$ *time.*

We can enumerate all triangles in a cograph in $O(\# \mathrm{T} + n + m)$ time. Solving $\triangle$-ENUM parameterized by distance to cograph using Lemma 13 then yields:

Proposition 14 ($\star$). $\triangle$-ENUM *parametrized by deletion set* K *to cographs is solvable in* $O(\# \mathrm{T} + n + m \cdot |K|)$ *time.*

Clique-Width. We next turn to the parameter clique-width as it is incomparable to the degeneracy and upper-bounded by two parameters allowing for linear-time FPT algorithms: the distance to cographs (Proposition 14), and treewidth (as treewidth upper-bounds the degeneracy). We show a quadratic-time FPT algorithm for $\triangle$-ENUM with respect to the clique-width of the input graph. We leave open whether it admits a linear FPT algorithm. Due to our results, the parameters clique-width and average degree form the border case between parameters admitting linear FPT algorithms and those that are GP-hard.

Theorem 15 ($\star$). *Given a k-expression of the graph G, $\triangle$-ENUM is solvable in* $O(n^2 + n \cdot k^2 + \# \mathrm{T})$ *time.*

5   Conclusion

Employing the framework of FPT-in-P analysis, we provided novel notions and insights concerning potentially faster solution algorithms for enumerating (and detecting) triangles in undirected graphs. It remains to be seen whether General-Problem-hardness is the appropriate notion for intractability within the field of FPT-in-P. Although data reduction is a theoretically and practically very promising concept, there is still little work in the context of enumeration problems. We hope that the notion of enum-advice kernels can be used to further develop this area of research.

In ongoing work we want to perform empirical studies with our algorithms (kernelization as well as solving algorithms). Moreover, it remains open to study whether our exponential-size kernel for parameter distance to d-degenerate graphs (see Table 1) can be improved in terms of size and running time. On a more general scale, note that triangles are both the smallest non-trivial cliques and cycles. Can we generalize our findings to these two different settings? Finally, we mention that following the FPT-in-P route might be an attractive way to "circumvent" lower bound results for other polynomial-time solvable problems.

References

1. Abboud, A., Williams, V.V.: Popular conjectures imply strong lower bounds for dynamic problems. In: Proceedings of the 55th FOCS, pp. 434–443. IEEE Computer Society (2014)
2. Abboud, A., Williams, V.V., Wang, J.R.: Approximation and fixed parameter subquadratic algorithms for radius and diameter in sparse graphs. In: Proceedings of the 27th SODA, pp. 377–391. SIAM (2016)
3. Bar-Yehuda, R., Geiger, D., Naor, J., Roth, R.M.: Approximation algorithms for the feedback vertex set problem with applications to constraint satisfaction and Bayesian inference. SIAM J. Comput. **27**(4), 942–959 (1998)
4. Becchetti, L., Boldi, P., Castillo, C., Gionis, A.: Efficient semi-streaming algorithms for local triangle counting in massive graphs. In: Proceedings of the 14th ACM KDD, pp. 16–24. ACM (2008)
5. Björklund, A., Pagh, R., Williams, V.V., Zwick, U.: Listing triangles. In: Esparza, J., Fraigniaud, P., Husfeldt, T., Koutsoupias, E. (eds.) ICALP 2014. LNCS, vol. 8572, pp. 223–234. Springer, Heidelberg (2014). doi:10.1007/978-3-662-43948-7_19
6. Bretscher, A., Corneil, D.G., Habib, M., Paul, C.: A simple linear time LexBFS cograph recognition algorithm. SIAM J. Discret. Math. **22**(4), 1277–1296 (2008)
7. Chiba, N., Nishizeki, T.: Arboricity and subgraph listing algorithms. SIAM J. Comput. **14**(1), 210–223 (1985)
8. Corneil, D.G., Perl, Y., Stewart, L.K.: A linear recognition algorithm for cographs. SIAM J. Comput. **14**(4), 926–934 (1985)
9. Creignou, N., Meier, A., Müller, J.S., Schmidt, J., Vollmer, H.: Paradigms for parameterized enumeration. Theory Comput. Syst. **60**(4), 737–758 (2017)
10. Doucha, M., Kratochvíl, J.: Cluster vertex deletion: a parameterization between vertex cover and clique-width. In: Rovan, B., Sassone, V., Widmayer, P. (eds.) MFCS 2012. LNCS, vol. 7464, pp. 348–359. Springer, Heidelberg (2012). doi:10.1007/978-3-642-32589-2_32

11. Ferrara, E.: Measurement and analysis of online social networks systems. In: Alhajj, R., Rokne, J. (eds.) Encyclopedia of Social Network Analysis and Mining, pp. 891–893. Springer, New York (2014). doi:10.1007/978-1-4614-6170-8_242

12. Fluschnik, T., Komusiewicz, C., Mertzios, G.B., Nichterlein, A., Niedermeier, R., Talmon, N.: When can graph hyperbolicity be computed in linear time? In: Ellen F., Kolokolova A., Sack J.R. (eds.) Proceedings of the 15th WADS. LNCS, vol. 10389, pp. 397–408. Springer, Heidelberg (2017). doi:10.1007/978-3-319-62127-2_34. ISBN 978-3-319-62126-5

13. Fomin, F.V., Lokshtanov, D., Pilipczuk, M., Saurabh, S., Wrochna, M.: Fully polynomial-time parameterized computations for graphs and matrices of low treewidth. In: Proceedings of the 28th SODA, pp. 1419–1432. SIAM (2017)

14. Giannopoulou, A.C., Mertzios, G.B., Niedermeier, R.: Polynomial fixed-parameter algorithms: a case study for longest path on interval graphs. In: Proceedings of the 10th IPEC, LIPIcs, vol. 43, pp. 102–113. Schloss Dagstuhl - Leibniz-Zentrum fuer Informatik (2015)

15. Grabow, C., Grosskinsky, S., Kurths, J., Timme, M.: Collective relaxation dynamics of small-world networks. Phys. Rev. E **91**, 052815 (2015)

16. Green, O., Bader, D.A.: Faster clustering coefficient using vertex covers. In: Proceedings of the 6th SocialCom, pp. 321–330. IEEE Computer Society (2013)

17. Habib, M., Paul, C., Viennoti, L.: A synthesis on partition refinement: a useful routine for strings, graphs, boolean matrices and automata. In: Morvan, M., Meinel, C., Krob, D. (eds.) STACS 1998. LNCS, vol. 1373, pp. 25–38. Springer, Heidelberg (1998). doi:10.1007/BFb0028546

18. Itai, A., Rodeh, M.: Finding a minimum circuit in a graph. SIAM J. Comput. **7**(4), 413–423 (1978)

19. Khamis, M.A., Ngo, H.Q., Ré, C., Rudra, A.: Joins via geometric resolutions: worst case and beyond. ACM Trans. Database Syst. **41**(4), 22:1–22:45 (2016)

20. Kopelowitz, T., Pettie, S., Porat, E.: Dynamic set intersection. In: Dehne, F., Sack, J.-R., Stege, U. (eds.) WADS 2015. LNCS, vol. 9214, pp. 470–481. Springer, Cham (2015). doi:10.1007/978-3-319-21840-3_39

21. Kopelowitz, T., Pettie, S., Porat, E.: Higher lower bounds from the 3SUM conjecture. In: Proceedings of the 27th SODA, pp. 1272–1287. SIAM (2016)

22. Lagraa, S., Seba, H.: An efficient exact algorithm for triangle listing in large graphs. Data Min. Knowl. Disc. **30**(5), 1350–1369 (2016)

23. Latapy, M.: Main-memory triangle computations for very large (sparse (power-law)) graphs. Theor. Comput. Sci. **407**(1–3), 458–473 (2008)

24. Lee, T., Magniez, F., Santha, M.: Improved quantum query algorithms for triangle detection and associativity testing. Algorithmica **77**(2), 459–486 (2017)

25. Lewis, J.M., Yannakakis, M.: The node-deletion problem for hereditary properties is NP-complete. J. Comput. Syst. Sci. **20**(2), 219–230 (1980)

26. Mertzios, G.B., Nichterlein, A., Niedermeier, R.: The power of linear-time datare-duction for maximum matching. In: Proceedings of the 42nd MFCS, LIPIcs, vol. 83, pp. 46:1–46:14. Schloss Dagstuhl - Leibniz-Zentrum fuer Informatik (2017)

27. Newman, M.E.J.: The structure and function of complex networks. SIAM Rev. **45**(2), 167–256 (2003)

28. Park, H., Silvestri, F., Kang, U., Pagh, R.: Mapreduce triangle enumeration with guarantees. In: Proceedings of CIKM 2014, pp. 1739–1748. ACM (2014)

29. Patrascu, M.: Towards polynomial lower bounds for dynamic problems. In: Proceedings of the 42nd STOC, pp. 603–610. ACM (2010)

30. Schank, T., Wagner, D.: Finding, counting and listing all triangles in large graphs, an experimental study. In: Nikoletseas, S.E. (ed.) WEA 2005. LNCS, vol. 3503, pp. 606–609. Springer, Heidelberg (2005). doi:10.1007/11427186_54

31. Sorge, M., Weller, M.: The graph parameter hierarchy, TU Berlin (2016). Unpublished Manuscript

32. Zhang, Y., Parthasarathy, S.: Extracting analyzing and visualizing triangle k-core motifs within networks. In: Proceedings of the 28th ICDE, pp. 1049–1060. IEEE Computer Society (2012)

Testing Polynomial Equivalence
by Scaling Matrices

Markus Bläser[1]([✉]), B.V. Raghavendra Rao[2], and Jayalal Sarma[2]

[1] Saarland Informatics Campus, Saarland University, Saarbrücken, Germany
mblaeser@cs.uni-saarland.de
[2] IIT Madras, Chennai, India
{bvrr,jayalal}@cse.iitm.ac.in

Abstract. In this paper we study the polynomial equivalence problem:
test if two given polynomials f and g are equivalent under a non-singular
linear transformation of variables.

We begin by showing that the more general problem of testing whether
f can be obtained from g by an arbitrary (not necessarily invertible) linear transformation of the variables is equivalent to the existential theory
over the reals. This strengthens an NP-hardness result by Kayal [9].

Two n-variate polynomials f and g are said to be equivalent up to scal-
ing if there are scalars $a_1, \ldots, a_n \in \mathbb{F} \setminus \{0\}$ such that $f(a_1 x_1, \ldots, a_n x_n) =
g(x_1, \ldots, x_n)$. Testing whether two polynomials are equivalent by scaling
matrices is a special case of the polynomial equivalence problem and is
harder than the polynomial identity testing problem.

As our main result, we obtain a randomized polynomial time algorithm for testing if two polynomials are equivalent up to a scaling of
variables with black-box access to polynomials f and g over the real
numbers.

An essential ingredient to our algorithm is a randomized polynomial
time algorithm that given a polynomial as a black box obtains coefficients and degree vectors of a maximal set of monomials whose degree
vectors are linearly independent. This algorithm might be of independent interest. It also works over finite fields, provided their size is large
enough to perform polynomial interpolation.

1   Introduction

The polynomial equivalence problem (PolyEq), i.e., testing if two given polynomials are equivalent under a non-singular change of coordinates is one of the fundamental computational tasks related to polynomials. More precisely, two polynomials $p(x_1, x_2, \cdots, x_n)$ and $q(x_1, x_2, \cdots, x_n)$ are said to be *linearly equivalent*
if there is an invertible linear transformation, A such that for $y_i = \sum_j A_{ij} x_j$,
$p(y_1, y_2, \ldots y_n) = q(x_1, \ldots x_n)$. When A is not restricted to be invertible, the
problem is referred to as polynomial projection problem (PolyProj).

Indeed, observing that only a polynomial with all coefficients equal to zero
can be equivalent to the zero polynomial, PolyEq is a generalization of the
well studied polynomial identity testing problem which has close connections to

© Springer-Verlag GmbH Germany 2017
R. Klasing and M. Zeitoun (Eds.): FCT 2017, LNCS 10472, pp. 111–122, 2017.
DOI: 10.1007/978-3-662-55751-8_10

arithmetic circuit lower bounds [7]. Further, since a non-singular change of coordinates is one of the fundamental geometric primitives, POLYEQ is of primary importance to computational algebraic geometry.

Saxena [14] showed that the graph isomorphism problem is polynomial time many one reducible to the case of POLYEQ where the polynomials are of degree three. Thus, the problem simultaneously generalizes the graph isomorphism problem and the polynomial identity testing problem. Further, if the change of coordinates (the matrix A) is not restricted to be invertible, that is, A need not be invertible, then the problem POLYPROJ is NP-hard under polynomial time many-one reductions [9]. We first strengthen this hardness result over $\mathbb{R}$ (in fact, also over any integral domain) as follows:

Theorem 1. *Given two sparse polynomials $f, g \in \mathbb{R}[x_1, \ldots, x_n]$, deciding whether there is a matrix A such that $f(x) = g(Ax)$ is as hard as the existential theory over the reals.*

Both POLYPROJ and POLYEQ can be solved in polynomial space over $\mathbb{R}$ and $\mathbb{C}$ in the Blum-Shub-Smale model of algebraic computation [4] using existential theories over these fields. Since the best upper bound for existential theory over reals is PSPACE, the above hardness result indicates that the POLYPROJ is possibly harder than just being NP-hard. However, the hardness result does not apply to when A is restricted to be non-singular, and the complexity of the POLYEQ problem remains elusive. Over finite fields the problem is in NP ∩ co-AM [17]. However, over the field of rational numbers, it is not known if the problem is decidable [14].

Given the lack of progress in the general problem POLYEQ, it is natural to solve special instances of the problem. A natural restriction is to study the problem when the input polynomials are restricted. When both polynomials are restricted to quadratic forms (homogeneous degree 2 polynomials), we know about the structure of equivalent polynomials and this also leads to a polynomial time algorithm for testing equivalence of such polynomials (see Witt's equivalence theorem [12]). As indicated above the problem already becomes harder when the degree is allowed to be even three. Agrawal and Saxena [1] showed that ring isomorphism testing problem, reduces to the POLYEQ problem when the degree of the polynomials is at most three. [1] Patarin [13] even designed a cryptosystem which assumes the hardness of the degree bounded (by three) version of the problem to prove security guarantees.

Instead of simultaneously restricting both of the polynomials in the problem, it is even interesting to study the problem when one of the polynomials is fixed to be a well-structured family and the other polynomial is allowed to be arbitrary. In this direction, Kayal [8] obtained randomized polynomial time algorithms to test if a given polynomial (as a black-box) is equivalent to either an elementary symmetric polynomial or to the power symmetric polynomial of a given degree.

---

[1] For a (partial) converse, they [1] also showed that deciding equivalence of degree k polynomials having n variables over $\mathbb{F}_q$ (such that k and $q - 1$ are co-prime), can be reduced to the ring isomorphism problem.

Further, Kayal [9] obtained similar algorithms when one of the polynomials is fixed to be either the permanent polynomial or the determinant polynomial. More recently, Kayal *et al.* [10] obtained randomized polynomial time algorithm for POLYEQ when one of the polynomials is the iterated matrix multiplication polynomial.

Another possibility of obtaining restrictions of POLYEQ is by restricting the structure of change of coordinates. Grigoriev [6] considered the problem of testing equivalence of polynomials under Taylor shifts[2]: given two polynomials f and g, are there $a_1, \ldots, a_n \in \mathbb{K}$ such that $f(x_1 + a_1, \ldots, x_n + a_n) = g$? Grigoriev obtained a polynomial time algorithm to the problem when the polynomial is given in the sparse representation. The algorithm is deterministic polynomial time if $\mathbb{K}$ is algebraically closed, randomized polynomial time if $\mathbb{K} = \mathbb{Q}$ and quantum polynomial time if $\mathbb{K}$ is finite. More recently, Dvir et al. [5] showed that the shift equivalence problem is polynomial time equivalent to the polynomial identity testing problem in the black-box as well as non-black-box setting.

In this paper, we restrict the structure of the matrices under which the equivalence is tested, to *diagonal matrices*. We obtain a randomized polynomial time algorithm for testing if two polynomials are equivalent up to a scaling of variables with black-box access to polynomials f and g. More precisely, we prove the following theorem:

Theorem 2 (Main). *Given $f, g \in \mathbb{R}[x_1, \ldots, x_n]$ as a blackbox, there exists a randomized algorithm that tests if there is an invertible diagonal matrix A such that $f(X) = g(AX)$. The algorithm runs in time* poly(n, Δ, L)*, where the degree of f and g is bounded by Δ and all of the coefficients of f and g can be represented by at most L bits.*

As mentioned above, Kayal [9] designed randomized polynomial time algorithms for testing equivalence if one of the polynomials comes from a well-structured family of polynomials like the permanent family or determinant family. These algorithms follow the following general scheme: First, the general problem is reduced to permutation and scaling equivalence testing by studying the Lie algebra of the input polynomial. Then permutation and scaling equivalence testing is reduced to scaling equivalence testing. Our result shows that this last step can always be done in randomized polynomial time, even when one of the polynomials does not come from a nice family but is arbitrary. Thus, the hardness of POLYEQ most likely lies in the first step, since we need a large enough Lie algebra to make the approach work. The Lie algebra of a random polynomial is trivial [9].

As an ingredient to our proof of Theorem 2, we obtain a randomized polynomial time algorithm that given a polynomial as a black box obtains coefficients of a maximal set of monomials whose degree vectors are linearly independent, this might be of independent interest.

---

[2] which is strictly speaking not a (homogeneous) linear change of coordinates.

Theorem 3. *There is a randomized algorithm, that given a polynomial $f \in$ $\mathbb{R}[x_1, \ldots, x_n]$ by black box access outputs a maximal collection of*

$$\{(m, \alpha) \mid \alpha \neq 0, \text{ and } \alpha \text{ is the coefficient of the monomial } m \text{ in } f\}$$

such that the set of degree vectors is linearly independent over $\mathbb{R}$. The running time is polynomial in the degree Δ of f, the number of variables n, and the bit size L of the representation of the coefficients.

We remark that the latter algorithm also works over finite fields provided they are large enough. Here large enough means $p(\Delta, n)$ for some polynomial p for small degree.

2 Preliminaries

In this section, we fix the notations that we use throughout the paper.

For a monomial $m = x_1^{d_1} \cdots x_n^{d_n}$ let $\mathsf{Deg}(m) = (d_1, \ldots, d_n)$ denote the degree vector of m. For a polynomial $f \in \mathbb{K}[x_1, \ldots, x_n]$, let $\mathsf{Mon}(f)$ denote the set of monomials that have non-zero coefficients in f. A **degree-basis** for f is a maximal collection $S = \{(m, \alpha) \mid \alpha \neq 0 \text{ is the coefficient of monomial } m \text{ in } f\}$ such that the set $\{\mathsf{Deg}(m) \mid (m, \alpha) \in S, \text{ for some } \alpha \neq 0\}$ is linearly independent over $\mathbb{Q}$ or equivalently $\mathbb{R}$.

Isolating Monomials: Klivans and Spielman [11] obtained a randomized polynomial time algorithm that tests if a polynomial given as a black-box is identically zero or not. Their algorithm involves a randomized polynomial time algorithm that isolates a monomial in the given polynomial if it is not identically zero. We state the result below:

Theorem 4 (Klivans and Spielman [11]). *There is a probabilistic algorithm that given a non-zero polynomial $f \in K[x_1, \ldots, x_n]$ (by blackbox access) outputs a monomial m of f, its degree vector $\mathsf{Deg}(m)$ and its coefficient α in f with probability $\geq 1 - \epsilon$ in time polynomial in n, Δ, and $1/\epsilon$.*

Theorem 4 is going to be a building block for our proof of Theorem 3. We need a bit more insight into the proof of Theorem 4 listed as follows:

- The algorithm in [11] first replaces the variables x_i by y^{a_i} where the a_i are numbers with $O(\log(n\Delta/\epsilon))$ bits. We get a new univariate polynomial $\hat{f}$. Monomials of f get mapped to monomials in $\hat{f}$ and are grouped together. The substitution has the property that with probability $\geq 1 - \epsilon$, there is only one monomial of f getting mapped to the (non-zero) monomial of $\hat{f}$ having minimum degree. Since we have only black-box access to f, this substitution is only conceptual and is simulated when later on plugging values into the blackbox.

- Then we interpolate $\hat{f}$ by evaluating it at $\text{poly}(n, \Delta, 1/\epsilon)$ many values. That is, we plug in values v^{a_i} for each x_i for polynomially many values v. The lowest nonzero coefficient of $\hat{f}$ is also a coefficient of f, however, we do not know the degree pattern of this monomial (yet).
- Then we modify the substituion in the first step by replacing x_1 by $2y^{a_1}$. In this way, the lowest nonzero coefficient of $\hat{f}$ will get multiplied by 2^{d_1} where d_1 is the x_1-degree of the unique monomial in f that is mapped to the lowest degree monomial of $\hat{f}$. Doing this for all x_i, we can also extract the degree vector of the monomial.

Tensors and Tensor Rank: We also fix notations and state the preliminary results about tensors that we need in the paper. We call $t = (t_{i,j,\ell}) \in \mathbb{R}^{n \times n \times n}$ a tensor. A rank one tensor is a tensor that can we written as $u \otimes v \otimes w$ with $u, v, w \in \mathbb{R}^n$. The minimum number of such rank-one-tensors such that t can be written as the sum of them is called the rank of t. For an introduction to this problem, the reader is referred to [2,3].

With a tensor t, we can associate the trilinear form

$$F(x, y, z) = \sum_{i,j,\ell=1}^{n} t_{i,j,\ell} x_i y_j z_\ell.$$

The so called unit tensor $e_r \in \mathbb{R}^{r \times r \times r}$ is given by the trilinear form

$$E_r = \sum_{i=1}^{r} x_i y_i z_i$$

The following fact is well known :

Proposition 1 (see [3]). Let $t = (t_{i,j,\ell}) \in \mathbb{R}^{n \times n \times n}$ be a tensor. The tensor rank of t is bounded by r if and only if there are matrices $S, T, U \in \mathbb{R}^{r \times n}$ such that

$$F(x, y, z) = E_r(Sx, Ty, Uz).$$

3   Hardness of the PolyProj Problem

Testing whether there is an arbitrary matrix A such that $f(x) = g(Ax)$ is a hard problem. In this section, we prove Theorem 1. As mentioned in the introduction, this improves the hardness result shown in [9].

Theorem 1. Given two polynomials $f, g \in \mathbb{R}[x_1, \ldots, x_n]$, as a list of monomials and their coefficients, deciding whether there is a matrix A (not necessarily nonsingular) such that $f(x) = g(Ax)$ is as hard as the existential theory over the reals.

Proof. We proceed by reducing the tensor rank problem to POLYPROJ problem. Given a tensor $t = (t_{i,j,\ell}) \in \mathbb{R}^{n \times n \times n}$, we observe that Proposition 1 suggests two polynomials of the form $f(x) = g(Ax)$. However, there are two issues. Firstly, we have three sets of variables and secondly, the matrices S, T, and U are not square matrices. The second problem is easy to circumvent. We consider F as a polynomial in the variables $x_1, \ldots, x_r$, $y_1, \ldots, y_r$, and $z_1, \ldots, z_r$ instead and extend the matrices S, T, and U by zero rows.

To address the first problem, we modify the problem we reduce from. A tensor is called *symmetric*, if $t_{i,j,\ell} = t_{i,\ell,j} = \ldots$ for all six permutations of the indices. In the same way as for general tensors, we can associate a trilinear form with symmetric tensors, too:

$$F'(x) = \sum_{i,j,\ell=1}^{n} t_{i,j,\ell} x_i x_j x_\ell.$$

Definition 1 (Symmetric Rank). *The symmetric or Waring rank of a symmetric tensor t is the smallest r such that there is an $r \times n$ matrix A with $F'(x) = E'_r(Ax)$ where*

$$E'_r = \sum_{i=1}^{r} x_i^3.$$

Shitov [16] recently proved that the problem of deciding whether a symmetric tensor t has symmetric rank r is as hard as the existential theory over the underlying ground field. The same is true for the ordinary tensor rank. (Independently, Schaefer and Stefankovic proved a similar result [15], but only for the tensor rank.)   □

We remark that, since Shitov's result [16] holds over any integral domain, the above theorem is also true for any integral domain.

4   Extracting a degree-basis of a Polynomial

In this section we obtain a randomized polynomial time algorithm that given a polynomial f as a black-box computes a **degree-basis** for f. We re-state Theorem 3 for readability:

Theorem 3. *There is a randomized algorithm, that given a polynomial $f \in \mathbb{R}[x_1, \ldots, x_n]$ by black box access outputs a maximal collection of*

$$\{(m, \alpha) \mid \alpha \neq 0, \text{ and } \alpha \text{ is the coefficient of the monomial } m \text{ in } f\}$$

such that the set of degree vectors is linearly independent over $\mathbb{R}$. The running time is polynomial in the degree Δ of f, the number of variables n, and the bit size L of the representation of the coefficients.

Proof. Algorithm 1 is our proposed algorithm. It starts with extracting a first monomial using the algorithm by Klivans and Spielman (Theorem 4). Then it proceeds iteratively and extends the set one by one.

Now assume we have already extracted monomials $m_1, \ldots, m_t$ of f such that the corresponding degree vectors $v_1, \ldots, v_t$ are linearly independent. Let $t < n$. We describe a procedure that finds a new monomial m_{t+1} such that its degree vector v_{t+1} is not contained in the span of $v_1, \ldots, v_t$ or reports that there does not exists such a v_{t+1}.

Let $v_1, \ldots, v_{\hat{t}}$ be a basis of the $\mathbb{R}$-vector space spanned by all degree vectors of f, that is, we extend $v_1, \ldots, v_t$ to a basis. Let p be a prime such that $v_1, \ldots, v_{\hat{t}}$ stay linearly independent over $\mathbb{F}_p$. By the Hadamard bound for the determinant, the matrix formed by $v_1, \ldots, v_t$ has a non vanishing minor whose absolute value is bounded by $(\Delta n)^n$. And by the prime number theorem, a prime number of size $O(n\mathsf{polylog}(\Delta n))$ will have the property stated above with high probability. Note that we only know the vectors $v_1, \ldots, v_t$ so far, but since we simply choose p uniformly at random, we do not need to know the vectors $v_1, \ldots, v_{\hat{t}}$ at all.

Let $u_1, \ldots, u_{n-t}$ be linearly independent vectors such that $v_i u_j = 0$ over $\mathbb{F}_p$ for all $1 \le i \le t$, $1 \le j \le n-t$, where $v_i u_j$ denotes the standard scalar product. If w is a vector not contained in the span of $v_1, \ldots, v_t$, then there is a j such that $w u_j \ne 0$ over $\mathbb{F}_p$. Consider the substitution $x_i \to y^{u_{j,i}} x_i$, $1 \le i \le n$, where $u_{j,i}$ are the entries of u_j. This substitution maps every monomial m of f to some monomial of the form $y^d m$. Let f_j be the resulting polynomial.

By construction, we have:

Lemma 1. *1. The degree of f_j is bounded by $O(\Delta n \mathsf{polylog}(\Delta n))$ for all j.*
2. If a monomial m is contained in the span of $v_1, \ldots, v_t$, then for every j, $p \mid d$ where $y^d m$ is the image of m in f_j.
3. If a monomial m is not contained in the span of $v_1, \ldots, v_t$, then there is a j such that $p \nmid d$ where $y^d m$ is the image of m in f_j. □

We continue with the proof of the theorem. The strategy is now clear: We treat each f_j as a univariate polynomial in y with coefficients from $K[x_1, \ldots, x_n]$. Then we use the algorithm from Theorem 4 to extract a monomial from the coefficient polynomial of a power y^d with $p \nmid d$. If we find a monomial then we set v_{t+1} to be its degree vector. If we do not find such a monomial, then $v_1, \ldots, v_t$ is a maximal linearly independent set.

Let $f_j = \sum_{d=0}^{\Delta_j} g_d \cdot y^d$. To be able to apply Theorem 4, we have to provide blackbox access to the g_d's but we have only blackbox access to f. We simulate this as follows:

- Given blackbox access to f, it is easy to simulate blackbox access to f_j.
- Now assume we want to evaluate g_d at a point $\xi \in K^n$.
- We evaluate f_j at the points $(\xi, \alpha_i) \in K^{n+1}$, $0 \le i \le \Delta_j$, where the α_i are pairwise distinct, that is, we compute values $f_j(\xi, \alpha_i) = \sum_{d=0}^{\Delta_j} g_d(\xi) \alpha_i^d$. From these values, we interpolate the coefficients of f_j, viewed as a univariate polynomial in y. The coefficient of y^d is $g_d(\xi)$.

It is clear from construction that Algorithm 1 returns the correct result if no errors occur in the randomized computations. Thus, if we make every error

---

Algorithm 1. Gen-Mon(f)

Input: Black box access to polynomial $f \in K[x_1, \ldots, x_n]$
Output: A degree-basis for f
  $S \leftarrow \emptyset$.
  $t \leftarrow 1$
  if f is not identically 0 **then**
    Extract a monomial m_1 of f (using Theorem 4) with coefficient α_1.
    Let v_1 be the degree vector.
    $S \leftarrow S \cup \{(m_1, \alpha_1)\}$.
    while TRUE do
      Randomly choose a prime p of size $O(n \mathsf{polylog}(\Delta n))$.
      Compute linearly independent vectors $u_1, \ldots, u_{n-t}$ such that $v_i u_j = 0$ over $\mathbb{F}_p$
      for all $1 \leq i \leq t$, $1 \leq j \leq n - t$.
      Let $f_j(x_1, \ldots, x_n) = f(x_1 y^{u_{j,1}}, \ldots, x_n y^{u_{j,n}})$, $1 \leq j \leq n - t$.
      Write $f_j(x, y) = \sum_{d=0}^{\Delta_j} h_{j,d} y^d$.
      Try to extract a monomial of every $h_{j,d}$ using Theorem 4, $1 \leq j \leq n - t$, $p \nmid d$.
      Let m_{t+1} be the first such monomial found, α_{t+1} be its coefficient and let v_{t+1}
      be its degree vector. Set $S = S \cup \{(m_{t+1}, \alpha_{t+1})\}$.
      If no such monomial m_{t+1} is found, then output S and HALT.
    end while
  end if

---

probability of every randomized subroutine polynomially small in Algorithm 1, then by the union bound, it will compute the correct result with high probability. For the running time observe that the while loop is executed at most $n - 1$ times. The degrees Δ_j are bounded by $\mathsf{poly}(n, \Delta)$ by the bound on p. All numbers occurring as coefficients have length bounded by $\mathsf{poly}(n, \Delta, L)$, since the degrees of all polynomials are bounded by $\mathsf{poly}(n, \Delta)$. $\qquad\square$

5 Testing for Equivalence by Scaling

Let $f(X)$ and $g(X)$ be polynomials in $\mathbb{R}[x_1, \ldots, x_n]$ given by black box access. We assume that the degree of f and g is bounded by Δ and that all coefficients of f and g can be represented by at most L bits.

Assume there is a non-singular diagonal matrix A such that $f(X) = g(AX)$. Let $(a_1, \ldots, a_n)$ denote the entries of A on the diagonal. Clearly, if $f(X) = g(AX)$ with A diagonal, f and g should have the same set of monomials. We first treat the case that the degree basis has maximum cardinality n.

Lemma 2. *Let* $S = \{(m_i, \alpha_i) \mid 1 \leq i \leq n\}$ *be a* degree-basis *of* f. *If* $f(X) = g(AX)$ *for a non-singular diagonal matrix* A, *then such an* A *can be computed deterministically in time polynomial in* n, Δ *and* L, *where the* a_i *are represented by polynomial size expressions with roots.*

Proof. Let $\alpha_i \neq 0$ and $\beta_i \neq 0$ be the coefficient of m_i in f and g, respectively. Suppose $f(X) = g(AX)$ for some non-singular diagonal matrix A with diagonal $(a_1, \ldots, a_n)$. We have n polynomial equations

$$\alpha_i = \beta_i \prod_{j=1}^{n} a_j^{d_{i,j}}$$

where $v_i = \text{Deg}(m_i) =: (d_{i,1}, \ldots, d_{i,n})$. Taking logarithms on both sides, we have

$$\log \alpha_i = \log \beta_i + \sum_{j=1}^{n} d_{i,j} \log a_j.$$

(Formally, you have to choose an appropriate branch of the complex algorithm, since the α_i or β_i can be negative. Since we exponentiate again in the end, the actual choice does not matter.)

Since the vectors $v_1, \ldots, v_n$ are linearly independent over $\mathbb{R}$, there are unique values for $\log a_1, \ldots, \log a_n$ satisfying the above equations. Now $a_1, \ldots, a_n$ can be obtained by inverse logarithms. This proves the uniqueness of $a_1, \ldots, a_n$.

Let $D = (d_{i,j})$. Then the a_i are given by

$$\begin{pmatrix} \log a_1 \\ \vdots \\ \log a_n \end{pmatrix} = D^{-1} \begin{pmatrix} \log(\alpha_1/\beta_1) \\ \vdots \\ \log(\alpha_n/\beta_n) \end{pmatrix}.$$

So each $a_i = \prod_{j=1}^{n} (\alpha_j/\beta_j)^{\bar{d}_{i,j}}$ where $D^{-1} = (\bar{d}_{i,j})$. □

When the set of degree vectors of f has cardinality less than n, we can still use Algorithm 1 to compute a linearly independent set of degree vectors $v_1, \ldots, v_t$ of maximal size. Let $v_i = (d_{i,1}, \ldots, d_{i,n})$, $1 \leq i \leq t$. Let $m_1, \ldots, m_t$ be the corresponding monomials with coefficients $\alpha_1, \ldots, \alpha_t$. Let $\beta_1, \ldots, \beta_t$ be the corresponding coefficients of the monomials of g. From these values, we can set up a system of equations as in Lemma 2, however, this time there might be more than one solution. The next lemma states that it actually does not matter which of these solutions we choose:

Lemma 3. *Let $a_1, \ldots, a_n$ be any solution to*

$$\log \alpha_i = \log \beta_i + \sum_{j=1}^{n} d_{i,j} \log a_j, \qquad 1 \leq i \leq t,$$

and let A be the corresponding diagonal matrix. Let $r(x)$ be a monomial with coefficient δ and degree vector $u = (e_1, \ldots, e_n)$ contained in the linear span of $v_1, \ldots, v_t$, i.e., $u = \lambda_1 v_1 + \cdots + \lambda_t v_t$. Then the coefficient of $r(Ax)$ is

$$\delta \cdot \left(\frac{\alpha_1}{\beta_1}\right)^{\lambda_1} \cdots \left(\frac{\alpha_t}{\beta_t}\right)^{\lambda_t},$$

in particular, it is independent of the chosen solution for $a_1, \ldots, a_n$.

Algorithm 2. Scaling equivalence test

Input: Black box access to polynomials $f, g \in K[x_1, \ldots, x_n]$
the degree vectors of f have full rank
Output: Nonsingular diagonal matrix A with $f(x) = g(Ax)$ if such an A exists
Apply Gen-Mon with polynomial f as the black-box to get a set S.
Apply Gen-Mon to g using the *same* random bits as above to get a set S'.
If the monomials in the set S is not the same as S' then REJECT.
Solve for the entries of A using Lemma 2 (choosing any solution if there is more than one).
ACCEPT if and only if $f(x) - g(Ax)$ is identically zero.

Proof. Let $u = \sum_{i=1}^{t} \lambda_i v_i$. Now

$$
\begin{aligned}
r(Ax) &= \delta \cdot (a_1 x_1)^{e_1} \cdots (a_n x_n)^{e_n} \\
&= \delta \cdot a_1^{e_1} \cdots a_n^{e_n} \cdot x_1^{e_1} \cdots x_n^{e_n} \\
&= \delta \cdot a_1^{\sum_{i=1}^{t} \lambda_i d_{i,1}} \cdots a_n^{\sum_{i=1}^{t} \lambda_i d_{i,n}} \cdot x_1^{e_1} \cdots x_n^{e_n} \\
&= \delta \cdot (a_1^{d_{1,1}} \cdots a_n^{d_{1,n}})^{\lambda_1} \cdots (a_1^{d_{t,1}} \cdots a_n^{d_{t,n}})^{\lambda_t} \cdot x_1^{e_1} \cdots x_n^{e_n} \\
&= \delta \cdot \left(\frac{\alpha_1}{\beta_1} \right)^{\lambda_1} \cdots \left(\frac{\alpha_t}{\beta_t} \right)^{\lambda_t} \cdot x_1^{e_1} \cdots x_n^{e_n}. \qquad \square
\end{aligned}
$$

Thus for testing if there is a diagonal matrix A with $f(X) = g(AX)$, it is enough to compute the non-zero coefficients of at most n monomials $m_1, \ldots, m_n$ in f and g the degree vectors of which are linearly independent.

We complete the correctness of Algorithm 2 in the following Theorem, which in turn completes the proof of Theorem 2.

Theorem 5. *Algorithm 2 returns correct the correct answer with high probability. It runs in time polynomial in Δ, n and L.*

Proof. The algorithm calls two times the routine Get-Mon and makes one call to a polynomial identity test. By making the error probabilities of these calls small enough, we can controll the error probability of Algorithm 2 by the union bound.

Now we need to argue that if the polynomials f and g have the same set of monomials, then the calls for Gen-Mon(f) and Gen-Mon(g) with same set of random bits (i.e., by re-using the random bits) will result in sets S and S' with the same set of monomials.

Consider parallel runs of Get-Mon with f and g as inputs respectively such that they use a common random string say R. If f and g have the same set of monomials, then clearly $h_{j,d}^f$ and $h_{j,d}^g$ both have the same set of monomials at every iteration of the two parallel instances of the algorithm.

Since the randomness is only used in the exponents, if f and g have the same set of monomials, then the algorithm applied to f and to g with the *same* random bits will result in the same set of degree vectors. Therefore, we get the appropriate coefficients of g.

Once we have the coefficients, we can find the the entries of a scaling matrix using the set of equations in Lemma 2. By Lemma 3, it does not matter which solution we choose. The algorithm is correct by construction. Each single step of the algorithm can be performed in polynomial time.   □

Acknowledgments. The work was supported by the Indo-German Max Planck Center for Computer Science.

References

1. Agrawal, M., Saxena, N.: Equivalence of $\mathbb{F}$-algebras and cubic forms. In: Durand, B., Thomas, W. (eds.) STACS 2006. LNCS, vol. 3884, pp. 115–126. Springer, Heidelberg (2006). doi:10.1007/11672142_8
2. Bläser, M.: Explicit tensors. In: Agrawal, M., Arvind, V. (eds.) Perspectives in Computational Complexity. PCSAL, vol. 26, pp. 117–130. Springer, Cham (2014). doi:10.1007/978-3-319-05446-9_6
3. Bürgisser, P., Clausen, M., Shokrollahi, M.A.: Algebraic Complexity Theory. Grundlehren der mathematischen Wissenschaften, vol. 315. Springer, Heidelberg (1997)
4. Canny, J.F.: Some algebraic and geometric computations in PSPACE. In: Proceedings of the 20th Annual ACM Symposium on Theory of Computing, Chicago, Illinois, USA, 2–4 May 1988, pp. 460–467 (1988)
5. Dvir, Z., Oliveira, R.M., Shpilka, A.: Testing equivalence of polynomials under shifts. In: Esparza, J., Fraigniaud, P., Husfeldt, T., Koutsoupias, E. (eds.) ICALP 2014. LNCS, vol. 8572, pp. 417–428. Springer, Heidelberg (2014). doi:10.1007/978-3-662-43948-7_35
6. Grigoriev, D.: Testing shift-equivalence of polynomials by deterministic, probabilistic and quantum machines. Theor. Comput. Sci. **180**(1), 217–228 (1997)
7. Kabanets, V., Impagliazzo, R.: Derandomizing polynomial identity tests means proving circuit lower bounds. Comput. Complex. **13**(1/2), 1–46 (2004)
8. Kayal, N.: Efficient algorithms for some special cases of the polynomial equivalence problem. In: Proceedings of the Twenty-Second Annual ACM-SIAM Symposium on Discrete Algorithms, SODA 2011, San Francisco, California, USA, 23–25 January 2011, pp. 1409–1421 (2011)
9. Kayal, N.: Affine projections of polynomials: extended abstract. In: Proceedings of the 44th Symposium on Theory of Computing Conference, STOC 2012, New York, NY, USA, 19–22 May 2012, pp. 643–662 (2012)
10. Kayal, N., Nair, V., Saha, C., Tavenas, S.: Reconstruction of full rank algebraic branching programs. In: 32nd IEEE Conference on Computational Complexity (CCC) (2017, to appear)
11. Klivans, A.R., Spielman, D.A.: Randomness efficient identity testing of multivariate polynomials. In: Proceedings on 33rd Annual ACM Symposium on Theory of Computing, 6–8 July 2001, Heraklion, Crete, Greece, pp. 216–223 (2001)
12. Lang, S.: Algebra. Springer, Heidelberg (2002)
13. Patarin, J.: Hidden fields equations (HFE) and Isomorphisms of polynomials (IP): two new families of asymmetric algorithms. In: Maurer, U. (ed.) EURO-CRYPT 1996. LNCS, vol. 1070, pp. 33–48. Springer, Heidelberg (1996). doi:10.1007/3-540-68339-9_4

14. Saxena, N.: Morphisms of rings and applications to complexity. Ph.D. thesis, Department of Computer Science, Indian Institute of Technology, Kanpur, India (2006)
15. Schaefer, M., Stefankovic, D.: The complexity of tensor rank. CoRR, abs/1612.04338 (2016)
16. Shitov, Y.: How hard is tensor rank? CoRR, abs/1611.01559 (2016)
17. Thierauf, T.: The isomorphism problem for read-once branching programs and arithmetic circuits. Chic. J. Theor. Comput. Sci. **1998**(1) (1998)

Strong Duality in Horn Minimization

Endre Boros[1], Ondřej Čepek[2]([✉]), and Kazuhisa Makino[3]

[1] MSIS Department and RUTCOR, Rutgers University, 100 Rockafellar Road,
Piscataway, NJ 08854, USA
`endre.boros@rutgers.edu`
[2] Department of Theoretical Computer Science, Charles University,
Malostranskénám. 25, 118 00 Prague 1, Czech Republic
`ondrej.cepek@mff.cuni.cz`
[3] Research Institute for Mathematical Sciences (RIMS),
Kyoto University, Kyoto 606-8502, Japan
`makino@kurims.kyoto-u.ac.jp`

Abstract. A pure Horn CNF is minimal if no shorter pure Horn CNF representing the same function exists, where the CNF length may mean several different things, e.g. the number of clauses, or the total number of literals (sum of clause lengths), or the number of distinct bodies (source sets). The corresponding minimization problems (a different problem for each measure of the CNF size) appear not only in the Boolean context, but also as problems on directed hypergraphs or problems on closure systems. While minimizing the number of clauses or the total number of literals is computationally very hard, minimizing the number of distinct bodies is polynomial time solvable. There are several algorithms in the literature solving this task.

In this paper we provide a structural result for this body minimization problem. We develop a lower bound for the number of bodies in any CNF representing the same Boolean function as the input CNF, and then prove a strong duality result showing that such a lower bound is always tight. This in turn gives a simple sufficient condition for body minimality of a pure Horn CNF, yielding a conceptually simpler minimization algorithm compared to the existing ones, which matches the time complexity of the fastest currently known algorithm.

1 Introduction

A Boolean function of n variables is a mapping from $\{0,1\}^n$ to $\{0,1\}$. Boolean functions naturally appear in many areas of mathematics and computer science and constitute a key concept in complexity theory. In this paper we shall study an important problem connected to Boolean functions, a so called Boolean minimization problem, which aims at finding a shortest possible representation of a given Boolean function. The formal statement of the Boolean minimization problem (BM) of course depends on (i) how the input function is represented, (ii) how it is represented on the output, and (iii) the way how the output size is measured.

© Springer-Verlag GmbH Germany 2017
R. Klasing and M. Zeitoun (Eds.): FCT 2017, LNCS 10472, pp. 123–135, 2017.
DOI: 10.1007/978-3-662-55751-8_11

One of the most common representations of Boolean functions are conjunctive normal forms (CNFs). There are two usual ways how to measure the size of a CNF: the number of clauses and the total number of literals (sum of clause lengths). It is easy to see that BM is NP-hard if both input and output is a CNF (for both above mentioned measures of the output size). This is an easy consequence of the fact that BM contains the CNF satisfiability problem (SAT) as its special case (an unsatisfiable formula can be trivially recognized from its shortest CNF representation). In fact, BM was shown to be in this case probably harder than SAT: while SAT is NP-complete (i.e. Σ_1^p-complete [9]), BM is Σ_2^p-complete [16] (see also the review paper [17] for related results). It was also shown that BM is Σ_2^p-complete when considering Boolean functions represented by general formulas of constant depth as both the input and output for BM [8].

Due to the above intractability result, it is reasonable to study BM for subclasses of Boolean functions for which SAT (or more generally consistency testing, if the function is not represented by a CNF) is solvable in polynomial time. An extensively studied example of such a class is the class of Horn functions. A CNF is Horn if every clause in it contains at most one positive literal, and it is pure Horn (or definite Horn in some literature) if every clause in it contains exactly one positive literal. A Boolean function is (pure) Horn, if it admits a (pure) Horn CNF representation. Pure Horn functions represent a very interesting concept which was studied in many areas of computer science and mathematics under several different names. The same concept appears as directed hypergraphs in graph theory and combinatorics, as implicational systems in artificial intelligence and database theory, and as lattices and closure systems in algebra and CLA (concept lattice analysis). Consider a pure Horn CNF $\Phi = (\overline{a} \vee b) \wedge (\overline{b} \vee a) \wedge (\overline{a} \vee \overline{c} \vee d) \wedge (\overline{a} \vee \overline{c} \vee e)$ on variables a, b, c, d, e, or its equivalent directed hypergraph $\mathcal{H} = (V, \mathcal{E})$ with vertex set $V = \{a, b, c, d, e\}$ and directed hyperarcs $\mathcal{E} = \{(\{a\}, b), (\{b\}, a), (\{a, c\}, d), (\{a, c\}, e)\}$. This latter can be expressed more concisely using adjacency lists (a generalization of adjacency lists for ordinary digraphs) in which all hyperarcs with the same source set are grouped together $\{a\} : b, \{b\} : a, \{a, c\} : d, e$, or can be represented as an implicational (closure) system on variables a, b, c, d, e defined by rules $a \longrightarrow b, b \longrightarrow a, ac \longrightarrow de$.

It is not difficult to see that all of these constitute identical relations among the five entities a, b, c, d, e, only using different terminology. Using these notions, the same concept has been traditionally studied within logic, combinatorics, database theory, artificial intelligence, and algebra using different techniques, different terminology, and often exploring similar questions with somewhat different emphasis corresponding to the particular area. Interestingly, in each of these areas the problem similar to BM, i.e. a problem of finding the shortest equivalent representation of the input data (CNF, directed hypergraph, set of rules) was studied. However, already the examples above suggest that a "natural" way how to measure the size of the representation depends on the area. Five different measures and corresponding concepts of minimality were introduced in [3] in the context of directed hypergraphs:

- (SM) source minimum hypergraph (no equivalent hypergraph with fewer source sets)
- (HM) hyperarc minimum hypergraph (no equivalent hypergraph with fewer hyperarcs)
- (SHM) source-hyperarc minimum hypergraph (no equivalent hypergraph with fewer hyperarcs plus source sets)
- (SAM) source-area minimum hypergraph (no equivalent hypergraph with smaller source area = sum of sizes of source sets)
- (O) optimum hypergraph (no equivalent hypergraph with smaller size = source area + number of hyperarcs).

(SM) minimizes the number of adjacency lists for a hypergraph or equivalently the number of rules for an implicational system, and hence it is a very natural measure for the size of the representation in these two contexts. On the other hand, it is not very natural in the CNF context and thus it is rarely used there. (HM) minimizes the number of hyperarcs or equivalently the number of clauses in the CNF, so it is one of the two common measures for CNFs. (SHM) and (SAM) are slightly strange mixed size measures and do not appear in many other papers. (O) is simply the total size of the adjacency list representation of a directed hyperarc and similarly a total size of an implicational system. It should be noted that the second common measure for CNFs, namely the total number of literals (sum of clause lengths) - let us denote it by (L) - is missing in the above list (in the hypergraph context (L) is the total size of the representation if all hyperarcs are listed individually). In our example the representations are already minimal and the sizes are as follows: (SM) three, (HM) four, (SHM) seven, (SAM) four, (O) eight, and (L) ten.

It can be shown that for five of these six measures it is NP-hard to find the shortest representation, the sole exception being (SM). There is an extensive literature on the intractability results for the described minimization problems, especially for (HM). Although these results originally appeared in various contexts, let us rephrase them here using the CNF terminology. (HM) was first proved to be NP-hard in [3] (hypergraph context) and later independently in [12] (CNF context). Both proofs construct high degree clauses (with the degree proportional to the number of all variables, where the degree of a clause is the number of literals in it), which left open the question, what is the complexity of (HM) when the clause degrees are bounded. It can be shown that (HM) stays NP-hard even when the inputs are limited to cubic (degree at most three) pure Horn CNFs [7]. It should be also noted that there exists a hierarchy of tractable subclasses of pure Horn CNFs for which (HM) is polynomial time solvable, namely acyclic and quasi-acyclic pure Horn CNFs [13], and CQ Horn CNFs [5]. There are also few heuristic minimization algorithms for pure Horn CNFs [4]. (SHM), (SAM), and (O) were proved to be NP-hard in [3] (hypergraph context), (O) independently also in [12] (CNF context). (L) was proved to be NP-hard in [15] (implicational system context) and this result was later strengtened by several results which limit the clause degrees.

The only one of the above six minimization problems for which a polynomial time procedure exists to derive a minimum representation is (SM). The first such algorithm appeared in database theory literature [15]. Different algorithms for the same task were then independently discovered in hypergraph theory [3], and in the theory of closure systems [11] (see also [1, 2] for more recent literature on the subject). It may be somewhat puzzling what makes (SM) so different (in terms of tractability of minimization) from the other five cases. In this paper, we will try to provide an explanation.

All the polynomial time algorithms developed for (SM) so far [3, 11, 15] use a similar method for proving its optimality. The algorithm takes an input representation (e.g., hypergraph, implicational system) and by a sequence of steps, called (1) *right-saturation,* (2) *left-saturation, and* (3) *redundancy elimination* in the implicational systems terminology, transforms it into an equivalent representation which is *unique,* i.e., the algorithm arrives to the same output for all equivalent inputs. This unique output is called the GD-basis of the implicational systems in the related literature. Each step of the algorithm either preserves or decreases the number of bodies (source sets) in the representation. Therefore, the unique output has the minimum possible number of bodies, since the algorithm can be applied also to a representation which is already body minimum without ever increasing the number of bodies. The algorithms and especially the proofs of their correctness are typically quite involved. In this paper we take a mathematical programming approach to (SM), which simplifies the proof of optimality.

For problem (HM), there exists a natural lower bound for the number of clauses in any CNF representation of a pure Horn function f denoted by $ess(f)$ [6]. The quantity $ess(f)$ is defined as the maximum number of pairwise disjoint essential sets, where an essential set is a set of implicates of f which evaluate to zero on a given false point of f. For some family of pure Horn functions f, this lower bound is tight, i.e., the lower bound $ess(f)$ is equal to the minimum number of clauses needed for a CNF representation of f, denoted by $cnf(f)$. However, in general, there is a gap, i.e., $ess(f) < cnf(f)$ [10] (more on the size of this gap can be found in [14]). In other words, there is only a weak duality between the maximization problem of computing $ess(f)$ and minimization problem of computing $cnf(f)$.

In this paper[1], we modify this idea for (SM) to compute the minimum number of bodies (source sets), denoted by $body(f)$, in any CNF representation of a given pure Horn function f. We introduce as a lower bound for $body(f)$, the maximum number of body-disjoint essential sets of implicates of f, denoted by $bess(f)$, and prove a strong duality between the maximization problem of computing $bess(f)$ and the minimization problem of computing $body(f)$. This min-max theorem is the main contribution of the paper. Moreover, our approach gives a simple sufficient condition for body minimality, namely right-saturation and body-irredundancy, which respectively correspond to Steps (1) and (3) of the existing algorithms. Note that left-saturation in Step (2) eliminated by our

---

[1] Due to space limitations we had to leave out most of our proofs.

algorithm is the conceptually most difficult step in the existing algorithms; see discussion in Sect. 4. Therefore, our approach yields a conceptually simpler minimization algorithm compared with the existing ones. This algorithm matches the time complexity of the fastest currently known algorithm for (SM).

2 Definitions

In this section we recall some of the necessary definitions, introduce notation, and state some basic properties.

2.1 Boolean Functions

We denote by V the set of variables, set $n = |V|$, and consider Boolean functions $h : \mathbb{B}^n \to \mathbb{B}$, where $\mathbb{B} = \{0, 1\}$. We shall write $h \leq g$ if for all $X \in \mathbb{B}^n$ we have $h(X) \leq g(X)$. We denote by $\mathbb{T}(h) = \{X \in \mathbb{B}^n \mid f(X) = 1\}$ the set of *true points* of h, and by $\mathbb{F}(h) = \mathbb{B}^n \setminus \mathbb{T}(h)$ its set of *false points*. To a subset $S \subseteq V$ we associate its *characteristic vector* $\chi_S = (x_1, \ldots, x_n) \in \mathbb{B}^n$ defined by $x_i = 1$ if and only if $i \in S$.

The components x_i, $i = 1, \ldots, n$ of a binary vector X can be viewed as Boolean variables (where truth values are represented by 0 and 1). The logical negation of these variables will be denoted by $\overline{x}_i = 1 - x_i$, $i = 1, \ldots, n$, and called *complemented variables*. Since variables and their complements frequently play a very symmetric role, we call them together as *literals*. An elementary disjunction of literals $C = \bigvee_{i \in I} \overline{x}_i \vee \bigvee_{j \in J} x_j$ is called a *clause*, if every propositional variable appears in it at most once, i.e. if $I \cap J = \emptyset$. It is a well-known fact that every Boolean function h can be represented by a conjunction of clauses. Such an expression is called a *conjunctive normal form* (or CNF) of the Boolean function h.

We say that a clause C_1 *subsumes* another clause C_2 if $C_1 \leq C_2$ (e.g. the clause $\overline{x} \vee z$ subsumes the clause $\overline{x} \vee \overline{y} \vee z$). A clause C is called an *implicate* of a function h if $h \leq C$. An implicate C is called *prime* if there is no distinct implicate C' subsuming C, or in other words, an implicate of a function is prime if dropping any literal from it produces a clause which is not an implicate of that function.

2.2 Pure Horn Functions

A clause is *pure Horn* if exactly one of its literals is an un-complemented variable. A Boolean function is *pure Horn* if it can be represented by a *pure Horn CNF*, that is, a conjunction of pure Horn clauses. It is well known that each prime implicate of a pure Horn function is pure Horn. Thus, in particular, any prime CNF representing a pure Horn function is pure Horn.

In the sequel we assume that we have a pure Horn function h given, and will relate all subsequent definitions to it. Whenever we speak about a CNF representation of h, we assume that this CNF is pure Horn, we never consider

CNFs containing clauses which are not pure Horn. For a pure Horn function h we denote by $\mathcal{I}(h)$ the set of pure Horn implicates of h and by $\mathcal{P}(h)\,(\subseteq \mathcal{I}(h))$ we denote the set of prime implicates of h.

For a subset S and variable $u \notin S$ we write $S \to u$ to denote the pure Horn clause $C = u \vee \bigvee_{v \in S} \bar{v}$, where S is called the *body* (or *source set*) of C and u the *head* of C, which we denote by $S = body(C)$ and $u = head(C)$. For two subsets A, B of the variables we write $A \to B$ to denote the conjunction (or set) of the clauses $\bigwedge_{u \in B} (A \to u)$. For a subset $\Phi \subseteq \mathcal{P}(h)$ we shall view Φ both as a set and as a conjunction of clauses. We also interpret any subset $\Phi \subseteq \mathcal{P}(h)$ as a function (represented by that CNF.) Furthermore, by writing $\Phi = h$, $\Phi = \Psi$ and $\Phi \neq \Sigma$ we mean that Φ represents the same function as h and Ψ, and that it does not represent the same function as Σ. This will never cause confusion, since we do not need to compare in the sequel by equality/non-equality subsets of implicates, as set families. We shall write $\Psi \subseteq \Phi$ if Ψ, as a set of clauses, is a subset of Φ. We shall write $\Psi \leq \Phi$ if the Boolean functions defined by these conjunctions have this relation, that is, if $\Psi(X) \leq \Phi(X)$ for all $X \in \mathbb{B}^n$.

2.3   Forward Chaining

In verifying that a given clause is an implicate of a given pure Horn function, a very useful and simple procedure is the following. Let Φ be a pure Horn CNF of a pure Horn function h. We shall define a *forward chaining* procedure which associates to any subset S of the propositional variables of h a set $F_\Phi(S)$ in the following way. The procedure takes as input the subset S of propositional variables, initializes the set $F_\Phi(S) = S$, and at each step it looks for a pure Horn clause $Y \to y$ in Φ such that $Y \subseteq F_\Phi(S)$, and $y \notin F_\Phi(S)$. If such a clause is found, the propositional variable y is included into $F_\Phi(S)$, and the search is repeated as many times as possible.

It is easy to see that the forward chaining operator satisfies the following properties, where $\Psi \subseteq \Phi$ and $A \subseteq B \subseteq V$:

$$F_\Phi(A) = F_\Phi(F_\Phi(A)), \quad F_\Phi(A) \subseteq F_\Phi(B), \quad \text{and} \quad F_\Psi(A) \subseteq F_\Phi(A) \qquad (1)$$

FORWARD CHAINING PROCEDURE(Φ, S)	
Input:	A pure Horn CNF Φ of h and $S \subseteq V$.
Initialization:	Set $Q = S$.
Main Step:	**While** $\exists\, C \in \Phi : body(C) \subseteq Q$ and $head(C) \notin Q$ **do** $Q = Q \cup \{head(C)\}$.
Output:	$F_\Phi(S) = Q$.

The following lemma, proved in [12], shows how the above procedure can help in determining whether a given clause is an implicate of a given CNF, or not.

Lemma 1. *Given a CNF Φ representing h, a subset S of its propositional variables, and its variable $y \notin S$, we have $y \in F_\Phi(S)$ if and only if $S \to y$ is an implicate of h.*

Let Φ and Ψ be two distinct pure Horn CNF representations of a given pure Horn function h. Then by Lemma 1 we have $F_\Phi(S) = F_\Psi(S)$ for an arbitrary subset S of the variables, since Φ and Ψ represent the same function f. Therefore, the set of variables reachable from S by forward chaining depends only on the underlying function rather than on a particular CNF representation. For this reason, we shall also use the expression $F_h(S)$ instead of $F_\Phi(S)$ whenever we do not want to refer to a specific CNF.

Corollary 1. *If h is a pure Horn function and $\Phi \subseteq \mathcal{P}(h)$, then Φ represents h if and only if $F_h(S) = F_\Phi(S)$ for all subsets $S \subseteq V$.*

Finally, it follows from the first equation in (1) that the forward chaining operator is in fact a closure operator. It is quite easy to see what are the closed sets.

Lemma 2. *Let h be a pure Horn function and let S be a subset of its propositional variables. Then $F_h(S) = S$ if and only if $\chi_S \in \mathbb{T}(h)$, that is, the characteristic vector of S is a true point of h.*

Let us add that the forward chaining procedure can be executed in linear time in the size of Φ. This and Lemma 1 imply that the equivalence of two pure Horn CNF-s can also be tested in polynomial time.

2.4 Essential Sets of Implicates

To a subset $S \subseteq V$ of the variables we associate the following subset of implicates

$$\mathcal{E}_S = \{C \in \mathcal{I}(h) \mid body(C) \subseteq S,\ head(C) \notin S\}. \tag{2}$$

We shall call $\mathcal{E}_S$ to be an *essential set* for h defined by the set S, or equivalently by the binary vector χ_S.

Lemma 3. *Let h be a pure Horn function and let S be a subset of its propositional variables. Then $\mathcal{E}_S \neq \emptyset$ if and only if $\chi_S \in \mathbb{F}(h)$, that is, the characteristic vector of S is a false point of h. In fact $\mathcal{E}_S$ is exactly the set of implicates of h that are falsified by χ_S.*

A key property of essential sets was shown in [6,10].

Lemma 4. *Let h be a pure Horn function. Then a pure Horn CNF Φ represents h if and only if $\Phi \cap \mathcal{E}_S \neq \emptyset$ for all nonempty essential sets $\mathcal{E}_S$.*

For a pure Horn CNF Φ we denote by $\mathcal{B}(\Phi) = \{body(C) \mid C \in \Phi\}$ the set of bodies in Φ. Let us now introduce two measures for the minimum size of a representation of h, namely the number of clauses and the number of bodies needed to represent h:

$$cnf(h) = \min_{\substack{\Phi \subseteq \mathcal{I}(h) \\ \Phi = h}} |\Phi| \quad \text{and} \quad body(h) = \min_{\substack{\Phi \subseteq \mathcal{I}(h) \\ \Phi = h}} |\mathcal{B}(\Phi)|.$$

By definition for any pure Horn function h, we have $body(h) \leq cnf(h)$.

Let us define $ess(h)$ as the maximum number of pairwise disjoint essential sets. Let us further call two essential sets $\mathcal{E}$ and $\mathcal{E}'$ *body-disjoint* if there are no implicates $(S \to u), (S \to v) \in \mathcal{I}(h)$ such that $(S \to u) \in \mathcal{E}$ and $(S \to v) \in \mathcal{E}'$. Clearly, body-disjoint essential sets are also disjoint, since $u = v$ is possible in the above definition. We define $bess(h)$ as the maximum number of pairwise body-disjoint essential sets. For any pure Horn function h, we have $bess(h) \leq ess(h) \leq cnf(h)$.

Let us note that disjointness or body-disjointness of essential sets can be tested efficiently.

Lemma 5. *Let h be a pure Horn function. For subsets $P, Q \subseteq V$, we have the following equivalences.*

(i) $\mathcal{E}_P$ and $\mathcal{E}_Q$ are disjoint if and only if $F_h(P \cap Q) \subseteq P \cup Q$.
(ii) $\mathcal{E}_P$ and $\mathcal{E}_Q$ are body-disjoint if and only if $F_h(P \cap Q) \subseteq P$ or $F_h(P \cap Q) \subseteq Q$.

Both properties can be tested in polynomial time in terms of the size of a pure Horn CNF representing h.

As a consequence, lower bounds on the quantities $ess(h)$ and $bess(h)$ have polynomial certificates. For instance, to prove that $K \leq bess(h)$ for a pure Horn function h represented by a pure Horn CNF Φ, it is enough to exhibit subsets $Q_i, i = 1, \ldots, K$ such that the essential sets $\mathcal{E}_{Q_i}, i = 1, \ldots, K$ are pairwise body-disjoint. By Lemma 5 the latter can be verified in polynomial time in terms of K and the size of Φ.

Let us finish this section by defining two more notions. Let h be a pure Horn function. For a subset $S \subseteq V$, we denote by $\mathcal{C}_S = S \to (F_h(S) \setminus S)$ the set of pure Horn implicates of h with body S. A pure Horn CNF Φ representing h is called *right-saturated* if we have $\mathcal{C}_S \subseteq \Phi$ for every clause $(S \to u) \in \Phi$. A CNF Φ is called *body-irredundant* if for every clause $C \in \Phi$, the CNF $\Phi \setminus \mathcal{C}_{body(C)}$ represents a function different from Φ.

3   Strong Duality

Our main result in this section is the min-max theorem claiming that the maximum number of pairwise body-disjoint essential sets is the same as the minimum number of bodies one needs in a representation of the function.

Let us show first a weak dual relation between these quantities.

Lemma 6 (Weak Duality). *Let h be a pure Horn function. Let $\mathbb{S}$ be an arbitrary family of pairwise body-disjoint nonempty essential sets for h, and let Φ be an arbitrary pure Horn CNF representation of h. Then we have*

$$|\mathbb{S}| \leq |\mathcal{B}(\Phi)|.$$

In particular, $bess(h) \leq body(h)$ holds.

For a pure Horn CNF Φ representing a function h, let $\mathcal{B}(\Phi) = \{S_1, \ldots, S_m\}$. We denote by $\Phi_{-i} = \Phi \backslash \mathcal{C}_{S_i}$ the truncated CNF obtained by removing all clauses from Φ with body S_i, and define

$$P_i = F_{\Phi_{-i}}(S_i)$$

as the forward chaining closure of S_i with respect to the truncated CNF Φ_{-i}, $i = 1, \ldots, m$.

Lemma 7. *For a pure Horn function h, let Φ be a pure Horn CNF representation of h that is right-saturated and body-irredundant. Let P_i be defined as above. Then, the essential sets $\mathcal{E}_{P_i}$, $i = 1, \ldots, m$ are pairwise body-disjoint and nonempty.*

Corollary 2. *If Φ is a body-irredundant right-saturated CNF of a pure Horn function h, then*

$$bess(h) \geq |\mathcal{B}(\Phi)|.$$

Theorem 1 (Strong Duality). *Let h be an arbitrary pure Horn function. Then, we have*

$$bess(h) = body(h).$$

Furthermore, any body-irredundant right-saturated CNF of h is body minimum.

Corollary 3. *For a pure Horn function h in n variables we have*

$$bess(h) \leq cnf(h) \leq n \cdot bess(h).$$

The above corollary shows that we may efficiently compute both a lower bound and an upper bound for $cnf(h)$ which is itself NP-hard to compute. The upper bound seems to be quite weak at a first glance, but the example

$$h(x_1, \ldots, x_n) = \bigwedge_{i=1}^{n} x_i$$

shows that the above inequalities are best possible, since we have $bess(h) = 1$ (the only body is the empty set) and $cnf(h) = n$.

4   Algorithmic Consequences

The main claim of the paper is the min-max theorem (Theorem 1) which states that the maximum number of pairwise body disjoint essential sets is always equal to the minimum number of bodies in a pure Horn CNF representation of the input function. Moreover, it shows that right-saturation and body-irredundancy are sufficient for body minimality. This provides a conceptually very simple algorithm for computing a body minimum CNF.

R-SATURATED&B-IRREDUNDANT CNF

Input: A pure Horn CNF Φ of h.

Step 1: /* Compute a right-saturated CNF Ψ from Φ */
For each body S in Φ, compute $F_\Phi(S)$
Construct $\Psi = \bigwedge_{\text{body } S \text{ of } \Phi}(S \to F_\Phi(S) \setminus S)$.

Step 2: /* Compute a body-irredundant CNF from Ψ */
For each body S of Ψ, if $F_{\Psi \setminus \mathcal{C}_S}(S) = F_\Phi(S)$, then replace Ψ by $\Psi \setminus \mathcal{C}_S$.

Step 3: Output a CNF Ψ.

It is obvious that right-saturation in Step 1 does not change the represented function, i.e., that Ψ also represents h. On the other hand, the correctness of Step 2 requires the following lemma.

Lemma 8. *Let Ψ be a pure Horn CNF representing h, and let S be a body in Ψ. Then $\Psi \setminus \mathcal{C}_S$ also represents h if and only if $F_{\Psi \setminus \mathcal{C}_S}(S) = F_h(S)$.*

Now let us turn our attention to the time complexity of R-SATURATED&B-IRREDUNDANT CNF. Let p be the number of distinct bodies in Φ. Steps 1 and 2 each require $O(p)$ invocations of the forward chaining procedure. Since the forward chaining procedure can be implemented to run in linear time with respect to the size of the input CNF, Steps 1 and 2 can be done in $O(p|\Phi|)$ and $O(p \max\{|\Phi|, |\Psi \setminus \mathcal{C}_S|\})$ time, respectively. Note that the size of $\Psi \setminus \mathcal{C}_S$ might be $\Theta(n|\Phi|)$ for certain input CNFs. For example, let $\Phi = \bigwedge_{i=1}^{n-1}(x_i \to x_{i+1})$. Then its right-saturated CNF is $\Psi = \bigwedge_{i=1}^{n-1}(x_i \to \{x_j \mid j > i\})$, implying that $|\Phi| = \Theta(n)$ and $|\Psi \setminus \mathcal{C}_S| = \Theta(n^2)$ for every body S in Φ. Therefore, a straightforward implementation of the algorithm requires $O(np|\Phi|)$ time. In order to reduce this time complexity to $O(p|\Phi|)$, we make use of the following two technical lemmas, which show that for testing the equality $F_{\Psi \setminus \mathcal{C}_S}(S) = F_\Phi(S)$ in Step 2 it suffices to run forward chaining on Φ instead of Ψ.

Let us start with a simple observation. Let Φ be a pure Horn CNF, let Ψ be its right-saturated CNF, and let $S, T \subseteq V$ be arbitrary two sets of variables. Then by (1) we have

$$F_{\Phi \setminus \mathcal{C}_S}(T) \subseteq F_{\Psi \setminus \mathcal{C}_S}(T) \subseteq F_\Psi(T) = F_\Phi(T). \tag{3}$$

Lemma 9. *Let Φ and Ψ be defined as above. If $S \subseteq V$ satisfies $F_{\Phi \setminus \Phi_S}(S) \subsetneq F_{\Psi \setminus C_S}(S)$, then there exists a body $T \neq S$ in Φ such that $T \subseteq F_{\Phi \setminus C_S}(S)$ and $F_\Phi(T) \supseteq S$.*

Lemma 10. *Let Φ and Ψ be defined as above, and let $S \subseteq V$. Then $F_{\Psi \setminus C_S}(S) = F_\Phi(S)$ if and only if at least one of the following conditions holds:*

(i) $F_{\Phi \setminus C_S}(S) = F_\Phi(S)$
(ii) there exists a body $T \neq S$ in Φ such that $T \subseteq F_{\Phi \setminus C_S}(S)$ and $F_\Phi(T) \supseteq S$.

By making use of (i) and (ii) in Lemma 10, we may implement Step 2 in R-SATURATED&B-IRREDUNDANT CNF in $O(p|\Phi|)$ time. For each ordered pair of bodies T and S in Φ, we define $D[T, S] = 1$ if $F_\Phi(T) \supseteq S$, and $D[T, S] = 0$ otherwise. The $p \times p$ matrix D can be computed in $O(p|\Phi|)$ time by applying forward chaining procedure on each body T, which can be obviously done at no extra cost in Step 1 where all forward chaining closures of bodies are computed anyway. After this preprocessing, we compute for each body S in Φ the set $F_{\Phi \setminus C_S}(S)$, and then we check (i) whether $F_{\Phi \setminus C_S}(S) = F_\Phi(S)$ (the right hand side was computed in Step 1), and if not then we check (ii) the existence of $T \subseteq F_{\Phi \setminus C_S}(S)$ such that $D[T, S] = 1$. This can be done in $O(|\Phi|)$ time for each S. Therefore, Step 2 in R-SATURATED&B-IRREDUNDANT CNF can be implemented to run in $O(p|\Phi|)$ time, which concludes the proof of the following statement.

Theorem 2. *Algorithm* R-SATURATED&B-IRREDUNDANT CNF *can be implemented to run in $O(p|\Phi|)$ time.*

As stated earlier in this paper, several body minimization algorithms for pure Horn CNFs are long known [3,11,15]. They differ in many aspects, but all of them define an equivalence relation on the set of bodies and perform left-saturation (not necessarily called this way). Let Φ be a pure Horn CNF. For two bodies $S, T \subseteq V$ in Φ we say that S and T are *equivalent with respect to* Φ if $F_\Phi(S) = F_\Phi(T)$. Define $\Phi_{[S]}$ be a CNF consisting of all clauses in Φ whose bodies are equivalent to S with respect to Φ. The left-saturation of CNF Φ is then performed by replacing every body S in Φ by $F_{\Phi \setminus \Phi_{[S]}}(S)$. For example, a more recent formulation of the minimization algorithm from the implicational systems literature described in [2] is as follows:

GD-BASIS (RL-SATURATED&B-IRREDUNDANT CNF)

Input: A pure Horn CNF Φ of h.

Step 1: Compute a right-saturated CNF Ψ from Φ.

Step 2: Compute a left-saturated CNF Ψ^* from Ψ.

Step 3: Compute a body-irredundant CNF from Ψ^*.

Step 4: Output the resulting CNF.

Clearly, it works the same as algorithm R-SATURATED&B-IRREDUNDANT CNF except of an extra Step 2 performing left-saturation, which is conceptually the most difficult step to understand. It is clear from Theorem 1 that in fact left-saturation is not necessary for body minimality (right-saturation and irredundancy suffice), it is only needed for the uniqueness of the output. However, uniqueness of the output is an essential part of all proofs of body minimality of the output that do not use a lower bound obtained from the strong duality relation presented in this paper. Such proofs typically proceed as follows: (i) prove the uniqueness of the output (called the GD-basis of the input function), (ii) note that neither right-saturation, nor left-saturation, nor redundancy elimination increases the number of bodies in the current CNF, and (iii) argue that the unique output has the minimum possible number of bodies, since the algorithm can also be applied to a representation which is already body minimum without ever increasing the number of bodies.

Although algorithm R-SATURATED&B-IRREDUNDANT CNF is conceptually simpler than the older body minimization algorithms, it does not achieve a better asymptotic complexity. Left-saturation, if implemented carefully using clever data structures, also runs in $O(p|\Phi|)$ time (see e.g. [15]), same as right-saturation and redundancy elimination. Finally, let us note that once left-saturation is omitted, the uniqueness of the output is lost. Consider a pure Horn CNF in four variables.

$$x_1 \to x_2, \quad x_2 \to x_1, \quad x_1 x_3 \to x_2 x_4, \quad x_2 x_3 \to x_1 x_4$$

It is easy to see that $body(h) = bess(h) = 3$, and in fact both of the following CNFs are body-irredundant and right-saturated (that is body minimum representations of the input CNF)

$$
\begin{array}{ll}
x_1 \to x_2 & x_1 \to x_2 \\
x_2 \to x_1 & x_2 \to x_1 \qquad (4) \\
x_1 x_3 \to x_2 x_4 & x_2 x_3 \to x_1 x_4
\end{array}
$$

Thus, body minimum representations of a pure Horn function are not unique, and any of the above presented two minimum CNFs may be the output of algorithm R-SATURATED&B-IRREDUNDANT CNF, where the output depends on the order in which bodies are tested in the Step 2 (which eliminates redundancy). If left-saturation is performed (prior to the redundancy check) the last implication is in both cases replaced by $x_1 x_2 x_3 \to x_4$ and the output becomes unique.

Acknowledgements. The second author gratefully acknowledges a support by the Czech Science Foundation (grant P202/15-15511S). The third author gratefully acknowledges partial support by Grant-in-Aid for Scientific Research 24106002, 25280004, 26280001 and JST CREST Grant Number JPMJCR1402, Japan.

References

1. Arias, M., Balcázar, J.L.: Canonical Horn representations and query learning. In: Gavaldà, R., Lugosi, G., Zeugmann, T., Zilles, S. (eds.) ALT 2009. LNCS, vol. 5809. Springer, Heidelberg (2009). doi:10.1007/978-3-642-04414-4_16
2. Arias, M., Balcázar, J.L.: Construction and learnability of canonical Horn formulas. Mach. Learn. **85**(3), 273–297 (2011)
3. Ausiello, G., D'Atri, A., Sacca, D.: Minimal representation of directed hypergraphs. SIAM J. Comput. **15**(2), 418–431 (1986)
4. Boros, E., Čepek, O., Kogan, A.: Horn minimization by iterative decomposition. Ann. Math. Artif. Intell. **23**(3–4), 321–343 (1998)
5. Boros, E., Čepek, O., Kogan, A., Kučera, P.: A subclass of Horn CNFs optimally compressible in polynomial time. Ann. Math. Artif. Intell. **57**(3–4), 249–291 (2009)
6. Boros, E., Čepek, O., Kogan, A., Kučera, P.: Exclusive and essential sets of implicates of Boolean functions. Discret. Appl. Math. **158**(2), 81–96 (2010)
7. Boros, E., Čepek, O., Kučera, P.: A decomposition method for CNF minimality proofs. Theoret. Comput. Sci. **510**, 111–126 (2013)
8. Buchfuhrer, D., Umans, C.: The complexity of Boolean formula minimization. J. Comput. Syst. Sci. **77**(1), 142–153 (2011). Celebrating Karp's Kyoto Prize
9. Cook, S.A.: The complexity of theorem-proving procedures. In: Proceedings of the Third Annual ACM Symposium on Theory of Computing, STOC 1971, pp. 151–158. ACM, New York (1971)
10. Čepek, O., Kučera, P., Savicky, P.: Boolean functions with a simple certificate for CNF complexity. Discret. Appl. Math. **160**(4–5), 365–382 (2012)
11. Guigues, J.L., Duquenne, V.: Familles minimales d'implications informatives résultant d'une tables de données binares. Math. Sci. Hum. **95**, 5–18 (1986)
12. Hammer, P.L., Kogan, A.: Optimal compression of propositional Horn knowledge bases: complexity and approximation. Artif. Intell. **64**(1), 131–145 (1993)
13. Hammer, P.L., Kogan, A.: Quasi-acyclic propositional Horn knowledge bases: optimal compression. IEEE Trans. Knowl. Data Eng. **7**(5), 751–762 (1995)
14. Hellerstein, L., Kletenik, D.: On the gap between ess(f) and cnf-size(f). Discret. Appl. Math. **161**(1), 19–27 (2013)
15. Maier, D.: Minimum covers in relational database model. J. ACM **27**(4), 664–674 (1980)
16. Umans, C.: The minimum equivalent DNF problem and shortest implicants. J. Comput. Syst. Sci. **63**(4), 597–611 (2001)
17. Umans, C., Villa, T., Sangiovanni-Vincentelli, A.L.: Complexity of two-level logic minimization. IEEE Trans. Comput.-Aided Des. Integr. Circ. Syst. **25**(7), 1230–1246 (2006)

Token Jumping in Minor-Closed Classes

Nicolas Bousquet[1(✉)], Arnaud Mary[2], and Aline Parreau[3]

[1] Laboratoire G-SCOP, CNRS, Univ. Grenoble Alpes, Grenoble, France
nicolas.bousquet@grenoble-inp.fr
[2] Univ Lyon, Université Lyon 1, LBBE CNRS UMR 5558, 69622 Lyon, France
[3] Univ Lyon, Université Lyon 1, LIRIS UMR CNRS 5205, 69621 Lyon, France

Abstract. Given two k-independent sets I and J of a graph G, one can ask if it is possible to transform the one into the other in such a way that, at any step, we replace one vertex of the current independent set by another while keeping the property of being independent. Deciding this problem, known as the Token Jumping (TJ) reconfiguration problem, is PSPACE-complete even on planar graphs. Ito et al. proved in 2014 that the problem is FPT parameterized by k if the input graph is $K_{3,\ell}$-free.

We prove that the result of Ito et al. can be extended to any $K_{\ell,\ell}$-free graphs. In other words, if G is a $K_{\ell,\ell}$-free graph, then it is possible to decide in FPT-time if I can be transformed into J. As a by product, the TJ-reconfiguration problem is FPT in many well-known classes of graphs such as any minor-free class.

1 Introduction

Reconfiguration problems arise when, given an instance of a problem and a solution to it, we make elementary changes to transform the current solution into another. The objective can be to sample a solution at random, to generate all possible solutions, or to reach a certain desired solution. Many types of reconfiguration problems have been introduced and studied in various fields. For instance reconfiguration of graph colorings [1,10], Kempe chains [4,11], shortest paths [5], satisfiability problems [13] or dominating sets [20] have been studied. For a survey on reconfiguration problems, the reader is referred to [25]. Our reference problem is the independent set problem.

In the whole paper, $G = (V, E)$ is a graph where n denotes the size of V, and k is an integer. For standard definitions and notations on graphs, we refer the reader to [9]. A k-*independent set* of G is a subset of vertices of size k such that no two elements of S are adjacent. The k-independent set reconfiguration graph

N. Bousquet—The author is partially supported by ANR project STINT (ANR-13-BS02-0007).

A. Mary—The author is partially supported by ANR project GraphEn (ANR-15-CE40-0009).

A. Parreau—The author is partially supported by ANR project GAG (ANR-14-CE25-0006).

© Springer-Verlag GmbH Germany 2017
R. Klasing and M. Zeitoun (Eds.): FCT 2017, LNCS 10472, pp. 136–149, 2017.
DOI: 10.1007/978-3-662-55751-8_12

is a graph where vertices are k-independent sets and two independent sets are adjacent if they are "close" to each other.

Three possible definitions of adjacency between independent sets have been introduced. In the *Token Addition Removal* (TAR) model [2,23], two independent sets I, J are adjacent if they differ on exactly one vertex (i.e. if there exists a vertex u such that $I = J \cup \{u\}$ or the other way round). In the *Token Sliding* (TS) model [3,8,14], vertices are moved along edges of the graph. In the *Token Jumping* (TJ) model [6,15,17,18], two k-independent sets I, J are adjacent if the one can be obtained from the other by replacing a vertex with another one. In other words there exist $u \in I$ and $v \in J$ such that $I = (J \setminus \{v\}) \cup \{u\}$. In this paper, we concentrate on the Token Jumping model.

The *k-TJ-reconfiguration graph of G*, denoted $TJ_k(G)$, is the graph whose vertices are all k-independent sets of G (of size exactly k), with the adjacency defined above. The TJ-RECONFIGURATION problem is defined as follows:

TOKEN JUMPING (TJ)-RECONFIGURATION
Input: A graph G, an integer k, two k-independent sets I and J.
Output: YES if and only if I and J are in the same connected component of $TJ_k(G)$.

The TJ-RECONFIGURATION problem is PSPACE-complete even for planar graphs with maximum degree 3 [14], for perfect graphs [18], and for graphs of bounded bandwidth [26]. On the positive side, Bonsma et al. [6] proved that it can be decided in polynomial time in claw-free graphs. Kamiński et al. [18] gave a linear-time algorithm on even-hole-free graphs.

Parameterized Algorithm. A problem Π is *FPT* parameterized by a parameter k if there exists a function f and a polynomial P such that for any instance $\mathcal{I}$ of Π of size n and of parameter k, the problem can be decided in $f(k) \cdot P(n)$. A problem Π *admits a kernel* parameterized by k (for a function f) if for any instance I of size n and parameter k, one can find in polynomial time, an instance I' of size $f(k)$ such that I' is positive if and only if I is positive. A folklore result ensures that the existence of a kernel is equivalent to the existence of an FPT algorithm, but the function f might be exponential. A kernel is *polynomial* if f is a polynomial function.

Ito et al. [17] proved that the TJ-reconfiguration problem is W[1]-hard[1] parameterized by k. On the positive side they show that the problem becomes FPT parameterized by both k and the maximum degree of G. Mouawad et al. [22] proved that the problem is W[1]-hard parameterized by the treewidth of the graph but is FPT parameterized by the length of the sequence plus the treewidth of the graph. In [16], the authors showed that the TJ-RECONFIGURATION problem is FPT on planar graphs parameterized by k. They actually remarked that their proof can be extended to $K_{3,\ell}$-free graphs, i.e. graphs that do not contain any copy of $K_{3,\ell}$ as a subgraph. In this paper (Sects. 2 and 3), we prove that the

---

[1] Under standard algorithmic assumptions, W[1]-hard problems do not admit FPT algorithms.

result of [16] can be extended to any $K_{\ell,\ell}$-free graphs. More formally we show the following:

Theorem 1. TJ-RECONFIGURATION *is FPT parameterized by* $k + \ell$ *in* $K_{\ell,\ell}$-*free graphs. Moreover there exists a function* h *such that* TJ-RECONFIGURATION *admits a kernel of size* $\mathcal{O}(h(\ell) \cdot k^{\ell 3^\ell})$ *which is polynomial if* ℓ *is a fixed constant.*

As a consequence, Theorem 1 ensures that TJ-RECONFIGURATION admits a polynomial kernel on many classical graph classes such as bounded degree graphs, bounded treewidth graphs, graphs of bounded genus or $\mathcal{H}$-(topological) minor free graphs where $\mathcal{H}$ is a finite collection of graphs.

The proof of [16] consists in partitioning the graph into classes according to its neighborhood in $I \cup J$ (two vertices lie in the same class if they have the same neighborhood in $I \cup J$). The authors showed that (i) some classes have bounded size (namely those with at least 3 neighbors in $I \cup J$); (ii) if some classes are large enough, one can immediately conclude (namely those with at most one neighbor in $I \cup J$); (iii) we can "reduce" classes with two neighbors in $I \cup J$ if they are too large. As they observed, this proof cannot be directly extended to $K_{\ell,\ell}$-free graphs for $\ell \geq 4$. In this paper, we develop new tools to "reduce" classes. Namely, we iteratively apply a lemma of Kővári et al. [19] to find a subset X of vertices such that X has size at most $f(k, \ell)$, contains $I \cup J$ and is such that for every $Y \subset X$, if the set of vertices with neighborhood Y in X is too large, then it can be replaced by an independent set of size k.

Note finally that the TJ-RECONFIGURATION problem is W[1]-hard parameterized only by ℓ since graphs of treewidth at most ℓ are $K_{\ell+1,\ell+1}$-free graphs. And the TJ-RECONFIGURATION problem is W[1]-hard parameterized by the treewidth [22]. We left the existence of a polynomial kernel parameterized by $k + \ell$ as an open question.

Hardness for Graphs of Bounded VC-Dimension. A natural way of extending our result would consist in proving it for graphs of bounded VC-dimension. The VC-dimension is a classical way of defining the complexity of a hypergraph that received considerable attention in various fields, from learning to discrete geometry. Informally, the VC-dimension is the maximum size of a set on which the hyperedges of the hypergraph intersect on all possible ways. A formal definition will be provided in Sect. 4. In this paper, we define the VC-dimension of a graph as the VC-dimension of its closed neighborhood hypergraph, which is the most classical definition used in the literature (see [7] for instance).

Bounded VC-dimension graphs generalize $K_{\ell,\ell}$-free graphs since $K_{\ell,\ell}$-free graphs have VC-dimension at most $\ell + \log \ell$. One can naturally ask if our results can be extended to graphs of bounded VC-dimension. Unfortunately the answer is negative since we can obtain as simple corollaries of existing results that the TJ-RECONFIGURATION problem is NP-complete on graphs of VC-dimension 2 and $W[1]$-hard parameterized by k on graphs of VC-dimension 3. We complete these results in Sect. 4 by showing that the problem is polynomial on graphs of VC-dimension 1. The parameterized complexity status remains open on graphs of VC-dimension 2.

2   Density of $K_{\ell,\ell}$-Free Graphs

Kövári et al. [19] proved that any $K_{\ell,\ell}$-free graph has a sub-quadratic number of edges. The initial bound of [19] was later improved, see e.g. [12].

Theorem 2 (Kövári et al. [19]). *Let G be a $K_{\ell,\ell}$-free graph on n vertices. Then G has at most $ex(n, K_{\ell,\ell})$ edges, with*

$$ex(n, K_{\ell,\ell}) \leq \left(\frac{\ell-1}{2}\right)^{1/\ell} \cdot n^{2-1/\ell} + \frac{1}{2}(\ell-1)n.$$

As a corollary, for every ℓ, there exists a polynomial function P_ℓ such that every $K_{\ell,\ell}$-free graph with at least $n \geq P_\ell(k)$ vertices contains a stable set of size at least k. Note that in the following statements, we did not make any attempt in order to optimize the functions. Due to space restriction, the proofs of both corollaries are not included in this extended abstract.

Corollary 1. *Every $K_{\ell,\ell}$-free graph with $k\ell(4k)^\ell$ vertices contains an independent set of size k.*

We will also need a "bipartite version" of both Theorem 2 and Corollary 1.

Theorem 3 (Kövári et al. [19]). *Let $G = ((A, B), E)$ be a $K_{\ell,\ell}$-free bipartite graph where $|A| = n$ and $|B| = m$. The number of edges of G is at most*

$$ex(n, m, K_{\ell,\ell}) \leq (\ell-1)^{1/\ell} \cdot (n - \ell + 1) \cdot m^{1-1/\ell} + (\ell-1)m.$$

Corollary 2. *Let $\ell \geq 3$. Let G be a $K_{\ell,\ell}$-free graph and C be a subset of vertices of size at least $(3\ell)^{4\ell}$. There are at most $(3\ell)^{2\ell}$ vertices of G incident to a fraction of at least $\frac{1}{8\ell}$ of the vertices of C.*

3   Polynomial Kernel on $K_{\ell,\ell}$-Free Graphs

In this section we prove the following that implies Theorem 1.

Theorem 4. *The* TJ-RECONFIGURATION *problem admits a kernel of size $h(\ell) \cdot k^{\ell \cdot 3^\ell}$.*

Let G be a $K_{\ell,\ell}$-free graph and k be an integer. Let I and J be two distinct independent sets of size k. Two vertices a and b are *similar* for a subset X of vertices if both a and b have the same neighborhood in X. A *similarity class* (for X) is a maximum subset of vertices of $V \setminus X$ with the same neighborhood in X.

In Sect. 3.1, we present basic facts and describe the kernel algorithm. In Sect. 3.2, we bound the size of the graph returned by the algorithm. It will be almost straightforward to see that the size of this graph is a function of k and ℓ. However, we will need additional lemmas to prove that its size is at most $h(\ell) \cdot k^{\ell \cdot 3^\ell}$ and that the algorithm is polynomial when ℓ is a fixed constant. Section 3.3 is devoted to prove that the algorithm is correct.

3.1   The Algorithm

Let us first briefly informally describe the behavior of Algorithm 1. During the algorithm, we will update a set X of important vertices. At the beginning of the algorithm we set $X = I \cup J$. At each step of the algorithm, at most $f(k, \ell)$ vertices will be added to X. So each similarity class of the previous step will be divided into at most $2^{f(k,\ell)}$ parts. The main ingredient of the proof (essentially) consists in showing that, at the end of the algorithm, either the size of a class is bounded or the whole class can be replaced by an independent set of size k. As a by-product, the size of the graph can be bounded by a function of k and ℓ.

Let X be a set of vertices containing $I \cup J$. The *rank* of a similarity class C for X is the number of neighbors of C in X. Our proof consists in applying different arguments depending on the rank of the similarity classes. We actually consider the 3 distinct types of classes: classes of rank at least ℓ, classes of rank at most 1 and classes of rank at least 2 and at most $\ell - 1$. The size of the class C in the first two cases can be bounded as shown in [16] for $\ell = 3$.

Lemma 1. *The size of a class C of rank at least ℓ is at most $\ell - 1$.*

Lemma 2. *Let X be a set of vertices containing $I \cup J$. If the size of a class C of rank at most 1 for X is at least $k\ell(4k)^\ell$, then it is possible to transform I into J.*

Our approach to deal with the remaining classes consists in adding vertices in X to increase their ranks. Since it is simple to deal with a class of rank at least ℓ, it provides a way to simplify classes. However, some vertices might not be incident to the new vertices of X, and then their ranks do not increase. The central arguments of the proof consists in proving that we can deal with these vertices if we repeat a "good" operation at least $2k + 1$ steps (see Lemma 5).

The set X is called the set of *important vertices*. Initially, $X = I \cup J$. We denote by X_t the set X at the beginning of step t and $X_0 = I \cup J$. A class D for $X_{t'}$ is *inherited* from a class C of X_t if $t' > t$ and $D \subseteq C$. We say that C is an *ancestor* of D. Note that the rank of C is at most the rank of D since when we add vertices in X the rank can only increase and the set X is increasing during the algorithm.

A similarity class is *big* for X_t if its size is at least $g(k, \ell) := 4 \cdot k\ell(4k)^\ell$. We say that we *reduce a class* C when we replace all the vertices of the class C by an independent set of size k with the same neighborhood in $V \setminus C$: $N(c) \cap X$ where c is any vertex of the class C.

The classes that are in $\mathcal{C}_t$ are said to be *treated* at round t. For these classes, we add in X_t all the vertices that are incident to a $1/8\ell$-fraction of the vertices of the class. When we say that we *refine* the classes at step t, it means that we partition the vertices of the classes according to their respective neighborhood in the new set X_{t+1}.

Algorithm 1. Kernel algorithm

Let $X_0 = I \cup J$ Initially important vertices are $I \cup J$
if a class of rank 1 is big **then**
  Return a YES instance   Valid operation, see Lemma 2
end if
for every $j = 2$ to $\ell - 1$ **do**
  for $s = 0$ to $2k$ **do**
    $\mathcal{C}_t :=$ big similarity classes of rank j for X_t.   Treat big classes of current rank
    $Z = \varnothing$
    for every $C \in \mathcal{C}_t$ **do**
      $Y := \{y \in V \setminus X_t$ such that $|N(y) \cap C| \geq \frac{|C|}{8\ell}\}$. $|Y|$ is bounded, Corollary 2
      if $|(N(Y) \cup Y) \cap C| \leq |C|/2$ **then**
        Reduce the class C.   Valid operation by Lemma 4
      else
        $Z = Z \cup Y$
      end if
    end for
    $X_{t+1} := X_t \cup Z$.   Update the important vertices (and the classes)
  end for
  Reduce all the big classes of rank j.   Valid operation by Lemma 5
end for
Return the reduced graph.

3.2 Size of the Reduced Graph

This part is devoted to prove that the size of the graph output by the algorithm is $h(\ell) \cdot k^{\ell \cdot 3^\ell}$. When we have finished to treat classes of rank j, $j < \ell$, (i.e. when the index of the first loop is at least $j + 1$) then either all the classes of rank j have size less than $g(k, \ell)$ or they are replaced by an independent set of size k. Since classes of rank j cannot be created further in the algorithm, any class of rank at most j has size at least $g(k, \ell)$ at the end of the algorithm. Moreover, any class of rank ℓ has size at most $\ell - 1$ by Lemma 1.

A *step* of the algorithm is an iteration of the second loop (variable s) in the algorithm. The value of j at a given step of the algorithm is called the *index* of the step and the value of s is called the *depth* of the step.

Note that at the step of index i and of depth $2k$, all the classes of rank i are reduced. So at the end of this step, no class of rank i is big anymore. Since the set X_{t+1} contains X_t for every t, the future classes are subsets of classes of rank t. So there is no big class of rank i at any step further in the algorithm. In particular we have the following:

Remark 1. At any step of index j, no class of rank $i < j$ is big. Moreover, at the end of the algorithm, no class is big.

The structure of the algorithm ensures that the algorithm ends. Actually, we have the following:

Remark 2. The number of steps is equal to $(2k + 1) \cdot (\ell - 2)$.

Using Corollary 2, it is simple to prove that the final size of X is bounded by a function of k and ℓ. Since the number of classes only depends on k and ℓ and each class has bounded size by Remark 1, the final size of the graph is bounded in terms of k and ℓ. The rest of this subsection is devoted to prove a better bound on the size of the final graph. We show that it is actually polynomial if ℓ is a fixed constant. The proof will be a consequence of the following lemma.

Lemma 3. *The size of X at the end of the algorithm is at most $h'(\ell) \cdot k^{3^\ell}$.*

Proof. Let us denote by N_j an upper bound on the maximum number of big classes of rank j at any step of the algorithm. Let us first give an upper bound on the number of vertices that are added in X during the steps of index j. Let t be a step of index j. The number of classes in $\mathcal{C}_t$ is at most N_j. Moreover, for each class in $\mathcal{C}_t$, Corollary 2 ensures that at most $(3\ell)^{2\ell}$ vertices are added in X. So the size of $X_{t+1} \setminus X_t$ is at most $N_j \cdot (3\ell)^{2\ell}$. Since there are $2k + 1$ steps of index j, the number of vertices that are added in X during all the steps of index j is at most $N_j \cdot (2k + 1) \cdot (3\ell)^{2\ell}$. Thus the set of important vertices X at the end of the algorithm satisfies the following:

$$|X| \leq 2k + \sum_{j=2}^{\ell-1} \left(N_j \cdot (3\ell)^{2\ell} \right) \cdot (2k + 1).$$

The remainder of the proof is devoted to find a bound on N_j that immediately provides an upper bound on $|X|$. Let us prove by induction on j that $N_j = f_j(\ell) \cdot k^{3^j}$, with $f_2(\ell) = 4$ and $f_j(\ell) = \left(f_{j-1}(\ell) \cdot (3\ell)^{2\ell} \right)^j$ is a valid upper bound.

Since classes are refinement of previous classes and by Remark 1, the number of big classes of rank r is non increasing when we are considering steps of index r. As an immediate consequence, there are at most $(2k)^2$ big classes of rank 2, which is the maximum number of classes of rank 2 when $X = I \cup J$. So the results holds for $j = 2$.

Let $j > 2$ and assume that $f_i(\ell) \cdot k^{3^i}$ is an upper bound on N_i for any $2 \leq i < j$. We say that a class of rank j is *created* at step t if it is inherited from a class (at step t) of rank smaller than j. The maximum number of big classes of rank j is at most the initial number of big classes of rank j plus the number of big classes of rank j created at any step of the algorithm. Let us count how many big classes of rank j can be created at step t. By Remark 1, if the index of the step is at least j, no new big class of rank j can be created. So if a class of rank j is created at step t, then the index of t is $j - i$ with $i > 0$.

Consider a big class C of rank $j - i$ at step t. Let us count how many big classes of rank j can be inherited from this class. Since C is big, the index r of the step t is at most $j - i$. As we already noticed, the set $Z = X_{t+1} \setminus X_t$ has size at most $N_r \cdot (3\ell)^{2\ell} \leq N_{j-i} \cdot (3\ell)^{2\ell}$ since $(N_r)_r$ is an increasing sequence. Each class of rank j inherited from C must have i neighbors in Z. Since there are at most $(N_{j-i} \cdot (3\ell)^{2\ell})^i$ ways of selecting i vertices in Z, the class C can lead to the creation of at most $(N_{j-i} \cdot (3\ell)^{2\ell})^i$ big classes of rank j. By induction hypothesis, the number of big classes of rank $j - i$ is at most N_{j-i}. So at step t, the number of classes of rank j that are created from classes of rank $j - i$ is at most $N_{j-i} \cdot (N_{j-i} \cdot (3\ell)^{2\ell})^i \leq (N_{j-i} \cdot (3\ell)^{2\ell})^{i+1}$.

The total number of rounds of the algorithm is at most $(2k + 1) \cdot (\ell - 2)$ by Remark 2. So the number of big classes of rank j that are created all along the algorithm from (big) classes of rank $j - i$ is at most $(N_{j-i} \cdot (3\ell)^{2\ell})^i \cdot (2k+1) \cdot (\ell-2)$. And then the number of big classes of rank j is at most:

$$(2k)^j + \sum_{i=1}^{j-2} \left(N_{j-i} \cdot (3\ell)^{2\ell} \right)^{i+1} \cdot (2k + 1) \cdot (\ell - 2) \leq f_j(\ell) \cdot k^{3^j}.$$

This inequality is a consequence of the induction hypothesis. Hence, $N_j = f_j(\ell) \cdot k^{3^j}$ is an upper bound on the number of big classes of rank j. Finally, the final size of X satisfies

$$|X| \leq 2k + \sum_{j=2}^{\ell-1} \left(N_j \cdot (3\ell)^{2\ell} \right) \cdot (2k + 1) = h'(\ell) \cdot k^{3^\ell}.$$

□

We have all the ingredients to determine the size of the graph at the end of the algorithm. Let us denote by X the set of important vertices at the end of the algorithm. For every subset of X of size ℓ, there exist at most $\ell - 1$ vertices incident to them by Lemma 1. So the number of vertices with at least ℓ neighbors on X is at most $(\ell - 1) \cdot |X|^\ell$. Moreover every class of rank 0 or 1 contains less than $g(k, \ell) := k \cdot \ell \cdot (4k)^\ell$ vertices by Lemma 2. And every class of rank between 2 and $\ell - 1$ has size at most $g(k, \ell)$ by Remark 1. Since there are $|X|^{\ell-1}$ classes of rank at most $\ell - 1$, Lemma 3 ensures that the size s of the graph returned by the algorithm is at most

$$s \leq (\ell - 1) \cdot \left(h'(\ell) \cdot k^{3^\ell} \right)^\ell + \left(h'(\ell) \cdot k^{3^\ell} \right)^{\ell-1} \cdot k \cdot \ell \cdot (4k)^\ell \leq h(\ell) \cdot k^{\ell \cdot 3^\ell}.$$

So the size of the reduced graph has the size of the claimed kernel.

Complexity of the Algorithm. Let us now briefly discuss the complexity of the algorithm. By Lemma 3, the size of X is bounded by a function of k and ℓ and is polynomial in k if ℓ is a fixed constant. The only possible non polynomial step of the algorithm would consist in maintaining an exponential number of classes. But Lemma 1 ensures that the number of classes of rank at least ℓ is at most $\ell \cdot \binom{\ell}{|X|}$ which is polynomial if ℓ is a constant. So the total number of classes is at most $(\ell + 1) \cdot |X|^\ell$ which ensures that this algorithm runs in polynomial time. Note moreover that the power of the algorithm does not depend on ℓ.

3.3   Equivalence of Transformations

This section is devoted to prove that Algorithm 1 is correct. To do it, we just have to prove that reducing classes does not modify the existence of a transformation. In other words we have to show that I can be transformed into J in the original graph if and only if I can be transformed into J in the reduced graph. In Algorithm 1, there are two cases where we reduce a class. Lemmas 4 and 5 ensure that in both cases these reductions are correct.

Lemma 4. *Let C be a big class of $\mathcal{C}_t$. Assume moreover that the set Y of vertices of $V \setminus X$ incident to a fraction $\frac{1}{8\ell}$ of the vertices of C satisfies $|N(Y) \cap C| \leq |C|/2$. Then there is a transformation of I into J in G if and only if there is a transformation in the graph where C is reduced.*

Proof. Let G be the original graph and G' be the graph where the class C has been replaced by an independent set of size k. We denote by C' the independent set of size k that replaces C in G'.

Assume that there exists a transformation from I to J in G. Let us prove that such a sequence also exists in G'. Either no independent set in the sequence contains a vertex of C, and then the sequence still exists in the graph G'. So we may assume that at least one independent set contains vertices of C. Let us denote by I' the last independent set of the sequence between I and J such that the sequence between I and I' does not contain any vertex of C. In other words, it is possible to move a vertex of I' to a vertex of C. Similarly let J' be the first independent set such that the sequence between J' and J does not contain any vertex of C. Note that in the graph G', the transformations of I into I' and of J' into J still exist since all the independent sets are in $G[V \setminus C]$ that is not modified.

Let us denote by c the vertex of C in the independent set after I' in the sequence and i_0 the vertex deleted from I'. No vertex of $(I' \setminus i_0) \cap X$ has a neighbor in C. Otherwise it would not be possible to move i_0 on c since sets have to remain independent. Thus in G' we can move the vertex i_0 to any vertex of C' and then move the remaining vertices of I' to C'. These operations are possible since for every vertex c' of C', we have $N(c') \subseteq N(c) \cap X$ and $(I \cup \{c\}) \setminus \{i_0\}$ is an independent set. Free to reverse the sequence, a similar argument holds for J'. So there is a transformation from I to J in the graph G'.

Assume now that there exists a transformation from I to J in G'. As in the previous case, we can assume that an independent set of the transformation sequence contain a vertex of C'. Let us denote by I' the last independent set such that the sequence between I and I' does not contain any vertex of C'. Similarly J' is the first independent set such that the sequence between J' and J does not contain any vertex of C'. Let us denote by i_0 and j_0 the vertices respectively deleted between I' and the next independent set and added between the independent before J' and J'.

Note that no vertex of $(I' \setminus i_0) \cap X$ has a neighbor in C. Otherwise the independent set after I' in the sequence would not be independent. Similarly no vertex of $(J' \setminus j_0) \cap X$ has a neighbor in C. Let us partition $F = (I' \cup J') \setminus \{i_0, j_0\}$ into two sets A and B. The set A is the subset of vertices of F incident to a fraction of at least $\frac{1}{8\ell}$ of the vertices of C in G. By hypothesis on C, $N(A) \cap C$ covers at most half of the vertices of C. Let B be the complement of A in F. Every vertex of B is incident to a fraction of at most $\frac{1}{8\ell}$ of the vertices of C in

G. So $N(B)$ covers at most a quarter of the vertices of C. Let us denote by D the set $C \setminus N(I' \cup J')$. The size of D is at least one quarter of the size of C. Since C is big, the size of D is at least $k \cdot \ell \cdot (4k)^\ell$. Theorem 1 ensures that exists an independent set of size I'' at least k in D. By construction of D, one can move i_0 to any vertex of I''. And then the remaining vertices of I' to I''. Similarly, one can transform I'' into J'. So there exists a transformation from I to J in the graph G that concludes this proof.   □

Lemma 5. *Let C be a class of rank j when the index of the step equals j and the depth of the step equals $2k$. Assume moreover that the size of C is at least $4k\ell \cdot (4k)^\ell$. Then there is a transformation of I into J in G if and only if there is a transformation in the graph where C is reduced.*

Proof. Let G be the original graph and G' be the graph where C is reduced. We will denote by C' the independent set of size k that replaces C in G'.

Assume that there exists a transformation from I to J in G. A sequence also exists in G'. The proof works exactly as the proof of the first part of Lemma 4.

Assume now that there exists a transformation from I to J in G'. Let us prove that a transformation from I to J also exists on G. If none of the independent sets of the sequence contains a vertex of C', the sequence still exists in G. So we can assume that an independent set of the sequence contains a vertex of C'. Let us denote by I' the last independent set such that the sequence between I and I' does not contain any vertex of C' and J' the first independent set such that the sequence between J' and J does not contain any vertex of C'. Let us denote by i_0 and j_0 the vertices respectively deleted between I' and the next independent set and added between the independent before J' and J'. We denote by I_0 and J_0 the sets $I' \setminus i_0$ and $J' \setminus j_0$. Note that no vertex of $(I' \setminus i_0) \cap X$ has a neighbor in C. Otherwise the independent set after I' in the sequence would not be independent. Similarly no vertex of $(J' \setminus j_0) \cap X$ has a neighbor in C.

Let us denote by t_0 the step of index j and depth 0. And let t be the step of index j and depth $2k$. In other words $t = t_0 + 2k$. Let $C_0, C_1, \ldots, C_{2k} = C$ be the ancestors of C at round $t_0, \ldots, t_0 + 2k$. All these classes have rank j and $C_{2k} \subseteq C_{2k-1} \subseteq \cdots C_0$. Since $|C_{2k}| \geq g(k, \ell)$, the same holds for any class C_i. In particular, the class C_i is big at step $t_0 + i$. Since the class C_i is not reduced at step $t_0 + i$, the subset of vertices incident to a $\frac{1}{8\ell}$-fraction of the vertices of C_i covers at least half of the vertices of C_i. In particular, for every $i < 2k$, we have

$$|C_{i+1}| \leq |C_i|/2. \tag{1}$$

Let $i \leq 2k$. Let us denote by Y_i the set of vertices of $V \setminus X_{t_0+i}$ that are incident to at least $\frac{1}{8\ell}$ of the vertices of C_i. Any vertex y of Y_i has no neighbor in C_h for $h > i$. Indeed the set Y_i is added in X_{t_0+i+1} at the end of step $t_0 + i$. And by definition of C_h, the rank of C_h is still j. Note moreover that, if $i \neq h$ then Y_i and Y_h are disjoint. Since a vertex in Y_h is not incident to a $\frac{1}{8\ell}$-fraction of the vertices of C_i, Eq. (1) ensures for every $i < 2k$ and every vertex $x \notin Y_i$

$$|N(x) \cap (C_i \setminus C_{i+1})| \leq \frac{|C_i \setminus C_{i+1}|}{4\ell}$$

Moreover, by definition of Y_{2k}, every vertex x which is not in Y_{2k} satisfies

$$|N(x) \cap C_{2k}| \leq \frac{|C_{2k}|}{8\ell}$$

Since the sets $Y_0, \ldots, Y_{2k}$ are disjoint, there exists an index i such that $I' \cup J'$ does not contain any vertex of Y_i. Let $C_i' = C_i \setminus C_{i+1}$ (or $C_i' = C_i$ if $i = 2k$). Every vertex of $I' \cup J'$ is incident to at most $\frac{|C_i'|}{4\ell}$ of the vertices of C_i'. So the complement of $N(I_0 \cup J_0)$ in C_i', denoted by C_i'' has size at least

$$|C_i''| \geq \frac{|C_i'|}{2} \geq \frac{|C_i|}{4} \geq \frac{|C_{2k}|}{4} \geq g(k, \ell) \geq k\ell \cdot (4k)^\ell.$$

By Corollary 1, C_i'' contains an independent set S of size k. Since I' and S are anticomplete (up to one vertex, namely i_0), one can transform the independent set from I' into S. Similarly, one can transform S into J' which completes the proof. $\qquad\square$

4 Bounded VC-Dimension

Let $H = (V, E)$ be a hypergraph. A set X of vertices of H is *shattered* if for every subset Y of X there exists a hyperedge e such that $e \cap X = Y$. An intersection between X and a hyperedge e of E is called a *trace* (on X). Equivalently, a set X is shattered if all its $2^{|X|}$ traces exist. The *VC-dimension* of a hypergraph is the maximum size of a shattered set.

Let $G = (V, E)$ be a graph. The *closed neighborhood hypergraph* of G is the hypergraph with vertex set V where $X \subseteq V$ is a hyperedge if and only if $X = N[v]$ for some vertex $v \in V$ (where $N[v]$ denotes the closed neighborhood of v). The *VC-dimension* of a graph is the VC-dimension of its closed neighborhood hypergraph. The VC-dimension of a class of graphs $\mathcal{C}$ is the maximum VC-dimension of a graph of $\mathcal{C}$.

There is a correlation between VC-dimension and complete bipartite subgraphs. Namely, a $K_{\ell,\ell}$-free graph has VC-dimension at most $\mathcal{O}(\ell)$. Since the TJ-RECONFIGURATION problem is W[1]-hard for general graphs and FPT on $K_{\ell,\ell}$-free graphs, one can naturally ask if this result can be extended to graphs of bounded VC-dimension. Let us remark that the problem is W[1]-hard even on graphs of VC-dimension 3. This is a corollary of two simple facts. First, to prove that the TJ-RECONFIGURATION problem is W[1]-hard on general graphs, Ito et al. [17] showed that if the Independent Set problem is W[1]-hard on a class of graph $\mathcal{G}$, then the TJ-RECONFIGURATION problem is W[1]-hard on the class $\mathcal{G}'$ where graphs of $\mathcal{G}'$ consist in the disjoint union of a graph of $\mathcal{G}$ and a complete bipartite graph. Note that the VC-dimension of a complete bipartite graph equals 1. Moreover, if $\mathcal{G}$ is a class closed by disjoint union, then the VC-dimension of the class $\mathcal{G}'$ is equal to the VC-dimension of $\mathcal{G}$. Hence we have the following:

Remark 3. If $\mathcal{C}$ is a class of graphs of VC-dimension at most d closed by disjoint union, then the TJ-RECONFIGURATION problem on graphs of VC-dimension at most d is at least as hard as the INDEPENDENT SET problem on $\mathcal{C}$.

So any hardness result for INDEPENDENT SET provides a hardness result for TJ-RECONFIGURATION. The INDEPENDENT SET problem is $W[1]$-hard on graphs of VC-dimension at most 3. Indeed, Marx proved in [21] that the INDEPENDENT SET problem is $W[1]$-hard on unit disk graphs, and unit disk graphs have VC-dimension at most 3 (see for instance [7]). To complete the picture, we have to determine the complexity of the problem for $k = 1$ and $k = 2$. For graphs of VC-dimension 2, the problem is NP-hard. Indeed the INDEPENDENT SET problem is NP-complete on graphs of girth at least 5 [24] and this class has VC-dimension at most 2 (see for instance [7]).

The remaining of this section is devoted to prove that TJ-RECONFIGURATION can be decided in polynomial time on graphs of VC-dimension at most 1.

Theorem 5. *The* TJ-RECONFIGURATION *problem can be solved in polynomial time on graphs of VC-dimension at most 1.*

Let us give the three lemmas that permits to prove Theorem 5 whose proofs are not included in this extended abstract.

Lemma 6. *Let G be a graph of VC-dimension at most 1 and let u and v be two vertices of G. Then one of the following holds:*

1. *The closed neighborhoods of u and v are disjoint.*
2. *One of the closed neighborhoods is included in the other.*
3. *u and v form a dominating pair.*

The following lemma ensures that if the graph contains a vertex satisfying the second point of Lemma 6, then it can be deleted.

Lemma 7. *Let G be a graph of VC-dimension at most 1 and let u and v be two vertices such that $N[u] \subseteq N[v]$. Let I, J be two independent set that do not contain v. Then there exists a TJ-transformation from I to J in G if and only if there exists a TJ-transformation from I to J in $G' := G \setminus \{v\}$.*

Note moreover that if I (or J) contains v then we can transform I into $I \cup \{u\} \setminus \{v\}$ in one step. Lemma 7 combined with this remark ensures that we can reduce the graph in such a way no vertex satisfies the second point of Lemma 6.

Lemma 8. *Let G be a graph of VC-dimension at most 1 such that no pair of vertices satisfies the second point of Lemma 6. Let I and J be two independent sets of size at least 3, then $I \cup J$ is an independent set.*

This completes the proof of Theorem 5 since either I and J have size at most 2 and the problem is obviously polynomial. Or $I \cup J$ is an independent set and one can simply move every vertex from I to J one by one.

Question 1. Is the TJ-RECONFIGURATION problem FPT on graphs of VC-dimension at most 2?

References

1. Bonamy, M., Bousquet, N.: Recoloring bounded treewidth graphs. Electron. Notes Discrete Math. (LAGOS 2013) **44**, 257–262 (2013)
2. Bonamy, M., Bousquet, N.: Reconfiguring Independent Sets in Cographs. CoRR, abs/1406.1433 (2014)
3. Bonamy, M., Bousquet, N.: Token sliding on chordal graphs. In: International Workshop on Graph-Theoretic Concepts in Computer Science (WG) (2017, to appear)
4. Bonamy, M., Bousquet, N., Feghali, C., Johnson, M.: On a conjecture of Mohar concerning Kempe equivalence of regular graphs. CoRR, abs/1510.06964 (2015)
5. Bonsma, P.: The complexity of rerouting shortest paths. Theor. Comput. Sci. **510**, 1–12 (2013)
6. Bonsma, P., Kamiński, M., Wrochna, M.: Reconfiguring independent sets in claw-free graphs. In: Ravi, R., Gørtz, I.L. (eds.) SWAT 2014. LNCS, vol. 8503, pp. 86–97. Springer, Cham (2014). doi:10.1007/978-3-319-08404-6_8
7. Bousquet, N., Lagoutte, A., Li, Z., Parreau, A., Thomassé, S.: Identifying codes in hereditary classes of graphs and VC-dimension. SIAM J. Discrete Math. **29**(4), 2047–2064 (2015)
8. Demaine, E.D., Demaine, M.L., Fox-Epstein, E., Hoang, D.A., Ito, T., Ono, H., Otachi, Y., Uehara, R., Yamada, T.: Polynomial-time algorithm for sliding tokens on trees. In: Ahn, H.-K., Shin, C.-S. (eds.) ISAAC 2014. LNCS, vol. 8889, pp. 389–400. Springer, Cham (2014). doi:10.1007/978-3-319-13075-0_31
9. Diestel, R.: Graph Theory. Graduate Texts in Mathematics, vol. 173, 3rd edn. Springer, Heidelberg (2005)
10. Feghali, C., Johnson, M., Paulusma, D.: A reconfigurations analogue of brooks' theorem and its consequences. CoRR, abs/1501.05800 (2015)
11. Feghali, C., Johnson, M., Paulusma, D.: Kempe equivalence of colourings of cubic graphs. CoRR, abs/1503.03430 (2015)
12. Fredi, Z.: An upper bound on Zarankiewicz' problem. Comb. Probab. Comput. **5**(1), 29–33 (1996)
13. Gopalan, P., Kolaitis, P., Maneva, E., Papadimitriou, C.: The connectivity of Boolean satisfiability: computational and structural dichotomies. SIAM J. Comput. **38**, 2330–2355 (2009)
14. Hearn, R., Demaine, E.: PSPACE-completeness of sliding-block puzzles and other problems through the nondeterministic constraint logic model of computation. Theor. Comput. Sci. **343**(1–2), 72–96 (2005)
15. Ito, T., Demaine, E., Harvey, N., Papadimitriou, C., Sideri, M., Uehara, R., Uno, Y.: On the complexity of reconfiguration problems. Theor. Comput. Sci. **412**(12–14), 1054–1065 (2011)
16. Ito, T., Kamiński, M., Ono, H.: Fixed-parameter tractability of token jumping on planar graphs. In: Ahn, H.-K., Shin, C.-S. (eds.) ISAAC 2014. LNCS, vol. 8889, pp. 208–219. Springer, Cham (2014). doi:10.1007/978-3-319-13075-0_17
17. Ito, T., Kamiński, M., Ono, H., Suzuki, A., Uehara, R., Yamanaka, K.: On the parameterized complexity for token jumping on graphs. In: Gopal, T.V., Agrawal, M., Li, A., Cooper, S.B. (eds.) TAMC 2014. LNCS, vol. 8402, pp. 341–351. Springer, Cham (2014). doi:10.1007/978-3-319-06089-7_24
18. Kamiński, M., Medvedev, P., Milanič, M.: Complexity of independent set reconfigurability problems. Theoret. Comput. Sci. **439**, 9–15 (2012)

19. Kővári, T., Sós, V., Turán, P.: On a problem of K. Zarankiewicz. Colloq. Math. **3**, 50–57 (1954)

20. Lokshtanov, D., Mouawad, A.E., Panolan, F., Ramanujan, M.S., Saurabh, S.: Reconfiguration on sparse graphs. In: Dehne, F., Sack, J.-R., Stege, U. (eds.) WADS 2015. LNCS, vol. 9214, pp. 506–517. Springer, Cham (2015). doi:10.1007/978-3-319-21840-3_42

21. Marx, D.: Efficient approximation schemes for geometric problems? In: Brodal, G.S., Leonardi, S. (eds.) ESA 2005. LNCS, vol. 3669, pp. 448–459. Springer, Heidelberg (2005). doi:10.1007/11561071_41

22. Mouawad, A.E., Nishimura, N., Raman, V., Wrochna, M.: Reconfiguration over tree decompositions. In: Cygan, M., Heggernes, P. (eds.) IPEC 2014. LNCS, vol. 8894, pp. 246–257. Springer, Cham (2014). doi:10.1007/978-3-319-13524-3_21

23. Mouawad, A.E., Nishimura, N., Raman, V., Simjour, N., Suzuki, A.: On the parameterized complexity of reconfiguration problems. In: Gutin, G., Szeider, S. (eds.) IPEC 2013. LNCS, vol. 8246, pp. 281–294. Springer, Cham (2013). doi:10.1007/978-3-319-03898-8_24

24. Murphy, O.J.: Computing independent sets in graphs with large girth. Discrete Appl. Math. **35**(2), 167–170 (1992)

25. van den Heuvel, J.: The complexity of change. In: Blackburn, S.R., Gerke, S., Wildon, M. (eds.) Surveys in Combinatorics 2013, pp. 127–160. Cambridge University Press (2013)

26. Wrochna, M.: Reconfiguration in bounded bandwidth and treedepth. CoRR, abs/1405.0847 (2014)

Expressive Power of Evolving Neural Networks Working on Infinite Input Streams

Jérémie Cabessa[1]([⊠]) and Olivier Finkel[2]([⊠])

[1] Laboratoire d'économie mathématique – LEMMA,
Université Paris 2, 4 Rue Blaise Desgoffe, 75006 Paris, France
jeremie.cabessa@u-paris2.fr
[2] Institut de Mathématiques de Jussieu - Paris Rive Gauche, CNRS et Université
Paris Diderot, UFR de mathématiques case 7012, 75205 Paris Cedex 13, France
finkel@math.univ-paris-diderot.fr

Abstract. Evolving recurrent neural networks represent a natural model of computation beyond the Turing limits. Here, we consider evolving recurrent neural networks working on infinite input streams. The expressive power of these networks is related to their attractor dynamics and is measured by the topological complexity of their underlying neural ω-languages. In this context, the deterministic and non-deterministic evolving neural networks recognize the (boldface) topological classes of $BC(\mathbf{\Pi}_2^0)$ and $\mathbf{\Sigma}_1^1$ ω-languages, respectively. These results can actually be significantly refined: the deterministic and nondeterministic evolving networks which employ $\alpha \in 2^\omega$ as sole binary evolving weight recognize the (lightface) relativized topological classes of $BC(\Pi_2^0)(\alpha)$ and $\Sigma_1^1(\alpha)$ ω-languages, respectively. As a consequence, a proper hierarchy of classes of evolving neural nets, based on the complexity of their underlying evolving weights, can be obtained. The hierarchy contains chains of length ω_1 as well as uncountable antichains.

Keywords: Neural networks · Attractors · Formal languages · ω-languages · Borel sets · Analytic sets · Effective Borel and analytic sets

1 Introduction

The theoretical approach to neural computation has mainly been focused on comparing the computational capabilities of diverse neural models with those of abstract computing machines. Nowadays, the computational capabilities of various models of neural networks have been shown to range from the finite automaton degree [14–16,18], up to the Turing [20,23] or even to the super-Turing level [4,6,19,21].

In particular, the real-weighted (or analog) neural networks are strictly more powerful than Turing machines. They decide the complexity class $P/poly$ in polynomial time of computation [19]. The precise computational capabilities of these networks can actually be characterized in terms of the Kolmogorov

© Springer-Verlag GmbH Germany 2017
R. Klasing and M. Zeitoun (Eds.): FCT 2017, LNCS 10472, pp. 150–163, 2017.
DOI: 10.1007/978-3-662-55751-8_13

complexity of their underlying synaptic real weights. As a consequence, a proper hierarchy of classes of analog neural nets employing real weights of increasing Kolmogorov complexity has been obtained [2]. On the other hand, the evolving neural networks (i.e., those employing time-dependent synaptic weights) turn out to be computationally equivalent to the analog ones, irrespectively of the nature (rational or real) of their underlying synaptic weights [4,6].

More recently, based on biological as well as theoretical considerations, these studies have been extended to the paradigm of infinite computation [3,5–10]. In this context, the expressive power of the networks is intrinsically related to their attractor dynamics, and is measured by the topological complexity of their underlying neural ω-languages. In this case, the Boolean recurrent neural networks provided with certain type specification of their attractors are computationally equivalent to Büchi or Muller automata [8]. The rational-weighted neural nets are equivalent to Muller Turing machines. The deterministic and nondeterministic analog and evolving neural networks recognize the (boldface) topological classes of $BC(\mathbf{\Pi}_2^0)$ and $\mathbf{\Sigma}_1^1$ ω-languages, respectively, and in this respect, are super-Turing [3,9].

Here, we refine the above mentioned results for the case of evolving neural networks. More precisely, we focus without loss of generality on evolving neural nets employing only one time-dependent binary weight throughout their computational process. We show that the deterministic and nondeterministic evolving networks using the sole changing weight $\alpha \in 2^\omega$ recognize the (lightface) relativized topological classes of $BC(\Pi_2^0)(\alpha)$ and $\Sigma_1^1(\alpha)$ ω-languages, respectively. As a consequence, a proper hierarchy of classes of evolving neural nets, based on the complexity of their underlying evolving weights, can be obtained. The hierarchy contains chains of length ω_1 as well as uncountable antichains. These achievements generalize the proper hierarchy of classes of analog networks obtained in the context of classical computation [2].

2 Preliminaries

Given a finite set X, usually referred to as an *alphabet*, we let X^* and X^ω denote the sets of finite sequences (or *finite words*) and infinite sequences (or *infinite words*) of elements of X. A set $L \subseteq X^*$ or $L \subseteq X^\omega$ is called a *language* or an ω-*language*, respectively. In the sequel, any space of the form X^ω will be assumed to be equipped with the product topology of the discrete topology on X. Accordingly, the basic open sets of X^ω are of the form $p \cdot X^\omega$, for some $p \in X^*$. The general open sets are countable unions of basic open sets. In particular, the space of infinite words of bits (Cantor space) and that of infinite words of N-dimensional Boolean vectors will be denoted by $2^\omega = \{0,1\}^\omega$ and $(\mathbb{B}^N)^\omega$, respectively. They are assumed to be equipped with the above mentioned topology.

Let $(\mathcal{X}, \mathcal{T})$ be one of the above topological spaces, or a product of such spaces. The class of *Borel subsets* of $\mathcal{X}$, denoted by $\mathbf{\Delta}_1^1$ (boldface), is the σ-algebra generated by $\mathcal{T}$, i.e., the smallest collection of subsets of $\mathcal{X}$ containing all open sets and closed under countable union and complementation. For every non-null countable ordinal $\alpha < \omega_1$, where ω_1 is the first uncountable ordinal, the Borel classes $\mathbf{\Sigma}^0_\alpha$, $\mathbf{\Pi}^0_\alpha$ and $\mathbf{\Delta}^0_\alpha$ of $\mathcal{X}$ are defined as follows:

- $\mathbf{\Sigma}^0_1$ is the class of open subsets of $\mathcal{X}$ (namely $\mathcal{T}$).
- $\mathbf{\Pi}^0_1$ is the class of closed subsets of $\mathcal{X}$, i.e., that of complements of open sets.
- $\mathbf{\Sigma}^0_\alpha$ is the class of countable unions of subsets of $\mathcal{X}$ in $\bigcup_{\gamma < \alpha} \mathbf{\Pi}^0_\gamma$.
- $\mathbf{\Pi}^0_\alpha$ is the class of countable intersections of subsets of $\mathcal{X}$ in $\bigcup_{\gamma < \alpha} \mathbf{\Sigma}^0_\gamma$.
- $\mathbf{\Delta}^0_\alpha = \mathbf{\Sigma}^0_\alpha \cap \mathbf{\Pi}^0_\alpha$.

The Borel classes $\mathbf{\Sigma}^0_\alpha$, $\mathbf{\Pi}^0_\alpha$ and $\mathbf{\Delta}^0_\alpha$ provide a stratification of the class of Borel sets known as the *Borel hierarchy*. One has $\mathbf{\Delta}^1_1 = \bigcup_{\alpha < \omega_1} \mathbf{\Sigma}^0_\alpha = \bigcup_{\alpha < \omega_1} \mathbf{\Pi}^0_\alpha$ [12]. The *rank* of a Borel set $A \subseteq \mathcal{X}$ is the smallest ordinal α such that $A \in \mathbf{\Sigma}^0_\alpha \cup \mathbf{\Pi}^0_\alpha$. It is commonly considered as a relevant measure of the topological complexity of Borel sets. The class of sets obtained as finite Boolean combinations (unions, intersections and complementations) of $\mathbf{\Pi}^0_2$-sets is denoted by $BC(\mathbf{\Pi}^0_2)$.

Analytic sets are more complicated than Borel sets. They are obtained as projections of either $\mathbf{\Pi}^0_2$-sets or general Borel sets [12]. More precisely, a set $A \subseteq \mathcal{X}$ is *analytic* if there exists some $\mathbf{\Pi}^0_2$-set $B \subseteq \mathcal{X} \times 2^\omega$ such that $A = \{x \in \mathcal{X} : (x, \beta) \in B, \text{for some } \beta \in 2^\omega\} = \pi_1(B)$ [12]. The class of analytic sets is denoted by $\mathbf{\Sigma}^1_1$. It strictly contains that of Borel sets, i.e., $\mathbf{\Delta}^1_1 \subsetneq \mathbf{\Sigma}^1_1$ [12].

The *effective* (lightface) counterpart of the Borel and analytic classes, denoted by $\Sigma^0_n, \Pi^0_n, \Delta^0_n$ as well as Δ^1_1 and Σ^1_1, are obtained by a similar effective construction, yet starting from the class Σ^0_1 of effective open sets [17]. The class of finite Boolean combinations of Π^0_2-sets, denoted by $BC(\Pi^0_2)$ (lightface), and that of effective analytic sets, denoted by Σ^1_1 (lightface), correspond to the collections of ω-languages recognizable by deterministic and nondeterministic Muller Turing machines, respectively [22]. One has $BC(\Pi^0_2) \subsetneq BC(\mathbf{\Pi}^0_2)$ and $\Sigma^1_1 \subsetneq \mathbf{\Sigma}^1_1$.

Any topological class Γ of the underlying topological space $\mathcal{X}$ will also be written as $\Gamma \restriction \mathcal{X}$, whenever we want $\mathcal{X}$ to be specified. In addition, for any point $x \in \mathcal{X}$, we will use the notation $x \in \Gamma$ to mean that $\{x\} \in \Gamma$. Besides, any product space $\mathcal{X} \times \mathcal{Y}$ is assumed to be equipped with the product topology. If $A \subseteq \mathcal{X} \times \mathcal{Y}$ and $y \in \mathcal{Y}$, the *y-section* of A is defined by $A_y = \{x \in \mathcal{X} : (x, y) \in A\}$. For any class Γ equal to Σ^0_1, $BC(\Pi^0_2)$, Σ^1_1, or Π^1_1 with underlying product space $\mathcal{X} \times \mathcal{Y}$, and for any $y \in \mathcal{Y}$, we consider the *relativization of* Γ *to* y, denoted by $\Gamma(y)$, which is the class of all y-sections of sets in Γ. In other words: $A \in \Gamma(y) \restriction \mathcal{X}$ if and only if there exists $B \in \Gamma \restriction \mathcal{X} \times \mathcal{Y}$ such that $A = B_y$. Moreover, we denote as usual $\Delta^1_1(y) = \Sigma^1_1(y) \cap \Pi^1_1(y)$ [17, p. 118].

For any $\alpha \in 2^\omega$, one can show that the relativized classes $BC(\Pi^0_2)(\alpha)$ and $\Sigma^1_1(\alpha)$ correspond to the collections of ω-languages recognizable by deterministic and nondeterministic Muller Turing machine with oracle α, respectively. In addition, it can be shown that $x \in \Sigma^0_1(\alpha)$ if and only if the successive letters of x can be produced step by step by some TM with oracle α. Besides, one has $x \in \Sigma^1_1(\alpha)$ iff $x \in \Delta^1_1(\alpha)$, for any $\alpha \in 2^\omega$ [17].

Finally, the spaces $(\mathbb{B}^M)^\omega \times 2^\omega$ and $(\mathbb{B}^{M+1})^\omega$ are isomorphic via the natural identification. Accordingly, subsets of these spaces will be identified without it being explicitly mentioned.

3 Recurrent Neural Networks on Infinite Input Streams

We consider first-order recurrent neural networks composed of Boolean input cells, Boolean output cells and sigmoidal internal cells. The sigmoidal internal neurons introduce the biological source of nonlinearity which is crucial to neural computation. They provide the possibility to surpass the capabilities of finite state automata, or even of Turing machines. The Boolean input and output cells carry out the exchange of discrete information between the network and the environment. When some infinite input stream is supplied, the output cells eventually enter into some attractor dynamics. The expressive power of the networks is related to the attractor dynamics of their Boolean output cells.

3.1 Deterministic Case

A *deterministic (first-order) recurrent neural network*, denoted by D-RNN, consists of a synchronous network of neurons related together in a general architecture. It is composed of M Boolean input cells $(u_i)_{i=1}^M$, N sigmoidal internal neurons $(x_i)_{i=1}^N$, and P Boolean output cells $(y_i)_{i=1}^P$. The dynamics of the network is computed as follows: given the activation values of the input and internal neurons $(u_j)_{j=1}^M$ and $(x_j)_{j=1}^N$ at time t, the activation values of each internal and output neuron x_i and y_i at time $t+1$ are updated by the following equations, respectively:

$$x_i(t+1) = \sigma \left(\sum_{j=1}^N a_{ij}(t) \cdot x_j(t) + \sum_{j=1}^M b_{ij}(t) \cdot u_j(t) + c_i(t) \right) \text{ for } i = 1, \ldots, N \quad (1)$$

$$y_i(t+1) = \theta \left(\sum_{j=1}^N a_{ij}(t) \cdot x_j(t) + \sum_{j=1}^M b_{ij}(t) \cdot u_j(t) + c_i(t) \right) \text{ for } i = 1, \ldots, P \quad (2)$$

where $a_{ij}(t)$, $b_{ij}(t)$, and $c_i(t)$ are the time dependent *synaptic weights* and *bias* of the network at time t, and σ and θ are the linear-sigmoid[1] and Heaviside step activation functions defined by

$$\sigma(x) = \begin{cases} 0, & \text{if } x < 0 \\ x, & \text{if } 0 \le x \le 1 \\ 1, & \text{if } x > 1 \end{cases} \quad \text{and} \quad \theta(x) = \begin{cases} 0, & \text{if } x < 1 \\ 1, & \text{if } x \ge 1 \end{cases}$$

A synaptic weight or a bias w will be called *static* if it remains constant over time, i.e., if $w(t) = c$ for all $t \ge 0$. It will be called *bi-valued evolving* if it varies among two possible values over time, i.e., if $w(t) \in \{0,1\}$ for all $t \ge 0$. A D-RNN is illustrated in Fig. 1.

---

[1] The results of the paper remain valid for any other kind of sigmoidal activation function satisfying the properties mentioned in [13, Sect. 4].

The dynamics of a D-RNN $\mathcal{N}$ is therefore given by the function $f_{\mathcal{N}} : \mathbb{B}^M \times \mathbb{B}^N \to \mathbb{B}^N \times \mathbb{B}^P$ defined by

$$f_{\mathcal{N}}(\boldsymbol{u}(t), \boldsymbol{x}(t)) = (\boldsymbol{x}(t+1), \boldsymbol{y}(t+1))$$

where the components of $\boldsymbol{x}(t+1)$ and $\boldsymbol{y}(t+1)$ are given by Eqs. (1) and (2), respectively.

Consider some D-RNN $\mathcal{N}$ provided with M Boolean input cells, N sigmoidal internal cells, and P Boolean output cells. For each time step $t \geq 0$, the *state* of $\mathcal{N}$ at time t consists of a pair of the form

$$\langle \boldsymbol{x}(t), \boldsymbol{y}(t) \rangle \in [0,1]^N \times \mathbb{B}^P.$$

The second element of this pair, namely $\boldsymbol{y}(t)$, is the *output state* of $\mathcal{N}$ at time t.

Assuming the initial state of the network to be $\langle \boldsymbol{x}(0), \boldsymbol{y}(0) \rangle = \langle \boldsymbol{0}, \boldsymbol{0} \rangle$, any infinite input stream

$$s = (\boldsymbol{u}(t))_{t \in \mathbb{N}} = \boldsymbol{u}(0)\boldsymbol{u}(1)\boldsymbol{u}(2) \cdots \in (\mathbb{B}^M)^{\omega}$$

induces via Eqs. (1) and (2) an infinite sequence of consecutive states

$$c_s = (\langle \boldsymbol{x}(t), \boldsymbol{y}(t) \rangle)_{t \in \mathbb{N}} = \langle \boldsymbol{x}(0), \boldsymbol{y}(0) \rangle \langle \boldsymbol{x}(1), \boldsymbol{y}(1) \rangle \cdots \in ([0,1]^N \times \mathbb{B}^P)^{\omega}$$

which is the *computation* of $\mathcal{N}$ induced by s. The corresponding infinite sequence of output states

$$bc_s = (\boldsymbol{y}(t))_{t \in \mathbb{N}} = \boldsymbol{y}(0)\boldsymbol{y}(1)\boldsymbol{y}(2) \cdots \in (\mathbb{B}^P)^{\omega}$$

is the *Boolean computation* of $\mathcal{N}$ induced by s. The computation of such a D-RNN is illustrated in Fig. 1.

Note that any D-RNN $\mathcal{N}$ with P Boolean output cells can only have 2^P – i.e., finitely many – possible distinct output states. Consequently, any Boolean computation bc_s necessarily consists of a finite prefix of output states followed by an infinite suffix of output states that repeat infinitely often – yet not necessarily in a periodic manner – denoted by $\mathrm{inf}(bc_s)$. A set of states of the form $\mathrm{inf}(bc_s) \subseteq \mathbb{B}^P$ will be called an *attractor*[2] of $\mathcal{N}$ [8], as illustrated in Fig. 1. A precise definition can be given as follows:

Definition 1. *Let $\mathcal{N}$ be some D-RNN. A set $A = \{\boldsymbol{y_0}, \ldots, \boldsymbol{y_k}\} \subseteq \mathbb{B}^P$ is an attractor for $\mathcal{N}$ if there exists some infinite input stream s such that the corresponding Boolean computation bc_s satisfies $\mathrm{inf}(bc_s) = A$.*

We suppose that the attractors are of two distinct types, either *accepting* or *rejecting*. The type specification of these attractors is not the subject of this work (cf. [8]), and from this point onwards, we assume that any D-RNN is equipped

---

[2] In words, an attractor of $\mathcal{N}$ is a set of output states into which the Boolean computation of the network could become forever trapped – yet not necessarily in a periodic manner.

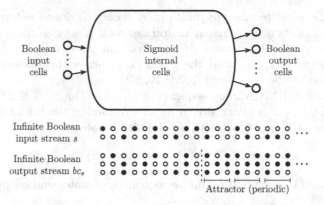

Attractor (periodic)

Fig. 1. Illustration of the computational process performed by some D-RNN. The infinite Boolean input stream $s = \boldsymbol{u}(0)\boldsymbol{u}(1)\boldsymbol{u}(2)\cdots \in (\mathbb{B}^M)^\omega$, represented by the first pattern, induces a corresponding Boolean output stream – or Boolean computation – $bc_s = \boldsymbol{y}(0)\boldsymbol{y}(1)\boldsymbol{y}(2)\cdots \in (\mathbb{B}^P)^\omega$, represented by the second pattern. The filled and empty circles represent active and quiet Boolean cells, respectively. From some time step onwards, a certain set of output states begins to repeat infinitely often, which corresponds to the attractor dynamics associated with input stream s.

with a corresponding classification of all of its attractors into accepting and rejecting types.

This classification of attractors yields the following Muller acceptance condition: given some D-RNN $\mathcal{N}$, an infinite input stream $s \in (\mathbb{B}^M)^\omega$ is *accepted* $\mathcal{N}$ if $\inf(bc_s)$ is an accepting attractor; it is *rejected* by $\mathcal{N}$ if $\inf(bc_s)$ is a rejecting attractor. The set of all accepted input streams of $\mathcal{N}$ is called the *neural ω-language recognized by* $\mathcal{N}$, denoted by $L(\mathcal{N})$. A set $L \subseteq (\mathbb{B}^M)^\omega$ is said to be *recognizable* by some D-RNN if there exists a network $\mathcal{N}$ such that $L(\mathcal{N}) = L$.

Two different models of D-RNNs can be considered according to the nature of their synaptic weights:

1. The class of *deterministic static rational neural nets*, denoted by D-St-RNN[$\mathbb{Q}$], which refers to the D-RNNs whose every weights are static and modelled by rational values.
2. The class of *deterministic bi-valued evolving rational neural nets*, denoted by D-Ev$_2$-RNN[$\mathbb{Q}$], which refers to the D-RNNs whose every evolving weights are bi-valued and every static weights are rational. In this case, the subclass of networks containing $\alpha_1, \ldots, \alpha_k \in 2^\omega$ as sole bi-valued evolving weights, all other ones being static rational, is denoted by D-Ev$_2$-RNN[$\mathbb{Q}, \alpha_1, \ldots, \alpha_k$].

3.2 Nondeterministic Case

We also consider nondeterministic recurrent neural networks, as introduced in [19,20]. The nondeterminism is expressed by means of an external binary guess stream processed via some additional Boolean guess cell.

Formally, a *nondeterministic (first-order) recurrent neural network*, denoted by N-RNN, consists of a recurrent neural network $\mathcal{N}$ as described in previous Sect. 3.1, except that it contains $M+1$ Boolean input cells $(u_i)_{i=1}^{M+1}$, rather than M. The cell u_{M+1}, called the *guess cell*, carries the Boolean source of nondeterminism to be considered [3,7,9,19,20].

Given some N-RNN $\mathcal{N}$, any sequence $g = g(0)g(1)g(2)\cdots \in 2^\omega$ submitted to guess cell u_{M+1} is a *guess stream* for $\mathcal{N}$. Assuming the initial state of the network to be $\langle \boldsymbol{x}(0), \boldsymbol{y}(0) \rangle = \langle \boldsymbol{0}, \boldsymbol{0} \rangle$, any infinite input and guess streams

$$s = (\boldsymbol{u}(t))_{t \in \mathbb{N}} \in (\mathbb{B}^M)^\omega \text{ and } g = (g(t))_{t \in \mathbb{N}} \in 2^\omega$$

induce via Eqs. (1) and (2) two infinite sequences of states and output states

$$c_{(s,g)} = (\langle \boldsymbol{x}(t), \boldsymbol{y}(t) \rangle)_{t \in \mathbb{N}} \in ([0,1]^N \times \mathbb{B}^P)^\omega$$

$$bc_{(s,g)} = (\boldsymbol{y}(t))_{t \in \mathbb{N}} \in (\mathbb{B}^P)^\omega$$

called the *computation* and *Boolean computation* of $\mathcal{N}$ induced by (s,g), respectively. Furthermore, Definition 1 of an *attractor* remains unchanged in this case.

We also assume that any N-RNN $\mathcal{N}$ is equipped with a corresponding classification of all of its attractors into accepting and rejecting types. An infinite input stream $s \in (\mathbb{B}^M)^\omega$ is *accepted* by $\mathcal{N}$ if there exists some guess stream $g \in 2^\omega$ such that $\inf(bc_{(s,g)})$ is an accepting attractor. It is *rejected* by $\mathcal{N}$ otherwise, i.e., if for all guess streams $g \in 2^\omega$, the set $\inf(bc_{(s,g)})$ is a rejecting attractor. The set of all accepted input streams is the *neural ω-language recognized by* $\mathcal{N}$, denoted by $L(\mathcal{N})$. A set $L \subseteq (\mathbb{B}^M)^\omega$ is said to be *recognizable* by some nondeterministic recurrent neural network if there exists a N-RNN $\mathcal{N}$ such that $L(\mathcal{N}) = L$.

As for the deterministic case, the following classes and subclasses of N-RNNs will be considered according to the nature of their synaptic weights:

1. The class of *nondeterministic static rational neural nets*, denoted by N-St-RNN[$\mathbb{Q}$].
2. The class of *nondeterministic bi-valued evolving rational neural nets*, denoted by N-Ev$_2$-RNN[$\mathbb{Q}$], which is stratified into the subclasses of N-Ev$_2$-RNN[$\mathbb{Q}, \alpha_1, \ldots, \alpha_k$], where $\alpha_1, \ldots, \alpha_k \in 2^\omega$.

4 Expressive Power of Neural Networks

We provide a precise characterization of the expressive power of the evolving neural networks, according to the specific evolving weights that they employ. As a consequence, a proper hierarchy of classes of evolving networks, related to the complexity of their underlying evolving weights, will be obtained in Sect. 5.

4.1 Deterministic Case

The expressive powers of the classes D-St-RNN[$\mathbb{Q}$] and D-Ev$_2$-RNN[$\mathbb{Q}$] have been established in [9, Theorems 1, 2]. We recall these results:

Theorem 1 [9, Theorems 1 and 2]. *Let $L \subseteq (\mathbb{B}^M)^\omega$ be some ω-language.*

(a) *L is recognizable by some D-St-RNN[$\mathbb{Q}$] iff L is recognizable by some deterministic Muller Turing machine iff $L \in BC(\Pi_0^2)$.*
(b) *L is recognizable by some D-Ev$_2$-RNN[$\mathbb{Q}$] iff $L \in BC(\mathbf{\Pi_0^2})$.*

Theorem 1 states that D-St-RNN[$\mathbb{Q}$]s are Turing equivalent and D-Ev$_2$-RNN[$\mathbb{Q}$]s are strictly more powerful than deterministic Muller Turing machines, since $BC(\Pi_0^2) \subsetneq BC(\mathbf{\Pi_0^2})$. In this sense, the deterministic evolving neural networks are *super-Turing*.

Remark 1. The proof of Theorem 1(b) [9, Theorem 2] shows that any ω-language $L \in BC(\mathbf{\Pi_0^2})$ can be recognized by some D-Ev$_2$-RNN[$\mathbb{Q}$] employing only one bi-valued evolving weight given in the form of a bias. In other words, any D-Ev$_2$-RNN[$\mathbb{Q}$] is expressively equivalent to some D-Ev$_2$-RNN[$\mathbb{Q}, \alpha$], for some $\alpha \in 2^\omega$. Hence, from this point onwards, we will without loss of generality focus on the latter subclass of networks.

A precise characterization of the expressive power of the subclass of D-Ev$_2$-RNN[$\mathbb{Q}, \alpha$] can be obtained, for any $\alpha \in 2^\omega$. The result is achieved by forthcoming Propositions 1 and 2.

Proposition 1. *Let $L \subseteq (\mathbb{B}^M)^\omega$ be some ω-language and $\alpha \in 2^\omega$. If $L \in BC(\Pi_0^2)(\alpha)$, then L is recognizable by some D-Ev$_2$-RNN[$\mathbb{Q}, \alpha$].*

Proof. If $L \in BC(\Pi_0^2)(\alpha) \upharpoonright (\mathbb{B}^M)^\omega$, then by definition, there exists $L' \in BC(\Pi_0^2) \upharpoonright (\mathbb{B}^{M+1})^\omega$ such that $L = L'_\alpha = \{s \in (\mathbb{B}^M)^\omega : (s, \alpha) \in L'\}$. Theorem 1 ensures that there exists a D-St-RNN[$\mathbb{Q}$] $\mathcal{N}'$ with $M+1$ input cells $u_1, \ldots, u_{M+1}$ such that $L(\mathcal{N}') = L'$.

Now, we consider the D-Ev$_2$-RNN[$\mathbb{Q}, \alpha$] $\mathcal{N}$ which consists in a slight modification of the D-St-RNN[$\mathbb{Q}$] $\mathcal{N}'$. More precisely, $\mathcal{N}$ contains the same cells and synaptic connections as $\mathcal{N}'$, it admits only $u_1, \ldots, u_M$ as its input cells, but u_{M+1} is transformed into an internal cell receiving the bi-valued evolving weight $\alpha \in 2^\omega$ in the form of a bias. Moreover, the attractors of $\mathcal{N}$ are the same as those of $\mathcal{N}'$. By construction, on every input $s \in (\mathbb{B}^M)^\omega$, $\mathcal{N}$ receives the bi-valued evolving weight α as bias and it works precisely like $\mathcal{N}'$ on input $(s, \alpha) \in (\mathbb{B}^{M+1})^\omega$. Consequently, $s \in L(\mathcal{N})$ if and only if $(s, \alpha) \in L(\mathcal{N}') = L'$. Therefore, $L(\mathcal{N}) = L'_\alpha = L$, meaning that L is recognized by the D-Ev$_2$-RNN[$\mathbb{Q}, \alpha$] $\mathcal{N}$. □

Proposition 2. *Let $L \subseteq (\mathbb{B}^M)^\omega$ be some ω-language and $\alpha \in 2^\omega$. If L is recognizable by some D-Ev$_2$-RNN[$\mathbb{Q}, \alpha$], then $L \in BC(\Pi_0^2)(\alpha)$.*

Proof. Let $\mathcal{N}$ be a D-Ev$_2$-RNN[$\mathbb{Q}, \alpha$] such that $L(\mathcal{N}) = L$. By Remark 1, we may assume without loss generality that the bi-valued evolving weight α of $\mathcal{N}$ is a bias related to some cell x. Let $\mathcal{N}'$ be the D-St-RNN[$\mathbb{Q}$] obtained by replacing in $\mathcal{N}$ the evolving bias $\alpha \in 2^\omega$ by a new input cell u_{M+1} related to x with a weight of 1. Hence, $\mathcal{N}'$ is a D-St-RNN[$\mathbb{Q}$] with $M+1$ input cells, and Theorem 1 ensures that $L(\mathcal{N}') \in BC(\Pi_0^2)$. By construction, if $\mathcal{N}'$ receives input $(s, \alpha) \in (\mathbb{B}^{M+1})^\omega$, then it works precisely like $\mathcal{N}$ on input $s \in (\mathbb{B}^M)^\omega$, which means that $(s, \alpha) \in L(\mathcal{N}')$ if and only if $s \in L(\mathcal{N})$. Thus $L(\mathcal{N}) = L(\mathcal{N}')_\alpha$. Since $L(\mathcal{N}') \in BC(\Pi_0^2)$, it follows that $L(\mathcal{N}) \in BC(\Pi_0^2)(\alpha)$. □

By combining Propositions 1 and 2, one obtains the following theorem:

Theorem 2. *Let $L \subseteq (\mathbb{B}^M)^\omega$ be some ω-language and $\alpha \in 2^\omega$. The following conditions are equivalent:*

(a) $L \in BC(\Pi_0^2)(\alpha)$;
(b) L is recognizable by some $D\text{-}Ev_2\text{-}RNN[\mathbb{Q}, \alpha]$.

From Theorem 2 and Remark 1, the following set-theoretical result can be retrieved: $BC(\mathbf{\Pi}_0^2) = \bigcup_{\alpha \in 2^\omega} BC(\Pi_0^2)(\alpha)$. Indeed, $L \in BC(\mathbf{\Pi}_0^2)$ if and only if, by Remark 1, L is recognizable by some $D\text{-}Ev_2\text{-}RNN[\mathbb{Q}, \alpha]$, for some $\alpha \in 2^\omega$, if and only if, by Theorem 2, $L \in BC(\Pi_0^2)(\alpha)$, for some $\alpha \in 2^\omega$. In words, the relativized classes $BC(\Pi_0^2)(\alpha)$ span the class $BC(\mathbf{\Pi}_0^2)$, when α varies over 2^ω.

4.2   Nondeterministic Case

The expressive power of the classes N-St-RNN[$\mathbb{Q}$] and N-Ev$_2$-RNN[$\mathbb{Q}$] has also been established in [3, Theorems 1 and 2]. We have the following results:

Theorem 3 [3, Theorems 1 and 2]. *Let $L \subseteq (\mathbb{B}^M)^\omega$ be some ω-language.*

(a) L is recognizable by some $N\text{-}St\text{-}RNN[\mathbb{Q}]$ iff $L \in \Sigma_1^1$;
(b) L is recognizable by some $N\text{-}Ev_2\text{-}RNN[\mathbb{Q}]$ iff $L \in \Sigma_1^1$.

Theorem 3 states that N-St-RNN[$\mathbb{Q}$]s are Turing equivalent and that N-Ev$_2$-RNN[$\mathbb{Q}$] are strictly more powerful than nondeterministic Muller Turing machines, since $\Sigma_1^1 \subsetneq \Sigma_1^1$. In this sense, the nondeterministic evolving neural networks are also *super-Turing*.

Remark 2. The nondeterministic counterpart of Remark 1 holds. More precisely, the proof of Theorem 3(b) [3, Theorem 2] shows that any ω-language $L \in \Sigma_1^1$ can be recognized by some N-Ev$_2$-RNN[$\mathbb{Q}$] employing only one bi-valued evolving weight given in the form of a bias. Consequently, from this point onwards, we will without loss of generality focus on the subclass of N-Ev$_2$-RNN[$\mathbb{Q}, \alpha$], for $\alpha \in 2^\omega$.

We now provide a precise characterization of the expressive power of the subclass of N-Ev$_2$-RNN[$\mathbb{Q}, \alpha$], for some given $\alpha \in 2^\omega$. The result is achieved via forthcoming Propositions 3 and 4, which are simple generalizations of Propositions 1 and 2, respectively.

Proposition 3. *Let $L \subseteq (\mathbb{B}^M)^\omega$ be some ω-language. If $L \in \Sigma_1^1(\alpha)$, with $\alpha \in 2^\omega$, then L is recognizable by some $N\text{-}Ev_2\text{-}RNN[\mathbb{Q}, \alpha]$.*

Proof. If $L \in \Sigma_1^1(\alpha) \upharpoonright (\mathbb{B}^M)^\omega$, then by definition, there exists $L' \in \Sigma_1^1 \upharpoonright (\mathbb{B}^{M+1})^\omega$ such that $L = L'_\alpha = \{s \in (\mathbb{B}^M)^\omega : (s, \alpha) \in L'\}$. Theorem 3 ensures that there exists a N-St-RNN[$\mathbb{Q}$] $\mathcal{N}'$ with $M + 1$ input cells such that $L(\mathcal{N}') = L'$. As in the proof of Proposition 1, one can modify network $\mathcal{N}'$ to obtain a N-Ev$_2$-RNN[$\mathbb{Q}, \alpha$] $\mathcal{N}_1$ such that $L(\mathcal{N}_1) = L'_\alpha = L$. $\qquad\square$

Proposition 4. *Let $L \subseteq (\mathbb{B}^M)^\omega$ be some ω-language. If, for some $\alpha \in 2^\omega$, L is recognizable by some N-Ev$_2$-RNN[$\mathbb{Q}, \alpha$], then $L \in \Sigma_1^1(\alpha)$.*

Proof. Let $\mathcal{N}$ be a N-Ev$_2$-RNN[$\mathbb{Q}, \alpha$] such that $L(\mathcal{N}) = L$. By Remark 2, we may assume without loss generality that the bi-valued evolving weight α of $\mathcal{N}$ is a bias. As in the proof of Proposition 2, there exists a N-St-RNN[$\mathbb{Q}$] $\mathcal{N}'$ with $P + 1$ input cells such that $(s, \alpha) \in L(\mathcal{N}')$ if and only if $s \in L(\mathcal{N})$. This means that $L(\mathcal{N}) = L(\mathcal{N}')_\alpha$. In addition, Theorem 3 ensures that $L(\mathcal{N}') \in \Sigma_1^1$. Therefore, $L(\mathcal{N}) \in \Sigma_1^1(\alpha)$. $\qquad\square$

By combining Propositions 3 and 4, the following theorem is obtained:

Theorem 4. *Let $L \subseteq (\mathbb{B}^M)^\omega$ be some ω-language and $\alpha \in 2^\omega$. The following conditions are equivalent:*

(a) $L \in \Sigma_1^1(\alpha)$;
(b) L is recognizable by some N-Ev$_2$-RNN[$\mathbb{Q}, \alpha$].

From Theorem 4 and Remark 2, the following set-theoretical result can be retrieved: $\Sigma_1^1 = \bigcup_{\alpha \in 2^\omega} \Sigma_1^1(\alpha)$. In other words, the relativized classes $\Sigma_1^1(\alpha)$ span the class Σ_1^1, when α varies over 2^ω.

5   The Hierarchy Theorem

Theorems 2 and 4 provide a precise characterization of the expressive power of the classes of D-Ev$_2$-RNN[$\mathbb{Q}, \alpha$] and N-Ev$_2$-RNN[$\mathbb{Q}, \alpha$], for $\alpha \in 2^\omega$. We first present some conditions that the evolving weights satisfy whenever their corresponding relativized classes are included one into the other.

Proposition 5. *Let $\alpha, \beta \in 2^\omega$. The following relations hold:*

$$BC(\Pi_2^0)(\alpha) \subseteq BC(\Pi_2^0)(\beta) \;\longrightarrow\; \alpha \in \Delta_1^1(\beta) \qquad (3)$$
$$\Sigma_1^1(\alpha) \subseteq \Sigma_1^1(\beta) \;\longleftrightarrow\; \alpha \in \Delta_1^1(\beta) \qquad (4)$$

Proof. We prove both left-to-right implications. Recall that $\alpha \in \Sigma_1^0(\alpha)$. In the first case, one has $\alpha \in \Sigma_1^0(\alpha) \subseteq BC(\Pi_2^0)(\alpha) \subseteq BC(\Pi_2^0)(\beta) \subseteq \Delta_1^1(\beta)$. In the second case, $\alpha \in \Sigma_1^0(\alpha) \subseteq \Sigma_1^1(\alpha) \subseteq \Sigma_1^1(\beta)$. Hence, $\alpha \in \Delta_1^1(\beta)$, by [17].

For the converse implication of Relation (4), suppose that $\alpha \in \Delta_1^1(\beta)$. Then $\alpha \in \Sigma_1^1(\beta)$, which means that the ω-language $\{\alpha\}$ is recognized by some non-deterministic Muller TM $\mathcal{M}_1$ with oracle β. Now, let $L \in \Sigma_1^1(\alpha)$. Then L is recognized by a nondeterministic Muller TM $\mathcal{M}_2$ with oracle α. Consider the nondeterministic Muller TM $\mathcal{M}$ with oracle β which works as follows: if x is written on its input tape, then $\mathcal{M}$ nondeterministically writes some $y \in 2^\omega$ bit by bit on one of its work tape, and concomitantly, simulates in parallel the behaviors of $\mathcal{M}_1$ on y as well as that of $\mathcal{M}_2$ with oracle y on x. The TM $\mathcal{M}$ is suitably programmed in order to always have enough bits of y being written on its work tape so that the next simulations steps of $\mathcal{M}_1$ with oracle y can be performed without fail. In addition, the machine $\mathcal{M}$ accepts input x iff both simulation processes of $\mathcal{M}_1$ and $\mathcal{M}_2$ are accepting, i.e., iff $y = \alpha$ and the simulation of $\mathcal{M}_2$ with oracle $y = \alpha$ accepts x, which is to say that $x \in L(\mathcal{M}_2) = L$. Hence, $\mathcal{M}$ recognizes L also, and thus $L \in \Sigma_1^1(\beta)$. This shows that $\Sigma_1^1(\alpha) \subseteq \Sigma_1^1(\beta)$.   □

We now show the existence of an infinite sequence of weights whose corresponding succession of relativized classes properly stratify the "super-Turing" classes of $BC(\Pi_2^0)$ and Σ_1^1 neural ω-languages. The hierarchy induced by the inclusion relation between the relativized classes possesses chains of length ω_1 as well as uncountable antichains.

Proposition 6. *There exists a sequence $(\alpha_i)_{i<\omega_1}$, where $\alpha_i \in 2^\omega$ for all $i < \omega_1$, such that*

(a) $BC(\Pi_2^0)(\alpha_0) = BC(\Pi_2^0)$ and $BC(\Pi_2^0)(\alpha_i) \subsetneq BC(\Pi_2^0)(\alpha_j)$, for all $i < j < \omega_1$;
(b) $\Sigma_1^1(\alpha_0) = \Sigma_1^1$ and $\Sigma_1^1(\alpha_i) \subsetneq \Sigma_1^1(\alpha_j)$, for all $i < j < \omega_1$.

Moreover, there exists some uncountable set $A \subseteq 2^\omega$ such that $BC(\Pi_2^0)(\alpha_i) \not\subseteq BC(\Pi_2^0)(\alpha_j)$ and $\Sigma_1^1(\alpha_i) \not\subseteq \Sigma_1^1(\alpha_j)$, for every distinct $\alpha_i, \alpha_j \in A$.

Proof. Take $\alpha_0 \in \Sigma_1^0$. Suppose that for $\gamma < \omega_1$, the sequence $(\alpha_i)_{i<\gamma}$ has been constructed and satisfies the required property. We build the next element α_γ of that sequence, i.e., the element such that $\Sigma_1^1(\alpha_i) \subsetneq \Sigma_1^1(\alpha_\gamma)$, for all $i < \gamma$. Note that, for each $i < \gamma$, the set $\Delta_1^1(\alpha_i)$ is countable. Since $\gamma < \omega_1$, the union $\bigcup_{i<\gamma} \Delta_1^1(\alpha_i)$ is countable too. Hence, there exists $\alpha \in 2^\omega \setminus \bigcup_{i<\gamma} \Delta_1^1(\alpha_i)$. Now, let $\{\beta_i : i < \omega\}$ be an enumeration of the countable set $\{\alpha\} \cup \{\alpha_i : i < \gamma\}$, and let $\alpha_\gamma \in 2^\omega$ be the encoding of $\{\beta_i : i < \omega\}$ given by $\alpha_\gamma(\langle i, n\rangle) = \beta_i(n)$, where $\langle .,.\rangle : \omega^2 \to \omega$ is a classical recursive pairing function. Each function $f_i : \alpha_\gamma \mapsto (\alpha_\gamma)_i = \beta_i$ is recursive, and therefore, $\beta_i \in \Sigma_1^0(\alpha_\gamma)$, for each $i < \omega$.

We show that $BC(\Pi_2^0)(\alpha_j) \subseteq BC(\Pi_2^0)(\alpha_\gamma)$, for all $j < \gamma$. Let $L \in BC(\Pi_2^0)(\alpha_j) = BC(\Pi_2^0)(\beta_i)$, for some $i < \omega$. This means that L is recognizable by some deterministic Muller TM $\mathcal{M}$ with oracle β_i. Since $\beta_i \in \Sigma_1^0(\alpha_\gamma)$, L is also recognized by the deterministic Muller TM $\mathcal{M}'$ with oracle α_γ which, in a suitable alternating manner, produces β_i bit by bit from α_γ, and works precisely like $\mathcal{M}$ with oracle β_i. Therefore, $L \in BC(\Pi_2^0)(\alpha_\gamma)$. By replacing in this argument every occurrences of "$BC(\Pi_2^0)$" by "Σ_1^1" and of "deterministic" by "nondeterministic", one obtains that $\Sigma_1^1(\alpha_j) \subseteq \Sigma_1^1(\alpha_\gamma)$, for all $j < \gamma$.

We now show that $BC(\Pi_2^0)(\alpha_j) \subsetneq BC(\Pi_2^0)(\alpha_\gamma)$ and $\Sigma_1^1(\alpha_j) \subsetneq \Sigma_1^1(\alpha_\gamma)$, for all $j < \gamma$. Towards a contradiction, suppose that $BC(\Pi_2^0)(\alpha_\gamma) \subseteq BC(\Pi_2^0)(\alpha_j)$ or $\Sigma_1^1(\alpha_\gamma) \subseteq \Sigma_1^1(\alpha_j)$, for some $j < \gamma$. Then Relations (3) and (4) ensure that $\alpha_\gamma \in \Delta_1^1(\alpha_j)$. But $\alpha = \beta_k$ for some $k < \omega$, and by the above stated fact,

$\alpha = \beta_k \in \Sigma_1^0(\alpha_\gamma)$. The two relations $\alpha \in \Sigma_1^0(\alpha_\gamma)$ and $\alpha_\gamma \in \Delta_1^1(\alpha_j)$ imply that $\alpha \in \Delta_1^1(\alpha_j)$. This contradicts the fact that $\alpha \in 2^\omega \setminus \bigcup_{i<\gamma} \Delta_1^1(\alpha_i)$.

We finally prove the existence of an uncountable antichain. There exists an uncountable set $A \subseteq 2^\omega$ such that $\alpha_i \notin \Delta_1^1(\alpha_j)$, for all distinct $\alpha_i, \alpha_j \in A$ [1]. By Relations (3) and (4), $BC(\Pi_2^0)(\alpha_i) \not\subseteq BC(\Pi_2^0)(\alpha_j)$ and $\Sigma_1^1(\alpha_i) \not\subseteq \Sigma_1^1(\alpha_j)$, for all distinct $\alpha_i, \alpha_j \in A$. $\qquad\square$

Let $\mathcal{L}(\text{D-Ev}_2\text{-RNN}[\mathbb{Q}, \alpha])$ and $\mathcal{L}(\text{N-Ev}_2\text{-RNN}[\mathbb{Q}, \alpha])$ denote the classes of neural ω-languages recognized by D-Ev$_2$-RNN$[\mathbb{Q}, \alpha]$ and N-Ev$_2$-RNN$[\mathbb{Q}, \alpha]$, respectively. Theorems 2 and 4 together with Proposition 6 imply the existence of two proper hierarchies of classes of deterministic and nondeterministic evolving neural networks of increasing expressive power.

Theorem 5. *There exists a sequence of binary evolving weights $(\alpha_i)_{i<\omega_1}$ such that*

(a) $\mathcal{L}(\text{D-Ev}_2\text{-RNN}[\mathbb{Q}, \alpha_i]) \subsetneq \mathcal{L}(\text{D-Ev}_2\text{-RNN}[\mathbb{Q}, \alpha_j])$, for all $i < j < \omega_1$;
(b) $\mathcal{L}(\text{N-Ev}_2\text{-RNN}[\mathbb{Q}, \alpha_i]) \subsetneq \mathcal{L}(\text{N-Ev}_2\text{-RNN}[\mathbb{Q}, \alpha_j])$, for all $i < j < \omega_1$.

Finally, let R be the equivalence relation defined by $R(\alpha, \beta)$ iff $\mathcal{L}(\text{N-Ev}_2\text{-RNN}[\mathbb{Q}, \alpha]) = \mathcal{L}(\text{N-Ev}_2\text{-RNN}[\mathbb{Q}, \beta])$. This relation represents the decision problem of whether two classes of nondeterministic evolving networks (determined by the evolving weights α and β) have the same expressive power. We show that this relation is undecidable and of complexity of $\Pi_1^1 \setminus \Sigma_1^1$.

Proposition 7. *The equivalence relation R is in the class $\Pi_1^1 \setminus \Sigma_1^1$.*

Proof. According to Theorem 4 and Relation (4), the relation $R \subseteq 2^\omega \times 2^\omega$ satisfies $R(\alpha_1, \alpha_2)$ iff $\alpha_1 \in \Delta_1^1(\alpha_2)$ and $\alpha_2 \in \Delta_1^1(\alpha_1)$. It is known that the relation "$\alpha \in \Delta_1^1(\beta)$" is a Π_1^1 relation which can be expressed by a Π_1^1-formula $\phi(\alpha, \beta)$, see [17, 4D.14 p. 171] and [11]. Thus R is a Π_1^1-relation. Towards a contradiction, assume now that R is Σ_1^1, and take $\beta \in \Sigma_1^0$. Then $R(., \beta) = \{\alpha : R(\alpha, \beta)\} = \{\alpha : \alpha \in \Delta_1^1(\beta) \,\&\, \beta \in \Delta_1^1(\alpha)\} = \{\alpha : \alpha \in \Delta_1^1(\beta)\} = \{\alpha : \alpha \in \Delta_1^1\}$ should also be in Σ_1^1. But it is known that the set $\{\alpha : \alpha \in \Delta_1^1\}$ is not Σ_1^1, see [17, 4D.16 p. 171]. This concludes the proof. $\qquad\square$

6 Conclusion

The expressive power of evolving neural networks working on infinite input streams has been finely characterized in terms of relativized topological classes. As a consequence, a proper hierarchy of classes of evolving neural nets, based on the complexity of their underlying evolving weights, has been obtained. The hierarchy contains chains of length ω_1 as well as uncountable antichains.

These results (together with [3,9]) show that evolving and analog neural networks represent a natural model for oracle-based ω-computation. For future work, a similar refined characterization of the expressive power of analog neural networks is expected to be studied. In fact, we prove in an extended version of this paper that if $r_\alpha \in \mathbb{R}$ is some recursive encoding of $\alpha \in 2^\omega$, then the analog networks employing $r_\alpha \in \mathbb{R}$ as sole real weight are computationally equivalent to the evolving networks employing α as sole evolving weight.

References

1. Apt, K.R.: ω-models in analytical hierarchy. Bulletin de l'académie polonaise des sciences **XX**(11), 901–904 (1972)
2. Balcázar, J.L., Gavaldà, R., Siegelmann, H.T.: Computational power of neural networks: a characterization in terms of Kolmogorov complexity. IEEE Trans. Inf. Theory **43**(4), 1175–1183 (1997)
3. Cabessa, J., Duparc, J.: Expressive power of nondeterministic recurrent neural networks in terms of their attractor dynamics. IJUC **12**(1), 25–50 (2016)
4. Cabessa, J., Siegelmann, H.T.: Evolving recurrent neural networks are super-Turing. In: Proceedings of IJCNN 2011, pp. 3200–3206. IEEE (2011)
5. Cabessa, J., Siegelmann, H.T.: The computational power of interactive recurrent neural networks. Neural Comput. **24**(4), 996–1019 (2012)
6. Cabessa, J., Siegelmann, H.T.: The super-turing computational power of plastic recurrent neural networks. Int. J. Neural Syst. **24**(8), 1450029 (2014)
7. Cabessa, J., Villa, A.E.P.: The expressive power of analog recurrent neural networks on infinite input streams. Theor. Comput. Sci. **436**, 23–34 (2012)
8. Cabessa, J., Villa, A.E.P.: An attractor-based complexity measurement for Boolean recurrent neural networks. PLoS ONE **9**(4), e94204+ (2014)
9. Cabessa, J., Villa, A.E.P.: Expressive power of first-order recurrent neural networks determined by their attractor dynamics. J. Comput. Syst. Sci. **82**(8), 1232–1250 (2016)
10. Cabessa, J., Villa, A.E.P.: Recurrent neural networks and super-turing interactive computation. In: Koprinkova-Hristova, P., Mladenov, V., Kasabov, N.K. (eds.) Artificial Neural Networks. SSB, vol. 4, pp. 1–29. Springer, Cham (2015). doi:10.1007/978-3-319-09903-3_1
11. Finkel, O.: Ambiguity of omega-languages of turing machines. Log. Methods Comput. Sci. **10**(3), 1–18 (2014)
12. Kechris, A.S.: Classical Descriptive Set Theory. Graduate Texts in Mathematics, vol. 156. Springer, New York (1995)
13. Kilian, J., Siegelmann, H.T.: The dynamic universality of sigmoidal neural networks. Inf. Comput. **128**(1), 48–56 (1996)
14. Kleene, S.C.: Representation of events in nerve nets and finite automata. In: Shannon, C., McCarthy, J. (eds.) Automata Studies, pp. 3–41. Princeton University Press, Princeton (1956)
15. McCulloch, W.S., Pitts, W.: A logical calculus of the ideas immanent in nervous activity. Bull. Math. Biophys. **5**, 115–133 (1943)
16. Minsky, M.L.: Computation: Finite and Infinite Machines. Prentice-Hall Inc., Englewood Cliffs (1967)
17. Moschovakis, Y.N.: Descriptive Set Theory. Mathematical Surveys and Monographs, 2nd edn. American Mathematical Society, Providence (2009)
18. Siegelmann, H.T.: Recurrent neural networks and finite automata. Comput. Intell. **12**, 567–574 (1996)
19. Siegelmann, H.T., Sontag, E.D.: Analog computation via neural networks. Theor. Comput. Sci. **131**(2), 331–360 (1994)
20. Siegelmann, H.T., Sontag, E.D.: On the computational power of neural nets. J. Comput. Syst. Sci. **50**(1), 132–150 (1995)
21. Síma, J., Orponen, P.: General-purpose computation with neural networks: a survey of complexity theoretic results. Neural Comput. **15**(12), 2727–2778 (2003)

22. Staiger, L.: ω-languages. In: Rozenberg, G., Salomaa, A. (eds.) Handbook of Formal Languages: Beyond Words, vol. 3, pp. 339–387. Springer, New York (1997)
23. Turing, A.M.: Intelligent machinery. Technical report, National Physical Laboratory, Teddington, UK (1948)

Minimal Absent Words in a Sliding Window and Applications to On-Line Pattern Matching

Maxime Crochemore[1,2], Alice Héliou[3(✉)], Gregory Kucherov[2],
Laurent Mouchard[4], Solon P. Pissis[1], and Yann Ramusat[5]

[1] Department of Informatics, King's College London, London, UK
{maxime.crochemore,solon.pissis}@kcl.ac.uk
[2] CNRS & Université Paris-Est, Champs-sur-Marne, France
gregory.kucherov@univ-mlv.fr
[3] LIX, École Polytechnique, CNRS, INRIA, Université Paris-Saclay,
Palaiseau, France
alice.heliou@polytechnique.org
[4] University of Rouen, LITIS EA 4108, TIBS, Rouen, France
laurent.mouchard@univ-rouen.fr
[5] DI ENS, ENS, CNRS, PSL Research University & INRIA, Paris, France
yann.ramusat@ens.fr

Abstract. An *absent* (or forbidden) word of a word y is a word that does not occur in y. It is then called *minimal* if all its proper factors occur in y. There exist linear-time and linear-space algorithms for computing all minimal absent words of y (Crochemore et al. in Inf Process Lett 67:111–117, 1998; Belazzougui et al. in ESA 8125:133–144, 2013; Barton et al. in BMC Bioinform 15:388, 2014). Minimal absent words are used for data compression (Crochemore et al. in Proc IEEE 88:1756–1768, 2000, Ota and Morita in Theoret Comput Sci 526:108–119, 2014) and for alignment-free sequence comparison by utilizing a metric based on minimal absent words (Chairungsee and Crochemore in Theoret Comput Sci 450:109–116, 2012). They are also used in molecular biology; for instance, three minimal absent words of the human genome were found to play a functional role in a coding region in Ebola virus genomes (Silva et al. in Bioinformatics 31:2421–2425, 2015). In this article we introduce a new application of minimal absent words for on-line pattern matching. Specifically, we present an algorithm that, given a pattern x and a text y, computes the distance between x and every window of size $|x|$ on y. The running time is $\mathcal{O}(\sigma|y|)$, where σ is the size of the alphabet. Along the way, we show an $\mathcal{O}(\sigma|y|)$-time and $\mathcal{O}(\sigma|x|)$-space algorithm to compute the minimal absent words of every window of size $|x|$ on y, together with some new combinatorial insight on minimal absent words.

1 Introduction

Pattern matching is the problem of finding a *pattern* in a usually much longer *text*. Both pattern and text are words (or strings) drawn over some alphabet. This problem has been studied for a long time and efficient solutions have been proposed (see for example [1,13,20,22] or also [9,16]). A related problem is the

© Springer-Verlag GmbH Germany 2017
R. Klasing and M. Zeitoun (Eds.): FCT 2017, LNCS 10472, pp. 164–176, 2017.
DOI: 10.1007/978-3-662-55751-8_14

approximate pattern matching problem: it is the same problem but allowing some *errors* in the matching process (see [9,16,27]). This problem depends mainly on how errors are interpreted and thus which metric is used for the comparison. Pattern matching algorithms are classified into on-line and off-line. With *off-line* algorithms the text can be processed before searching; a survey of such algorithms was written by Navarro et al. [26]. A more recent algorithm based on a bidirectional index has been proposed by Kucherov et al. [21]. With *on-line* algorithms the text cannot be processed before searching. A famous such algorithm is *bitap*, one of the underlying algorithms of Unix utility *agrep*; it was first invented by Dömölki [12] and it underwent several improvements among them the last one was done by Myers [24]. A survey on on-line algorithms for approximate pattern matching was written by Navarro [25] (see also [27]).

In this article we propose a new on-line pattern matching scheme using a metric that is based on minimal absent words. This notion of negative information has first been coined as minimal forbidden words by Béal et al. [5]. A *minimal absent word* of word y is a word absent from y whose all proper factors occur in y. A tight upper bound on the number of minimal absent words of a word y of length n over an alphabet of size σ is known to be $\mathcal{O}(\sigma n)$ [10,23]. Moreover it was shown that the set of all minimal absent words of y is sufficient to uniquely reconstruct y [10,14]. The notion has been used in data compression [11,29] and in molecular biology [2,8,17–19,32,34], where authors often focus on the computation of the shortest absent words (sometimes called *unwords*).

Chairungsee and Crochemore introduced the Length Weighted Index (LWI), a metric based on the symmetric difference of minimal absent words sets [7]. The LWI was then applied by Crochemore et al. [8] to devise an $\mathcal{O}(m+n)$-time and $\mathcal{O}(m+n)$-space algorithm for alignment-free comparison of two sequences of length m and n on a constant-sized alphabet. More recently, different such indices have been studied for sequence comparison and phylogeny reconstruction [30]. We base our new pattern matching algorithm on this LWI. To maintain the LWI across the word y for a pattern x, we need to compute the set of minimal absent words in a sliding window of size $m = |x|$ of y. Several linear-time and linear-space algorithms have been proposed to compute the set of minimal absent words [3,4,6,10,15]. Ota et al. presented an on-line algorithm that requires linear time and linear space [28]. However, to the best of our knowledge, the problem of computing minimal absent words in a sliding window has not been addressed.

Our Contributions. Here we present the *first* algorithm to compute minimal absent words in a sliding window. For a window of size m and a word of length n on an alphabet of size σ, our algorithm performs $\mathcal{O}(\sigma n)$ insert and delete operations on the set of minimal absent words. With a careful implementation of the data structures, it requires $\mathcal{O}(\sigma n)$ time overall using $\mathcal{O}(\sigma m)$ space. We apply this algorithm for on-line approximate pattern matching using the LWI for a pattern of length m over every window of size m of the text. This yields the *first* algorithm for the classical on-line exact pattern matching problem that uses some form of negative information (minimal absent words) for the comparison.

Definitions and Notation

Let $y = y[0]y[1] \cdots y[n-1]$ be a *word* of length $n = |y|$ on a finite ordered *alphabet* of size $\sigma = |\Sigma|$. We denote by $y[i \mathinner{.\,.} j] = y[i] \cdots y[j]$ the *factor* of y whose occurrence *starts* at position i and *ends* at position j on y, and by ε the *empty word*, the word of length 0. The set of all possible words on Σ (including the empty word) is denoted by Σ^*. A *prefix* of y is a factor that starts at position 0 ($y[0 \mathinner{.\,.} j]$) and a *suffix* is a factor that ends at position $n-1$ ($y[i \mathinner{.\,.} n-1]$). A factor x of y is *proper* if $x \neq y$.

Let u be a non-empty word. An integer p such that $0 < p \leq |u|$ is called a *period* of u if $u[i] = u[i+p]$, for $i = 0, 1, \ldots, |u| - p - 1$. For every word u and every natural number k, we define the kth *power* of the word u, denoted by u^k, by $u^0 = \varepsilon$ and $u^k = u^{k-1}u$, for $k = 1, 2, \ldots, n$.

Let x be a word of length $m \leq n$. We say that there exists an *occurrence* of x in y when x is a factor of y. Opposingly, we say that the word x is an *absent* word of y if it does not occur in y. We consider absent words of length at least 2 only. An absent word x of length m, $m \geq 2$, of y is *minimal* if and only if all its proper factors occur in y. This is equivalent to saying that a minimal absent word (MAW) of y is of the form aub, $a, b \in \Sigma, u \in \Sigma^*$, such that au and ub are factors of y but aub is not. We can easily see that, if x is a MAW of y, then $2 \leq |x| \leq |y| + 1$. Note that $|x| = |y| + 1$ if and only if $y = a^{|y|}$ for some $a \in \Sigma$.

Example 1. Let $y = $ ABAACA. Its factors of lengths 1 and 2 are A, B, C, AA, AB, AC, BA, and CA. The set of MAWs of y is obtained by combining the aforementioned factors: $\{$BB, BC, CB, CC, AAA, AAB, BAB, BAC, CAA, CAB, CAC$\}$.

Let U and V be two sets. We denote by $U \triangle V$ their symmetric difference, that is, $U \triangle V = (U \setminus V) \cup (V \setminus U)$. We consider the LWI, a distance on Σ^*, for two words x and y on Σ^* [7]. It is based on the set $M(x) \triangle M(y)$, where $M(x)$ is the set of minimal absent words of x, and it is defined by:

$$\mathsf{LWI}(x, y) = \sum_{w \in M(x) \triangle M(y)} \frac{1}{|w|^2}.$$

2   Combinatorial Results

In this section we consider a word z of fixed length m on an alphabet Σ of size σ and denote by $M(z)$ its set of MAWs. The word z essentially represents the content of the window on word y used in the algorithm of Sect. 3. We first discuss changes to be done on the set of MAWs when appending and removing letters on the word of interest. Then we show bounds on the number of changes on the set of MAWs when moving forward the current window by one position.

2.1   Changes When Appending One Letter to the Window

We denote by $M(z)|\alpha$, $\alpha \in \Sigma$, the operation on the set of MAWs when concatenating the letter α to the, possibly empty, word z. The operation creates $M(z\alpha)$

from $M(z)$. We introduce some bounds on the number of insertions/deletions for the on-line computation of the set of MAWs. These results have already been shown in [28] and we briefly present them for completeness.

We denote by s the starting position of the longest suffix of z that repeats in z; when this suffix is empty we set $s = |z|$. We also denote by s_α the starting position of the longest suffix that occurs in z followed by α; when this suffix is empty we set $s_\alpha = |z|$. Note that we have $s \le s_\alpha$ because the latter suffix obviously repeats in z. This is illustrated in Fig. 1.

The next two lemmas state bounds of the number of insert and delete operations performed by $M(z)|\alpha$.

Lemma 1. $M(z)|\alpha$ deletes exactly one MAW from $M(z)$, namely, $z[s_\alpha - 1 .. |z| - 1]\alpha$.

Proof. Let $w = aub$, $a, b \in \Sigma$ and $u \in \Sigma^*$, be a MAW to be removed. This means that aub is absent in z but present in $z\alpha$. Thus $b = \alpha$ and au is a suffix of z that does not occur followed by α in z. The word $ub = u\alpha$ is also present in z, so u is a suffix of z that occurs in z followed by α. Then the starting position of the suffix occurrence of u in z is s_α and $w = z[s_\alpha - 1 .. |z| - 1]\alpha$.   □

To establish an upper bound on the number of MAWs added by the operation $M(z)|\alpha$, we first divide the new MAWs of the form aub, $a, b \in \Sigma$ and $u \in \Sigma^*$, into three types (see also Fig. 1):

1. au and ub are absent in z.
2. au is absent in z and ub is present in z.
3. au is present in z and ub is absent in z.

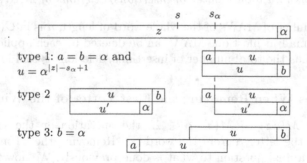

Fig. 1. Illustration of the three different types of MAWs that are added when letter α is appended to z.

Lemma 2. There are at most one MAW of type 1, σ MAWs of type 2, and $(s_\alpha - s)(\sigma - 1)$ MAWs of type 3 added by the operation $M(z)|\alpha$.

Proof. We consider a new MAW $w = aub$, $a, b \in \Sigma$ and $u \in \Sigma^*$, created by the operation. Let w be of type 1, that is, au and ub do not occur in z. Then they are both suffixes of $z\alpha$, and because they have same length, are equal. This implies that u is both a prefix and a suffix of $ub = u\alpha$. Thus the latter has period 1, w is of the form $\alpha^{|w|}$, and $u = \alpha^{|w|-2}$. But then $u\alpha$ is absent in z. Therefore, $\alpha^{|w|-3}$ is the longest repeated suffix of z that occurs followed by α in z. Consequently $|w| = |z| - s_\alpha + 3$.

Let w be of type 2, that is, ub occurs in z and au occurs in $z\alpha$ but not in z. Then au is a suffix of $z\alpha$ and u can be written $u'\alpha$. As ub occurs in z, u' is a suffix of z that occurs in z followed by α. Moreover, since $au = au'\alpha$ does not occur in z, u' is the longest suffix of z that occurs in z followed by α, therefore its starting position as a suffix is s_α. The letter b can be any letter of the alphabet of z that occurs after an occurrence of u in z. Consequently there are at most σ such MAWs.

Let w be of type 3, that is, au occurs in z and ub occurs in $z\alpha$ but not in z. This implies that $b = \alpha$, u is a suffix of z not preceded by a, and au occurs elsewhere in z. Since no occurrence of u in z is followed by α, we have that the starting position k of u as a suffix satisfies $s \le k < s_\alpha$. Therefore, there are at most $s_\alpha - s$ possible words u and for each of them, there are at most $\sigma - 1$ possibilities for the letter a to obtain a MAW. Consequently, there are at most $(s_\alpha - s)(\sigma - 1)$ such MAWs. $\square$

The previous lemma shows that during one step of the computation of MAWs for a sliding window of size m we may have to handle $\mathcal{O}(\sigma m)$ new MAWs. However, the total number of insertions when computing the set of MAWs for a word y of length n get amortized to $\mathcal{O}(\sigma n)$ in an on-line computation.

Proposition 1 [28]. *Starting with the empty word, and applying n times the operation $|$ leads to a total number of insertions/deletions of MAWs in $\mathcal{O}(\sigma n)$.*

Proof. The number of MAWs of the whole word of length n is in $\mathcal{O}(\sigma n)$ [10]. As stated by Lemma 1 at most one MAW can be deleted by each application of the operation $|$. Thus the total number of insertions/deletions is still in $\mathcal{O}(\sigma n)$. $\square$

2.2   Changes When Removing the First Letter of the Window

We denote by $M(\alpha z) \rightarrow M(z)$, $\alpha \in \Sigma$, the operation on the set of MAWs when deleting the letter α from the word αz. Removing the leftmost letter of the window is a dual question to what is done previously. We now focus on the longest repeated prefix instead of the longest repeated suffix.

Let us denote by p the ending position of the longest repeated prefix of z and by p_α the ending position of the longest prefix of z that occurs in z preceded by α. We set them to 0 when the prefixes are empty. Note that $p_\alpha \le p$. Similar to Lemma 1, removing a letter from the left creates exactly one MAW.

Lemma 3. *The operation $M(\alpha z) \rightarrow M(z)$ creates exactly one MAW, which is $\alpha z[0 \mathrel{..} p_\alpha + 1]$.*

Similar to Sect. 2.1, we distinguish among three types of MAWs to be deleted by the operation:

1. au and ub are absent in z.
2. au is absent in z and ub present in z.
3. ub is absent in z and au present in z.

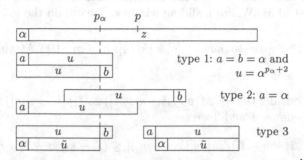

Fig. 2. Illustration of the three different types of MAWs that are deleted when removing α, the letter before z.

We note that types 1, 2, and 3 behave respectively similarly to type 1, 3, and 2 in Sect. 2.1; see Fig. 2 for an illustration. The following result is similar to that stated in Lemma 2.

Lemma 4. *There are at most one MAW of type 1, $(\sigma - 1)(p - p_\alpha)$ MAWs of type 2, and σ MAWs of type 3 to be deleted by the operation $M(\alpha z) \rightarrow M(z)$.*

2.3 Changes When Sliding a Window over a Text

We now focus on our main problem: MAWs in a sliding window. For $m < n$ and for all i, $0 \le i \le n - m$, we consider the window $y[i \mathinner{.\,.} i + m - 1]$ and define:

- s_i the starting position of its longest repeated suffix,
- $\tilde{s}_i$ the starting position of its longest suffix that occurs followed by $y[i + m]$,
- ss_i the starting position of its longest suffix that is a power,
- p_i the ending position of its longest repeated prefix,
- $\tilde{p}_i$ the ending position of its longest prefix that occurs preceded by $y[i - 1]$,
- pp_i the ending position of its longest prefix that is a power.

In what follows, we make use of this notation considering the case of a sliding window. The following lemma shows that we cannot output in linear time the set of MAWs in the sliding window at each step of the process.

Lemma 5. *The upper bound of $\sum_{i=0}^{n-m} |M(y[i \mathinner{.\,.} i + m - 1])|$ is $\mathcal{O}(\sigma n m)$ and this bound is tight.*

Proof. For every factor z of length m of y, $|M(z)|$ is $\mathcal{O}(\sigma m)$. Thus the upper bound of their sum is $\mathcal{O}(\sigma nm)$. Now consider $y = (\mathsf{A}^{m-1}\mathsf{C}^{m-1})^{\frac{n}{2m-2}}$ of length n and its factors of length $2m$. In each factor w of length $2m$, this kind of pattern occurs: $XY^{m-1}X$, with $\{X, Y\} = \{\mathsf{A}, \mathsf{C}\}$. Thus $\{XY^iX | 1 \leq i \leq m-1\} \subseteq M(w)$, so $|M(w)| \geq m-1$. Consequently the bound is tight. One can generalize this construction of y to obtain a tight bound for larger alphabets (Lemma 1 in [2]). $\square$

However, as shown below, we can bound the number of changes necessary to maintain the set of MAWs for a sliding window. We obtain the following result.

Theorem 1. *The upper bound of* $\sum_{i=0}^{n-m-1} |M(y[i..i+m-1]) \triangle M(y[i+1..i+m])|$ *is in* $\mathcal{O}(\sigma n)$.

Proof. Let us consider the set $M(y[i..i+m-1]) \triangle M(y[i..i+m])$ with $0 \leq i < n-m$. From Lemmas 1 and 2 we get

$$|M(y[i..i+m-1]) \triangle M(y[i..i+m])| \leq (\tilde{s}_i - s_i)(\sigma - 1) + \sigma + 2.$$

Then,

$$\sum_{i=0}^{n-m-1} |M(y[i..i+m-1]) \triangle M(y[i..i+m])| \leq \sum_{i=0}^{n-m-1} (\tilde{s}_i - s_i)(\sigma - 1) + n\sigma + 2n.$$

We note that $\tilde{s}_i \leq s_{i+1} \leq \tilde{s}_i + 1$ and we have $s_i \leq \tilde{s}_i$ thus

$$0 \leq \sum_{i=0}^{n-m-1} (\tilde{s}_i - s_i) = \sum_{i=0}^{n-m-1} \tilde{s}_i - \sum_{i=0}^{n-m-1} s_i$$

$$0 \leq \sum_{i=0}^{n-m-1} (\tilde{s}_i - s_i) = \tilde{s}_{n-m-1} - s_0 + \sum_{i=0}^{n-m-2} (\tilde{s}_i - s_{i+1}) \leq n$$

Then $\sum_{i=0}^{n-m-1} |M(y[i..i+m-1]) \triangle M(y[i..i+m])| \leq 2n\sigma + n$. Now, we consider the set $M(y[i..i+m]) \triangle M(y[i+1..i+m])$. From Lemmas 3 and 4 we obtain a similar inequality: $\sum_{i=0}^{n-m-1} |M(y[i..i+m]) \triangle M(y[i+1..i+m])| \leq 2n\sigma + n$. Thus we obtain the desired bound by the triangle inequality. $\square$

3   Minimal Absent Words in a Sliding Window

For a general introduction to suffix trees, see [9]. The *suffix tree* T of a non-empty word w of length n is a compact trie representing all suffixes of w. The nodes of the trie which become nodes of the suffix tree (i.e., branching nodes and leaves) are called *explicit* nodes, while the other nodes are called *implicit*. We use $L(v)$ to denote the path-label of a node v, i.e., the concatenation of the edge labels along the path from the root to v. Node v is a *terminal* node if and only if $L(v) = w[i..n-1]$, $0 \leq i < n$; here v is also labelled with index i. The *suffix link* of a node v with path-label $L(v) = \alpha s$ is a pointer to the node path-labelled s, where $\alpha \in \Sigma$ is a single letter and s is a word. The suffix link of v exists if v is a non-root internal node of T. Our algorithm relies on Senft's on-line construction algorithm of the suffix tree for a sliding window [31] that is itself based on Ukkonen's on-line construction algorithm of the suffix tree [33].

3.1 An Overview of Senft's Algorithm

The algorithm of Ukkonen constructs the suffix tree on-line in $\mathcal{O}(n)$ time for a word of length n on a constant-sized alphabet by processing the word from left to right. To adapt it for a sliding window with amortized constant time per one window shift, two additional problems need to be resolved: (i) deleting the leftmost letter of a window; and (ii) maintaining edge labels under window shifts.

Deleting the Leftmost Letter. Consider the longest repeated prefix of the current window. When the leftmost letter is deleted, all prefixes that are longer than this prefix need to be removed from the tree but the longest repeated prefix and all shorter prefixes will remain in the tree. To remove these prefixes we delete the leaf corresponding to the whole window and its incoming edge as follows:

- If the longest repeated prefix corresponds to an explicit node, this node is the parent of the leaf to be deleted. If this node has only one child remaining, we delete the node and merge the two edges. Otherwise, we do nothing.
- If the longest repeated prefix corresponds to an implicit node, it is equal to the longest repeated suffix. We create a new leaf in the place of the one we have deleted. We label it with the starting position of what was the longest repeated suffix and its incoming edge is labelled accordingly.

Maintaining Edge Labels. Assume by induction that all edge labels are correctly positioned relative to the current window. For the next m shifts of the window, we still maintain the same relative positioning of edge labels. After the m shifts, edge labels are recomputed by a bottom-up traversal of the tree. Since m shifts create at most $2m$ nodes, the amortized time spent on one shift is $O(1)$.

3.2 Our Algorithm

Consider a word y of length n on an alphabet Σ of size σ. Our goal is to maintain the set of MAWs for a sliding window of size m. That is, for all successive $i \in [0, \ldots, n - m]$, we want to compute $M_m(i) = M(y[i \mathinner{.\,.} i + m - 1])$.

For a word z, by $\Sigma(z)$ we denote the alphabet of z and by $V(z)$ the set of explicit nodes in the suffix tree of z. Consider a mapping $f : M(z) \to \Sigma(z) \times V(z)$ defined by $f(aub) = (a, v_{ub})$, where $a \in \Sigma$ and v_{ub} is either the explicit node corresponding to the factor ub or the immediate explicit descendant node if this node is implicit.

Lemma 6. *Mapping f is an injection.*

Proof. Let $w, w' \in M(z)$, $w \neq w'$, $w = aub$ and $w' = a'u'b'$, with $a, b, a', b' \in \Sigma(z)$ and $u, u' \in \Sigma(z)^*$.

Suppose that $f(w) = f(w')$, then $a = a'$ and $v_{ub} = v_{u'b'}$. Thus ub and $u'b'$ are distinct prefixes of the factor corresponding to v_{ub}, consequently one is prefix of the other, without loss of generality ub is prefix of $u'b'$. Then aub is a prefix of $au'b'$, this is impossible as they are both MAWs of z. Thus two distinct elements of $M(z)$ cannot share the same image by f, so f is an injection.   □

Lemma 6 allows us to represent all MAWs by storing a set of letters in each explicit node of the tree. We will call this set the *maw*-set. Moreover, a letter a in the *maw*-set will be *tagged* if and only if u corresponds to an implicit node in the tree. Observe that a can become tagged only when u is a repeated suffix of y. This is because factors au and ub define distinct occurrences of u, and the occurrence of au must be a suffix, otherwise u would be followed by two distinct letters and would then be an explicit node. Besides *maw*-sets, we will also need to store at each explicit node another set of letters: the set of all letters preceding the occurrences of the factor corresponding to the node.

By induction, assume we are at position i, the suffix tree $T_m(i)$ for $y[i .. i + m - 1]$ is built and the set of MAWs $M_m(i)$ has been computed. We now explain how to update $T_m(i)$ and $M_m(i)$ to obtain $T_m(i + 1)$ and $M_m(i + 1)$. The tree is updated based on Senft's algorithm, by first adding a letter to the right of the current window and then deleting the leftmost letter. The set of MAWs is updated using Lemmas 1, 2 and 3, 4 respectively. The algorithm will maintain positions $s_i, p_i, \tilde{s}_i, \tilde{p}_i, ss_i, pp_i$ as defined in Sect. 2.3. We store the leaf nodes in a list so that the last created leaf and the "oldest" leaf currently in the tree can be accessed in constant time.

Adding a Letter to the Right. We follow Ukkonen's algorithm for updating the suffix tree. Recall that Ukkonen's algorithm proceeds by updating the *active node* in the tree. At the beginning of each iteration, the active node corresponds to the longest repeated suffix, i.e. to factor $y[s_i .. i + m - 1]$. The node corresponding to the longest repeated prefix is called the *head node*.

The algorithm starts from the active node and updates it following the suffix links until reaching a node with an outgoing edge starting with $y[i + m]$ – this node corresponds to the suffix starting at $\tilde{s}_i$. At the same time, we compute MAWs of type 3 that are created. For each $s_i \leq j < \tilde{s}_i$, we perform the following.

- If the active node is implicit we make it explicit. We set its set of preceding letters equal to its child's set. We move the untagged letters of the *maw*-set of its child to the *maw*-set of the active node. We untag the tagged letters of the *maw*-set of its child. If the last node created at this window shift does not have a suffix link, we add a suffix link from this node to the active node. We add the letter corresponding to this suffix link to the set of preceding letters of the active node.
- We create a leaf labelled j, with $y[j - 1]$ in its set of preceding letters. We create an edge from the active node to this leaf with the label $y[i + m]$.

- For each letter $a \neq y[j-1]$ in the set of preceding letters of the active node, $ay[s_i + j \mathinner{\ldotp\ldotp} i + m] \in M_{m+1}(i) \backslash M_m(i)$ (type 3 in Lemma 2), therefore we add a in the maw-set of the leaf.

The current active node corresponds to the factor $y[\tilde{s}_i \mathinner{\ldotp\ldotp} i + m - 1]$. According to Lemma 1, there is exactly one MAW to be deleted which is $y[\tilde{s}_i - 1 \mathinner{\ldotp\ldotp} i + m]$. This MAW is stored in the child of the active node by following the edge starting with $y[i+m]$; we remove $y[\tilde{s}_i - 1]$ (tagged or not) from its maw-set.

Then we update the active node by following the edge starting with $y[i+m]$; now it corresponds to the factor $y[\tilde{s}_i \mathinner{\ldotp\ldotp} i + m]$. If the head node was also corresponding to the factor $y[\tilde{s}_i \mathinner{\ldotp\ldotp} i + m - 1]$, we move it down with the active node; we have $\tilde{p}_{i+1} = p_i + 1$, otherwise we have $\tilde{p}_{i+1} = p_i$. If the active node is explicit, we update its set of preceding letters by adding $y[\tilde{s}_i - 1]$.

Then, for each letter b occurring after an occurrence of $y[\tilde{s}_i \mathinner{\ldotp\ldotp} i+m]$ in $y[i \mathinner{\ldotp\ldotp} i+m-1]$, $y[\tilde{s}_i - 1 \mathinner{\ldotp\ldotp} i+m]b \in M_{m+1}(i) \backslash M_m(i)$ (type 2 in Lemma 2). These MAWs are stored in their corresponding child of the active node. If the active node is implicit, there is only one of them and we tag the letter.

By Lemma 2, if $ss_i = \tilde{s}_i - 1$, then $y[i+m]y[\tilde{s}_i - 1 \mathinner{\ldotp\ldotp} i+m]$ is the new MAW of type 1. We store it in the maw-set of the child of the active node by following the edge starting with $y[i+m]$.

Deleting the Leftmost Letter. We note that the longest repeated prefix of $y[i \mathinner{\ldotp\ldotp} i+m]$ is $y[i \mathinner{\ldotp\ldotp} \tilde{p}_{i+1}]$, and its longest repeated suffix is $y[\tilde{s}_i \mathinner{\ldotp\ldotp} i+m]$. At the beginning of this step they correspond respectively to the head node and the active node. Consider the parent of the oldest leaf of the tree, similarly as in Senft's algorithm two cases are distinguished.

- If the head node is an explicit node, then it is the parent of the oldest leaf. We remove the leaf and its incoming edge. If the head node has only one remaining child, we delete the node and merge the two edges; the maw-set associated to the node is added to the leaf.
- Otherwise, the head node is on the edge leading to the oldest leaf. We replace the leaf with a new one labelled by $\tilde{s}_i$, with $y[\tilde{s}_i - 1]$ as the only preceding letter, and the edge is relabelled by $y[\tilde{s}_i - 1]$. We add $y[\tilde{s}_i - 1]$ to the set of preceding letters of the parent of the leaf.

The MAWs associated to the leaf we have deleted were those of type 3 (Lemma 4). We now update the tree and compute the other MAWs to remove and add.

We visit the oldest leaf in the tree and empty its set of preceding letters. Then we move up in the tree following back the edges until we have covered $\tilde{p}_{i+1} - i$ letters. We move the head node to this node: it corresponds to the factor $y[i+1 \mathinner{\ldotp\ldotp} \tilde{p}_{i+1}]$. If the active node was equal to the head node, we move the active node to this node; we have $s_{i+1} = \tilde{s}_i - 1$, otherwise we have $s_{i+1} = \tilde{s}_i$. Each of the explicit nodes visited on the path from the oldest leaf to the head node corresponds to a factor $y[i+1 \mathinner{\ldotp\ldotp} j]$, with $p_{i+1} \geq j > \tilde{p}_{i+1}$. For each of them, we remove $y[i]$ from their set of preceding letters. For each of their children, we remove letter $y[i]$ (tagged or not) from their *maw*-set (type 2 Lemma 4).

There is at most one MAW of type 1 that has to be deleted (Lemma 4). It exists if and only if $y[i] = y[i+1]$ and $pp_{i+1} = \tilde{p}_{i+1} + 1$, in which case we remove it from the *maw*-set of the child of the head node by following the edge starting with $y[i]$. According to Lemma 3, removing the leftmost letter creates one MAW, which is $y[i]y[i+1 \mathinner{.\,.} \tilde{p}_{i+1} + 1]$, thus we add $y[i]$ to the *maw*-set of the child of the head node by following the edge starting with $y[\tilde{p}_{i+1} + 1]$. If the head node is implicit and thus equal to the active node we tag the letter $y[i]$.

Finally if the head node is above the parent of the oldest leaf of the tree, we move it down to this node. If the active node is implicit and on the edge leading to the oldest leaf of tree we set the head node equal to the active node.

Complexity. The algorithm extends Senft's algorithm for the construction of the suffix tree in a sliding window. For both addition and deletion of a letter, the number of operations is $\mathcal{O}(\sigma(\tilde{s}_i - s_i))$ and $\mathcal{O}(\sigma(p_{i+1} - \tilde{p}_{i+1}))$. Similar to the proof of Theorem 1, we obtain that the total number of operations is $\mathcal{O}(\sigma n)$. We use $\mathcal{O}(\sigma m)$ space to store the suffix tree for the factor inside the window. The σ factor is to store an array of size σ at each explicit node for constant-time child queries. We also use up to $4m$ arrays of size σ each to store the two sets of letters – the suffix tree has no more than $2m$ explicit nodes. We also store the word itself over two windows. Thus the total space complexity is bounded by $\mathcal{O}(\sigma m)$. We thus obtain the following result.

Theorem 2. *Given a word of length n on an alphabet of size σ, our algorithm computes the set of minimal absent words in a sliding window of size m in $\mathcal{O}(\sigma n)$ time and $\mathcal{O}(\sigma m)$ space.*

4   Applications to On-Line Pattern Matching

As a consequence of Theorem 2 we obtain the following result.

Theorem 3. *Given a word x of length m on an alphabet Σ of size σ, one can find on-line all occurrences of x in a word y of length $n \geq m$ on alphabet Σ in $\mathcal{O}(\sigma n)$ time and $\mathcal{O}(\sigma m)$ space. Within the same complexities, one can also compute on-line $LWI(x, y[i \mathinner{.\,.} i + m - 1])$, for all $0 \leq i \leq n - m$.*

Proof. As a pre-processing step, we build the suffix tree of x and compute the MAWs of x. At the same time, by Lemma 6, we represent all MAWs of x by storing a set of letters in each explicit node of the tree. This can be done in $\mathcal{O}(\sigma m)$ time and space [10]. We then apply Theorem 2 to build the suffix tree for a sliding window of size m over y on top of the suffix tree of x. This way when a MAW is created or deleted we can update LWI in $\mathcal{O}(1)$ time as we can check if it is a MAW of x or not. For the first part, note that two words x and z are equal if and only if $LWI(x, z) = 0$ [10,14]. We thus obtain the result.   □

References

1. Aho, A.V., Corasick, M.J.: Efficient string matching: an aid to bibliographic search. Commun. ACM **18**, 333–340 (1975)
2. Almirantis, Y., Charalampopoulos, P., Gao, J., Iliopoulos, C.S., Mohamed, M., Pissis, S.P., Polychronopoulos, D.: On avoided words, absent words, and their application to biological sequence analysis. Algorithms Mol. Biol. **12**(1), 5:1–5:12 (2017)
3. Barton, C., Heliou, A., Mouchard, L., Pissis, S.P.: Linear-time computation of minimal absent words using suffix array. BMC Bioinform. **15**, 11 (2014)
4. Barton, C., Heliou, A., Mouchard, L., Pissis, S.P.: Parallelising the computation of minimal absent words. In: Wyrzykowski, R., Deelman, E., Dongarra, J., Karczewski, K., Kitowski, J., Wiatr, K. (eds.) PPAM 2015. LNCS, vol. 9574, pp. 243–253. Springer, Cham (2016). doi:10.1007/978-3-319-32152-3_23
5. Béal, M.-P., Mignosi, F., Restivo, A.: Minimal forbidden words and symbolic dynamics. In: Puech, C., Reischuk, R. (eds.) STACS 1996. LNCS, vol. 1046, pp. 555–566. Springer, Heidelberg (1996). doi:10.1007/3-540-60922-9_45
6. Belazzougui, D., Cunial, F., Kärkkäinen, J., Mäkinen, V.: Versatile succinct representations of the bidirectional Burrows-wheeler transform. In: Bodlaender, H.L., Italiano, G.F. (eds.) ESA 2013. LNCS, vol. 8125, pp. 133–144. Springer, Heidelberg (2013). doi:10.1007/978-3-642-40450-4_12
7. Chairungsee, S., Crochemore, M.: Using minimal absent words to build phylogeny. Theoret. Comput. Sci. **450**, 109–116 (2012)
8. Crochemore, M., Fici, G., Mercas, R., Pissis, S.P.: Linear-time sequence comparison using minimal absent words. In: Kranakis, E., Navarro, G., Chávez, E. (eds.) LATIN 2016. LNCS, vol. 9644, pp. 334–346. Springer, Heidelberg (2016). doi:10.1007/978-3-662-49529-2_25
9. Crochemore, M., Hancart, C., Lecroq, T.: Algorithms on Strings. Cambridge University Press, Cambridge (2007)
10. Crochemore, M., Mignosi, F., Restivo, A.: Automata and forbidden words. Inf. Process. Lett. **67**(3), 111–117 (1998)
11. Crochemore, M., Mignosi, F., Restivo, A., Salemi, S.: Data compression using antidictionaries. Proc. IEEE **88**(11), 1756–1768 (2000)
12. Dömölki, B.: An algorithm for syntactical analysis. Comput. Linguist. **3**, 29–46 (1964)
13. Ferragina, P., Manzini, G.: Opportunistic data structures with applications. In: FOCS, pp. 390–398. IEEE Computer Society (2000)
14. Fici, G.: Minimal Forbidden Words and Applications. Thèse, Université de Marne la Vallée (2006)
15. Fujishige, Y., Tsujimaru, Y., Inenaga, S., Bannai, H., Takeda, M.: Computing DAWGs and minimal absent words in linear time for integer alphabets. In: MFCS. LIPIcs, vol. 58, pp. 38:1–38:14. Schloss Dagstuhl - Leibniz-Zentrum fuer Informatik (2016)
16. Gusfield, D.: Algorithms on Strings, Trees and Sequences: Computer Science and Computational Biology. Cambridge University Press, Cambridge (1997)
17. Hampikian, G., Andersen, T.L.: Absent sequences: nullomers and primes. In: PSB, pp. 355–366. World Scientific (2007)
18. Heliou, A., Pissis, S.P., Puglisi, S.J.: emMAW: computing minimal absent words in external memory. Bioinformatics (2017)

19. Herold, J., Kurtz, S., Giegerich, R.: Efficient computation of absent words in genomic sequences. BMC Bioinform. **9**, 167 (2008)
20. Knuth, D.E., Morris Jr., J.H., Pratt, V.R.: Fast pattern matching in strings. SIAM J. Comput. **6**(2), 323–350 (1977)
21. Kucherov, G., Salikhov, K., Tsur, D.: Approximate string matching using a bidirectional index. Theoret. Comput. Sci. **638**, 145–158 (2016)
22. Landau, G.M., Myers, E.W., Schmidt, J.P.: Incremental string comparison. SIAM J. Comput. **27–2**, 557–582 (1998)
23. Mignosi, F., Restivo, A., Sciortino, M.: Words and forbidden factors. Theoret. Comput. Sci. **273**(1–2), 99–117 (2002)
24. Myers, G.: A fast bit-vector algorithm for approximate string matching based on dynamic programming. J. ACM **46**(3), 395–415 (1999)
25. Navarro, G.: A guided tour to approximate string matching. ACM Comput. Surv. **33**(1), 31–88 (2001)
26. Navarro, G., Baeza-Yates, R.A., Sutinen, E., Tarhio, J.: Indexing methods for approximate string matching. IEEE Data Eng. Bull. **24**(4), 19–27 (2001)
27. Navarro, G., Raffinot, M.: Flexible Pattern Matching in Strings: Practical Online Search Algorithms for Texts and Biological Sequences. Cambridge University Press, Cambridge (2008)
28. Ota, T., Fukae, H., Morita, H.: Dynamic construction of an antidictionary with linear complexity. Theor. Comput. Sci. **526**, 108–119 (2014)
29. Ota, T., Morita, H.: On a universal antidictionary coding for stationary ergodic sources with finite alphabet. In: ISITA, pp. 294–298. IEEE (2014)
30. Rahman, M.S., Alatabbi, A., Athar, T., Crochemore, M., Rahman, M.S.: Absent words and the (dis)similarity analysis of DNA sequences: an experimental study. BMC Bioinform. Notes **9**(1), 1–8 (2016)
31. Senft, M.: Suffix tree for a sliding window: an overview. In: WDS, pp. 41–46. Matfyzpress (2005)
32. Silva, R.M., Pratas, D., Castro, L., Pinho, A.J., Ferreira, P.J.S.G.: Three minimal sequences found in Ebola virus genomes and absent from human DNA. Bioinformatics **31**(15), 2421–2425 (2015)
33. Ukkonen, E.: On-line construction of suffix trees. Algorithmica **14**(3), 249–260 (1995)
34. Wu, Z., Jiang, T., Su, W.: Efficient computation of shortest absent words in a genomic sequence. Inf. Process. Lett. **110**(14–15), 596–601 (2010)

Subquadratic Non-adaptive Threshold Group Testing

Gianluca De Marco[1]([✉]), Tomasz Jurdziński[2], Michał Różański[2], and Grzegorz Stachowiak[2]

[1] Dipartimento di Informatica, University of Salerno, Fisciano, Italy
demarco@dia.unisa.it
[2] Institute of Computer Science, University of Wrocław, Wrocław, Poland

Abstract. We consider *threshold group testing* – a generalization of a well known and thoroughly examined problem of *combinatorial group testing*. In the classical setting, the goal is to identify a set of *positive* individuals in a population, by performing tests on pools of elements. The output of each test is an answer to the question: *is there at least one positive element inside a query set Q?* The threshold group testing is a natural generalization of this classical setting which arises when the answer to a test is positive if *at least* $t > 0$ elements under test are positive.

We show that there exists a testing strategy for the threshold group testing consisting of $O(d^{3/2} \log(N/d))$ tests, for d positive items in a population of size N. For any value of the threshold t, we also provide a lower bound of order $\Omega\left(\min\left\{\left(\frac{d}{t}\right)^2, \frac{N}{t}\right\}\right)$. Our subquadratic bound shows a complexity separation with the classical group testing (which corresponds to $t = 1$) where $\Omega(d^2 \log_d N)$ tests are needed [25].

Next, we introduce a further generalization, the *multi-threshold group testing* problem. In this setting, we have a set of $s > 0$ thresholds, $t_1, t_2, \ldots, t_s$. The output of each test is an integer between 0 and s which corresponds to which thresholds get passed by the number of positives in the queried pool. Here, one may be interested in minimizing not only the number of tests, but also the number of thresholds which is related to the accuracy of the tests. We show the existence of two strategies for this problem. The first one of size $O(d^{3/2} \log(N/d))$ is an extension of the above-mentioned result. The second strategy is more general and works for a range of parameters. As a consequence, we show that $O(\frac{d^2}{t} \log(N/d))$ tests are sufficient for $t \le d/2$. Both strategies use respectively $O(\sqrt{d})$ and $O(\sqrt{t})$ thresholds.

Keywords: Group testing · Threshold group testing · Non-adaptive strategies · Randomized algorithms

This work was supported by the Polish National Science Centre grants DEC-2012/06/M/ST6/00459 and 2014/13/N/ST6/01850.

© Springer-Verlag GmbH Germany 2017
R. Klasing and M. Zeitoun (Eds.): FCT 2017, LNCS 10472, pp. 177–189, 2017.
DOI: 10.1007/978-3-662-55751-8_15

1   Introduction

In this paper we discuss a generalization to *group testing* – a problem introduced by Dorfman [24]. Its classical version consists of discovering a set of up to d *positive elements* in a population of N individuals in a series of group tests. The output of each test is an answer to the question: *is there at least one positive element inside a query set Q?* Typically, it is assumed that d is much smaller than N. A naive solution is to query each individual in the population separately, which gives a solution with N tests. However, knowing that the set of positives is sparse it is possible to arrange a strategy of testing in which the number of tests depends, to a large extent, on the sparsity d rather than the size of the population N.

The problem has been defined under many variants. At the two extremes, adaptive and non-adaptive strategies can be defined. While in the former, one can adjust the query sets basing on the results of previous tests, in the non-adaptive case, the query pattern has to be defined in advance. In the present paper we consider only the latter.

Since its introduction [24], group testing and its variations have been extensively studied. Surprisingly, many applications have been found in apparently unrelated areas. In particular, there are applications in molecular biology and DNA library screening (cf. the survey [31] and the references therein), pattern matching [10,28], compressed sensing [11], multiple access communications [4–6,18–22,30,34], radio networks [8,9,16,17] and streaming algorithms [12]. Other natural variants include the case where only a partial discovery of the positives [1,8,9,15,23] is required. Another variant of the problem arises in the area of molecular biology [26,27].

The generalization of the problem we consider in this paper was first addressed by Damaschke [13]. In the *threshold group testing* the output of each test responds to the question: *are there at least t positive elements in the query set?* The classical version of the problem corresponds to the choice $t = 1$. Moreover, we extend the setting to the *multi-threshold group testing*, where the output tells if the number of positives under test falls into one of the predefined "buckets".

1.1   Problem Definition

The *population* is identified with the set $[N] = \{1, 2, \ldots, N\}$, i.e., *an individual* is a natural number from the set. We consider a set $P \subseteq [N]$ of distinguished individuals, called *positives*. In our considerations we assume that P contains, depending on the variant of the problem, at most d or exactly d elements.

A single *testing pool* is a subset of individuals $Q \subseteq [N]$ to be tested. Tests are modeled by the *feedback functions*. We consider two types of feedback functions. The first type relates to the usual threshold group testing with threshold $t > 0$. Let Q be the pool to be tested, then the output of the threshold function $f_{t,P}$ satisfies

$$f_{t,P}(Q) = \begin{cases} 0 \text{ if } |Q \cap P| < t \\ 1 \text{ if } |Q \cap P| \geq t \end{cases}.$$

The second type of feedbacks is a natural generalization of this idea. Let $t_1 \leq \ldots \leq t_s$ be thresholds in a multi-threshold setting. Then the multi-threshold feedback function $f_{\mathcal{T},P}$ with s thresholds satisfies the following,

$$f_{\mathcal{T},P}(Q) = \begin{cases} 0 & \text{if } |Q \cap P| < t_1 \\ i & \text{if } |Q \cap P| \in [t_i, t_{i+1}) \text{ for } i < s \\ s & \text{if } |Q \cap P| \geq t_s \end{cases}$$

where Q is a test pool and $\mathcal{T} = \{t_1, \ldots, t_s\}$.

A sequence $\mathcal{Q} = (Q_1, \ldots, Q_m)$, where $Q_i \subseteq [N]$, is called a *pooling strategy* of size m. For a given feedback function f and pooling strategy $\mathcal{Q} = (Q_1, \ldots, Q_m)$ we define a *measurement* to be a vector $y = f(\mathcal{Q}) = (y_1, \ldots, y_m)$ such that $y_i = f(Q_i)$.

Our goal in the *threshold group testing* with threshold $t > 0$ is to provide a pooling strategy $\mathcal{Q}$ of minimal size that allows to decode the set of positives P based on the measurement vector for $\mathcal{Q}$. Formally, there exists a mapping φ such that: for any set of positives $P \subseteq [N]$ satisfying $|P| \leq d$, we have $\varphi(\mathcal{Q}, y) = P$ provided that $y = f_{t,P}(\mathcal{Q})$. We call the pooling strategies satisfying the above condition to be *correct*.

Multi-threshold group testing is a generalization of threshold group testing in which measurement vectors are no longer binary. Our goal is to propose a set of thresholds $\mathcal{T}$ and a pooling strategy $\mathcal{Q}$ such that there exists a mapping ψ such that for any $P \subseteq [N]$ of size at most d we have $\psi(\mathcal{Q}, \mathcal{T}, y) = P$ provided that $y = f_{\mathcal{T},P}(\mathcal{Q})$.

1.2 Previous Results

As explained previously, the threshold group testing is a natural generalization, of the classical group testing, introduced by Damaschke [13]. Namely, the model introduced by Damaschke can be defined as follows. Let l and u be integer parameters with $0 \leq l < u$. Each test with a query set Q outputs *Yes* if Q contains at least u positives, and *No* if there are at most l positives. If, on the other hand, the number of positives in Q is between l and u, the result of the test is arbitrary. It is supposed that l and u are constant and previously known. As in the classical group testing, the question is to determine the set of positives P by using as few tests as possible. The classical group testing corresponds to the special case with $u = 1$ (and $l = 0$).

Damaschke [13] showed that the set of positives P can be identified only when the number of positives is at least u. Moreover, he proved that in general the set of positives can be identified only approximately, i.e. up to g wrongly identified items, where $g = u - l - 1$ is the gap between the two thresholds. In other words, regardless the number of tests, the identified set may contain up to g false positives/negatives. A special case is when the gap $g = 0$, which corresponds to the problem considered in the present paper ($t = u = l + 1$). In this case, as a consequence of the above mentioned Damaschke's result, a precise identification of the set of positives is possible.

As far as adaptive strategies are allowed, in [13] efficient upper bounds are also given. Namely, for $g = 0$ a nearly optimal scheme requiring only $O(d \log N)$ tests is presented.

The non-adaptive case, which is the topic of the present work, has been studied by Chen and Fu [2]. They give a generalization of the standard notion of disjunct matrix which is suitable for the threshold model.

Using their strongly disjunct matrices, they show that $O(d^{u+1} \log(N/d))$ tests are sufficient to identify non-adaptively the set of positives. This result is almost tight with respect to the number of rows of their definition of strongly disjunct matrix. So the main question left open from their work was to understand if a different approach could beat their complexity, which, because of the d^{u+1} term, becomes quite disadvantageous for non-constant values of the threshold.

Cheraghchi [3] showed that a weaker version of disjunct matrix can be used to solve the threshold group testing. With an efficient probabilistic construction of such a weaker disjunct matrix, Cheraghchi showed that, assuming that the gap g and the upper threshold u are constants, it is possible to non-adaptively find the positives in $O(d^{g+2} \log d \log(N/d))$ tests. This implies that, in our case when $g = 0$, i.e., when one is interested to identify exactly the set of positives, *assuming that the threshold is constant*, the number of tests becomes $O(d^2 \log d \log(N/d))$. This is a remarkable improvement over the previous bound, as the complexity now does not depend on the value of the threshold. In [3] it is assumed that the upper threshold u (which in our case of $g = 0$ corresponds to threshold t in our setting) is a constant. This assumption allows to get rid of the occurrences of parameter u within the asymptotic bound. One of the main questions left open by Cheraghchi's work is whether the number of tests can be reasonably controlled even for large values of the threshold.

We answer to this question in the positive with even an improvement over the above mentioned quadratic bound. Namely, we show an interesting dependence between the number of tests and the value of the threshold. While for small values of the threshold the number of tests is still quadratic in the number of positives, it becomes subquadratic as the value of the threshold grows. In particular, we get a subquadratic $O(d^{3/2} \log(N/d))$ bound for a threshold group testing with threshold $t = d/2$ for gap $g = 0$ and d positives. Our result is existential and relies on the assumption that the number of positives is exactly d.

1.3 Our Results

Our first result concerns the standard threshold group testing model. We show that there exists a pooling design of size $O(\frac{d^2}{q(d,t)} \log \frac{N}{d})$ for a single-threshold group testing, where $q(d,t) = \Omega\left(\sqrt{\frac{dt}{d-t}}\right)$. As a consequence, we have a subquadratic number of tests for a range of values of t. In particular, there is a pooling strategy revealing d positives, of size $O(d^{3/2} \log(N/d))$ for $t = d/2$. Then we present a lower bound of order $\Omega\left(\min\left\{\left(\frac{d}{t}\right)^2, \frac{N}{t}\right\}\right)$. Our subquadratic bound shows a complexity separation with the classical group testing where $\Omega(d^2 \log_d n)$ tests are needed [25].

Then, we introduce the multi-threshold variant of group testing and present a solution with $O(d^{3/2} \log(N/d))$ tests and $O(\sqrt{d})$ thresholds for up to d positives. We conclude the paper with a result using $O(s)$ thresholds arranged around $t > 0$, i.e., $\mathcal{T} = \{t - s, ..., t + s\}$, to resolve the multi-threshold group testing, with exactly d positives, within $O(\frac{d^2}{q} \log(N/d))$ tests for $q = \Omega(t(1 - e^{s^2/t})^2)$. A special case of this result gives a pooling strategy of size $O(\frac{d^2}{t} \log(N/d))$ for $O(\sqrt{t})$ thresholds for any $t \leq d/2$.

2   Preliminaries

Throughout the paper we use the following notation:

- $\mathcal{B}(n, p)$ – binomial distribution with parameters n and p,
- $\mathcal{N}(\mu, \sigma^2)$ – normal distribution with mean μ and variance σ^2.

For the rest of this section we denote μ and σ^2 to be expectation and variance of $\mathcal{B}(n, p)$, i.e., $\mu = np$ and $\sigma^2 = np(1 - p)$.

Consider a random process that picks a subset $Q \subseteq [N]$ such that each $x \in [N]$ is chosen to be in Q with probability p. We introduce a series of random variables χ_S, for all $S \subseteq [N]$, to denote the size of the intersection of sets S and Q, i.e., $\chi_S = |Q \cap S|$. Whenever we use this notation, the support is always $[N]$ and probability p follows directly from the context. Note that $\chi_S \sim \mathcal{B}(|S|, p)$. Finally, we will be using the following mathematical fact which results from the observation that $f(x) = (1 - x)^{1/x}$ is a decreasing function.

Fact 1. *For any $0 < x \leq 1/2$ we have $(1 - x)^{1/x} \geq 1/4$.*

2.1   Technical Results

We devote this section to discuss the properties of binomial distribution. We use a normal approximation to the binomial distribution to provide two-sided bounds on the probability of $X \sim \mathcal{B}(n, p)$ being at least k, for values of k that are close to the expectation of X — meaning that $|k - \mu| = O(1)$. Because of the Central Limit Theorem, one expects that this value should be close to some constant around $\frac{1}{2}$, for sufficiently large n. We use Berry-Esseen theorem, in a variant stated in [14], which provides quantitative guarantees of normal approximation to binomial distributions. We use them to show in Lemma 1 that the above-mentioned probability $P(X \geq k)$ is indeed constant for a range of parameters.

We refer the reader to the full version of the paper for the proofs of the following technical results.

Proposition 1. *Let $X \sim \mathcal{B}(n, p)$ with $p \leq 1/2$, $\beta > 0$, and k be a positive integer satisfying*

$$(i) \ |k - \mu| \leq \beta \ and \ (ii) \ 2\beta \leq k \leq n - 2\beta.$$

Then we have the following

$$\hat{\gamma}\sqrt{\frac{n}{k(n-k)}} \geq \mathrm{P}\left(X=k\right) \geq \gamma_\beta \sqrt{\frac{n}{k(n-k)}}\,,$$

where $\gamma_\beta = \Theta\left(\frac{1}{4^\beta}\right)$ *and* $\hat{\gamma} > 0$ *is a constant.*

Proposition 2. *Let* $X \sim \mathcal{B}(n,p)$ *with* $p \leq 1/2$, $n \geq 2$, *and* k *satisfying* $|k - \mu| \leq 1$. *Then*

- $e^{-1/2} \geq \mathrm{P}\left(X=k\right) \geq np/16 \qquad$ *for* $k=1$,
- $\mathrm{P}\left(X=k\right) \geq 1/16 \qquad\qquad$ *for* $k=n-1$.

The following lemma concludes the section.

Lemma 1. *Let* $X \sim \mathcal{B}(n,p)$ *where* $p \leq 1/2$ *and* $k \in \mathrm{N}$ *is such that* $k-1 \leq \mu \leq k$ *then the following conditions hold:*

1. $3/4 \geq \mathrm{P}\left(X \geq k\right) \geq np/16 \qquad\qquad\qquad\qquad$ *for* $k=1$,
2. $1-c \geq \mathrm{P}\left(X \geq k\right) \geq c \qquad\qquad\qquad\qquad\qquad$ *for* $k>1$,

where $c = \Theta(1)$.

3   Single Threshold

We devote this section to prove the existence of a pooling design for threshold group testing with d positives with number of tests which is $o(d^2)$ (for a suitable choice of threshold t). We also sketch a lower bound on the number of tests in a threshold design by generalizing a result from [25]. Throughout this section we assume that $t \leq d/2$.

3.1   Upper Bound

We consider random tests Q constructed by picking each element of $[N]$ with probability $p = \frac{t}{d}$. The assumption that $t \leq d/2$ implies $p \leq 1/2$.

Let A, B be d-element subsets of $[N]$ and let $s = |A \cap B|$. Recalling the definition of χ_S, for any set $S \subseteq [N]$, we can observe that $\chi_{A\cap B} \sim \mathcal{B}(s, \frac{t}{d})$. Let γ_β be defined like in Proposition 1, i.e., $\gamma_\beta = \Theta\left(\frac{1}{4^\beta}\right)$, for any $\beta > 0$, and let $\mu_s = \lfloor st/d \rfloor$.

Proposition 3. *The following inequalities hold:*

- $\mathrm{P}\left(\chi_{A\cap B} = \mu_s\right) \geq \gamma_1 \sqrt{\frac{d}{t(d-t)}} \quad$ *for* $\mu_s \in \{2, \ldots, s-2\}$,
- $\mathrm{P}\left(\chi_{A\cap B} = \mu_s\right) \geq 1/16 \quad$ *for* $\mu_s \in \{0, 1, s-1\}$.

Proof. Let us prove the first inequality. Let $\tilde{\lambda}(s) = \gamma_1 \sqrt{\frac{s}{\mu_s(s-\mu_s)}}$. Proposition 1, for $k = \mu_s$ and $\beta = 1$, implies $\mathrm{P}\left(\chi_{A \cap B} = \mu_s\right) \geq \tilde{\lambda}(s)$ for $\mu_s \in \{2, \ldots, s-2\}$.

The size of the intersection of A and B, namely s, can take values from 0 to $d-1$. Function $\tilde{\lambda}$ is non-increasing, thus we can write

$$\mathrm{P}\left(\chi_{A \cap B} = \mu_s\right) \geq \tilde{\lambda}(d-1) = \Omega\left(\gamma_1 \sqrt{\frac{d}{t(d-t)}}\right).$$

This proves the first inequality.

Let us prove the second part of the proposition, i.e. for $\mu_s \in \{0, 1, s-1\}$. First, let us observe that $\mathrm{E}[\chi_{A \cap B}] = \frac{st}{d}$ and, therefore, $\mathrm{E}[\chi_{A \cap B}] - 1 < \mu_s \leq \mathrm{E}[\chi_{A \cap B}]$. We have $\mathrm{P}\left(\chi_{A \cap B} = \mu_s\right) = \binom{s}{\mu_s} \cdot p^{\mu_s}(1-p)^{s-\mu_s}$.

For $\mu_s = 0$, we have $\mathrm{P}\left(\chi_{A \cap B} = \mu_s\right) = (1-p)^s \geq (1-p)^{1/p}$. The last inequality follows from $st/d < 1$ (since $\mu_s = 0$) which in turn implies $s < 1/p$.

For $\mu_s = 1$, we have $s \cdot p = st/d \geq 1$ and therefore $\mathrm{P}\left(\chi_{A \cap B} = \mu_s\right) = s \cdot p(1-p)^{s-1} \geq (1-p)^{s-1} \geq (1-p)^s \geq (1-p)^{2/p}$, where the last inequality follows from $st/d < 2$, which is a consequence of $\mu_s = 1$.

Finally, for $\mu_s = s-1$, we can assume that $s \geq 2$, otherwise for $s = 1$ we obtain $\mu_s = 0$, which we have already considered. But we have also that $s \leq 2$, indeed $sp = \mu_s = s-1$, which, recalling that $p \leq 1/2$, implies $s \leq 2$. Hence, we must have $s = 2$ in this case. We can also observe that, in order to have $sp = \mu_s = s-1 = 1$, we need also that $p = 1/2$. So, the binomial formula becomes $\mathrm{P}\left(\chi_{A \cap B} = \mu_s\right) = s \cdot p^{s-1}(1-p) = 2p(1-p) = 1/2$.

In the first two cases we can apply Fact 1 and get that $\mathrm{P}\left(\chi_{A \cap B} = \mu_s\right)$ is greater than or equal to $1/4$ in the first case and to $1/16$ in the second case. In the last case it is $1/2$. This concludes the proof of the second inequality. □

Recalling that A and B are d-element sets and that $s = |A \cap B|$, we can observe that $\chi_{A \setminus B}, \chi_{B \setminus A} \sim \mathcal{B}(d-s, \frac{t}{d})$.

Proposition 4. *The following inequalities hold:*

- $\mathrm{P}\left(\chi_{A \setminus B} \geq t - \mu_s\right) \mathrm{P}\left(\chi_{B \setminus A} < t - \mu_s\right) \geq \frac{p}{4}$ *for $\mu_s = t-1$,*
- $\mathrm{P}\left(\chi_{A \setminus B} \geq t - \mu_s\right) \mathrm{P}\left(\chi_{B \setminus A} < t - \mu_s\right) \geq c^2$ *for $\mu_s < t-1$.*

Proof. For $\mu_s = t-1$, we have $st/d \geq t-1$ from which we derive $s \geq d(t-1)/t$ and then

$$d - s \leq d - d \cdot \frac{t-1}{t} = \frac{d}{t} = \frac{1}{p}.$$

Hence, the left-hand side of the first inequality becomes

$$\mathrm{P}\left(\chi_{A \setminus B} \geq 1\right) \mathrm{P}\left(\chi_{B \setminus A} < 1\right) = \mathrm{P}\left(\chi_{A \setminus B} \geq 1\right) \mathrm{P}\left(\chi_{B \setminus A} = 0\right)$$
$$\geq (d-s)p \cdot (1-p)^{d-s}$$
$$\geq p \cdot (1-p)^{1/p}$$
$$\geq p/4,$$

where the last inequality follows from Fact 1. This proves the first inequality.

Let us show the second inequality. For $\mu_s < t - 1$, we have $t - \mu_s > 1$. We can now apply Lemma 1 (second condition) for $k = t - \mu_s$. It follows that

$$P\left(\chi_{A\backslash B} \geq t - \mu_s\right) P\left(\chi_{B\backslash A} < t - \mu_s\right) = P\left(\chi_{A\backslash B} \geq t - \mu_s\right)\left(1 - P\left(\chi_{B\backslash A} \geq t - \mu_s\right)\right)$$
$$\geq c \cdot (1 - (1 - c)) = c^2.$$

$\square$

Definition 1. *We say that a test $Q \subseteq [N]$ separates A and B if $f_{t,A}(Q) \neq f_{t,B}(Q)$.*

Lemma 2. *We have*

$$P\left(Q \text{ separates } A \text{ and } B\right) = \Omega\left(\sqrt{\frac{t}{d(d-t)}}\right).$$

Proof. Observe that variables $\chi_{A\backslash B}, \chi_{B\backslash A}, \chi_{A\cap B}$ are independent. Thus, we have

$$P\left(Q \text{ separates } A \text{ and } B\right) \geq P\left(\chi_A \geq t \wedge \chi_B < t\right)$$
$$= \sum_{i=0}^{\min\{t-1,s\}} P\left(\chi_{A\cap B} = i \wedge \chi_{A\backslash B} + i \geq t \wedge \chi_{B\backslash A} + i < t\right)$$
$$\geq P\left(\chi_{A\cap B} = \mu_s\right) P\left(\chi_{A\backslash B} \geq t - \mu_s\right) P\left(\chi_{B\backslash A} < t - \mu_s\right).$$

From Proposition 3 we get that $P\left(\chi_{A\cap B} = \mu_s\right) = \Omega(\sqrt{\frac{d}{t(d-t)}})$. Proposition 4 yields $P\left(\chi_{A\backslash B} \geq t - \mu_s\right) P\left(\chi_{B\backslash A} < t - \mu_s\right) = \Omega(p) = \Omega(t/d)$. Thus, we get $P\left(Q \text{ separates } A \text{ and } B\right) = \Omega\left(\sqrt{\frac{t}{d(d-t)}}\right)$. This concludes the proof. $\square$

Theorem 2. *There exists a pooling design of size $O(\frac{d^2}{q(d,t)}\log\frac{N}{d})$ for a single-threshold group testing, where $q(d,t) = \Omega\left(\sqrt{\frac{dt}{d-t}}\right)$.*

Proof. Let $\mathcal{Q} = Q_1, ..., Q_m$ be a sequence of random queries. For each Q_i we choose each element in $[N]$ to be contained in Q_i with probability $p = t/d$.

Now, we show that $\mathcal{Q}$ is a correct pooling strategy with positive probability. Let $\mathcal{F}$ be the family of unordered pairs of sets of size d. We say that $\mathcal{Q}$ separates A and B if there is a test $Q \in \mathcal{Q}$ that separates them. Let $p(d,t)$ be the lower bound on the probability of separating two sets resulting from Lemma 2.

$$P\left(\mathcal{Q} \text{ is not correct pooling strategy}\right) = P\left(\mathcal{Q} \text{ does not separate two sets in } \mathcal{F}\right)$$
$$\leq \sum_{\{A,B\}\in\mathcal{F}} P\left(\mathcal{Q} \text{ does not separate } \{A,B\}\right)$$
$$\leq |\mathcal{F}|\left(1 - p(d,t)\right)^m$$
$$\leq \exp\left(O(d\log(\frac{N}{d})) - m \cdot p(d,t)\right),$$

where the last inequality is due to $|\mathcal{F}| \leq \binom{N}{d}^2$ and $1 - x \leq e^{-x}$. The upper bound formula obtained above shows that it is sufficient for m to be $\Omega(\frac{d}{p(d,t)} \log(\frac{N}{k})) = \Omega(\frac{d^2}{q(d,t)} \log(\frac{N}{k}))$ in order to make $P(\mathcal{Q}$ is not correct pooling strategy) less than 1. □

Corollary 1. *For any* $t = \Omega(d)$ *such that* $t \leq \frac{d}{2}$ *there exists a pooling design of size* $O(d^{3/2} \log \frac{N}{d})$ *that discovers any set of positives of size* d.

3.2 Lower Bound

We now get a lower bound on the length of any pooling testing strategy revealing *up to d positives.*

Theorem 3. *For a threshold* $t > 0$ *the size of the pooling strategy discovering up to* d *positives is of order* $\Omega\left(\min\left\{\left(\frac{d}{t}\right)^2, \frac{N}{t}\right\}\right)$.

Proof. In the proof we will consider a generalization of a pooling strategy with a threshold. Let $\mathcal{Q} = Q_1, ..., Q_m$ be a family of m tests. We admit a situation in which each test Q_i uses a possibly different threshold t_i. If $t_i = 0$, then ith test is always positive and in fact can be omitted. Let $\phi(S, N, d)$ for a given t, m denote, that the following assertion is true:

> There exist integer thresholds $t_1, t_2, \ldots, t_k \in [0, t]$ such that $t_1 + t_2 + \cdots + t_k \leq S$ for which there is a generalized pooling strategy $\mathcal{Q}$ which has m tests and discovers up to d positives amongst N individuals.

The theorem will follow from the following claim

Claim. If $\phi(S, N, d)$, then $m \geq N/t$ or $\phi(S - d/t, N - 1, d - 1)$.

We consider two cases. In the first case, each individual $n \in [N]$ participates in less than d/t tests. We will prove that, in this case, for any arbitrary individual $n \in [N]$, there is a test in which only n and at most $t - 1$ other individuals participate. This implies that the total number of tests $m \geq N/t$.

Assume by contradiction that in each test Q_i in which there is n there are also at least t other individuals. Let, for such a test, the set $A_i \subseteq Q_i$ consist of t individuals other than n. The set A being the union of all such sets A_i has less than d elements since there are less than d/t such sets. Note that A and $A \cup \{n\}$ have the same test results in the generalized pooling strategy $\mathcal{Q}$ which contradicts its validity.

The second case occurs when there is an individual n participating in at least d/t tests. Note that the generalized pooling strategy consisting of tests $Q_i \setminus \{n\}$ is valid for detecting at most $d - 1$ positives in $[N] \setminus \{n\}$ if thresholds t_i are decreased by 1 for i such that $n \in Q_i$. Thus in this case $\phi(S, N, d)$ implies $\phi(S - d/t, N - 1, d - 1)$.

Now having the Claim we can prove the Theorem. Note that if there is a pooling strategy with threshold t in each round, then $\phi(kt, N, d)$. Unless $m \geq (N - d)/t$ the following chain of implications holds

$$\phi(mt, N, d) \implies \phi(mt - d/t, N - 1, d - 1) \implies \cdots$$
$$\cdots \implies \phi(mt - (d - 1)d/t, N - (d - 1), d - (d - 1)).$$

Since in the last formula $mt - (d - 1)d/t \geq 0$, then $m \geq (d - 1)d/t^2$. So

$$m \geq \min\left\{\left(\frac{d(d - 1)}{t^2}\right), \frac{N - d}{t}\right\} = \Omega\left(\min\left\{\left(\frac{d}{t}\right)^2, \frac{N}{t}\right\}\right).$$

$\square$

4   Multiple Thresholds

In this section we consider feedback functions $f = f_{T,P}$ with multiple thresholds. We prove a result for thresholds arranged consecutively, i.e., $T = \{a, a+1, ..., b\}$ for $a < b$. Thus, the result of the test $f(Q)$ tells the number of positives in Q if it is in a range (a, b). Otherwise its results are "at most a positives" or "at least b positives".

We prove a result for the following class of thresholds. Let $T = [t - 2\alpha, t + 2\alpha]$ for $t, \alpha \in \{1, ..., d\}$. The line of the proof is similar to the case of single threshold designs. Namely, we bound the probability of separating a fixed pair of individuals and then we apply the probabilistic method.

Theorem 4. *There exists a multi-threshold pooling strategy with 4α thresholds centered around t of size $O\left(\frac{d^2}{q(d,t,\alpha)} \log \frac{N}{d}\right)$, where $q(d, t, \alpha) = \Omega\left(t(1 - e^{-\alpha^2/t})^2\right)$.*

In the remainder of this section we prove Theorem 4 and formulate a corollary with a slightly simplified formula.

Let $Q = Q_1, ..., Q_m$ be a sequence of random pools. For each Q_i we choose each element in $[N]$ to be contained in Q_i with probability $p = t/d$.

Now, we state the definition of separation analogous to Definition 1.

Definition 2. *Let A and B be subsets of $[N]$. We say that a test $Q \subseteq [N]$ separates A and B if $f_{T,A}(Q) \neq f_{T,B}(Q)$.*

Lemma 3. *Let A, B be d-element subsets of $[N]$ and Q be a random query constructed by picking each element of $[N]$ with probability $p = \frac{t}{d}$. Recalling the notation χ_S from Sect. 2, we have*

$$P(Q \text{ separates } A \text{ and } B) = \Omega\left(\frac{t}{d}\left(1 - e^{-\alpha^2/t}\right)^2\right).$$

Proof. Let us denote $A' = A \setminus B$ and $B' = B \setminus A$. We exploit the observation that for A and B to be separated by test Q it is sufficient that $\chi_A \in \mathcal{T}$ and $\chi_B \neq \chi_A$ and proceed by using independence of random variables $\chi_{A \cap B}, \chi_{A'}, \chi_{B'}$.

$$P\,(Q \text{ separates A and B}) \geq P\,(\chi_A \in \mathcal{T} \wedge \chi_B \neq \chi_A)$$
$$\geq P\,(|\chi_{A \cap B} - E[\chi_{A \cap B}]| \leq \alpha) \cdot P\,(|\chi_{A'} - E[\chi_{A'}]| \leq \alpha \wedge \chi_{B'} \neq \chi_{A'}).$$

Let $p_i = P\,(\chi_{A'} = i)$, $r = |A'|$ and $R = [E[\chi_{A'}] - \alpha, E[\chi_{A'}] + \alpha]$. We have,

$$P\,(|\chi_{A'} - E[\chi_{A'}]| \leq \alpha \wedge \chi_B \neq \chi_A) = \sum_{i \in R} p_i(1 - p_i) \geq (1 - p^*)P\,(\chi_{A'} \in R),$$

where $p^* = \max_{i \in R} p_i$. We are interested in bounding p^* from above. For $i > 0$ we have $p_i \leq 1/2$ (see Propositions 1 and 2), but p_0 may be arbitrarily close to 1 as p approaches 0. Observe, however, that Proposition 2 guarantees that, in such a case, we have $p_1 \geq p/16$ and thus $p_0 \leq 1 - p/16$. We bound the other factor using the Chernoff bound. This gives us

$$P\,(|\chi_{A \cap B} - E[\chi_{A \cap B}]| > \alpha) \leq 2\exp\left(-\frac{\alpha^2}{(d-s)p}\right) \leq 2\exp\left(-\frac{\alpha^2}{dp}\right).$$

The same observation applies to the value $P\,(\chi_{A'} \in R)$. Thus, we have

$$P\,(Q \text{ separates A and B}) \geq (1 - p^*)P\,(|\chi_{A \cap B} - E[\chi_{A \cap B}]| \leq \alpha)$$
$$\cdot P\,(|\chi_{A'} - E[\chi_{A'}]| \leq \alpha) \geq p(1 - e^{-\alpha^2/(dp)})^2/16.$$

$$\square$$

Now we use the probabilistic method to show that there exists a pooling strategy od certain size. We show that $\mathcal{Q}$ has probability greater than zero of being a correct multi-threshold pooling strategy for suitable choice of m. The idea behind this is similar to that used in proof of existence of single-threshold testing strategy.

Let $p(d, t, \alpha) = \frac{t}{d}(1 - e^{-\alpha^2/t})^2/16$. By Lemma 3 we have

$$P\,(Q \text{ separates A and B}) \geq p(d, t, \alpha).$$

$P\,(\mathcal{Q} \text{ is not correct pooling strategy}) = P\,(\mathcal{Q} \text{ does not separate two sets} \in \mathcal{F})$
$$\leq \sum_{\{A,B\} \in \mathcal{F}} P\,(\mathcal{Q} \text{ does not separate } \{A, B\})$$
$$\leq |\mathcal{F}|\,(1 - p(d, t, \alpha))^m$$
$$\leq \exp\left(O(d\log(\frac{N}{d})) - m \cdot p(d, t, \alpha)\right),$$

Thanks to the above inequalities we know that it is sufficient to choose $m = d\log(\frac{N}{d})/p(d, t, \alpha) = O\left(\frac{d^2}{q(d,t,\alpha)}\log\frac{N}{d}\right)$. This concludes the proof of Theorem 4.

Corollary 2. *There exists a multi-threshold pooling strategy of size* $O(\frac{d^2}{t}\log\frac{N}{d})$ *provided that* $\alpha = \Omega(\sqrt{t})$.

5   Concluding Remarks

In this work we have studied the threshold group testing model and introduced the multi-threshold group testing. We have presented various upper bounds indicating that, in the model with non-constant threshold(s), the group testing can be done much faster than in the case of one constant threshold. We believe that techniques developed for efficient construction and decoding in the standard model of group testing (e.g., [3,7,29,30,32,33]) could help in developing explicit constructions for threshold group testing considered in this paper.

Acknowledgments. The authors would like to thank Darek Kowalski for his comments to the paper.

References

1. Alon, N., Hod, R.: Optimal monotone encodings. IEEE Trans. Inf. Theory **55**(3), 1343–1353 (2009)
2. Chen, H.-B., Fu, H.-L.: Nonadaptive algorithms for threshold group testing. Discrete Appl. Math. **157**(7), 1581–1585 (2009)
3. Cheraghchi, M.: Improved constructions for non-adaptive threshold group testing. Algorithmica **67**(3), 384–417 (2013)
4. Chlebus, B.S., De Marco, G., Kowalski, D.R.: Scalable wake-up of multi-channel single-hop radio networks. Theoret. Comput. Sci. **615**, 23–44 (2016)
5. Chlebus, B.S., De Marco, G., Kowalski, D.R.: Scalable wake-up of multi-channel single-hop radio networks. In: Aguilera, M.K., Querzoni, L., Shapiro, M. (eds.) OPODIS 2014. LNCS, vol. 8878, pp. 186–201. Springer, Cham (2014). doi:10.1007/978-3-319-14472-6_13
6. Chlebus, B.S., De Marco, G., Talo, M.: Naming a channel with beeps. Fundam. Inf. **153**(3), 199–219 (2017)
7. Chlebus, B.S., Kowalski, D.R.: Almost optimal explicit selectors. In: Liśkiewicz, M., Reischuk, R. (eds.) FCT 2005. LNCS, vol. 3623, pp. 270–280. Springer, Heidelberg (2005). doi:10.1007/11537311_24
8. Chrobak, M., Gasieniec, L., Rytter, W.: Fast broadcasting and gossiping in radio networks. In: FOCS 2009, pp. 575–584 (2000)
9. Clementi, A.E.F., Monti, A., Silvestri, R.: Distributed broadcast in radio networks of unknown topology. Theor. Comput. Sci. **302**(1–3), 337–364 (2003)
10. Clifford, R., Efremenko, K., Porat, E., Rothschild, A.: Pattern matching with don't cares and few errors. J. Comput. Syst. Sci. **76**(2), 115–124 (2010)
11. Cormode, G., Muthukrishnan, S.: Combinatorial algorithms for compressed sensing. In: 40th Annual Conference on Information Sciences and Systems, pp. 198–201 (2006)
12. Cormode, G., Muthukrishnan, S.: What's hot and what's not: tracking most frequent items dynamically. ACM Trans. Database Syst. **30**(1), 249–278 (2005)
13. Damaschke, P.: Threshold group testing. Electron. Notes Discrete Math. **21**, 265–271 (2005)
14. DasGupta, A.: Fundamentals of Probability: A First Course. Springer Texts in Statistics. Springer, New York (2010). doi:10.1007/978-1-4419-5780-1
15. De Bonis, A., Gasieniec, L., Vaccaro, U.: Optimal two-stage algorithms for group testing problems. SIAM J. Comput. **34**(5), 1253–1270 (2005)

16. De Marco, G.: Distributed broadcast in unknown radio networks. In: SODA 2008, pp. 208–217 (2008)
17. De Marco, G.: Distributed broadcast in unknown radio networks. SIAM J. Comput. **39**(6), 2162–2175 (2010)
18. De Marco, G., Kowalski, D.R.: Contention resolution in a non-synchronized multiple access channel. Theor. Comput. Sci. (2017). https://doi.org/10.1016/j.tcs.2017.05.014
19. De Marco, G., Kowalski, D.R.: Fast nonadaptive deterministic algorithm for conflict resolution in a dynamic multiple-access channel. SIAM J. Comput. **44**(3), 868–888 (2015)
20. De Marco, G., Kowalski, D.R.: Contention resolution in a non-synchronized multiple access channel. In: IPDPS 2013, pp. 525–533 (2013)
21. De Marco, G., Kowalski, D.R.: Towards power-sensitive communication on a multiple-access channel. In: 30th International Conference on Distributed Computing Systems (ICDCS 2010), Genoa, Italy, May 2010
22. De Marco, G., Pellegrini, M., Sburlati, G.: Faster deterministic wakeup in multiple access channels. Discrete Appl. Math. **155**(8), 898–903 (2007)
23. De Marco, G., Kowalski, D.R.: Searching for a subset of counterfeit coins: randomization vs determinism and adaptiveness vs non-adaptiveness. Random Struct. Algorithms **42**(1), 97–109 (2013)
24. Dorfman, R.: The detection of defective members of large populations. Ann. Math. Stat. **14**(4), 436–440 (1943)
25. Dyachkov, A.G., Rykov, V.V.: Bounds on the length of disjunctive codes. Probl. Pereda. Informatsii **18**(3), 7–13 (1982)
26. Farach, M., Kannan, S., Knill, E., Muthukrishnan, S.: Group testing problems with sequences in experimental molecular biology. In: Proceedings of the Compression and Complexity of Sequences, SEQUENCES 1997, p. 357 (1997)
27. Fu, H.-L., Chang, H., Shih, C.-H.: Threshold group testing on inhibitor model. J. Comput. Biol. **20**(6), 464–470 (2013)
28. Indyk, P.: Deterministic superimposed coding with applications to pattern matching. In: FOCS 1997, pp. 127–136 (1997)
29. Indyk, P., Ngo, H.Q., Rudra, A.: Efficiently decodable non-adaptive group testing. In: SODA, pp. 1126–1142 (2010)
30. Kautz, W., Singleton, R.: Nonrandom binary superimposed codes. IEEE Trans. Inf. Theory **10**(4), 363–377 (1964)
31. Ngo, H.Q., Du, D.-Z.: A survey on combinatorial group testing algorithms with applications to DNA library screening. In: Discrete Mathematical Problems with Medical Applications. DIMACS Series in Discrete Mathematics and Theoretical Computer Science, vol. 55, pp. 171–182. American Mathematical Society (2000)
32. Ngo, H.Q., Porat, E., Rudra, A.: Efficiently decodable error-correcting list disjunct matrices and applications. In: Aceto, L., Henzinger, M., Sgall, J. (eds.) ICALP 2011. LNCS, vol. 6755, pp. 557–568. Springer, Heidelberg (2011). doi:10.1007/978-3-642-22006-7_47
33. Porat, E., Rothschild, A.: Explicit nonadaptive combinatorial group testing schemes. IEEE Trans. Inf. Theory **57**(12), 7982–7989 (2011)
34. Wolf, J.: Born again group testing: multiaccess communications. IEEE Trans. Inf. Theory **31**(2), 185–191 (1985)

The Snow Team Problem
(Clearing Directed Subgraphs by Mobile Agents)

Dariusz Dereniowski[1], Andrzej Lingas[2], Mia Persson[3], Dorota Urbańska[1],
and Paweł Żyliński[4(✉)]

[1] Faculty of Electronics, Telecommunications and Informatics,
Gdańsk University of Technology, 80-233 Gdańsk, Poland
[2] Department of Computer Science, Lund University, 221 00 Lund, Sweden
[3] Department of Computer Science, Malmö University, 205 06 Malmö, Sweden
[4] Institute of Informatics, University of Gdańsk, 80-309 Gdańsk, Poland
zylinski@inf.ug.edu.pl

Abstract. We study several problems of clearing subgraphs by mobile agents in digraphs. The agents can move only along directed walks of a digraph and, depending on the variant, their initial positions may be pre-specified. In general, for a given subset S of vertices of a digraph D and a positive integer k, the objective is to determine whether there is a subgraph $H = (\mathcal{V}_H, \mathcal{A}_H)$ of D such that (a) $S \subseteq \mathcal{V}_H$, (b) H is the union of k directed walks in D, and (c) the underlying graph of H includes a Steiner tree for S. We provide several results on parameterized complexity and hardness of the problems.

Keywords: Graph searching · FPT-algorithm · NP-hardness · Monomial

1 Introduction

Consider a city, after a snowstorm, where all streets have been buried in snow completely, leaving a number of facilities disconnected. For snow teams, distributed within the city, the main battle is usually first to re-establish connectedness between these facilities. This motivates us to introduce a number of (theoretical) *snow team* problems in graphs. Herein, in the introduction section, for simplicity of presentation, we shall formalize only one of them, leaving the other variants to be stated and discussed subsequently.

Let $D = (\mathcal{V}, \mathcal{A}, F, B)$ be a vertex-weighted digraph of order n and size m, with two vertex-weight functions $F: \mathcal{V} \rightarrow \{0, 1\}$ and $B: \mathcal{V} \rightarrow \mathbb{N}$, such that its underlying graph is connected. (Recall that the *underlying graph* of D is a simple graph with the same vertex set and its two vertices u and v being adjacent if and only if there is an arc between u and v in D.) The model is that vertices of D correspond to street crossings while its arcs correspond to (one-way) streets,

Research partially supported by National Science Centre, Poland, grant number 2015/17/B/ST6/01887.

R. Klasing and M. Zeitoun (Eds.): FCT 2017, LNCS 10472, pp. 190–203, 2017.
DOI: 10.1007/978-3-662-55751-8_16

the set $\mathcal{F} = F^{-1}(1)$ corresponds to locations of facilities, called also *terminals*, and the set $\mathcal{B} = B^{-1}(\mathbb{N}^+)$ corresponds to vertices, called from now on *snow team bases*, where a (positive) number of snow ploughs is placed (so we shall refer to the function B as a *plough-quantity* function). Let $\mathbf{k}_B = \sum_{v \in \mathcal{V}} B(v)$ be the total number of snow ploughs placed in the digraph.

The Snow Team problem (ST)

Do there exist $\mathbf{k}_B$ directed walks in D, with exactly $B(v)$ starting points at each vertex $v \in \mathcal{V}$, whose edges induce a subgraph H of D such that all vertices in $F^{-1}(1)$ belong to one connected component of the underlying graph of H?

The ST problem may be understood as a question, whether for $\mathbf{k}_B$ snow ploughs, initially located at snow team bases in $\mathcal{B} = B^{-1}(\mathbb{N}^+)$, where the number of snow ploughs located at $v \in \mathcal{B}$ is equal to $B(v)$, it is possible to follow $\mathbf{k}_B$ walks in D clearing their arcs so that the underlying graph of the union of cleared walks includes a Steiner tree for all facilities in $F^{-1}(1)$.

Related Work. The Snow Team problem is related to the problems of clearing connections by mobile agents placed at some vertices in a digraph, introduced by Levcopoulos et al. in [32]. In particular, the ST problem is a generalized variant of the Agent Clearing Tree (ACT) problem where one wants to determine a placement of the minimum number of mobile agents in a digraph D such that agents, allowed to move only along directed walks, can simultaneously clear some subgraph of D whose underlying graph includes a spanning tree of the underlying graph of D. In [32], the authors provided a simple 2-approximation algorithm for solving the Agent Clearing Tree problem, leaving its complexity status open.

All the aforementioned clearing problems are themselves variants of the path cover problem in digraphs, where the objective is to find a minimum number of directed walks that cover all vertices (or edges) of a given digraph. Without any additional constraints, the problem was shown to be polynomially tractable by Ntafos and Hakimi in [36]. Several other variants involve additional constraints on walks as the part of the input, see [5,22,27,30,35–37] to mention just a few, some of them combined with relaxing the condition that all vertices of the digraph have to be covered by walks.

A wider perspective locates our snow team problems as variants of graph searching problems. The first formulations by Parsons [39] and Petrov [40] of the first studied variant of these problems, namely the *edge search*, were inspired by a work of Breisch [11]. In [11], the problem was presented as a search (conducted by a team of agents/rescuers) of a person lost in a system of caves. The differences between the problem we study in this work and the edge search lie in the fact that in edge search the entity that needs to be found (usually called a *fugitive*) changes its location quickly while in our case each entity is static and its position is known. Also in edge search, an agent can be removed from the graph and placed on any node (which is often referred as *jumping*) while in our problem it needs to follow a directed path. A variant of the edge search that shares certain characteristics with the problem we study is the *connected search*. In the latter, the connectivity restriction is expressed by requiring that at any time point, the subgraph that is ensured not to contain the fugitive is connected; for some recent algorithmic and structural results see e.g. [4,7,16,17]. We also remark a different cleaning problem introduced in [33] and related to the variants we study: cleaning a graph with *brushes*—for some recent works, see e.g. [10,12,24,34]. (Two restrictions from the original problem of cleaning a graph with brushes, namely, enforcing that each edge is traversed once and each cleaning entity must follow a walk in the graph appear in a variant of edge search called *fast search* [18].) All aforementioned searching games are defined for simple graphs; for some works on digraphs, see e.g. [2,3,6,15,25].

Finally, the ST problem is related to the directed Steiner tree problem, where for a given edge-weighted directed graph $D = (\mathcal{V}, \mathcal{A})$, a root $r \in \mathcal{V}$ and a set of terminals $X \subseteq \mathcal{V}$, the objective is to find a minimum cost arborescence rooted at r and spanning all terminals in X (equivalently, there exists a directed path from r to each terminal in X) [13,46]. For some recent works and results related to this problem, see e.g. [1,21,26]. We also point out to a generalization of the Steiner tree problem in which pairs of terminals are given as an input and the goal is to find a minimum cost subgraph which provides a connection for each pair [13,19]. For some other generalizations, see e.g. [14,31,41–43].

Our Results. We show that the Snow Team problem as well as some of its variants are fixed-parameter tractable. In particular, we prove that the ST problem admits a fixed-parameter algorithm with respect to the total number l of facilities and snow team bases, running in $2^{O(l)} \cdot \mathrm{poly}(n)$ time, where $\mathrm{poly}(n)$ is a polynomial in the order n of the input graph (Sect. 2). The proof relies on the algebraic framework introduced by Koutis in [28]. On the other hand, we show that the ST problem (as well as some of its variants) is NP-complete, by a reduction from the Set Cover problem [23] (Sect. 3). Our result on NP-completeness of the ST problem implies NP-completeness of the Agent Clearing Tree problem studied in [32], where the complexity status of the latter has been posed as an open problem. Because of space consideration for the complete proof of the NP-completeness result the reader is referred to a full version of the paper.

Remark. Note that a weaker version of the ST problem with the connectivity requirement removed, that is, we require each facility only to be connected to some snow team base, admits a polynomial-time solution by a straightforward reduction to the minimum path cover problem in directed graphs [36].

Notation. The set of all source vertices in a directed graph D is denoted by $s(D)$. For a directed walk π in D, the set of vertices (arcs) of π is denoted by $V(\pi)$ (resp. $A(\pi)$). For two directed walks π_1 and π_2 in D, where π_2 starts at the ending point of π_1, the concatenation of π_1 and π_2 is denoted by $\pi_1 \circ \pi_2$.

Observe that in a border case, all non-zero length walks of snow ploughs start at the same vertex of the input digraph $D = (\mathcal{V}, \mathcal{A}, F, B)$. Therefore, we may assume that the number of snow ploughs at any vertex is at most $n - 1$, that is, $B(v) \leq n - 1$ for any $v \in \mathcal{V}$, and so the description of any input requires $O(n \log n + m)$ space (recall $m \geq n - 1$).

2 The ST Problem is Fixed-Parameter Tractable

In this section, we prove that the Snow Team problem is fixed-parameter tractable with respect to the number of facilities and snow team bases. The proof relies on the key fact (see Lemmas 1 and 2 below) that to solve the ST problem with the input D, by considering a restricted variant of the problem, we may shift with it to the transitive closure $\mathrm{TC}(D)$ of D and try to detect a particular directed subtree of 'small' order. We solve the latter tree detection problem by a reduction to the problem of testing whether some properly defined multivariate polynomial has a monomial with specific properties, essentially modifying the construction in [29] designed for undirected trees/graphs.

Let us consider the variant of the ST problem, which we shall refer to as the All-ST problem, where we restrict the input only to digraphs $D = (\mathcal{V}, \mathcal{A}, F, B)$ that satisfy $\mathcal{B} = B^{-1}(\mathbb{N}^+) \subseteq \mathcal{F} = F^{-1}(1)$. (In other words, snow team bases can be located only at some facilities.) We have the following lemma.

Lemma 1. *Suppose that the All-ST problem can be solved in $2^{O(k)} \cdot \mathrm{poly}(n)$ time, where k is the number of facilities in the input (restricted) digraph of order n. Then, the ST problem can be solved in $2^{O(l)} \cdot \mathrm{poly}(n)$ time, where l is the total number of facilities and snow team bases in the input digraph of order n.*

Proof. It follows from the fact that a digraph $D = (\mathcal{V}, \mathcal{A}, F, B)$ admits a positive answer to the ST problem if and only if there exists a subset $\mathcal{B}'$ of $\mathcal{B} \setminus (\mathcal{F} \cap \mathcal{B})$ such that the digraph $D' = (\mathcal{V}, \mathcal{A}, F', B')$, where $F'(v) = 1$ for $v \in \mathcal{F} \cup \mathcal{B}'$ and $F'(v) = 0$ otherwise, and $B'(v) = B(v)$ for $v \in \mathcal{B}' \cup (\mathcal{F} \cap \mathcal{B})$ and $B'(v) = 0$ otherwise, admits a positive answer to the All-ST problem. □

Taking into account the above lemma, we now focus on constructing an efficient fixed-parameter algorithm for the All-ST problem, with the restricted input digraph $D = (\mathcal{V}, \mathcal{A}, F, B)$ satisfying $\mathcal{B} = B^{-1}(\mathbb{N}^+) \subseteq \mathcal{F} = F^{-1}(1)$. Let $\mathcal{W}$ be a set of walks (if any) that constitute a positive answer to the All-ST problem in D. We say that $\mathcal{W}$ is *tree-like* if all walks in $\mathcal{W}$ are arc-distinct and the underlying graph of their union includes a Steiner tree for $\mathcal{F}$. Notice that if $\mathcal{W}$ is tree-like, then all walks in $\mathcal{W}$ are just (simple) paths.

Lemma 2. *A (restricted) instance $D = (\mathcal{V}, \mathcal{A}, F, B)$ admits a positive answer to the All-ST problem if and only if the transitive closure $\mathrm{TC}(D) = (\mathcal{V}, \mathcal{A}', F, B)$ of D, with the same vertex-weight functions F and B, admits a positive answer to the All-ST problem with a tree-like set of walks whose underlying graph is of order at most $2|\mathcal{F}| - 1$.*

Since the transitive closure $\mathrm{TC}(D) = (\mathcal{V}, \mathcal{A}', F, B)$ inherits the functions F and B from the restricted instance D, we emphasize that $\mathrm{TC}(D)$ is a proper (restricted) instance to the All-ST problem.

Proof (of Lemma 2). ($\Leftarrow$) It follows from the fact that a directed walk in the transitive closure $\mathrm{TC}(D)$ corresponds to a directed walk in D.

($\Rightarrow$) Assume that the snow ploughs initially located at vertices in $\mathcal{B}$, according to the plough-quantity function B, can simultaneously follow $\mathbf{k}_B$ directed walks $\pi_1, \ldots, \pi_{\mathbf{k}_B}$ whose edges induce a subgraph H of D such that the underlying graph of H includes a Steiner tree of $\mathcal{F}$. Consider now the same walks in the transitive closure $\mathrm{TC}(D)$. To prove the existence of a tree-like solution of 'small' order, the idea is to transform these $\mathbf{k}_B$ walks (if ever needed) into another arc-disjoint $\mathbf{k}_B$ walks. The latter walks have the same starting points as the original ones (so in $\mathcal{B}$ and preserving the plough-quantity function B) and the underlying graph of their union is a Steiner tree of $\mathcal{F}$ (in the underlying graph of $\mathrm{TC}(D)$) having at most $|\mathcal{F}| - 1$ non-terminal vertices.

Our transforming process is based upon the following 2-step modification. First, assume without loss of generality that the walk $\pi_1 = (v_1, \ldots, v_{|\pi_1|})$ has an arc (v_t, v_{t+1}) that is shared with another walk or corresponds to an edge in the underlying graph H of $\bigcup_{i=1}^{\mathbf{k}_B} \pi_i$ that belongs to a cycle (in H). If $t = |\pi_1| - 1$, then we shorten π_1 by deleting its last arc (v_t, v_{t+1}). Otherwise, if $t < |\pi_1| - 1$, then we replace arcs (v_t, v_{t+1}) and (v_{t+1}, v_{t+2}) in π_1 with the arc (v_t, v_{t+2}). One can observe that the underlying graph of the new set of walks is connected, includes a Steiner tree of $\mathcal{F}$, and the vertex v_1 remains the starting vertex of (the new) π_1. But, making walk arc-disjoint or cycle-free may introduce another arc that is shared with at least two walks or another cycle in the underlying graph. However, the length of the modified walk always decreases by one. Consequently, since the initial walks are of the finite lengths, we conclude that applying the above procedure eventually results in a tree-like set $\Pi = \{\pi_1, \ldots, \pi_{\mathbf{k}_B}\}$ of walks, being (simple) paths.

Assume now that in this set Π of arc disjoint paths, there is a non-terminal vertex v of degree at most two in the underlying graph H of $\bigcup_{i=1}^{\mathbf{k}_B} \pi_i$. Without loss of generality assume that v belongs to the path π_1. Similarly as above, if $\deg_H(v) = 1$, then we shorten π_1 by deleting its last arc. Otherwise, if $\deg_H(v) = 2$ and v is not the endpoint of π_1, then modify π_1 be replacing v together with the two arcs of π_1 incident to it by the arc connecting the predecessor and successor of v in π_1, respectively. Observe that since v was a non-terminal vertex in the underlying graph, the underlying graph of (the new) $\bigcup_{i=1}^{\mathbf{k}_B} \pi_i$ is another Steiner tree of $\mathcal{F}$. Moreover, the above modification keeps paths arc-disjoint and does not change the starting vertex of π_1. Therefore, by subsequently replacing all such degree at most two non-terminal vertices, we obtain a tree-like set of k_B paths in the transitive closure $\mathrm{TC}(D)$ such that the underlying graph of their union is a Steiner tree of $\mathcal{F}$ with no degree two vertices except those either belonging to $\mathcal{F}$ or being end-vertices of exactly two paths (in $\mathrm{TC}(D)$). Therefore, we conclude that the number of non-terminal vertices in this underlying graph is at most $|\mathcal{F}| - 1$, which completes our proof of the lemma. □

Now, taking into account the above lemma, a given (restricted) instance $D = (\mathcal{V}, \mathcal{A}, F, B)$ of the All-ST problem can be transformed (in polynomial time) into the answer-equivalent (restricted) instance $\mathrm{TC}(D) = (\mathcal{V}, \mathcal{A}', F, B)$

of the *tree-like-restricted* variant of the All-ST problem in which only tree-like plough paths that together visit at most $2|\mathcal{F}| - 1$ vertices are allowed. Observe that $\mathrm{TC}(D) = (\mathcal{V}, \mathcal{A}', F, B)$ admits a positive answer to the tree-like-restricted All-ST problem if and only if $\mathrm{TC}(D)$ has a subtree $T = (\mathcal{V}_T, \mathcal{A}_T)$ of order at most $2|\mathcal{F}| - 1$ and such that $\mathcal{F} \subseteq \mathcal{V}_T$ and all edges of T can be traversed by at most $\mathbf{k}_B$ snow ploughs following arc-distinct paths starting at vertices in $\mathcal{B}$ (obeying the plough-quantity function B). This motivates us to consider the following problem.

Let $D = (\mathcal{V}, \mathcal{A}, F, B)$ be a directed graph of order n and size m, with two vertex-weight functions $F \colon \mathcal{V} \to \{0, 1\}$ and $B \colon \mathcal{V} \to \mathbb{N}$ such that $B^{-1}(\mathbb{N}^+) \subseteq F^{-1}(1)$, and let $T = (V, A, L)$ be a directed vertex-weighted tree of order t, with a vertex-weight function $L \colon V \to \mathbb{N}$.

The Tree Pattern Embedding problem (TPE)

Does D have a subgraph $H = (\mathcal{V}_H, \mathcal{A}_H)$ isomorphic to T such that $F^{-1}(1) \subseteq \mathcal{V}_H$ and $L(v) \le B(h(v))$ for any vertex v of T, where h is an isomorphism of T and H?

In Subsect. 2.2, we prove Theorem 1 given below which states that there is a randomized algorithm that solves the TPE problem in $O^*(2^t)$ time, where the notation O^* suppresses polynomial terms in the order n of the input graph D. We point out that if the order t of T is less than $|F^{-1}(1)|$ or at least $n + 1$, then the problem becomes trivial, and so, in the following, we assume $|F^{-1}(1)| \le t \le n$.

Theorem 1. *There is a randomized algorithm that solves the* TPE *problem in* $O^*(2^t)$ *time.* □

Suppose that for each vertex $v \in V$, the value $L(v)$ corresponds to the number of snow ploughs located at v that are required to simultaneously traverse (clear) all arcs of T, in an arc-distinct manner, and T admits a positive answer to the TPE problem in the transitive closure $\mathrm{TC}(D) = (\mathcal{V}, \mathcal{A}', F, B)$. Then $\mathrm{TC}(D)$ admits a positive answer to the tree-like-restricted All-ST problem, which immediately implies that D admits a positive answer to the All-ST problem (by Lemma 2). Therefore, taking into account Theorem 1, we are now ready to present the main theorem of this section. For simplicity of presentation, we now assume that a (restricted) directed graph $D = (\mathcal{V}, \mathcal{A}, F, B)$ itself (not its transitive closure) is an instance of the tree-like-restricted All-ST problem.

Theorem 2. *There is a randomized algorithm that solves the tree-like-restricted* All-ST *problem for* $D = (\mathcal{V}, \mathcal{A}, F, B)$ *in* $O^*(144^{|\mathcal{F}|})$ *time, where* $\mathcal{F} = F^{-1}(1)$.

Proof. Keeping in mind Lemma 2, we enumerate all undirected trees of order t, where $|\mathcal{F}| \le t \le 2|\mathcal{F}| - 1$ (and $t \le n$); there are $O(9^{|\mathcal{F}|})$ such candidates [38]. For each such a t-vertex candidate tree, we enumerate all orientations of its edges, in order to obtain a directed tree; there are 2^{t-1} such orientations. Therefore, we have $O(36^{|\mathcal{F}|})$ candidates for a directed oriented tree T of order t, where $|\mathcal{F}| \le t \le 2|\mathcal{F}| - 1$.

For each candidate $T = (V, A)$, we determine in $O(t)$ time how many (at least) snow ploughs, together with their explicit locations at vertices in V, are needed to traverse all arcs of T, in an arc-disjoint manner. This problem can be solved in linear time just by noting that the number of snow ploughs needed at a vertex v is equal to $\max\{0, \deg_{out}(v) - \deg_{in}(v)\}$ (since arcs must be traversed in an arc-disjoint manner). The locations of snow ploughs define a vertex-weight function $L: V \to \mathbb{N}$. We then solve the TPE problem with the instance D and $T = (V, A, L)$ in $O^*(2^t)$ time by Theorem 1.

As already observed, if T admits a positive answer to the TPE problem for D, then D admits a positive answer to the tree-like-restricted All-ST problem. Therefore, by deciding the TPE problem for each of $O(36^{|\mathcal{F}|})$ candidates, taking into account the independence of any two tests, we obtain a randomized algorithm for the restricted ST problem with a running time $O^*(144^{|\mathcal{F}|})$.

Taking into account Lemma 1, we immediately obtain the following corollary.

Corollary 1. *The* ST *problem admits a fixed-parameter algorithm with respect to the total number l of facilities and snow team bases, running in $2^{O(l)} \cdot \text{poly}(n)$ time, where n is the order of the input graph.* □

2.1   Variations on the Snow Team Problem

The first natural variation on the Snow Team problem is its minimization variant, which we shall refer to as the min-ST problem, where for a given input n-vertex digraph $D = (\mathcal{V}, \mathcal{A}, F, B)$, we wish to determine the minimum number of snow ploughs among those available at snow team bases in $\mathcal{B} = B^{-1}(\mathbb{N}^+)$ that admits a positive answer to the (original) Snow Team problem in D. We claim that this problem also admits a fixed-parameter algorithm with respect to the total number l of facilities and snow bases, running in time $2^{O(l)}\text{poly}(n)$, and the solution is concealed in our algorithm for the ST problem. Namely, observe that by enumerating all directed trees of order at most $|\mathcal{F}|$, see the proof of Theorem 2, together with the relevant function L, and checking their embeddability in D, we accidentally solve this minimization problem: the embeddable tree with the minimum sum $\sum_{v \in V} L(v)$ constitutes the answer to min-ST problem.

Corollary 2. *The* min-ST *problem admits a fixed-parameter algorithm with respect to the total number l of facilities and snow team bases, running in $2^{O(l)} \cdot \text{poly}(n)$ time, where n is the order of the input graph.* □

In the case when for the input digraph $D = (\mathcal{V}, \mathcal{A}, F, B)$, not all facilities can be re-connected into one component, that is, D admits a negative answer to the Snow Team problem, one can ask about the maximum number of facilities in $F^{-1}(1)$ that can be re-connected by snow ploughs located with respect to the plough-quantity function B [45]; we shall refer to this problem as the max-ST problem. Since we can enumerate all subsets of $\mathcal{F} = F^{-1}(1)$ in $O^*(2^{|\mathcal{F}|})$ time, taking into account Theorem 2, we obtain the following corollary.

Corollary 3. *The* max-ST *problem admits a fixed-parameter algorithm with respect to the total number l of facilities and snow team bases, running in* $2^{O(l)} \cdot \mathrm{poly}(n)$ *time, where n is the order of the input graph.*   □

Finally, we consider the following variant of the Snow Team problem, called the *Snow Team problem with Unspecified snow team bases* (STU). Given a weight function $F \colon \mathcal{V} \to \{0,1\}$ and an integer $k \geq 1$, do there exist k directed walks in a digraph $D = (\mathcal{V}, \mathcal{A})$ whose edges induce a subgraph H of D such that the set $F^{-1}(1)$ is a subset of the vertex set of H and the underlying graph of H is connected? We claim that for the STU problem, there is also a randomized algorithm with the running time $2^{O(k+l)} \cdot \mathrm{poly}(n)$, where $l = |F^{-1}(1)|$ is the number of facilities, and n is the order of the input graph. The solution is analogous to that for the ST problem. Namely, one can prove a counterpart of Lemma 2 which allows us to restrict ourselves to the restricted variant where only order $O(k + l)$ tree-like solutions are allowed. Then, the restricted variant is solved also using the algorithm for the TPE problem as a subroutine: the function B is the constant function $B(v) = n$, and among all directed tree candidates, we check only those with $\sum_{v \in V} L(v) \leq k$. We omit details.

Corollary 4. *The* STU *problem admits a fixed-parameter algorithm with respect to the number l of facilities and the number k of snow ploughs, running in* $2^{O(k+l)} \cdot \mathrm{poly}(n)$ *time, where n is the order of the input graph.*   □

Observe that if the number k of available snow ploughs is not the part of the input, that is, we ask about the minimum number of walks whose underlying graph includes a Steiner tree for the set of facilities, then this problem seems to be non-fixed-parameter tractable with respect only to the number of facilities. This follows from the fact that the minimum number of snow ploughs is unrelated to the number of facilities in the sense that even for two facilities to be connected, a lot of snow ploughs may be required, see Fig. 1 for an illustration.

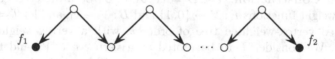

Fig. 1. Two facilities f_1 and f_2 require $n - 1$ snow ploughs, where n is the order of the digraph.

2.2   The Tree Pattern Embedding Problem

In this section, we solve the TPE problem by providing a randomized polynomial-time algorithm when the parameter t is fixed. Our algorithm is based upon the recent algebraic technique using the concepts of monotone arithmetic circuits and monomials, introduced by Koutis in [28], developed by Williams and Koutis in [29, 44], and adapted to some other graph problems, e.g., [8, 9, 20].

A (monotone) *arithmetic circuit* is a directed acyclic graph where each leaf (i.e., vertex of in-degree 0) is labeled either with a variable or a real non-negative constant (*input gates*), each non-leaf vertex is labeled either with + (an *addition gate* with an unbounded fan-in) or with × (a *multiplication gate* with fan-in two), and where a single vertex is distinguished (the *output gate*). Each vertex (gate) of the circuit represents (computes) a polynomial—these are naturally defined by induction on the structure of the circuit starting from its input gates—and we say that a polynomial is *represented* (*computed*) *by an arithmetic circuit* if it is represented (computed) by the output gate of the circuit. Finally, a polynomial that is just a product of variables is called a *monomial*, and a monomial in which each variable occurs at most once is termed a *multilinear monomial* [28, 44].

We shall use a slight generalization of the main results of Koutis and Williams in [28, 44], provided by them in Lemma 1 in [29], which, in terms of our notation, can be expressed as follows.

Fact 1 [29]. *Let $P(x_1, \ldots, x_n, z)$ be a polynomial represented by a monotone arithmetic circuit of size $s(n)$. There is a randomized algorithm that for the input P runs in $O^*(2^k t^2 s(n))$ time and outputs "YES" with high probability if there is a monomial of the form $z^t Q(x_1, \ldots, x_n)$, where $Q(x_1, \ldots, x_n)$ is a multilinear monomial of degree at most k, in the sum-product expansion of P, and always outputs "NO" if there is no such monomial $z^t Q(x_1, \ldots, x_n)$ in the expansion.* □

Taking into account the above fact, for the input digraph $D = (\mathcal{V}, \mathcal{A}, F, B)$ and directed tree $T = (V, A, L)$, the idea is to construct an appropriate polynomial $Q(X, z)$ such that $Q(X, z)$ contains a monomial of the form $z^{|S|} b(X)$, where $b(X)$ is a multilinear polynomial with exactly $|V|$ variables in X and $S = F^{-1}(1) \cup B^{-1}(\mathbb{N}^+)$, if and only if the TPE problem has a solution for the input D and T (see Lemma 3 below).

Polynomial Construction. Let $D = (\mathcal{V}, \mathcal{A}, F, B)$ be a directed graph, with two vertex-weight functions $F \colon \mathcal{V} \to \{0, 1\}$ and $B \colon \mathcal{V} \to \mathbb{N}$, and let $T = (V, A, L)$ be a directed vertex-weighted tree of order t, with a vertex-weight function $L \colon V \to \mathbb{N}$. We consider T to be rooted at a vertex $r \in V$, and for a non-root vertex v of T, we denote the parent of v in T by $p(v)$. Now, for $v \in V$, define two sets $N_T^+(v)$ and $N_T^-(v)$:

$$N_T^+(v) = \{u \in V \mid (u, v) \in A \text{ and } u \neq p(v)\},$$
$$N_T^-(v) = \{u \in V \mid (v, u) \in A \text{ and } u \neq p(v)\}.$$

The idea is to treat T as a 'pattern' that we try to embed into the digraph D, with respect to functions F, B and L. Denote $S = F^{-1}(1) \cup B^{-1}(\mathbb{N}^+)$ for brevity. We say that T has an *S-embedding into* D if the following holds (these are the formal conditions that need to be satisfied for the embedding to be correct):

(E1) There exists an injective function (homomorphism) $f \colon V \to \mathcal{V}$ such that if $(u, v) \in A$, then $(f(u), f(v)) \in \mathcal{A}$.

(E2) $S \subseteq f(V)$, where $f(V) = \{f(v) \,|\, v \in V\}$.
(E3) $L(v) \leq B(f(v))$ for any $v \in V$.

First, for $S \subseteq \mathcal{V}$, $w \in \mathcal{V}$ and $u \in V$, we introduce a particular indicator function, used for fulfilling Conditions (E2) and (E3):

$$
\mathbf{z}_S(u,w) = \begin{cases} z, & \text{if } w \in S \text{ and } L(u) \leq B(w), \\ 1, & \text{if } w \notin S \text{ and } L(u) \leq B(w), \\ 0, & \text{otherwise, i.e., if } L(u) > B(w). \end{cases}
$$

Next, following [29], we define a polynomial $Q(X,T)$ that we then use to test existence of a desired S-embedding of T in D. Namely, we root T at any vertex $r \in V$. Now, a polynomial $Q_{u,w}(X)$, for a subtree T_u of T rooted at $u \in V$ and for a vertex $w \in \mathcal{V}$, is defined inductively (in a bottom up fashion on T) as follows. For each $u \in V$ and for each $w \in \mathcal{V}$: if v is a leaf in T, then

$$
Q_{u,w}(X) = \mathbf{z}_S(u,w) \cdot x_w, \tag{1}
$$

and if u is not a leaf in T, then

$$
Q_{u,w}(X) = \begin{cases} \mathbf{z}_S(u,w) \cdot x_w \cdot Q_{u,w}^+(X) \cdot Q_{u,w}^-(X), & \text{if } N_T^-(u) \neq \emptyset \wedge N_T^+(u) \neq \emptyset, \\ \mathbf{z}_S(u,w) \cdot x_w \cdot Q_{u,w}^+(X), & \text{if } N_T^-(u) = \emptyset, \\ \mathbf{z}_S(u,w) \cdot x_w \cdot Q_{u,w}^-(X), & \text{if } N_T^+(u) = \emptyset, \end{cases} \tag{2}
$$

where

$$
Q_{u,w}^+(X) = \prod_{v \in N_T^+(u)} \left(\sum_{(w',w) \in \mathcal{A}} Q_{v,w'}(X) \right), \tag{3}
$$

$$
Q_{u,w}^-(X) = \prod_{v \in N_T^-(u)} \left(\sum_{(w,w') \in \mathcal{A}} Q_{v,w'}(X) \right). \tag{4}
$$

Finally, the polynomial $Q(X,z)$ is as follows:

$$
Q(X,z) = \sum_{w \in \mathcal{V}} Q_{r,w}(X). \tag{5}
$$

Lemma 3. *The polynomial $Q(X,z)$ contains a monomial of the form $z^{|S|}b(X)$, where $b(X)$ is a multilinear polynomial with exactly t variables in X, if and only if the t-vertex tree T has an S-embedding into D.*

Proof. Consider a vertex u of T and assume that the subtree T_u is of order j. Observe that, by a straightforward induction on the size of a subtree, a monomial $z^q x_{w_1} \cdots x_{w_j}$, where $w_i \in \mathcal{V}$ for each $i \in \{1, \ldots, j\}$, is present in $Q_{u,w_1}(X)$ if and only if the three following conditions hold.

(i) There exists a homomorphism f_u from the vertices of T_u to $w_1, \ldots, w_j$ such that $f_u(u) = w_1$.

(ii) $|\mathcal{S} \cap \{w_1, \ldots, w_j\}| \leq q$ and the equality holds if $w_1, \ldots, w_j$ are pairwise distinct.

(iii) $L(v) \leq B(f_u(v))$ for any vertex v of T_u.

The fact that f_u is a homomorphism follows from the observation that, during construction of $Q_{u,w_1}(X)$ in (3) and (4), a neighbor v of u is mapped to a node w' of D in such a way that if $(v, u) \in A$ then $(w', w) \in \mathcal{A}$ (see (3)), and if $(u, v) \in A$ then $(w, w') \in \mathcal{A}$ (see (4)). Conditions (ii) and (iii) are ensured by appropriate usage of the indicator function in (1), namely, if u is mapped to w in a homomorphism corresponding to $Q_{u,w}(X)$, then we add the multiplicative factor of z to $Q_{u,w}(X)$ provided that $L(v) \leq B(w)$.

Thus, we obtain that $Q(X, z)$ has a multilinear polynomial $z^{|\mathcal{S}|} x_{w_1} \cdots x_{w_t}$ if and only if T has an $\mathcal{S}$-embedding into D. $\qquad \square$

Observe that the polynomial $Q(X, z)$ and the auxiliary polynomials $Q_{u,w}^+(X)$, $Q_{u,w}^-(X)$ can be represented by a monotone arithmetic circuit of size polynomial in the order n of the input digraph D. To start with, we need $n + 1$ input gates for the variables corresponding to vertices of D, and the auxiliary variable z. With each of the aforementioned polynomials, we associate a gate representing it, in total $O(tn)$ gates. In order to implement the recurrences defining the polynomials, assuming unbounded fan-in of addition gates, we need $O(n)$ auxiliary gates for each recurrence involving large products. Thus, the resulting circuit is of size $O(n^3)$. Hence, by Fact 1 combined with Lemma 3, we conclude that the existence of an $\mathcal{S}$-embedding of the t-vertex tree T into D can be decided in $O^*(2^{|\mathcal{S}|})$ time. Consequently, since $|\mathcal{S}| \leq t$, we obtain Theorem 1 by the definition of an $\mathcal{S}$-embedding.

2.3   Embedding Directed Forests

We observe that the above approach can be adapted to the case when we want to embed a directed forest $T = (\mathcal{V}, \mathcal{A}, F, B)$ of order t into a directed graph. All we need is to build a relevant polynomial for each rooted directed tree-component of T, and then to consider the product $S(X, T)$ of these polynomials, asking about the existence of a monomial of the form $z^{|\mathcal{S}|} b(X)$, where $b(X)$ is a multilinear polynomial with exactly t variables in X. Also, by similar approach, we may consider and can solve (simpler) variants of our embedding problem without the weight function F or without the weight functions B and L; details are omitted.

3   The ST Problem is Hard

Based upon a polynomial-time reduction from the Set Cover problem [23], we can prove the following theorem. (Because of space consideration for the complete proof of the NP-completeness result the reader is referred to a full version of the paper.)

Theorem 3. *The* ST *problem is strongly NP-complete even for directed acyclic graphs* $D = (\mathcal{V}, \mathcal{A}, F, B)$ *with* $F^{-1}(1) = \mathcal{V}$ *and* $B(v) = 1$ *if* v *is a source vertex in* D *and* $B(v) = 0$ *otherwise.* $\qquad \square$

No Pre-specified Positions of Snow Ploughs. We claim that the Snow Team problem with Unspecified snow team bases is also NP-complete. The reduction is exactly the same as for the ST problem. All we need is to observe that if facilities are located at all vertices of the input digraph, then the number of snow ploughs sufficient to solve the STU problem is bounded from below by the number of source vertices in the digraph, since there must be at least one snow plough at each of its source vertices. Furthermore, without loss of generality we may assume that in any feasible solution of k walks, all snow ploughs are initially located at source vertices.

Corollary 5. *The* STU *problem is strongly NP-complete even for directed acyclic graphs* $D = (\mathcal{V}, \mathcal{A}, F)$ *with* $F^{-1}(1) = \mathcal{V}$ *and* k *equals the number of source vertices in* D. ☐

Since by setting $F(v) = 1$ for each vertex v of the input digraph, the STU problem becomes just the Agent Clearing Tree problem (ACT) studied in [32]. Hence, we immediately obtain the following corollary resolving the open problem of the complexity status of ACT posed in [32].

Corollary 6. *The* ACT *problem is NP-complete.* ☐

4   Open Problem

In all our variants of the Snow Team problem, we assumed that a snow plough can traverse arbitrary number of arcs. However, from a practical point of view, it is more natural to assume that each snow plough, called an *s-plough*, can traverse and clear only the fixed number s of arcs [35]. Observe that in this case, the key Lemma 2 does not hold, which immediately makes our algebraic approach unfeasible for the Snow Team problem with s-ploughs, so this variant requires further studies.

References

1. Abdi, A., Feldmann, A.E., Guenin, B., Könemann, J., Sanità, L.: Lehman's theorem and the directed Steiner tree problem. SIAM J. Discret. Math. **30**(1), 141–153 (2016)
2. Alspach, B., Dyer, D., Hanson, D., Yang, B.: Arc searching digraphs without jumping. In: Dress, A., Xu, Y., Zhu, B. (eds.) COCOA 2007. LNCS, vol. 4616, pp. 354–365. Springer, Heidelberg (2007). doi:10.1007/978-3-540-73556-4_37
3. Amiri, S.A., Kaiser, L., Kreutzer, S., Rabinovich, R., Siebertz, S.: Graph searching games and width measures for directed graphs. In: 32nd International Symposium on Theoretical Aspects of Computer Science (STACS 2015), pp. 34–47 (2015)
4. Barrière, L., Flocchini, P., Fomin, F.V., Fraigniaud, P., Nisse, N., Santoro, N., Thilikos, D.M.: Connected graph searching. Inf. Comput. **219**, 1–16 (2012)
5. Beerenwinkel, N., Beretta, S., Bonizzoni, P., Dondi, R., Pirola, Y.: Covering pairs in directed acyclic graphs. Comput. J. **58**(7), 1673–1686 (2015)

6. Berwanger, D., Dawar, A., Hunter, P., Kreutzer, S., Obdržálek, J.: The dag-width of directed graphs. J. Comb. Theory Ser. B **102**(4), 900–923 (2012)
7. Best, M.J., Gupta, A., Thilikos, D.M., Zoros, D.: Contraction obstructions for connected graph searching. Discret. Appl. Math. **209**, 27–47 (2016)
8. Björklund, A., Husfeldt, T., Taslaman, N.: Shortest cycle through specified elements. In: Twenty-Third Annual ACM-SIAM Symposium on Discrete Algorithms (SODA 2012), pp. 1747–1753 (2012)
9. Björklund, A., Kaski, P., Kowalik, L.: Constrained multilinear detection and generalized graph motifs. Algorithmica **74**(2), 947–967 (2016)
10. Borowiecki, P., Dereniowski, D., Pralat, P.: Brushing with additional cleaning restrictions. Theor. Comput. Sci. **557**, 76–86 (2014)
11. Breisch, R.L.: An intuitive approach to speleotopology. Southwest. Cavers (A Publ. Southwest. Region Natl. Speleol. Soc.) **6**, 72–78 (1967)
12. Bryant, D., Francetic, N., Gordinowicz, P., Pike, D.A., Pralat, P.: Brushing without capacity restrictions. Discret. Appl. Math. **170**, 33–45 (2014)
13. Charikar, M., Chekuri, C., Cheung, T.-Y., Dai, Z., Goel, A., Guha, S., Li, M.: Approximation algorithms for directed Steiner problems. J. Algorithms **33**(1), 73–91 (1999)
14. Chitnis, R.H., Esfandiari, H., Hajiaghayi, M.T., Khandekar, R., Kortsarz, G., Seddighin, S.: A tight algorithm for strongly connected steiner subgraph on two terminals with demands (extended abstract). In: Cygan, M., Heggernes, P. (eds.) IPEC 2014. LNCS, vol. 8894, pp. 159–171. Springer, Cham (2014). doi:10.1007/978-3-319-13524-3_14
15. de Oliveira Oliveira, M.: An algorithmic metatheorem for directed treewidth. Discret. Appl. Math. **204**, 49–76 (2016)
16. Dereniowski, D.: Connected searching of weighted trees. Theor. Comp. Sci. **412**, 5700–5713 (2011)
17. Dereniowski, D.: From pathwidth to connected pathwidth. SIAM J. Discret. Math. **26**(4), 1709–1732 (2012)
18. Dyer, D., Yang, B., Yaşar, Ö.: On the fast searching problem. In: Fleischer, R., Xu, J. (eds.) AAIM 2008. LNCS, vol. 5034, pp. 143–154. Springer, Heidelberg (2008). doi:10.1007/978-3-540-68880-8_15
19. Feldmann, A.E., Marx, D.: The complexity landscape of fixed-parameter directed Steiner network problems. In: 43rd International Colloquium on Automata, Languages, Programming (ICALP), pp. 27:1–27:14 (2016)
20. Fomin, F.V., Lokshtanov, D., Raman, V., Saurabh, S., Rao, B.V.R.: Faster algorithms for finding and counting subgraphs. J. Comput. Syst. Sci. **78**(3), 698–706 (2012)
21. Friggstad, Z., Könemann, J., Kun-Ko, Y., Louis, A., Shadravan, M., Tulsiani, M.: Linear programming hierarchies suffice for directed steiner tree. In: Lee, J., Vygen, J. (eds.) IPCO 2014. LNCS, vol. 8494, pp. 285–296. Springer, Cham (2014). doi:10.1007/978-3-319-07557-0_24
22. Gabow, H.N., Maheshwari, S.N., Osterweil, L.J.: On two problems in the generation of program test paths. IEEE Trans. Softw. Eng. **2**(3), 227–231 (1976)
23. Garey, M.R., Johnson, D.S.: Computers and Intractability: A Guide to the Theory of NP-Completeness. W. H. Freeman & Co., New York (1979)
24. Gaspers, S., Messinger, M.-E., Nowakowski, R.J., Pralat, P.: Parallel cleaning of a network with brushes. Discret. Appl. Math. **158**(5), 467–478 (2010)
25. Hunter, P., Kreutzer, S.: Digraph measures: Kelly decompositions, games, and orderings. Theor. Comput. Sci. **399**(3), 206–219 (2008)

26. Jones, M., Lokshtanov, D., Ramanujan, M.S., Saurabh, S., Suchý, O.: Parameterized complexity of directed Steiner tree on sparse graphs. In: Bodlaender, H.L., Italiano, G.F. (eds.) ESA 2013. LNCS, vol. 8125, pp. 671–682. Springer, Heidelberg (2013). doi:10.1007/978-3-642-40450-4_57

27. Kolman, P., Pangrác, O.: On the complexity of paths avoiding forbidden pairs. Discret. Appl. Math. **157**(13), 2871–2876 (2009)

28. Koutis, I.: Faster algebraic algorithms for path and packing problems. In: Aceto, L., Damgård, I., Goldberg, L.A., Halldórsson, M.M., Ingólfsdóttir, A., Walukiewicz, I. (eds.) ICALP 2008. LNCS, vol. 5125, pp. 575–586. Springer, Heidelberg (2008). doi:10.1007/978-3-540-70575-8_47

29. Koutis, I., Williams, R.: Limits and applications of group algebras for parameterized problems. ACM Trans. Algorithms **12**(3), 31 (2016)

30. Ková̂c̄, J.: Complexity of the path avoiding forbidden pairs problem revisited. Discret. Appl. Math. **161**(10–11), 1506–1512 (2013)

31. Laekhanukit, B.: Approximating directed Steiner problems via tree embedding. In: 43rd International Colloquium on Automata, Languages, Programming (ICALP), 74:1–74:13 (2016)

32. Levcopoulos, C., Lingas, A., Nilsson, B.J., Żyliński, P.: Clearing connections by few agents. In: Ferro, A., Luccio, F., Widmayer, P. (eds.) FUN 2014. LNCS, vol. 8496, pp. 289–300. Springer, Cham (2014). doi:10.1007/978-3-319-07890-8_25

33. Messinger, M.-E., Nowakowski, R.J., Pralat, P.: Cleaning a network with brushes. Theor. Comput. Sci. **399**(3), 191–205 (2008)

34. Messinger, M.-E., Nowakowski, R.J., Pralat, P.: Cleaning with brooms. Graphs Comb. **27**(2), 251–267 (2011)

35. Ntafos, S.C., Gonzalez, T.: On the computational complexity of path cover problems. J. Comput. Syst. Sci. **29**(2), 225–242 (1984)

36. Ntafos, S.C., Hakimi, S.L.: On path cover problems in digraphs and applications to program testing. IEEE Trans. Softw. Eng. **5**(5), 520–529 (1979)

37. Ntafos, S.C., Hakimi, S.L.: On structured digraphs and program testing. IEEE Trans. Comput. **C–30**(1), 67–77 (1981)

38. Otter, R.: The number of trees. Ann. Math. Second Ser. **49**(3), 583–599 (1948)

39. Parsons, T.D.: Pursuit-evasion in a graph. In: Alavi, Y., Lick, D.R. (eds.) Theory and Applications of Graphs. LNM, vol. 642, pp. 426–441. Springer, Heidelberg (1978). doi:10.1007/BFb0070400

40. Petrov, N.N.: A problem of pursuit in the absence of information on the pursued. Differentsial'nye Uravneniya **18**, 1345–1352 (1982)

41. Suchý, O.: On directed Steiner trees with multiple roots. In: Heggernes, P. (ed.) WG 2016. LNCS, vol. 9941, pp. 257–268. Springer, Heidelberg (2016). doi:10.1007/978-3-662-53536-3_22

42. Watel, D., Weisser, M.-A., Bentz, C., Barth, D.: Directed Steiner tree with branching constraint. In: Cai, Z., Zelikovsky, A., Bourgeois, A. (eds.) COCOON 2014. LNCS, vol. 8591, pp. 263–275. Springer, Cham (2014). doi:10.1007/978-3-319-08783-2_23

43. Watel, D., Weisser, M.-A., Bentz, C., Barth, D.: Directed Steiner trees with diffusion costs. J. Comb. Optim. **32**(4), 1089–1106 (2016)

44. Williams, R.: Finding paths of length k in $O^*(2^k)$ time. Inf. Process. Lett. **109**(6), 315–318 (2009)

45. Zientara, M.: Personal communication (2016)

46. Zosin, L., Khuller, S.: On directed Steiner trees. In: Thirteenth Annual ACM-SIAM Symposium on Discrete Algorithms (SODA 2002), pp. 59–63 (2002)

FO Model Checking on Map Graphs

Kord Eickmeyer[1]([✉]) and Ken-ichi Kawarabayashi[2]

[1] Department of Mathematics, Technical University Darmstadt,
Schlossgartenstr. 7, 64289 Darmstadt, Germany
eickmeyer@mathematik.tu-darmstadt.de
[2] Tokyo and JST, ERATO, Kawarabayashi Large Graph Project,
National Institute of Informatics,
Hitotsubashi 2-1-2, Chiyoda-ku, Tokyo 101-8430, Japan
k_keniti@nii.ac.jp

Abstract. For first-order logic model checking on monotone graph classes the borderline between tractable and intractable is well charted: it is tractable on all nowhere dense classes of graphs, and this is essentially the limit. In contrast to this, there are few results concerning the tractability of model checking on general, i.e. not necessarily monotone, graph classes.

We show that model checking for first-order logic on map graphs is fixed-parameter tractable, when parameterised by the size of the input formula. Map graphs are a geometrically defined class of graphs similar to planar graphs, but here each vertex of a graph is drawn homeomorphic to a closed disk in the plane in such a way that two vertices are adjacent if, and only if, the corresponding disks intersect. Map graphs may contain arbitrarily large cliques, and are not closed under edge removal.

Our algorithm works by efficiently transforming a given map graph into a nowhere dense graph in which the original graph is first-order interpretable. As a by-product of this technique we also obtain a model checking algorithm for FO on squares of trees.

1 Introduction

Starting with Courcelle's groundbreaking result [2] that model checking for monadic second-order logic (MSO) is fixed-parameter tractable on graphs of bounded tree width, efficient algorithms for model checking on restricted classes of structures have been thoroughly investigated. Since many well-known algorithmic problems on graphs (such as finding cliques, dominating sets, or vertex covers of a given size) can be rephrased as model checking problems, efficient algorithms for model checking immediately yield efficient algorithms for these problems as well. Therefore results showing the existence of such model checking algorithms are commonly referred to as *algorithmic meta theorems*.

For first-order logic (FO), model checking has been shown to be fixed-parameter tractable on a wide range of graph classes, cf. [4,6,9,14]. These results hinge on the fact that FO has very strong locality properties, and clever graph-theoretic tools for small-diameter graphs. In particular, the methods used in

© Springer-Verlag GmbH Germany 2017
R. Klasing and M. Zeitoun (Eds.): FCT 2017, LNCS 10472, pp. 204–216, 2017.
DOI: 10.1007/978-3-662-55751-8_17

proving these results are well-behaved under edge-removal. A graph class which is closed under taking (not necessarily induced) subgraphs is called *monotone*, and for monotone graph classes, FO model checking is fixed-parameter tractable if, and only if, the graph class is nowhere dense [14] (modulo some minor technicalities).

Thus to overcome the barrier of sparse graphs, entirely different algorithmic techniques are necessary. Previous results for model checking on non-sparse graph classes are few. In particular, Courcelle's result has been generalised to graphs of bounded clique width [3], and there are results for FO model checking on partially ordered sets of bounded width [10] and on certain interval graphs, if these graphs are given as an interval representation [12]. Recently, Gajarský et al. obtained an efficient model checking algorithmic for FO on graphs that are FO-interpretable in graphs of bounded degree [11].

In this work we obtain a new algorithmic meta theorem for first-order logic:

Theorem 1. *The model checking problem for first-order logic on vertex coloured map graphs is fixed-parameter tractable, parameterised by the size of the input formula.*

Map graphs have been introduced by Chen et al. [1] as a generalisation of planar graphs. They are defined as graphs which can be drawn in the plane in a way such that to every vertex of the graph a region homeomorphic to a closed disk is drawn, and the regions corresponding to vertices u and v touch if, and only if, uv is an edge of the graph. Here, two regions are considered to touch already if they intersect (as point sets) in a single point. If instead one insists that regions intersect in a set containing a homeomorphic image of a line segment, one obtains the familiar notion of planar graphs.

Note that unlike planar graphs, map graphs may contain arbitrarily large cliques, and the class of map graphs is not closed under taking arbitrary subgraphs. The recognition problem for map graphs, i.e. deciding for a given an abstract graph $G = (V, E)$ whether it can be realised as a map graph, has been shown to be feasible in polynomial time by Thorup in the extended abstract [22]. However, Thorup's algorithm has a running time of roughly $O(|V|^{120})$, and no complete description of it has been published. Moreover, it does not produce a witness graph (which is a combinatorical description of a map drawing) if the input graph is found to be a map graph. Recently, Mnich et al. [19] have given a linear algorithm that decides whether a map graph has an outerplanar witness graph, and computes one if the answer is yes.

The graph input to our algorithm is given as an abstract graph (and not as, say, a geometric representation as a map), and we do not rely on Thorup's algorithm nor any results from [22]. Instead, we use Chen et al.'s classification of cliques in a map graph and show how to efficiently compute, given a map graph G, a graph R in which G is first-order interpretable and such that the class of all graphs arising in this way is nowhere dense. In fact, G is an induced subgraph of the *square* of R, i.e. the graph with the same vertex set as R in which two vertices are adjacent if, and only if, they have distance at most 2 in R. In Sect. 7

we show how know results on squares and square roots of graphs can be used to obtain further algorithmic meta theorems.

2 Preliminaries

2.1 Logic

We use standard definitions for first-order logic (FO), cf. [7,8,16]. In particular, $\perp$ and $\top$ denote false and true, respectively. We will only be dealing with finite, vertex coloured graphs as logical structures, i.e. finite structures with vocabularies of the form $\{E, P_1, \ldots, P_k\}$, with a binary edge relation E and unary predicates $P_1, \ldots, P_k$.

2.2 Graphs

We will be dealing with finite simple (i.e. loop-free and without multiple edges) undirected graphs, cf. [5,24] for an in-depth introduction. Thus a *graph* $G = (V, E)$ consists of some finite set V of *vertices* and a set $E \subseteq \binom{V}{2}$ of *edges*. A *clique* $C \subseteq V$ is a set of pairwise adjacent vertices, i.e. such that $uv \in E$ for all $u, v \in C$, $u \neq v$. The *neighbourhood* of a vertex $v \in V$ is defined as

$$N(v) := \{w \in V \mid vw \in E\}.$$

For a set $W \subseteq V$ of vertices we denote by $E[W] \subseteq E$ the set of edges that have both endpoints in W.

A *topological embedding* of a graph $H = (W, F)$ into a graph $G = (V, E)$, is an injective mapping $\iota : W \to V$ together with a set $\{p_{xy} \mid xy \in F\}$ of paths in G such that

- each path p_{xy} connects $\iota(x)$ to $\iota(y)$ and
- the paths p_{xy} share no internal vertices, and no $\iota(z)$ is an internal vertex of any of these paths.

If a topological embedding of H into G exists we say that H is a *topological minor* of G, written $H \preceq G$.

If all paths p_{xy} of a topological embedding have length at most r then the embedding is said to be *r-shallow*. The notion of an *r-shallow topological minor*, written $\preceq_r$, is defined accordingly. A class $\mathcal{C}$ of graphs is called *nowhere dense* if for every r there is an m with $K_m \not\preceq_r G$ for any $G \in \mathcal{C}$.

We relax these notions by allowing vertices of G to be used more than once but at most c times, for a constant c. Thus H is a topological minor *of complexity* $\leq c$ of G (written $H \preceq^c G$) if there is a mapping $\iota : W \to V$ and paths p_{xy} connecting $\iota(x)$ to $\iota(y)$ for every $xy \in F$ such that no $v \in V$ is used more than c times as an internal vertex of some p_{xy} or as $\iota(x)$. Similarly for $H \preceq_r^c G$.

It is well known that $K_5 \not\preceq G$ for any planar graph G. While for every graph H and every $c \geq 2$ there is a planar graph G with $H \preceq^c G$, for every $c, r \in N$ there is some $m = m(c, r) \in \mathbb{N}$ such that $K_m \not\preceq_r^c G$ for any planar graph G (cf. [21, Sect. 4.8]).

2.3   Map Graphs

A graph $G = (V, E)$ is a *map graph* if there are sets $D_v \subseteq \mathbb{R}^2$, one for each $v \in V$, such that

- each D_v is homeomorphic to a closed disc (i.e. homeomorphic to $\{(x, y) \in \mathbb{R}^2 \mid x^2 + y^2 \le 1\}$,
- D_v and D_w intersect only on their boundaries, for $v \ne w$, and
- $D_v \cap D_w \ne \emptyset$ if, and only if, $vw \in E$.

Chen et al. showed that G is a map graph if, and only if, there is a planar bipartite graph $H = (V \cup P, F)$ having the vertices of G as one side of its bipartition and such that $uv \in E$ iff $up, vp \in F$ for some $p \in P$; moreover we may assume that $|P| \le 4|V|$ [1, Theorem 2.2, Lemma 2.3]. Such a graph H is called a *witness* for G. We call the elements of P the *points* of the witness, and refer the term vertex to elements of V.[1]

By [1, Theorem 3.1], every clique C in a map graph is of one (or more) of the following types (cf. Fig. 1):

pizza there is a $p \in P$ such that $pv \in F$ for all $v \in C$, or

pizza-with-crust there is a $v \in C$ and a $p \in P$ such that $pv \notin F$ but $pw \in F$ for all $w \in C \setminus \{v\}$, or

hamantasch there are $p, q, r \in P$ such that every $v \in C$ is adjacent to at least two of these points, or

rice ball $|C| \le 4$ and any $p \in P$ is adjacent to at most two vertices in C.

Furthermore, the number of *maximal* cliques in a map graph with n vertices is bounded by $27n$ [1, Theorem 3.2].

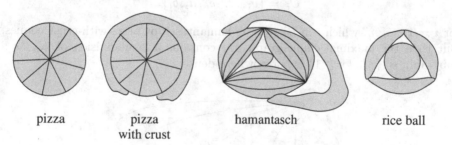

pizza   pizza   hamantasch   rice ball
                with crust

Fig. 1. The possible types of cliques in map graphs.

3   The Maximal Clique Graph

Let $G = (V, E)$ be a map graph and $H = (V \cup P, F)$ a planar witness graph for it. Let $C_1, \ldots, C_m \subseteq V$ be the maximal cliques in G. Then $m \le 27 \cdot |V|$. We

---

[1] Elements of V are referred to as *nations* by Chen et al.

define the maximal clique graph $M = (V \cup W, F_M)$ as the bipartite graph with $W = \{w_C \mid C = C_1, \ldots, C_m\}$ and $v \in V$ and $w_C \in W$ adjacent if, and only if, $v \in C$.

Note that

- any witness graph H is, by definition, planar, but we do not know how to efficiently compute one from G,
- we can recover G from coloured versions of both H and M by first-order interpretations,
- we can compute M from G in polynomial time, because we can enumerate the maximal cliques of G in output-polynomial time [23] and there are only linearly many.

In Sect. 5 we define a graph similar to M which will indeed be nowhere dense. Before doing so we give a sequence of a map graphs G_n for which $K_n \preceq_2 M_n$, i.e. the class of maximal clique graphs of map graphs is *not* nowhere dense.

Let $G_n := (V_n, E_n)$ with

$$V_n := \underbrace{\{v_1, \ldots, v_n\}}_{=:V} \cup \underbrace{\{a_1, \ldots, a_m\}}_{=:A} \cup \underbrace{\{b_1, \ldots, b_m\}}_{=:B},$$

$$E_n := \binom{V}{2} \cup \binom{A}{2} \cup \binom{B}{2} \cup \{a_i b_i \mid 1 \leq i \leq m\} \cup$$
$$\{v_i a_j, v_i b_j \mid 1 \leq i \leq n, 1 \leq j \leq m\},$$

where $m := \binom{n}{2}$. This is a map graph, as witnessed by Fig. 2. The maximal cliques are the sets $V \cup A$, $V \cup B$, and

$$C_i := \{v_1, \ldots, v_n, a_i, b_i\}$$

for $i = 1, \ldots, m$, which can be seen as hamantasch or pizza-with-crust cliques. But then the maximal clique graph M_n contains a 2-subdivision of K_n as a subgraph, so $\{M_n \mid n \geq 1\}$ is not nowhere dense.

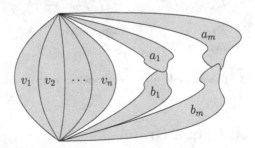

Fig. 2. The map graphs G_n whose maximal clique graphs contain large cliques as (even topological) 2-shallow minors.

We will deal with this by making sure that

- hamantasch cliques are of bounded size, and
- pizza-with-crust cliques with identical centre points are treated only once.

This will be the content of Sects. 4 and 5.

4 Neighbourhood Equivalence

We call two vertices $v, w \in V$ in a graph $G = (V, E)$ *neighbourhood equivalent*, written $v \sim w$, if

$$N(v) \setminus \{v, w\} = N(w) \setminus \{v, w\}.$$

This defines an equivalence relation on V; for transitivity note that $u \sim v$ and $v \sim w$ imply that $\{u, v, w\}$ is either a clique or an independent set. This equivalence relation has been studied before, e.g. in [11,15], where it is called twin relation. For the purpose of model checking we may prune a graph by removing neighbourhood equivalent vertices, as long as we keep track of their number, up to a given threshold.

Define the graph $G/\sim$ to be the graph with vertex set

$$V/\sim = \{[v] \mid v \in V\}$$

and vertices $[v] \neq [w]$ adjacent if, and only if, $vw \in E$. Note that this is independent of the particular choice of representatives, since $v \sim v'$ and $w \sim w'$ imply that $(vw \in E) \Leftrightarrow (v'w' \in E)$.

In $G/\sim$ no two vertices have identical neighbourhoods:

$$N([v]) = N([w]) \quad \Rightarrow \quad [v] = [w]$$

To see this, assume $[v] \neq [w]$. Wlog there is some $u \in V$ such that $uv \in E$ but $uw \notin E$. But then $[u][v] \in E/\sim$ but $[u][w] \notin E/\sim$, so $N([v]) \neq N([w])$.

We add information on the size of $[v]$ to the graph $G/\sim$ using unary predicates as follow: for $i \in \mathbb{N}$ we set

$$P_i(G) := \{[v] \mid \|[v]\| \geq i\} \subseteq V/\sim,$$

and we let $G_{\sim,m}$ be the graph $G/\sim$ together with the unary predicates $P_1, \ldots, P_m$. Note that $G_{\sim,m}$ can be computed on input G and m in polynomial time, e.g. using colour refinement techniques. Our definition of $G_{\sim,m}$ is motivated by the following lemma, whose (straight-forward) proof we omit here:

Lemma 1. *For every $\varphi \in FO$ of quantifier rank m there is a $\psi \in FO$ of the same quantifier rank such that*

$$G \models \varphi \quad \Leftrightarrow \quad G_{\sim,m} \models \psi$$

for every graph G.

5   3-Connected Map Graphs

In the following we show how to perform FO model checking on 3-connected map graphs. Using Lemma 1 we may, and will, assume that no two vertices of the input graph have identical neighbourhoods. We start by showing that this together with 3-connectedness implies that hamantasch cliques have size at most 9.

Lemma 2. *Let $G = (V, E)$ be a 3-connected map graph, $H = (V \cup P, F)$ a witness graph for G and $p, q \in P$. Consider the vertices of G adjacent to both p and q:*

$$N_{p,q} := N(p) \cap N(q) = \{v \in V \mid pv, qv \in F\}.$$

Then all but (possibly) two of the vertices in $N_{p,q}$ have identical neighbourhoods in G: There are vertices $u, w \in V$ such that

$$N(v) = N(v')$$

for all $v, v' \in N_{p,q} \setminus \{u, w\}$.

Proof. We fix an arbitrary drawing of H. If $|N_{p,q}| \leq 2$ we are done. Otherwise we may order the vertices of $N_{p,q}$ as $\{v_1, v_2, \ldots, v_\ell\}$ in such a way that for any $1 \leq i < j < k \leq \ell$, the vertex v_j is inside the region bounded by the cycle pv_iqv_kp in the drawing of H.

Now let $v \in V$ be a neighbour of v_j in G, for $1 < j < \ell$. Then there is a point $r \in P$ such that both v_j and v are adjacent to r in H. Then r must be inside the (closed) region bounded by $pv_{j-1}qv_{j+1}p$. If either $r = p$ or $r = q$, then v is adjacent to all vertices in $N_{p,q}$ and we are done.

Otherwise the point r must either be inside $pv_{j-1}qv_jp$ or inside $pv_jqv_{j+1}p$, but then either $\{v_{j-1}, v_j\}$ or $\{v_j, v_{j+1}\}$ disconnect v from v_{j+1} or v_{j-1}, contradicting the 3-connectedness of G. The lemma follows by choosing $u = v_1$ and $w = v_\ell$. ∎

Since we assume all vertices of our graph to have unique neighbourhoods, it follows that

$$|N_{p,q}| \leq 3$$

for all $p, q \in P$. Thus if $C \subseteq V$ is a hamantasch-clique, then $|C| \leq 9$, because

$$C = N_{p,q} \cup N_{q,r} \cup N_{p,r}$$

for a suitable choice of points $p, q, r \in P$ in a witness $H = (V \cup P, F)$.

The Reduced Maximal Clique Graphs

Starting from a given 3-connected map graph G we now compute a graph which we call *reduced maximal clique graph* $M(G)$. Let $C_1, \ldots, C_m \subseteq V$ be the maximal cliques of G, ordered in decreasing size:

$$|C_1| \geq |C_2| \geq \cdots \geq |C_m|,$$

and cliques of the same size may appear in arbitrary order. By a result of Chen et al. [1, Theorem 3.2] we know that $m \leq 27 \cdot |V|$, and using an algorithm of Tsukiyama et al. [23] we may compute all of these in polynomial time.

We construct a bipartite graph $R = (V \cup U, A)$ such that for every $v, w \in V$ there is a $u \in U$ adjacent to both v and w if, and only if, $vw \in E$. In this case we say that the edge vw is *covered* by u. We process the cliques in descending size, keeping a set $S_i \subseteq E$ of edges which are already covered, a set $T_i \subseteq E$ of edges which will be covered by individual vertices, and sets U_i, A_i of vertices and edges in the graph that is created. Initially we have

$$S_0 := \emptyset, \quad T_0 := \emptyset, \quad U_0 := \emptyset, \text{ and } \quad A_0 := \emptyset$$

and do the following for $i = 1, \ldots, m$:

(R1) $E[C_i] \subseteq S_{i-1}$ then all edges of C_i are already covered and we ignore C_i (setting $S_i = S_{i-1}$, $T_i = T_{i-1}$ and so on),

(R2) otherwise, if there is a vertex $v \in C_i$ such that $E[C_i \setminus \{v\}] \subseteq S_{i-1}$ we set

$$S_i := S_{i-1}, \qquad\qquad T_i := T_{i-1} \cup \{vw \mid w \in C_i, w \neq v\},$$
$$U_i := U_{i-1}, \text{ and} \qquad A_i := A_{i-1},$$

(R3) otherwise, if $|C_i| \leq 9$, we treat all edges in C_i as special edges:

$$S_i := S_{i-1}$$
$$T_i := T_{i-1} \cup E[C_i]$$
$$U_i := U_{i-1}, \text{ and}$$
$$A_i := A_{i-1}$$

(R4) In all other cases we introduce a new vertex u_i connected to all vertices in C_i:

$$S_i := S_{i-1} \cup E[C_i] \qquad\qquad T_i := T_{i-1}$$
$$U_i := U_{i-1} \cup \{u_i\}, \text{ and} \qquad A_i := A_{i-1} \cup \{u_i v \mid v \in C_i\}$$

At the end of this process we have a bipartite graph $(V \cup U_m, A_m)$, plus a set T_m of edges. For any edge $vw \in T_m \setminus S_m$ we add a new vertex u to U_m and connect it only to v and w. We call the resulting graph $R = R(G) := (V \cup U, A)$ the *reduced maximal clique graph* for G. Note that R is not uniquely determined by G but is also influenced by choices the algorithm makes at various stages.

By construction, the graph G is a half-square of R, i.e. for any $v, w \in V$ the edge vw is in E if, and only if, there is a $u \in U$ such that $uv, uw \in A$. Since R is not necessarily planar, it need not be a witness graph of G, but we will now show that the class of graphs arising in this way from map graphs is nowhere dense. Since we can easily recover G in a coloured version of R by a first-order interpretation, and since R can be constructed from G in polynomial time, we may use Grohe et al.'s model checking algorithm for first-order logic on nowhere dense classes of graphs [14] to obtain a model checking algorithm for first-order logic on 3-connected map graphs.

To show that the class

$$\{R(G) \mid G \text{ is a 3-connected map graph}\}$$

is indeed nowhere dense we choose an $r \in \mathbb{N}$ and assume that $K_{2m} \preceq_r R$. We will now show that in this case $K_m \preceq_r^c H'$ for some planar graph H', for an absolute constant c whose value will become apparent during the proof. Since this is not possible for large enough values of m, we conclude that for every r there is an m with $K_{2m} \npreceq_r R$, and thus the class of reduced maximal clique graphs is nowhere dense.

We fix a witness graph $H = (V \cup P, F)$ of G with an arbitrary drawing. Suppose $K_{2m} \preceq_r R$. This means that there are vertices $x_1, \ldots, x_{2m} \in V \cup U$ and pairwise internally vertex-disjoint paths p_{ij} connecting x_i and x_j, for $1 \le i < j \le 2m$. If we could map these vertices and paths *injectively* into H, we would obtain a topological K_{2m}-minor in H, contradicting the fact that H is planar if $2m \ge 5$. We can map the vertices in V to their respective counterparts in H. However,

(i) some maximal cliques (pizza-with-crust and hamantasch) do not correspond to single points in P, and

(ii) we may need to pass through points in P more than once.

We first deal with (i). This concerns vertices $u \in U$ that have been introduced to cover the edges of pizza-with-crust and hamantasch maximal cliques. Each $x_i \in V \cup U$ has degree $2m - 1$, which is > 9 if we choose m large enough. If $x_i = u \in U$ then u has been added by rule (R4) to R to cover the edges of a maximal clique C in G of size > 9. This clique must be either a pizza or pizza-with-crust, because all hamantasch cliques have size ≤ 9. Therefore there is a point $p \in P$ that is adjacent to all but at most one of the vertices in C. If there is a vertex $v \in V$ adjacent to x_i in R but not adjacent to p in H, we remove the x_j which is connected to x_i via the path containing v.

We do this for all the $2m$ vertices of the topological K_{2m} minor in R and, after relabelling the vertices, are left with a topological K_m-minor in R and a mapping of its vertices $x_1, \ldots, x_m \in V \cup U$ to $y_1, \ldots, y_m \in V \cup P$ such that:

− If $x_i \in V$, then $y_i = x_i$, and
− if $x_i \in U$, then $y_i \in P$, and all neighbours of x_i on paths of the K_m minor are also neighbours of y_i in H.

Furthermore, no $p \in P$ appears as y_i for more than one i: Obviously, any $p \in P$ can only be the centre vertex of at most one maximal clique of pizza type. It may be the centre vertex of more than one (in fact, an unbounded number of) maximal cliques of pizza-with-crust-type, but in this case it is also the centre vertex of a *larger* clique of pizza-type. It is precisely the purpose of rule (R2) in the construction of R to guarantee that only one of these maximal cliques results in a vertex in U.

It remains to map vertices on the paths connecting the x_i to vertices in H. Again we map vertices in V to their identical counterparts. For the remaining vertices, we do not need to preserve all adjacencies, but only their two neighbours on the path belonging to the topological minor. In the following, let xuy be a part of one of the paths connecting the x_i, with $x, y \in V$ and $u \in U$.

We make a case distinction, depending on how the vertex u was introduced to the graph R: If u was introduced using rule (R4) then there is a maximal clique C in G of size >9 containing both x and y. We make a case distinction on the type of C:

- If C is a maximal pizza-clique, then there is a $p \in P$ connected to exactly the elements of C, and we may map u to p.
- If C is a maximal pizza-with-crust-clique, there are two possibilities: If both x and y are connected to the centre point $p \in P$ of the pizza-with-crust, then we may map u to p. Using the same reasoning as above, we can ensure that no $p \in P$ is used more than once, because if it is the centre vertex of two or more pizzas-with-crust, then is also the centre vertex of an even larger pizza-clique, and by rule (R2) u could not have been introduced in this case. Finally, C may be a pizza-with-crust-clique, and x and y connected by a vertex $p \in P$ that is not the centre vertex of C. We can not bound the number of pairs x, y for which this happens, i.e. there may be arbitrarily many $x_1, \dots, x_k, y_1, \dots, y_k \in V$ such that
  - all x_i, y_i are adjacent to p,
  - each pair x_i, y_i belongs to some maximal clique C_i in G,
  - no $C_i \cup C_j$ is a clique for $1 \leq i < j \leq k$.
  However, in this case the paths $x_i p y_i$ do not cross but only touch at p, i.e. in the drawing of H, the pairs $x_i y_i$ are consecutive in the cyclic order of the neighbours $\{x_1, \dots, x_k, y_1, \dots, y_k\}$ of p. Therefore we may split p into vertices $p_1, \dots, p_k$, with each p_i adjacent to x_i and y_i, and still obtain a planar graph H'.

Otherwise, x and y are the endpoints of some edge $xy \in T_m$, and u was introduced to cover this edge. Then there is some $p \in P$ adjacent to both x and y in H. This p has degree at most 9, for otherwise the neighbours of p (plus possibly one other vertex) would form a maximal clique of size larger than 9, and there would be a vertex $u' \in U$ for this clique. Since p has degree ≤ 9 we may safely map u to p, because there can be at most $\binom{9}{2} = 36$ pairs of vertices that get routed through p in this way.

Thus after possibly splitting some vertices of H, we end up with a planar graph H' and an r-shallow topological embedding of K_m into H' of complexity at most 38, which gives the desired contradiction if m is large enough.

6   General Map Graphs

We briefly sketch how our algorithm can be adapted to map graphs that are not necessarily 3-connected. Recall that we needed 3-connectedness to bound the size of hamantasch cliques, which followed from Lemma 2.

Using Feferman and Vaught's composition theorem [18] we may treat connected components individually. Similarly, we may build a tree of 2-connected components (blocks) and process the blocks one by one. We are left with the case of 2-connected but not necessarily 3-connected graphs.

These can be tree-decomposed into parts which are cycles, parallel edges, or 3-connected map graphs, and such that these parts are glued together along edges (cf. [24], it is easy to see that the 3-connected parts in this decomposition are again map graphs). We could colour the edges of these component graphs with the FO[q]-types of the graphs attached to them, but this would result in 3-connected parts that are not necessarily map graphs. In fact, any graph can be encoded in a clique (which is a map graph) of the same size by colouring its edges with two colours.

Instead we introduce coloured vertices of degree 2 rather than colouring the edges. Essentially as in the proof of Lemma 2 we can then show that in any hamantasch clique, there can be only 9 different neighbourhood types if we neglect vertices of degree 2. Again using Feferman-Vaught, we can prune vertices from hamantasch cliques.

7 Squares of Trees

Algorithmic meta theorems for a logic L on a class $\mathcal{C}$ of structures immediately carry over to a structure $\mathcal{D}$ if

- every structure $A \in \mathcal{C}$ can be interpreted in a structure $A' \in \mathcal{D}$ using a L-interpretation that depends only on the classes $\mathcal{C}$ and $\mathcal{D}$, and
- the structure A' can be efficiently computed from A.

Courcelle's result for MSO model checking on graphs of bounded tree-width can be seen as an example of this, since for every graph G of tree-width k there is a tree T such that G is MSO-interpretable in T, using an interpretation that only depends on k, and T can be efficiently computed from G.

Our proof of Theorem 1 also uses this approach, with a very specific kind of FO-interpretation: The input graph G was interpreted as an induced subgraph of the square of the bipartite graph R computed in Sect. 5. Squares and square-roots of graphs have been studied in graph theory, cf. e.g. [13,20]. In particular, Lin and Skiena [17] showed that checking whether a given graph is a square of a tree, and computing such a tree, can be done in polynomial time. The key observation towards this algorithm is that if $G = (V, E)$ is a square of some tree $T = (V, F)$, then $v \in V$ is simplicial (i.e. $N(v)$ is a clique) if, and only if, v is a leaf of T.

Using Lin et al.'s result we immediately get:

Theorem 2. *Model checking for first-order logic on the class of (coloured) squares of trees is fixed-parameter tractable.*

References

1. Chen, Z.Z., Grigni, M., Papadimitriou, C.H.: Map graphs. J. ACM **49**(2), 127–138 (2002)
2. Courcelle, B.: Graph rewriting: an algebraic and logic approach. In: van Leeuwen, J. (ed.) Handbook of Theoretical Computer Science, vol. 2, pp. 194–242. Elsevier, Amsterdam (1990)
3. Courcelle, B., Makowsky, J., Rotics, U.: Linear time solvable optimization problems on graphs of bounded clique-width. Theory Comput. Syst. **33**(2), 125–150 (2000)
4. Dawar, A., Grohe, M., Kreutzer, S.: Locally excluding a minor. In: Logic in Computer Science (LICS), pp. 270–279 (2007)
5. Diestel, R.: Graph Theory. GTM, vol. 173, 4th edn. Springer, Heidelberg (2012)
6. Dvořák, Z., Král, D., Thomas, R.: Testing first-order properties for subclasses of sparse graphs. J. ACM (JACM) **60**(5), 36 (2013)
7. Ebbinghaus, H.D., Flum, J.: Finite Model Theory. Perspectives in Mathematical Logic, 2nd edn. Springer, Heidelberg (1999)
8. Ebbinghaus, H.D., Flum, J., Thomas, W.: Mathematical Logic, 2nd edn. Springer, Heidelberg (1994)
9. Frick, M., Grohe, M.: Deciding first-order properties of locally tree-decomposable structures. J. ACM **48**, 1148–1206 (2001)
10. Gajarský, J., Hliněný, P., Lokshtanov, D., Obdržálek, J., Ordyniak, S., Ramanujan, M., Saurabh, S.: FO model checking on posets of bounded width. In: 2015 IEEE 56th Annual Symposium on Foundations of Computer Science (FOCS), pp. 963–974. IEEE (2015)
11. Gajarský, J., Hliněný, P., Obdržálek, J., Lokshtanov, D., Ramanujan, M.S.: A new perspective on FO model checking of dense graph classes. In: Proceedings of the 31st Annual ACM/IEEE Symposium on Logic in Computer Science, LICS 2016, pp. 176–184. ACM (2016)
12. Ganian, R., Hliněný, P., Král', D., Obdržálek, J., Schwartz, J., Teska, J.: FO model checking of interval graphs. In: Fomin, F.V., Freivalds, R., Kwiatkowska, M., Peleg, D. (eds.) ICALP 2013. LNCS, vol. 7966, pp. 250–262. Springer, Heidelberg (2013). doi:10.1007/978-3-642-39212-2_24
13. Geller, D.P.: The square root of a digraph. J. Combin. Theory **5**, 320–321 (1968)
14. Grohe, M., Kreutzer, S., Siebertz, S.: Deciding first-order properties of nowhere dense graphs. In: Proceedings of the 46th Annual ACM Symposium on Theory of Computing, STOC 2014, pp. 89–98. ACM (2014)
15. Lampis, M.: Algorithmic meta-theorems for restrictions of treewidth. Algorithmica **64**(1), 19–37 (2012)
16. Libkin, L.: Elements of Finite Model Theory. Texts in Theoretical Computer Science. Spinger, Heidelberg (2004)
17. Lin, Y.-L., Skiena, S.S.: Algorithms for square roots of graphs. In: Hsu, W.-L., Lee, R.C.T. (eds.) ISA 1991. LNCS, vol. 557, pp. 12–21. Springer, Heidelberg (1991). doi:10.1007/3-540-54945-5_44
18. Makowsky, J.: Algorithmic uses of the Feferman-Vaught theorem. Annals Pure Appl. Log. **126**(1–3), 159–213 (2004)
19. Mnich, M., Rutter, I., Schmidt, J.M.: Linear-time recognition of map graphs with outerplanar witness. In: LIPIcs-Leibniz International Proceedings in Informatics, vol. 53. Schloss Dagstuhl-Leibniz-Zentrum fuer Informatik (2016)
20. Mukhopadhyay, A.: The square root of a graph. J. Comb. Theory **2**, 290–295 (1967)

21. Nešetřil, J., de Mendez, P.O.: Sparsity – Graphs, Structures, and Algorithms. Algorithms and Combinatorics, vol. 28. Springer, Heidelberg (2012). doi:10.1007/978-3-642-27875-4

22. Thorup, M.: Map graphs in polynomial time. In: Proceedings of 39th Annual Symposium on Foundations of Computer Science, pp. 396–405. IEEE (1998)

23. Tsukiyama, S., Ide, M., Ariyoshi, H., Shirakawa, I.: A new algorithm for generating all the maximal independent sets. SIAM J. Comput. **6**(3), 505–517 (1977)

24. Tutte, W.T.: Graph Theory. Encyclopedia of Mathematics and its Applications, vol. 21. Cambridge University Press, Cambridge (2001)

Multiple Context-Free Tree Grammars and Multi-component Tree Adjoining Grammars

Joost Engelfriet[1] and Andreas Maletti[2([⊠])]

[1] LIACS, Leiden University, P.O. Box 9512, 2300 RA Leiden, The Netherlands
j.engelfriet@liacs.leidenuniv.nl
[2] Universität Leipzig, P.O. Box 100 920, 04009 Leipzig, Germany
maletti@informatik.uni-leipzig.de

Abstract. Strong lexicalization is the process of turning a grammar generating trees into an equivalent one, in which all rules contain a terminal leaf. It is known that tree adjoining grammars cannot be strongly lexicalized, whereas the more powerful simple context-free tree grammars can. It is demonstrated that *multiple* simple context-free tree grammars are as expressive as *multi-component* tree adjoining grammars and that both allow strong lexicalization.

1 Introduction

In computational linguistics several grammar formalisms [7] have been proposed that generate semilinear superclasses of the context-free languages, are able to model cross-serial dependencies, but remain parsable in polynomial time. Among the most well known are the (set-local) multi-component tree adjoining grammar (MCTAG) [5,18], which is an extension of the tree adjoining grammar (TAG), and the multiple context-free (string) grammar (MCFG) [16], which was independently discovered as (string-based) linear context-free rewriting system (LCFRS) [17]. In both cases the ability to synchronously rewrite multiple components was added to a classical model (TAG and CFG). In the same spirit, the multiple context-free *tree* grammar (MCFTG) was introduced in [8, Sect. 5] as the context-free graph grammar in tree generating normal form of [1], but was implicitly envisioned as *tree-based* LCFRS already in [17].

We define the MCFTG as a straightforward generalization of both the MCFG and the classical simple (i.e., linear and nondeleting) context-free tree grammar (CFTG). Intuitively, an MCFTG G is a CFTG, in which several nonterminals are rewritten in one derivation step. Thus every rule of G is a sequence of rules of a CFTG, and the left-hand side nonterminals of these rules are rewritten synchronously. However, a sequence of nonterminals can only be rewritten if (earlier in the derivation) they were introduced explicitly as such by the application of a rule of G, which is called "locality" in [14,18]. Therefore, each rule of G must also specify the sequences of (occurrences of) nonterminals in its right-hand side that may later be rewritten. Although such derivations can easily be formalized, we prefer to define the semantics of G as a least fixed point (just as for an MCFG).

© Springer-Verlag GmbH Germany 2017
R. Klasing and M. Zeitoun (Eds.): FCT 2017, LNCS 10472, pp. 217–229, 2017.
DOI: 10.1007/978-3-662-55751-8_18

Two tree-generating grammars are *strongly* (resp. *weakly*) equivalent if they generate the same tree (resp. string) language, where the string language consists of the yields of the generated trees. It is not difficult to see that for every MCTAG there is a strongly equivalent MCFTG, just as for every TAG there is a strongly equivalent CFTG [5,13]. Our main contribution is that, vice versa, for every MCFTG there is a strongly equivalent MCTAG, generalizing the result of [9] that relates monadic CFTGs and non-strict TAGs. It also settles a problem stated in [18, Sect. 4.5]: "It would be interesting to investigate whether there exist LCFRS's with object level tree sets that cannot be produced by any MCTAG." We prove that such LCFRSs do not exist. It is proved in the cited section that MCTAGs are weakly equivalent to string-based LCFRSs, so MCFTGs are weakly equivalent to MCFGs.

Secondly, we consider lexicalized grammars [6] in which each rule contains a lexical item (i.e., a terminal symbol that appears in the yield of the generated tree). Lexicalized grammars are of importance because they are often more understandable and allow easier parsing (cf. the Introduction of [12]); moreover, a lexicalized grammar defines a so-called dependency structure on the lexical items of each generated string, allowing to investigate certain aspects of the grammatical structure of that string, see [10]. We investigate lexicalization, which is the process that transforms a grammar into an equivalent lexicalized one. Corresponding to the two notions of equivalence we obtain strong and weak lexicalization. Although TAGs can be weakly lexicalized [3], they cannot be strongly lexicalized, as unexpectedly shown in [11]. However, the more powerful CFTGs can be strongly lexicalized [12], and the used lexicalization procedure can easily be generalized to MCFTGs. Since our transformation of an MCFTG into an MCTAG preserves the property of being lexicalized, we obtain that MCTAGs can be strongly lexicalized in contrast to classical TAGs.

The *multiplicity* (or *fan-out*) of an MCFTG G is the maximal number of nonterminals that can be rewritten simultaneously in one derivation step. Our strong lexicalization of G preserves the multiplicity of G, but our transformation of G into a strongly equivalent MCTAG increases it polynomially, and so the same is true for the strong lexicalization of MCTAGs.

2   Preliminaries

The set $\{1, 2, 3, \dots\}$ of positive integers is denoted by $\mathbb{N}$, and $\mathbb{N}_0 = \mathbb{N} \cup \{0\}$. For all $k \in \mathbb{N}_0$ we write $[k]$ for $\{i \in \mathbb{N} \mid i \le k\}$. The cardinality of a set A is $|A|$, and we let $A^* = \bigcup_{n \in \mathbb{N}_0} A^n$ and $A^+ = \bigcup_{n \in \mathbb{N}} A^n$, where A^n is the n-fold Cartesian product of A. Note that $A^0 = \{\varepsilon\}$, where ε is the empty sequence. If A is finite, then the elements of A and A^* are also called symbols and strings, respectively. The length $|w|$ of $w \in A^*$ is such that $w \in A^{|w|}$. For a sequence $w = (a_1, \dots, a_n) \in A^n$, the set $\mathrm{occ}(w) = \{a_1, \dots, a_n\}$ contains the elements of A that occur in w, and w is *repetition-free* if no element occurs more than once (i.e., $|\mathrm{occ}(w)| = n$). The concatenation $w \cdot v$ (or just wv) of w with a sequence $v = (b_1, \dots, b_m)$ is $(a_1, \dots, a_n, b_1, \dots, b_m)$.

As usual, we let $w^0 = \varepsilon$ and $w^{n+1} = ww^n$ for every $n \in \mathbb{N}_0$. For a subset $B \subseteq A$, the *yield* of w with respect to B is the sequence $\mathrm{yd}_B(w)$ in B^* that is obtained from w by removing all symbols outside B. Formally, $\mathrm{yd}_B(\varepsilon) = \varepsilon$ and for all $v \in A^*$ we have $\mathrm{yd}_B(bv) = b \cdot \mathrm{yd}_B(v)$ for all $b \in B$ and $\mathrm{yd}_B(av) = \mathrm{yd}_B(v)$ for all $a \in A \setminus B$.

A ranked alphabet is a finite set Σ with a ranking $\mathrm{rk} \colon \Sigma \to \mathbb{N}_0$. For every $k \in \mathbb{N}_0$ we let $\Sigma^{(k)} = \{\sigma \in \Sigma \mid \mathrm{rk}(\sigma) = k\}$ be the set of k-ary symbols, and let mrk_Σ be the minimal $k \in \mathbb{N}_0$ such that $\bigcup_{n=0}^{k} \Sigma^{(n)} = \Sigma$. In examples we introduce a symbol σ of rank k as $\sigma^{(k)}$. With every string $\bar{\sigma} = (\sigma_1, \ldots, \sigma_n) \in \Sigma^*$ we associate a multiple rank $\mathrm{rk}^*(\bar{\sigma}) = (\mathrm{rk}(\sigma_1), \ldots, \mathrm{rk}(\sigma_n)) \in \mathbb{N}_0^*$. We fix the countably infinite set $X = \{x_1, x_2, \ldots\}$ of variables and let $X_k = \{x_i \mid i \in [k]\}$ for every $k \in \mathbb{N}_0$. For every set $Z \subseteq X$ of variables, the set $T_\Sigma(Z)$ of *trees* over Σ and Z is the smallest set $T \subseteq (\Sigma \cup Z)^*$ such that $Z \subseteq T$ and $\sigma t_1 \cdots t_k \in T$ for all $k \in \mathbb{N}_0$, $\sigma \in \Sigma^{(k)}$, and $t_1, \ldots, t_k \in T$. As usual we also write the term $\sigma(t_1, \ldots, t_k)$ to denote $\sigma t_1 \cdots t_k$. We denote $T_\Sigma(X_0) = T_\Sigma(\emptyset)$ by T_Σ. The nodes of a tree are formalized as "positions". The root is at position ε, and the position pi with $p \in \mathbb{N}^*$ and $i \in \mathbb{N}$ refers to the i-th child of the node at position p. Thus, the set $\mathrm{pos}(t) \subseteq \mathbb{N}^*$ of positions of a tree $t \in T_\Sigma(X)$ is defined by $\mathrm{pos}(x) = \{\varepsilon\}$ for $x \in X$ and $\mathrm{pos}(t) = \{\varepsilon\} \cup \{ip \mid i \in [k], p \in \mathrm{pos}(t_i)\}$ for $t = \sigma(t_1, \ldots, t_k)$. The label and subtree of t at $p \in \mathrm{pos}(t)$ are $t(p)$ and $t|_p$, respectively, so $x(\varepsilon) = x = x|_\varepsilon$, $t(\varepsilon) = \sigma$, $t|_\varepsilon = t$, $t(ip) = t_i(p)$, and $t|_{ip} = t_i|_p$.

A *forest* $t = (t_1, \ldots, t_m)$ is a sequence of trees $t_1, \ldots, t_m \in T_\Sigma(X)$. A single tree is a forest of length 1. The nodes of the forest t are addressed by positions from $(\mathbb{N} \cup \{\#\})^*$, where $\#$ is a special symbol. Intuitively, these positions are of the form $\#^{j-1}p$, in which $\#^{j-1}$ selects the tree t_j and $p \in \mathrm{pos}(t_j)$ is a position in t_j. Formally, $\mathrm{pos}(t) = \{\#^{j-1}p \mid j \in [m], p \in \mathrm{pos}(t_j)\}$. The label and subtree of t at position $\#^{j-1}p$ are $t(\#^{j-1}p) = t_j(p)$ and $t|_{\#^{j-1}p} = t_j|_p$, respectively. For every set $\Omega \subseteq \Sigma \cup X$, the set $\mathrm{pos}_\Omega(t) = \{p \in \mathrm{pos}(t) \mid t(p) \in \Omega\}$ contains the Ω-labeled positions of t. We let $\mathrm{occ}_\Omega(t) = \{t(p) \mid p \in \mathrm{pos}_\Omega(t)\}$ be the symbols of Ω that occur in t. The forest t is *uniquely Ω-labeled* if all symbols of Ω occur at most once in t; i.e., $|\mathrm{pos}_{\{\sigma\}}(t)| \leq 1$ for every $\sigma \in \Omega$. The set $P_\Sigma(X_k)$ of k-ary *patterns* is $P_\Sigma(X_k) = \{t \in T_\Sigma(X_k) \mid \forall x \in X_k \colon |\mathrm{pos}_{\{x\}}(t)| = 1\}$. The rank $\mathrm{rk}(t)$ of a k-ary pattern t is k. Clearly, $P_\Sigma(X_0) = T_\Sigma$. We let $P_\Sigma(X) = \bigcup_{k \in \mathbb{N}_0} P_\Sigma(X_k)$, and we associate the multiple rank $\mathrm{rk}^*(t) = (\mathrm{rk}(t_1), \ldots, \mathrm{rk}(t_m)) \in \mathbb{N}_0^*$ with every forest $t = (t_1, \ldots, t_m)$ of $P_\Sigma(X)^*$. For all $\theta \colon X \to T_\Sigma(X)$, the *first-order substitution* $t\theta$ is inductively defined by $x\theta = \theta(x)$, $t\theta = \sigma(t_1\theta, \ldots, t_k\theta)$, and $u\theta = (u_1\theta, \ldots, u_m\theta)$ for every $x \in X$, $t = \sigma(t_1, \ldots, t_k) \in T_\Sigma(X)$, and $u = (u_1, \ldots, u_m) \in T_\Sigma(X)^*$. Thus, each occurrence of a variable $x \in X$ is replaced by the tree $\theta(x)$. If there exists $n \in \mathbb{N}_0$ with $\theta(x_i) = x_i$ for all $i > n$, then we also write $t[\theta(x_1), \ldots, \theta(x_n)]$ instead of $t\theta$. In second-order substitution we replace nodes that are labeled by symbols of Σ. Let $\theta \colon \Sigma \to P_\Sigma(X)$ be such that $\mathrm{rk}(\theta(\sigma)) = \mathrm{rk}(\sigma)$ for all $\sigma \in \Sigma$. The *second-order substitution* $t\theta$ is inductively defined by $x\theta = x$, $t\theta = \theta(\sigma)[t_1\theta, \ldots, t_k\theta]$, and $u\theta = (u_1\theta, \ldots, u_m\theta)$ with x, t, and u as above. Intuitively, the second-order substitution $t\theta$ replaces each σ-labeled subtree of t by the tree $\theta(\sigma)$, into which the (recursively processed) direct subtrees are first-order substituted. If there exist distinct $\sigma_1, \ldots, \sigma_n \in \Sigma$ such that $\theta(\sigma) = \sigma(x_1, \ldots, x_k)$ for all $\sigma \in \Sigma^{(k)} \setminus \{\sigma_1, \ldots, \sigma_n\}$, we also write $t[(\sigma_1, \ldots, \sigma_n) \leftarrow (\theta(\sigma_1), \ldots, \theta(\sigma_n))]$ instead of $t\theta$. Finally, let $\mathcal{L} = \{\bar{\sigma}_1, \ldots, \bar{\sigma}_k\}$ be a subset of Σ^* such that $\bar{\sigma}_1 \cdots \bar{\sigma}_k$ is repetition-free. A (second-order) *substitution function* for $\mathcal{L}$ is a mapping $f \colon \mathcal{L} \to P_\Sigma(X)^*$ such that $\mathrm{rk}^*(f(\bar{\sigma})) = \mathrm{rk}^*(\bar{\sigma})$ for every $\bar{\sigma} \in \mathcal{L}$. For a forest $t \in P_\Sigma(X)^*$, the *simultaneous second-order substitution* $t[f]$ is defined by $t[f] = t[\bar{\sigma}_1 \cdots \bar{\sigma}_k \leftarrow f(\bar{\sigma}_1) \cdots f(\bar{\sigma}_k)]$. For a complete exposition of tree language theory, we refer the reader to [4].

3   Multiple Context-Free Tree Grammars

We define the (simple) multiple context-free tree grammar as a straightforward generalization of both the (simple) context-free tree grammar [2,15] and the multiple context-free (string) grammar [16,17]. We obtain essentially a tree-based linear context-free rewriting system.

Definition 1. A *(simple) multiple context-free tree grammar* (MCFTG) is a system $G = (N, \mathcal{N}, \Sigma, I, R)$ such that

- N is a ranked alphabet of *nonterminals*,
- $\mathcal{N} \subseteq N^+$ is a finite set of *big nonterminals*, which are nonempty repetition-free nonterminal sequences with $\mathrm{occ}(A) \neq \mathrm{occ}(A')$ for all distinct $A, A' \in \mathcal{N}$,
- Σ is a ranked alphabet of *terminals* such that $\Sigma \cap N = \emptyset$ and $\mathrm{mrk}_\Sigma \geq 1$,
- $I \subseteq \mathcal{N} \cap N^{(0)}$ is the set of *initial (big) nonterminals*, and
- R is a finite set of *rules* of the form $A \to (u, \mathcal{L})$, where $A \in \mathcal{N}$ is a big nonterminal, $u \in P_{N \cup \Sigma}(X)^+$ is a uniquely N-labeled forest (of patterns) such that $\mathrm{rk}^*(u) = \mathrm{rk}^*(A)$, and $\mathcal{L} \subseteq \mathcal{N}$ is a set of big nonterminals, called *links*, such that $\{\mathrm{occ}(B) \mid B \in \mathcal{L}\}$ is a partition of $\mathrm{occ}_N(u)$; i.e., $\mathrm{occ}(B) \cap \mathrm{occ}(B') = \emptyset$ for all distinct $B, B' \in \mathcal{L}$ and $\mathrm{occ}_N(u) = \bigcup_{B \in \mathcal{L}} \mathrm{occ}(B)$.

The *multiplicity* (or *fan-out*) of G, denoted by $\mu(G)$, is the maximal length of its big nonterminals. The *width* of G, denoted by $\omega(G)$, is the maximal rank of its nonterminals. A (simple) context-free tree grammar (CFTG) is an MCFTG of multiplicity 1.   □

For a given rule $\rho = A \to (u, \mathcal{L})$, its left-hand side is A, its right-hand side is u, and its set of links is $\mathcal{L}$. Since $\mathrm{rk}^*(A) = \mathrm{rk}^*(u)$, the rule ρ is of the form

$$(A_1, \ldots, A_n) \to ((u_1, \ldots, u_n), \{B_1, \ldots, B_k\}) \ ,$$

where $n \in \mathbb{N}$, $A_i \in N$, $u_i \in P_{N \cup \Sigma}(X_{\mathrm{rk}(A_i)})$, $k \in \mathbb{N}_0$, and $B_j \in \mathcal{N}$ for all $i \in [n]$ and $j \in [k]$. Intuitively, the application of the rule ρ consists of the simultaneous application of the n rules $A_i(x_1, \ldots, x_{\mathrm{rk}(A_i)}) \to u_i$ of an ordinary CFTG to occurrences of the nonterminals $A_1, \ldots, A_n$, and the introduction of all the nonterminals that occur in the big nonterminals $B_1, \ldots, B_k$. Every $B_j = (C_1, \ldots, C_m) \in N^+$ can be viewed as a link between the (unique) positions of u with labels $C_1, \ldots, C_m$ as well as a link between the corresponding positions after the application of ρ. The rule ρ can only be applied to positions

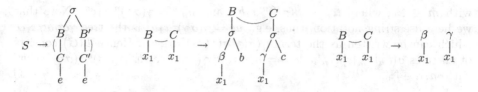

Fig. 1. The first three rules of Example 3.

with labels $A_1, \ldots, A_n$ that are joined by such a link. Thus, rule applications are "local" in the sense that a rule can rewrite only nonterminals that were previously introduced together in a single step of the derivation, just as for the local unordered scattered context grammar of [14], which is equivalent to the multiple context-free (string) grammar. Instead of defining derivation steps between trees in $T_{N \cup \Sigma}$, it is technically more convenient to define the generation of trees recursively. In an ordinary CFTG, a nonterminal A of rank k can be viewed as a generator of trees in $P_\Sigma(X_k)$ using derivations that start with $A(x_1, \ldots, x_k)$. In the same fashion, a big nonterminal A of an MCFTG generates forests in $P_\Sigma(X)^+$ of the same multiple rank as A.

Definition 2. Let $G = (N, \mathcal{N}, \Sigma, I, R)$ be an MCFTG. For every big nonterminal $A \in \mathcal{N}$ we recursively define the set $L(G, A) \subseteq P_\Sigma(X)^+$ of *forests generated* by A as follows. For every rule $\rho = A \to (u, \mathcal{L}) \in R$ and every substitution function $f \colon \mathcal{L} \to P_\Sigma(X)^+$ for $\mathcal{L}$ such that $f(B) \in L(G, B)$ for every $B \in \mathcal{L}$, the forest $u[f]$ is in $L(G, A)$. The *tree language* $L(G)$ *generated by* G is defined by $L(G) = \bigcup_{S \in I} L(G, S) \subseteq T_\Sigma$. ☐

Note that $u[f]$ is a simultaneous second-order substitution (see Sect. 2). Since $\mathrm{rk}^*(f(B)) = \mathrm{rk}^*(B)$ for all $B \in \mathcal{L}$, we have $\mathrm{rk}^*(t) = \mathrm{rk}^*(A)$ for every forest $t \in L(G, A)$. Two MCFTGs G and G' are (strongly) equivalent if $L(G) = L(G')$.

Example 3. We consider the MCFTG $G = (N, \mathcal{N}, \Sigma, \{S\}, R)$ with nonterminals $N = \{S^{(0)}, B^{(1)}, C^{(1)}, B'^{(1)}, C'^{(1)}\}$, big nonterminals $\mathcal{N} = \{S, (B, C), (B', C')\}$, terminals $\Sigma = \{\sigma^{(2)}, \beta^{(1)}, \gamma^{(1)}, b^{(0)}, c^{(0)}, e^{(0)}\}$, and the rules R (see Fig. 1):

$$S \to \sigma\big(B(C(e)), B'(C'(e))\big)$$
$$(B, C) \to \big(B(\sigma(\beta(x_1), b)), C(\sigma(\gamma(x_1), c))\big) \qquad (B, C) \to \big(\beta(x_1), \gamma(x_1)\big)$$
$$(B', C') \to \big(B(\sigma(\beta(x_1), b)), C(\sigma(\gamma(x_1), c))\big) \qquad (B', C') \to \big(\beta(x_1), \gamma(x_1)\big) \ ,$$

where we write a rule $A \to (u, \mathcal{L})$ as $A \to u$. In this example, and the next, the sets $\mathcal{L}$ of links are unique. Here they are $\{(B, C), (B', C')\}$ for the first rule, $\{(B, C)\}$ for the second and fourth rule, and $\emptyset$ for the third and fifth rule. Since the rules for (B, C) and (B', C') have the same right-hand sides and links, they are *aliases*. They represent essentially the same big nonterminal, but must be different because they occur together in the right-hand side of the first rule. It is easy to see that $L(G, (B, C)) = L(G, (B', C'))$ consists of all forests (t_m, u_m)

with $m \in \mathbb{N}_0$, where $t_m = \beta(\sigma\beta)^m x_1 b^m$ and $u_m = \gamma(\sigma\gamma)^m x_1 c^m$. Note that we here use string notation, thus, e.g., $u_2 = \gamma(\sigma\gamma)^2 x_1 c^2$ is the tree $\gamma\sigma\gamma\sigma\gamma x_1 cc$ which can be written as the term $\gamma(\sigma(\gamma(\sigma(\gamma(x_1), c)), c))$. Hence $L(G)$ consists of all trees $\sigma(t_m[u_m[e]], t_n[u_n[e]]) = \sigma\beta(\sigma\beta)^m\gamma(\sigma\gamma)^m ec^m b^m \beta(\sigma\beta)^n\gamma(\sigma\gamma)^n ec^n b^n$ with $m, n \in \mathbb{N}_0$.   □

4   Lexicalization

For a given terminal alphabet Σ we fix a subset $\Delta \subseteq \Sigma^{(0)}$ of *lexical* symbols. We say that an MCFTG G is *lexicalized* if each rule contains at least one lexical symbol; i.e., if $\mathrm{pos}_\Delta(u) \neq \emptyset$ for every rule $A \to (u, \mathcal{L})$ of G. Clearly, if G is lexicalized, then $L(G)$ has finite ambiguity, in the following sense. Let the yield $\mathrm{yd}(t)$ of a tree $t \in T_\Sigma$ be the string of lexical symbols that label its leaves, from left to right. So, $\mathrm{yd}(t) = \mathrm{yd}_\Delta(t) \in \Delta^*$ (as defined in Sect. 2). We say that a tree language $L \subseteq T_\Sigma$ has *finite ambiguity* if $\{t \in L \mid \mathrm{yd}(t) = w\}$ is finite for every $w \in \Delta^+$ and $\{t \in L \mid \mathrm{yd}(t) = \varepsilon\} = \emptyset$. We can lexicalize MCFTGs, which means that for each MCFTG G of which $L(G)$ has finite ambiguity, we can construct an equivalent lexicalized MCFTG. This is called strong lexicalization [6,11] because we require strong equivalence.

Theorem 4. *For each MCFTG G such that $L(G)$ has finite ambiguity there is an equivalent lexicalized MCFTG G' with $\mu(G') = \mu(G)$ and $\omega(G') = \omega(G) + 1$.*

The construction is essentially the same as the one in [12] for CFTGs. First, all nonlexicalized rules of rank 0 and rank 1 are removed, where the rank of a rule $A \to (u, \mathcal{L})$ is $|\mathcal{L}|$. This is similar to the removal of rules $A \to \varepsilon$ and $A \to B$ from a context-free grammar. Since $L(G)$ has finite ambiguity, such rules can only generate finitely many trees. Second, all rules of rank 0 with exactly one lexical symbol are removed. That can be done by applying all such rules to the other rules, in all possible ways. Finally, we guess a lexical symbol for every application of a nonlexicalized rule and put the guessed symbol in a new argument of a nonterminal (thus turning the rule into a lexicalized one). It is passed from nonterminal to nonterminal until a rule of rank 0 is applied, where we exchange the same lexical symbol for the new argument. The resulting rule is still lexicalized because we made sure that rules of rank 0 contain at least two lexical symbols. Lexical symbols that are guessed for distinct rule applications are transported to distinct applications of rules of rank 0.

5   MCFTG and MCTAG

Next we prove that MCTAGs have the same tree generating power as MCFTGs. It is shown in [9, Sect. 4] that "non-strict" TAGs have the same tree generating power as "footed" CFTGs. Since the translation from one formalism to the other is straightforward, we avoid the introduction of the formal machinery that is needed to define MCTAGs in the usual way. Rather we first define non-strict

MCTAGs to be footed MCFTGs, which generalize footed CFTGs in an obvious way. After that we define (strict) MCTAGs as a special type of non-strict MCTAGs. The main result of [9] is that non-strict TAGs have the same tree generating power as monadic CFTGs, where a CFTG G is monadic if $\omega(G) \leq 1$. Our result shows that the monadic restriction is not needed in the multi case.

According to [9], a CFTG is footed if for every rule $A(x_1, \ldots, x_k) \rightarrow u$ with $k \in \mathbb{N}$ there is a unique position of u with exactly k children that are labeled $x_1, \ldots, x_k$ from left to right.

Definition 5. Let Ω be a ranked alphabet. A pattern $t \in P_\Omega(X_k)$ is *footed* if either $k = 0$, or $k \in \mathbb{N}$ and there exists $p \in \mathrm{pos}_\Omega(t)$, called the *foot node* of t, such that $t|_p = \sigma(x_1, \ldots, x_k)$ for some $\sigma \in \Omega^{(k)}$. Let $G = (N, \mathcal{N}, \Sigma, I, R)$ be an MCFTG. A rule $A \rightarrow ((u_1, \ldots, u_n), \mathcal{L}) \in R$ is footed if $u_i{}'$ is footed for every $i \in [n]$. The MCFTG G is footed if every rule in R is footed. $\square$

Note that, by definition, every tree $t \in T_\Omega = P_\Omega(X_0)$ is footed. For a footed MCFTG G, it is straightforward to show that the trees $t_1, \ldots, t_n$ are footed for every forest $(t_1, \ldots, t_n) \in L(G, A)$. This implies that $\omega(G) \leq \mathrm{mrk}_\Sigma$.

Based on the close relationship between non-strict TAGs and footed CFTGs as shown in [9, Sect. 4], we here define a non-strict TAG to be a footed CFTG and, similarly, a non-strict MCTAG to be a footed MCFTG. To convince the reader familiar with TAGs of this definition, we add some more terminology. Let $A \rightarrow (u, \mathcal{L})$ be a rule with $A = (A_1, \ldots, A_n)$ and $u = (u_1, \ldots, u_n)$. If the rule is initial (i.e., $A \in I$), then the right-hand side u together with the set $\mathcal{L}$ of links is called an *initial forest*, and otherwise it is called an *auxiliary forest*. Application of the rule consists of adjunctions and substitutions. The replacement of the nonterminal A_i by u_i is called an *adjunction* if $\mathrm{rk}(A_i) \geq 1$ and a *substitution* otherwise. An occurrence of a nonterminal $C \in N$ in u with $\mathrm{rk}(C) \geq 1$ has an obligatory adjunction (OA) constraint, whereas an occurrence of a terminal $\sigma \in \Sigma$ in u with $\mathrm{rk}(\sigma) \geq 1$ has a null adjunction (NA) constraint. In the same manner we handle obligatory and null substitution (OS and NS) constraints. Each big nonterminal $B \in \mathcal{L}$ can be viewed as a selective adjunction/substitution (SA/SS) constraint, which restricts the auxiliary forests that can be adjoined/substituted for B to the right-hand sides of the rules with left-hand side B.

Given a footed pattern $t \in P_{N \cup \Sigma}(X_k)$ with $k \geq 1$, we define $\mathrm{rlab}(t) = t(\varepsilon)$ and $\mathrm{flab}(t) = t(p)$, where p is the foot node of t. Thus, $\mathrm{rlab}(t)$ and $\mathrm{flab}(t)$ are the labels of the root and the foot node of t, respectively. For $k = 0$, we let $\mathrm{rlab}(t) = t(\varepsilon)$ and $\mathrm{flab}(t) = t(\varepsilon)$ for technical convenience.

Definition 6. Let Ω, Σ be ranked alphabets and $\varphi \colon \Omega \rightarrow \Sigma$ be a fixed mapping. A pattern $t \in P_\Omega(X_k)$ is *adjoining* if it is footed and $\varphi(\mathrm{rlab}(t)) = \varphi(\mathrm{flab}(t))$. $\square$

Definition 7. A *(strict and set-local) multi-component tree adjoining grammar* (MCTAG) is an MCFTG $G = (N, \mathcal{N}, \Sigma, I, R)$, for which there exists a rank-preserving mapping $\varphi \colon (N \cup \Sigma) \rightarrow \Sigma$ such that $\varphi(\sigma) = \sigma$ for every $\sigma \in \Sigma$, and moreover, for every rule $(A_1, \ldots, A_n) \rightarrow ((u_1, \ldots, u_n), \mathcal{L}) \in R$ and every $i \in [n]$, u_i is an adjoining pattern and $\varphi(\mathrm{rlab}(u_i)) = \varphi(A_i)$.

A *tree adjoining grammar* (TAG) is an MCTAG of multiplicity 1. $\square$

The MCFTG G of Example 3 is an MCTAG with respect to the mapping φ such that $\varphi(S) = \sigma$, $\varphi(B) = \varphi(B') = \beta$, and $\varphi(C) = \varphi(C') = \gamma$.

Each nonterminal C with $\varphi(C) = \sigma$ can be viewed as the terminal symbol σ together with some information that is relevant to SA and SS constraints. The requirements in Definition 7 mean that the root and foot node of u_i represent the same terminal symbol as A_i. Thus, intuitively, adjunction always replaces a (constrained) terminal symbol by a tree with that same symbol as root label and foot node label. Thus, if $(t_1, \ldots, t_n) \in L(G, (A_1, \ldots, A_n))$ then $\mathrm{rlab}(t_i) = \mathrm{flab}(t_i) = \varphi(A_i)$ for every $i \in [n]$. Our MCTAGs and TAGs are slightly more general than the usual ones, because the roots of the generated trees need not have the same label; in other words, the underlying syntax may have more than one "sentence symbol" $\varphi(S)$ with $S \in I$. We view this as an irrelevant technicality.

Let MCFTL and MCTAL denote the classes of tree languages generated by MCFTGs and MCTAGs, respectively. We now prove that MCTAL = MCFTL. By definition, we have MCTAL $\subseteq$ MCFTL. The next theorem shows that for every MCFTG G there is an equivalent MCTAG, which is also lexicalized if G is lexicalized. Roughly speaking, the transformation of an MCFTG into an MCTAG is realized by decomposing each tree u_i in the right-hand side of a rule $A \to (u, \mathcal{L})$ with $A = (A_1, \ldots, A_n)$ and $u = (u_1, \ldots, u_n)$ into a bounded number of parts, to replace u_i in u by the sequence of these parts, and to replace A_i in A by a corresponding sequence of new nonterminals that simultaneously generate these parts.

Theorem 8. *For every MCFTG G with terminal alphabet Σ there is an equivalent MCTAG G' such that $\mu(G') \leq \mu(G) \cdot \mathrm{mrk}_\Sigma \cdot |\Sigma| \cdot (2 \cdot \omega(G) - 1)$ if $\omega(G) \neq 0$, and $\mu(G') = \mu(G)$ otherwise. Moreover, if G is lexicalized, then so is G'.*

Proof. The basic fact used in this proof is that, for any ranked alphabet Ω and mapping $\varphi \colon \Omega \to \Sigma$, every tree $u \in T_\Omega(X)$ with $u \notin X$ and $\mathrm{pos}_X(u) \neq \emptyset$ can be decomposed into at most $\mathrm{mrk}_\Omega \cdot |\Sigma| \cdot (2k - 1)$ adjoining patterns, where k is the number $|\mathrm{pos}_X(u)|$ of occurrences of variables in u. This decomposition can be obtained inductively as follows. Let $p \in \mathrm{pos}_\Omega(u)$ be the longest position such that $\varphi(u(p)) = \varphi(u(\varepsilon))$ and $|\mathrm{pos}_X(u|_p)| = |\mathrm{pos}_X(u)|$. Then there are an adjoining pattern $u_\varepsilon \in P_\Omega(X_m)$ and trees $u_1, \ldots, u_m \in T_\Omega(X)$ such that $m = \mathrm{rk}(u(p)) \geq 1$, $u = u_\varepsilon[u_1, \ldots, u_m]$, and p is the foot node of u_ε. In other words, u is decomposed as $u_\varepsilon[u_1, \ldots, u_m]$ where u_ε is an adjoining pattern. For every $i \in [m]$ with $u_i \notin X$, either $u_i \in T_\Omega$ and so u_i is an adjoining pattern of rank 0, or $\mathrm{pos}_X(u_i) \neq \emptyset$, in which case the tree u_i can be decomposed further. It should be clear that, in this inductive process, there are at most $|\Sigma| \cdot (2k - 1)$ such positions p. The factor mrk_Ω is due to the adjoining patterns of rank 0. As an example, let $\Omega = \{\sigma^{(2)}, \tau^{(2)}, \beta^{(1)}, a^{(0)}, b^{(0)}\}$ and let φ be the identity on Ω. Then the tree $u = \sigma(a, \sigma(v, \sigma(x_3, b)))$ with $v = \sigma(a, \tau(a, \sigma(a, \tau(x_1, \beta(\beta(x_2))))))$ is decomposed as $u = u_\varepsilon[u_1[u_{11}, u_{12}[x_1, u_{122}[x_2]]], u_2[x_3, u_{22}]]$ into the adjoining patterns $u_\varepsilon = \sigma(a, \sigma(x_1, x_2))$, $u_1 = \sigma(a, \tau(a, \sigma(x_1, x_2)))$, $u_{11} = a$, $u_{12} = \tau(x_1, x_2)$, $u_{122} = \beta(\beta(x_1))$, $u_2 = \sigma(x_1, x_2)$, and $u_{22} = b$. Using new symbols C_p^α such that $\alpha \in \Omega$, $p \in \mathbb{N}^*$, and $\mathrm{rk}(C_p^\alpha) = \mathrm{rk}(\alpha)$, we can also express this decomposition as $u = K[\gamma]$, where K is the tree $C_\varepsilon^\sigma(C_1^\sigma(C_{11}^a, C_{12}^\tau(x_1, C_{122}^\beta(x_2))), C_2^\sigma(x_3, C_{22}^b))$, which can be viewed as the skeleton of the decomposition, and γ is the substitution function such that $\gamma(C_p^\alpha) = u_p$. Note that the superscript α of C_p^α is equal to $\varphi(\mathrm{rlab}(u_p))$. This decomposition is formalized below and applied to (variants of) the trees in the right-hand sides of the rules of G.

Let $G = (N, \mathcal{N}, \Sigma, I, R)$ be an MCFTG. Provided that $\omega(G) \neq 0$, then we have $\mathrm{mrk}_\Sigma \cdot (2 \cdot \omega(G) - 1) \geq 1$ because $\mathrm{mrk}_\Sigma \geq 1$ by Definition 1. By straightforward constructions we may assume that G is "permutation-free" and "nonerasing". This means that if $(A_1, \ldots, A_n) \to ((u_1, \ldots, u_n), \mathcal{L})$ is a rule in R, then the pattern u_i is in $PF_{N \cup \Sigma}(X_{\mathrm{rk}(A_i)}) \setminus X$ for every $i \in [n]$, where $PF_\Omega(X_k)$ denotes the set of permutation-free k-ary patterns over Ω; i.e., patterns $t \in P_\Omega(X_k)$ such that $\mathrm{yd}_X(t) = x_1 \cdots x_k$. The nonerasing requirement that $u_i \notin X$ is only relevant when $\mathrm{rk}(A_i) = 1$, meaning that $u_i \neq x_1$.

We define $G' = (N', \mathcal{N}', \Sigma, I', R')$. The set N' of nonterminals consists of all quadruples $\langle C, \sigma, m, p \rangle$ with $C \in N$, $\sigma \in \Sigma$, $m \in \{0, \mathrm{rk}(\sigma)\}$, and $p \in \mathbb{N}^*$ such that $|p| \leq |\Sigma| \cdot \omega(G)$. The rank of $\langle C, \sigma, m, p \rangle$ is m. The set of initial nonterminals is $I' = \{\langle S, \sigma, 0, \varepsilon \rangle \mid S \in I, \sigma \in \Sigma\}$. We will define $\mathcal{N}'$ and R' in such a way that G' is an MCTAG with respect to the mapping $\varphi \colon (N' \cup \Sigma) \to \Sigma$ such that $\varphi(\langle C, \sigma, m, p \rangle) = \varphi(\sigma) = \sigma$. For every nonterminal $C \in N$, a *skeleton* of C is a pattern $K \in PF_{N'}(X_{\mathrm{rk}(C)}) \setminus X$ such that

(1) for every $p \in \mathrm{pos}_{N'}(K)$ there exist a symbol $\sigma \in \Sigma$ and $m \in \{0, \mathrm{rk}(\sigma)\}$ such that $K(p) = \langle C, \sigma, m, p \rangle$;
(2) for all $p, q \in \mathrm{pos}_{N'}(K)$, if position q is a proper descendant of position p, then $\varphi(K(q)) \neq \varphi(K(p))$ or $|\mathrm{pos}_X(K|_q)| < |\mathrm{pos}_X(K|_p)|$;
(3) for every $p \in \mathrm{pos}_{N'}(K)$, if $K|_p \in T_{N'}$ then $\mathrm{rk}(K(p)) = 0$.

For such a skeleton K, we let $\mathrm{seq}(K) = \mathrm{yd}_{N'}(K)$, which is in $(N')^+$. There are only finitely many skeletons K of C because $|\mathrm{pos}_{N'}(K)| \leq \mathrm{mrk}_\Sigma \cdot |\Sigma| \cdot (2k - 1)$, if $k = \mathrm{rk}(C) \geq 1$. If $\mathrm{rk}(C) = 0$, then each skeleton of C is of the form $\langle C, \sigma, 0, \varepsilon \rangle$ with $\sigma \in \Sigma$. Note that K can be reconstructed from $\mathrm{seq}(K)$ because K is permutation-free. In the example above, the tree K is a skeleton of C, provided C_p^α denotes $\langle C, \alpha, \mathrm{rk}(\alpha), p \rangle$, and $\mathrm{seq}(K) = (C_\varepsilon^\sigma, C_1^\sigma, C_{11}^a, C_{12}^\tau, C_{122}^\beta, C_2^\sigma, C_{22}^b)$.

We apply the above basic fact to patterns over $N' \cup \Sigma$. Let K be a skeleton of $C \in N$. A substitution function γ for $\mathrm{occ}_{N'}(K)$ is *adjoining* if, for every $C' \in \mathrm{occ}_{N'}(K)$, the pattern $\gamma(C') \in P_{N' \cup \Sigma}(X)$ is adjoining and we have $\varphi(\mathrm{rlab}(\gamma(C'))) = \varphi(C')$. We say that the pair $\langle K, \gamma \rangle$ is an *adjoining C-decomposition* of the tree $K[\gamma]$. By a straightforward induction, following the above basic fact, we can prove that every pattern u over $N' \cup \Sigma$ has an adjoining C-decomposition $\mathrm{dec}_C(u)$. More precisely, for every $C \in N$ and every $u \in PF_{N' \cup \Sigma}(X_{\mathrm{rk}(C)}) \setminus X$ there is a pair $\mathrm{dec}_C(u) = \langle K, \gamma \rangle$ such that K is a skeleton of C, γ is an adjoining substitution function for $\mathrm{occ}_{N'}(K)$, and $K[\gamma] = u$.

A *skeleton function* for $A \in \mathcal{N}$ is a substitution function κ for $\mathrm{occ}(A)$ that assigns a skeleton $\kappa(C)$ of C to every nonterminal $C \in \mathrm{occ}(A)$. The string homomorphism h_κ from $\mathrm{occ}(A)^*$ to $(N')^*$ is defined by $h_\kappa(C) = \mathrm{seq}(\kappa(C))$ for every $C \in \mathrm{occ}(A)$. We define the set $\mathcal{N}'$ of big nonterminals to be the set of all $h_\kappa(A)$, where $A \in \mathcal{N}$ and κ is a skeleton function for A.

We finally define the set R' of rules of G'. Let $\rho = A \to (u, \mathcal{L})$ be a rule in R such that $A = (A_1, \ldots, A_n)$, $u = (u_1, \ldots, u_n)$, and $\mathcal{L} = \{B_1, \ldots, B_k\}$. Moreover, let $\overline{\kappa} = (\kappa_1, \ldots, \kappa_k)$, where κ_j is a skeleton function for B_j for every $j \in [k]$. Intuitively, $\overline{\kappa}$ guesses for every nonterminal C that occurs in $B_1, \ldots, B_k$ the skeleton of an adjoining C-decomposition of the tree generated by C. Let f be the substitution function for $\mathrm{occ}_N(u)$ such that $f = \bigcup_{j \in [k]} \kappa_j$; i.e., $f(C) = \kappa_j(C)$ if $C \in \mathrm{occ}(B_j)$. Obviously, $u_i[f] \in PF_{N' \cup \Sigma}(X_{\mathrm{rk}(A_i)}) \setminus X$ for every $i \in [n]$. For every $i \in [n]$, let $u_i' = u_i[f]$ and let $\mathrm{dec}_{A_i}(u_i') = \langle K_i, \gamma_i \rangle$ (the adjoining A_i-decomposition of u_i'); moreover, if $\mathrm{seq}(K_i) = (C_1', \ldots, C_\ell')$ with $C_1', \ldots, C_\ell' \in N'$, then let $v_i' = (\gamma_j(C_1'), \ldots, \gamma_j(C_\ell'))$. Then we construct the rule

$$\langle \rho, \overline{\kappa} \rangle = \mathrm{seq}(K_1) \cdots \mathrm{seq}(K_n) \to (v_1' \cdots v_n', \mathcal{L}')$$

with $\mathcal{L}' = \{h_{\kappa_1}(B_1), \ldots, h_{\kappa_k}(B_k)\}$ in R'. We also define the skeleton function $\kappa_{\rho, \overline{\kappa}}$ for A by $\kappa_{\rho, \overline{\kappa}}(A_i) = K_i$ for every $i \in [n]$. Intuitively, K_i is the skeleton of an adjoining A_i-decomposition of the tree generated by A_i, resulting from the skeletons guessed by $\overline{\kappa}$.

It should be clear that G' is an MCTAG with respect to φ. Moreover, since the right-hand sides of the rules ρ and $\langle \rho, \overline{\kappa} \rangle$ contain the same terminal symbols, G' is lexicalized if G is lexicalized. The intuition underlying the correctness of G' is that for every A_i, the skeleton K_i generates the same terminal tree in G' as A_i generates in G, provided that the skeleton $\kappa_j(C)$ generates the same terminal tree in G' as C generates in G for every $j \in [k]$ and $C \in \mathrm{occ}(B_j)$.   □

Example 9. We consider the footed CFTG $G_1 = (N_1, \mathcal{N}_1, \Sigma, \{S\}, R_1)$ such that $N_1 = \mathcal{N}_1 = \{S^{(0)}, A^{(1)}, A'^{(1)}\}$, $\Sigma = \{\tau^{(3)}, \ell^{(1)}, r^{(1)}, a^{(0)}, b^{(0)}, e^{(0)}\}$, and R_1 contains the rules $S \to \ell A A' r e$, $A \to \ell A A' r x_1$, and $A \to \ell \ell \tau(a, b, r r x_1)$, plus the two rules for the alias A' of A. Note that, for the sake of readability, we omit here and in what follows the parentheses around the arguments of unary symbols; e.g., the right-hand side of the third rule is $\ell(\ell(\tau(a, b, r(r(x_1))))) = \ell \ell \tau a b r r x_1$. Let $\Delta = \{a, b\}$ be the set of lexical symbols. Clearly, $L(G_1)$ has finite ambiguity. However, there is no equivalent lexicalized footed CFTG. In fact, G_1 is a variant of the TAG of [11], for which there is no (strongly) equivalent lexicalized TAG. Thus, since we defined non-strict TAGs to be footed CFTGs, G_1 is a non-strict TAG that cannot be lexicalized (as non-strict TAG). We will construct an equivalent lexicalized MCTAG for G_1.

From Theorem 4, or rather from [12], we obtain an equivalent lexicalized CFTG G_2 with $\omega(G_2) = 2$. It has the new nonterminals $B^{(2)}$ and $B'^{(2)}$, where B' is an alias of B. Its rules are

$$\rho_1: \; S \to \ell A B(b, re) \quad \rho_2: \; A \to \ell A B(b, r x_1) \quad \rho_4: \; B \to \ell B(x_1, B'(b, r x_2))$$
$$\rho_3: \; A \to \ell \ell \tau(a, b, r r x_1) \quad \rho_5: \; B \to \ell \ell \tau(a, x_1, r r x_2)$$

plus the rules ρ_4' and ρ_5' for B'. Clearly, the tree $B(b, x_1)$ generates the same terminal trees as $A(x_1)$.

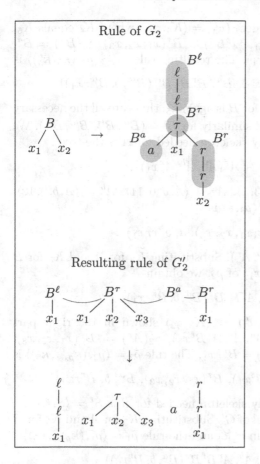

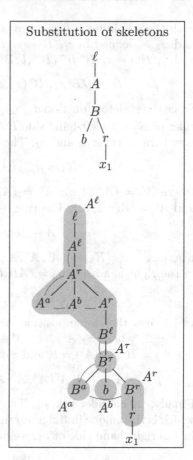

Fig. 2. Left part: the adjoining decomposition $\langle K_5, \gamma_5 \rangle$ of the right-hand side of rule ρ_5, with the resulting rule $\tilde{\rho}_5$. Right part: substitution of the skeletons K_3 of A and K_5 of B into the right-hand side of rule ρ_2, with the adjoining decomposition $\langle K_3, \gamma_2 \rangle$.

We now turn G_2 into an equivalent lexicalized MCTAG G_2' using the construction in the proof of Theorem 8. For the rule $\rho_5 = B \to u_5$ and $\overline{\kappa} = \varepsilon$, we obtain the adjoining B-decomposition $\mathrm{dec}_B(u_5) = \langle K_5, \gamma_5 \rangle$, in which we have $K_5 = B^\ell(B^\tau(B^a, x_1, B^r(x_2)))$, $\gamma_5(B^\ell) = \ell\ell x_1$, $\gamma_5(B^\tau) = \tau(x_1, x_2, x_3)$, $\gamma_5(B^a) = a$, and $\gamma_5(B^r) = rr x_1$, where $B^\ell = \langle B, \ell, 1, \varepsilon \rangle$, $B^\tau = \langle B, \tau, 3, 1 \rangle$, $B^a = \langle B, a, 0, 11 \rangle$, and $B^r = \langle B, r, 1, 13 \rangle$. The resulting rule $\tilde{\rho}_5 = \langle \rho_5, \varepsilon \rangle$ is

$$\tilde{\rho}_5: \quad \bar{B} \to (\ell\ell x_1, \ \tau(x_1, x_2, x_3), \ a, \ rr x_1) \ ,$$

where $\bar{B} = \mathrm{seq}(K_5) = (B^\ell, B^\tau, B^a, B^r)$, and the corresponding skeleton function for B is $\kappa_5 = \kappa_{\rho_5, \varepsilon}$ such that $\kappa_5(B) = K_5$. The construction of this rule is illustrated in the left part of Fig. 2. Of course we obtain similar primed results for B'. For the rule $\rho_4 = B \to u_4$ and $\overline{\kappa} = (\kappa_5, \kappa_5')$, we substitute K_5 for B and K_5' for B' in u_4 and obtain the tree $u_4' = \ell B^\ell B^\tau(B^a, x_1, B^r B'^\ell B'^\tau(B'^a, b, B'^r rr x_2))$

with the adjoining B-decomposition $\mathrm{dec}_B(u_4') = \langle K_4, \gamma_4 \rangle$, where K_4 equals K_5, and γ_4 is defined by $\gamma_4(B^\ell) = \ell B^\ell x_1$, $\gamma_4(B^\tau) = B^\tau(x_1, x_2, x_3)$, $\gamma_4(B^a) = B^a$, and $\gamma_4(B^r) = B^r B'^\ell B'^\tau(B'^a, b, B''^r x_1)$. The resulting rule $\tilde\rho_4 = \langle \rho_4, (\kappa_5, \kappa_5') \rangle$ is

$$\tilde\rho_4: \quad \bar B \to (\ell B^\ell x_1, B^\tau(x_1, x_2, x_3), B^a, B^r B'^\ell B'^\tau(B'^a, b, B''^r x_1)) .$$

Since the skeleton function $\kappa_{\rho_4, (\kappa_5, \kappa_5')}$ for B is again κ_5, these are all the necessary rules of G_2' with left-hand side $\bar B$, and similarly for $\bar B' = (B'^\ell, B'^\tau, B'^a, B'^r)$. We now turn to rules ρ_3 and ρ_2. The only skeleton needed for A is the tree

$$K_3 = \kappa_{\rho_3, \varepsilon}(A) = A^\ell A^\tau(A^a, A^b, A^r x_1) ,$$

where $A^\ell = \langle A, \ell, 1, \varepsilon \rangle$, $A^\tau = \langle A, \tau, 3, 1 \rangle$, $A^a = \langle A, a, 0, 11 \rangle$, $A^b = \langle A, b, 0, 12 \rangle$, and $A^r = \langle A, r, 1, 13 \rangle$. The rule $\tilde\rho_3 = \langle \rho_3, \varepsilon \rangle$ is

$$\tilde\rho_3: \quad \bar A \to (\ell\ell x_1, \tau(x_1, x_2, x_3), a, b, rr x_1) ,$$

where $\bar A = \mathrm{seq}(K_3) = (A^\ell, A^\tau, A^a, A^b, A^r)$. Substituting K_3 for A and K_5 for B in the right-hand side $u_2 = \ell A B(b, r x_1)$ of ρ_2, we obtain

$$u_2' = \ell A^\ell A^\tau(A^a, A^b, A^r B^\ell B^\tau(B^a, b, B^r r x_1)) ,$$

which has the decomposition $\mathrm{dec}_A(u_2') = \langle K_3, \gamma_2 \rangle$ shown in the right part of Fig. 2, where $\gamma_2(A^\ell) = \ell A^\ell A^\tau(A^a, A^b, A^r B^\ell x_1)$, $\gamma_2(A^\tau) = B^\tau(x_1, x_2, x_3)$, $\gamma_2(A^a) = B^a$, $\gamma_2(A^b) = b$, and $\gamma_2(A^r) = B^r r x_1$. The rule $\tilde\rho_2 = \langle \rho_2, (\kappa_{\rho_3, \varepsilon}, \kappa_5) \rangle$ is

$$\tilde\rho_2: \quad \bar A \to (\ell A^\ell A^\tau(A^a, A^b, A^r B^\ell x_1), B^\tau(x_1, x_2, x_3), B^a, b, B^r r x_1) .$$

Finally, we consider rule ρ_1. The only skeleton needed for S is $S^\ell = \langle S, \ell, 0, \varepsilon \rangle$, which is the unique initial nonterminal of G'. Substituting K_3 for A and K_5 for B in the right-hand side of ρ_1, we obtain $u_2'[e]$ and the rule $\tilde\rho_1 = \langle \rho_1, (\kappa_{\rho_3, \varepsilon}, \kappa_5) \rangle$:

$$\tilde\rho_1: \quad S^\ell \to \ell A^\ell A^\tau(A^a, A^b, A^r B^\ell B^\tau(B^a, b, B^r re)) .$$

Thus, G_2' has the rules $\{\tilde\rho_1, \tilde\rho_2, \tilde\rho_3, \tilde\rho_4, \tilde\rho_5, \tilde\rho_4', \tilde\rho_5'\}$. Clearly, the tree K_3 generates the same terminal trees as $A(x_1)$ and the tree K_5 generates the same terminal trees as $B(x_1, x_2)$, and hence the tree $B^\ell B^\tau(B^a, b, B^r(x_1))$ also generates the same terminal trees as $A(x_1)$. It is easy to check that G_2' is a lexicalized MCTAG with respect to the mapping φ such that $\varphi(C^x) = x$ for every $C \in \{S, A, B, B'\}$ and every $x \in \{\tau, \ell, r, a, b\}$. The multiplicity of G_2' is $\mu(G_2') = 5$. □

It follows from Theorems 4 and 8 that MCTAGs can be strongly lexicalized as opposed to TAGs.

Corollary 10. *For every finitely ambiguous MCTAG G with terminal alphabet Σ there is an equivalent lexicalized MCTAG G' such that*

$$\mu(G') \leq \mu(G) \cdot \mathrm{mrk}_\Sigma \cdot |\Sigma| \cdot (2 \cdot \omega(G) + 1) .$$
□

We do not know whether the multiplicity bounds in Theorem 8 and Corollary 10 are optimal.

Acknowledgment. We are grateful to the reviewers for their constructive comments.

References

1. Engelfriet, J., Maneth, S.: Tree languages generated by context-free graph grammars. In: Ehrig, H., Engels, G., Kreowski, H.-J., Rozenberg, G. (eds.) TAGT 1998. LNCS, vol. 1764, pp. 15–29. Springer, Heidelberg (2000). doi:10.1007/978-3-540-46464-8_2
2. Engelfriet, J., Schmidt, E.M.: IO and OI I. J. Comput. Syst. Sci. **15**(3), 328–353 (1977)
3. Fujiyoshi, A.: Epsilon-free grammars and lexicalized grammars that generate the class of the mildly context-sensitive languages. In: Proceedings of the 7th International Workshop on Tree Adjoining Grammar and Related Formalisms, pp. 16–23 (2005)
4. Gécseg, F., Steinby, M.: Tree languages. In: Rozenberg, G., Salomaa, A. (eds.) Handbook of Formal Languages, vol. 3, chap. 1, pp. 1–68. Springer, Heidelberg (1997). doi:10.1007/978-3-642-59126-6_1
5. Joshi, A.K., Levy, L.S., Takahashi, M.: Tree adjunct grammars. J. Comput. Syst. Sci. **10**(1), 136–163 (1975)
6. Joshi, A.K., Schabes, Y.: Tree-adjoining grammars and lexicalized grammars. In: Nivat, M., Podelski, A. (eds) Tree Automata and Languages, pp. 409–431. North-Holland (1992)
7. Kallmeyer, L.: Parsing Beyond Context-Free Grammars. Cognitive Technologies. Springer, Heidelberg (2010). doi:10.1007/978-3-642-14846-0
8. Kanazawa, M.: Second-order abstract categorial grammars as hyperedge replacement grammars. J. Log. Lang. Inf. **19**(2), 137–161 (2010)
9. Kepser, S., Rogers, J.: The equivalence of tree adjoining grammars and monadic linear context-free tree grammars. J. Log. Lang. Inf. **20**(3), 361–384 (2011)
10. Kuhlmann, M.: Dependency Structures and Lexicalized Grammars: An Algebraic Approach. LNCS (LNAI), vol. 6270. Springer, Heidelberg (2010)
11. Kuhlmann, M., Satta, G.: Tree-adjoining grammars are not closed under strong lexicalization. Comput. Linguist. **38**(3), 617–629 (2012)
12. Maletti, A., Engelfriet, J.: Strong lexicalization of tree adjoining grammars. In: Proceedings of the 50th Annual Meeting of the Association for Computational Linguistics, pp. 506–515. ACL (2012)
13. Mönnich, U.: Adjunction as substitution: an algebraic formulation of regular, context-free and tree adjoining languages. In: Proceedings of the 3rd International Conference on Formal Grammar, pp. 169–178. Université de Provence, France (1997)
14. Rambow, O., Satta, G.: Independent parallelism in finite copying parallel rewriting systems. Theor. Comput. Sci. **223**(1–2), 87–120 (1999)
15. Rounds, W.C.: Context-free grammars on trees. In: Proceedings of the 1st ACM Symposium on Theory of Computation, pp. 143–148. ACM (1969)
16. Seki, H., Matsumura, T., Fujii, M., Kasami, T.: On multiple context-free grammars. Theor. Comput. Sci. **88**(2), 191–229 (1991)
17. Vijay-Shanker, K., Weir, D.J., Joshi, A.K.: Characterizing structural descriptions produced by various grammatical formalisms. In: Proceedings of the 25th Annual Meeting of the Association for Computational Linguistics, pp. 104–111. ACL (1987)
18. Weir, D.J.: Characterizing mildly context-sensitive grammar formalisms. Ph.D. thesis, University of Pennsylvania (1988)

On $\Sigma \wedge \Sigma \wedge \Sigma$ Circuits: The Role of Middle Σ Fan-In, Homogeneity and Bottom Degree

Christian Engels[1], B.V. Raghavendra Rao[2(✉)], and Karteek Sreenivasaiah[3]

[1] Kyoto University, Kyoto, Japan
christian.engels@gmail.com
[2] IIT Madras, Chennai, India
bvrr@cse.iitm.ac.in
[3] Saarland University, Saarbrücken, Germany
karteek@mpi-inf.mpg.de

Abstract. We study polynomials computed by depth five $\Sigma \wedge \Sigma \wedge \Sigma$ arithmetic circuits where 'Σ' and '$\wedge$' represent gates that compute sum and power of their inputs respectively. Such circuits compute polynomials of the form $\sum_{i=1}^{t} Q_i^{\alpha_i}$, where $Q_i = \sum_{j=1}^{r_i} \ell_{ij}^{d_{ij}}$ where ℓ_{ij} are linear forms and r_i, α_i, $t > 0$. These circuits are a natural generalization of the well known class of $\Sigma \wedge \Sigma$ circuits and received significant attention recently. We prove an exponential lower bound for the monomial $x_1 \cdots x_n$ against depth five $\Sigma \wedge \Sigma^{[\leq n]} \wedge^{[\geq 21]} \Sigma$ and $\Sigma \wedge \Sigma^{[\leq 2^{\sqrt{n}/1000}]} \wedge^{[\geq \sqrt{n}]} \Sigma$ arithmetic circuits where the bottom Σ gate is homogeneous.

Our results show that the fan-in of the middle Σ gates, the degree of the bottom powering gates and the homogeneity at the bottom Σ gates play a crucial role in the computational power of $\Sigma \wedge \Sigma \wedge \Sigma$ circuits.

1 Introduction

Arithmetic circuits were introduced by Valiant [18] as a natural model for algebraic computation and conjectured that the permanent polynomial, perm_n, does not have polynomial size arithmetic circuits. Following Valiant's work, there have been intensive research efforts towards the resolution of Valiant's hypothesis. Further, obtaining super polynomial size lower bounds for arithmetic circuits computing explicit polynomials is a pivotal problem in Algebraic Complexity Theory. However, for general classes of arithmetic circuits, the best known lower bound is barely superlinear [2].

Lack of progress on lower bounds against general arithmetic circuits lead researchers to explore restricted classes of circuits. Grigoriev and Karpinski [5] proved an exponential size lower bound for depth three circuits computing the permanent over finite fields of fixed size. However, extending these results to infinite fields or depth four arithmetic circuits remains elusive. Agrawal and

K. Sreenivasaiah—This work was done while the author was working at Max Planck Institute for Informatics, Saarbrücken supported by IMPECS post doctoral fellowship.

© Springer-Verlag GmbH Germany 2017
R. Klasing and M. Zeitoun (Eds.): FCT 2017, LNCS 10472, pp. 230–242, 2017.
DOI: 10.1007/978-3-662-55751-8_19

Vinay [1] (see also [10,17]) explained this lack of progress by establishing that proving exponential lower bounds against depth four arithmetic circuits is enough to resolve Valiant's conjecture. This was strengthened further to depth three circuits over infinite fields by Gupta et al. [6].

Gupta et al. [7] obtained a $2^{\Omega(\sqrt{n})}$ size lower bound for depth four homogeneous circuits computing perm_n where the fan-in of the bottom product gate is bounded by $O(\sqrt{n})$. Following this, Fournier et al. [4] obtained a super polynomial lower bound against depth four homogeneous circuits computing a polynomial in VP. Further, the techniques in [7,8] have been generalized and applied to prove lower bounds against various classes of constant depth arithmetic circuits for polynomials in VP as well as in VNP (see e.g., [15] and references therein).

Most of the lower bound proofs against arithmetic circuits follow a common framework: (1) define a measure for polynomials that is sub-additive and/or sub-multiplicative, (2) show that the circuit class of interest has small measure and (3) show that the target polynomial has high measure. See [15] for a detailed survey of these measures.

Apart from the complexity measure based framework mentioned above, there have been two other prominent approaches towards a resolution of Valiant's hypothesis: A geometric approach by Mulmuley and Sohoni [14] and an approach based on the real τ conjecture proposed by Shub and Smale [16].

The geometric approach to complexity theory [14] involves the study of class of varieties associated with each of the complexity classes and studying their representations.

The real τ conjecture of Koiran [9] states that the number of real roots of a univariate polynomial computed by an arithmetic circuit of size s is bounded by a polynomial in s. Koiran [9] showed that any resolution of the real τ-conjecture or an integer variant of it, would imply a positive resolution of Valiant's hypothesis. There has been several approaches towards the resolution of the real τ-conjecture and its variants by Koiran et al. [11,12].

Circuit Model. We consider the class of depth five powering circuits, i.e., $\Sigma \wedge \Sigma \wedge \Sigma$ circuits. It was shown in [6] that any homogeneous polynomial f of degree d over a sufficiently large field computed by a circuit of size s can also be computed by a homogeneous $\Sigma \wedge^{[a]} \Sigma \wedge^{[d/a]} \Sigma$ circuit of size $s^{\sqrt{d \log n \log(sd)}}$ for suitably chosen a. Here the superscript $[a]$ for a gate denotes the fan-in (degree in the case of $\wedge$ gates) at that level. This was an intermediary step in [6] which went on to obtain a depth three $\Sigma \Pi \Sigma$ circuit of size $2^{O(\sqrt{d \log n \log(sd)})}$ for f.

Thus, combined with the results in [17], to prove Valiant's hypotheses over infinite fields, it is enough to prove a $2^{\omega(\sqrt{n} \log n)}$ size lower bound against any one of the following classes of circuits: (1) homogeneous depth four $\Sigma \Pi^{[\sqrt{n}]} \Sigma \Pi^{[O(\sqrt{n})]}$ circuits, (2) homogeneous depth five $\Sigma \wedge^{[\sqrt{n}]} \Sigma \wedge^{[O(\sqrt{n})]} \Sigma$ circuits or (3) depth three $\Sigma \Pi \Sigma$ circuits.

Models (1) and (3) have received extensive attention in the literature compared to model (2). It follows that obtaining a $2^{\omega(\sqrt{n} \log n)}$ lower bound for any one of the models above would give a similar lower bound to the other. However, known lower bounds for model (1) so far do not even imply a super polynomial lower bound for model (2) which leaves obtaining super polynomial lower bounds against this model wide open.

In this article, we prove lower bounds against two restrictions of model (2) mentioned above: $\Sigma \wedge \Sigma^{[\leq n]} \wedge^{[\geq 21]} \Sigma$ circuits and $\Sigma \wedge \Sigma^{[\leq 2^{\sqrt{n}/1000}]} \wedge^{[\geq \sqrt{n}]} \Sigma$ circuits with bottom gates computing homogeneous linear forms. Since the transformation from depth four $\Sigma \Pi^{[\sqrt{n}]} \Sigma \Pi^{[O(\sqrt{n})]}$ to depth five $\Sigma \wedge^{[a]} \Sigma \wedge^{[d/a]} \Sigma$ in [6], in contrast to their result from general circuits, works against any chosen parameter $a < d$, the restrictions on the degree of the bottom $\wedge$ gates in the models we consider are general enough.

Throughout, it helps to interpret the polynomials computed by $\Sigma \wedge \Sigma \wedge \Sigma$ as sums of powers of projections of power symmetric polynomials where the n variate power symmetric polynomial of degree d is given by $p_d(x_1, \ldots, x_n) = x_1^d + \cdots + x_n^d$.

Our Results. We prove lower bounds against the restrictions of depth five $\Sigma \wedge \Sigma \wedge \Sigma$. We show:

Theorem 1. *Let* $g = \sum_{i=1}^{s} f_i^{\alpha_i}$ *where* $f_i = p_{d_i}(\ell_{i_1}, \ldots, \ell_{i_n}) + \beta_i$ *for some scalars* β_i *and for every* i, *either* $d_i = 1$ *or* $d_i \geq 21$ *and* $\ell_{i_1}, \ldots, \ell_{i_n}$ *are homogeneous linear forms. If* $g = x_1 \cdot x_2 \cdots x_n$ *then* $s = 2^{\Omega(n)}$.

The proof of Theorem 1 involves the dimension of the space of projected multilinear derivatives as a complexity measure for a polynomial f. It is computed by first projecting the partial derivative space of f to its multilinear subspace and then setting a subset of variables to 0. The dimension of the resulting space of polynomials is our measure of complexity for polynomials. Further, the method of projected multilinear derivatives also gives our second important result of the paper: An exponential lower bound against depth five powering circuits where the middle Σ layers have fan-in at most $2^{\sqrt{n}/1000}$ with the degree of the bottom $\wedge$ gates at least $\sqrt{n}$:

Theorem 2. *Let* $g = \sum_{i=1}^{s} f_i^{\alpha_i}$ *where* $f_i = p_{d_i}(\ell_{i_1}, \ldots, \ell_{i_{N_i}}) + \beta_i$, *for some scalars* β_i *and* $\sqrt{n} \leq d_i \leq n$, $N_i \leq 2^{\sqrt{n}/1000}$, *and* $\ell_{i_1}, \ldots, \ell_{i_{N_i}}$ *are homogeneous linear forms. If* $g = x_1 \cdot x_2 \cdots x_n$ *then* $s = 2^{\Omega(n)}$.

It is not difficult to see that the polynomial $x_1 \cdots x_n$ has a homogeneous $\Sigma \wedge^{[\sqrt{n}]} \Sigma^{[O(2^{\sqrt{n}})]} \wedge^{[\sqrt{n}]} \Sigma$ circuit of size $2^{O(\sqrt{n})}$ (see Lemma 16). Theorem 2 shows that reducing the middle Σ gate fan-in by a constant factor in the exponent leads to an exponential lower bound.

The homogeneity condition on the lower Σ and $\wedge$ gates seems to be necessary to our proofs of Theorems 1 and 2. In fact, Saptharishi [15], in a result attributed to Forbes, showed that $x_1 \cdots x_n$ can be computed by $\Sigma \wedge \Sigma \wedge$ circuits of size $2^{O(\sqrt{n})}$ where the lower Σ gates are not necessarily homogeneous.

Thus, it is important to study depth five powering circuits where the bottom Σ gates are not necessarily homogeneous. Towards this, in Sect. 4, we consider the widely used measure of the dimension of the shifted partial derivatives of a polynomial. We show:

Theorem 3. *Let* $g = \sum_{i=1}^{s} f_i^{\alpha_i}$ *where* $f_i = p_{d_i}(x_{i_1}, \ldots, x_{i_{m_i}}, \ell_{i_1}, \ldots, \ell_{i_{r_i}})$, $m_i \leq \frac{1}{40}n$, $r_i \leq n^\epsilon$, $d \leq 2^{o(n)}$, $\alpha_i \leq 2^{o(n)}$ *for all* i *where* $0 < \epsilon < 1$. *If* $g = x_1 x_2 \ldots x_n$ *then* $s = 2^{\Omega(n)}$.

It should be noted that Theorem 3 is much weaker than Theorems 1 and 2, however, it allows non-homogeneous Σ gates at the bottom. It seems that the restrictions on r_i in the above theorem are necessary if the lower bound argument uses the method of shifted partial derivatives. In particular, we show:

Lemma 4. *Let* $k \leq \min\{l, d\}$ *and* $\alpha > 0$ *be large enough. Then*

$$\dim \left(\mathbb{F}\text{-}Span \left\{ x^{\leq l} \partial^{=k} \left(p_d(x_1, \ldots, x_n)^\alpha \right) \right\} \right) = \Omega \left(\frac{\binom{n}{k} \binom{n+l}{l}}{l^{l/(d-1)}} \right).$$

In the cases where $l/(d-1) = O(1)$ and $l = n^{O(1)}$ the above bound is tight up to a polynomial factor since $\dim \left(\mathbb{F}\text{-}Span \left\{ x^{\leq l} \partial^{=k} \left(p_d(x_1, \ldots, x_n)^\alpha \right) \right\} \right) \leq \binom{n}{k} \binom{n+l}{l}$ and hence indicating that the restrictions on the r_is in Theorem 3 would be necessary if the dimension of shifted partial derivatives is used as the measure of complexity.

2 Preliminaries

An *arithmetic circuit* is a labelled directed acyclic graph. Vertices of zero indegree are called *input* gates and are labelled by elements in $\mathbb{F} \cup \{x_1, \ldots, x_n\}$. Vertices of in-degree two or more are called *internal* gates and have their labels from $\{\times, +\}$. An arithmetic circuit has at least one vertex of zero out-degree called an *output* gate. We assume that an arithmetic circuit has exactly one output gate. A polynomial p_g in $\mathbb{F}[x_1, \ldots, x_n]$ can be associated with every gate g of an arithmetic circuit defined in an inductive fashion. Input gates compute their label. Let g be an internal gate with children $f_1, \ldots, f_m$ then $p_g = p_{f_1}$ op $\cdots$ op p_{f_m} where op $\in \{+, \times\}$ is the label of g. The polynomial computed by the circuit is the polynomial at one of the output gates and denoted by p_C. The size of an arithmetic circuit is the number of gates in it and is denoted by size(C). We will denote a fan-in/degree bound on a layer as a superscript to the corresponding gate e.g., $\Sigma \wedge \Sigma^{[\leq n]} \wedge^{[\geq 21]} \Sigma$ denotes the class of families of polynomials computed by depth five circuits with powering and sum gates, where the middle layer of sum gates have fan-in bounded from above by n and the bottom most powering gates have degree at least 21.

The following bound on the binomial coefficient is useful throughout the paper:

Proposition 5 ([13]). *Let* $r \leq n$. *Then* $\log_2 \binom{n}{r} \approx nH(r/n)$, *where* H *is the binary entropy function,* $H(p) = -p \log_2(p) - (1-p) \log_2(1-p)$, *and* $\approx$ *is equality up to an additive* $o(n)$ *error.*

We denote by $[n]$ the set $\{1, \ldots, n\}$. For a set of polynomials S, let $\mathcal{M}_{\leq d}(S)$ $(\mathcal{M}_{=d}(S))$ denote the set of all products of at most (exactly) d not necessarily distinct elements from S. Note that when S is a set of variables, $|\mathcal{M}_{\leq d}(S)| = \binom{|S|+d}{d}$. When the set S is clear from the context, we use $\mathcal{M}_{\leq d}$ $(\mathcal{M}_{=d})$ instead of $\mathcal{M}_{\leq d}(S)$ $(\mathcal{M}_{=d}(S))$.

For a subset S of variables, let $\mathcal{X}_a^b(S)$ denote the set of all multilinear monomials of degree $a \leq d \leq b$ in variables from the set S, i.e.,

$$\mathcal{X}_a^b(S) = \{ \prod_{x_i \in S} x_i^{\delta_i} \mid a \leq \sum_{i=1}^n \delta_i \leq b, \delta_i \in \{0,1\}\}.$$

For two sets A and B, define $A \odot B \triangleq \{a \cdot b \mid a \in A, b \in B\}$. Additionally, we define $A \cdot f$ for some polynomial f to be the set $\{a \cdot f \mid a \in A\}$.

The notion of *shifted partial derivatives* is given as follows: For $k \geq 0$ and $f \in \mathbb{F}[x_1, \ldots, x_n]$ let $\partial^{=k} f$ denote the set of all partial derivatives of f of order k. For $l \geq 0$, the (k, l) shifted partial derivative space of f, denoted by $\mathbb{F}\text{-Span}\{x^{\leq l}\partial^{=k} f\}$, is defined as:

$$\mathbb{F}\text{-Span}\{x^{\leq l}\partial^{=k} f\} \triangleq \mathbb{F}\text{-Span}\{\mathbf{m} \cdot \partial^{=k} f \mid \mathbf{m} \in \mathcal{M}_{\leq \ell}(x_1, \ldots, x_n)\}$$

where $\mathbb{F}\text{-Span}\{S\} \triangleq \{\alpha_1 f_1 + \cdots + \alpha_m f_m \mid f_i \in S \text{ and } \alpha_i \in \mathbb{F} \text{ for all } i \in [m]\}$. We restate the well known lower bound for the dimension of the space of shifted partial derivatives $x_1 \cdots x_n$:

Proposition 6 ([8]).

$$\dim\left(\mathbb{F}\text{-}Span\{x^{\leq l}\partial_{\mathsf{ML}}^{=k} x_1 \cdots x_n\}\right) = \dim\left(\mathbb{F}\text{-}Span\{x^{\leq l}\partial^{=k} x_1 \cdots x_n\}\right)$$

$$\geq \binom{n}{k} \cdot \binom{n-k+l}{l}.$$

In the above, $\partial_{\mathsf{ML}}^{=k} f$ denotes the set of kth order multilinear derivative space of f, i.e., $\partial_{\mathsf{ML}}^{=k} f \triangleq \{ \frac{\partial^k f}{\partial x_{i_1} \cdots \partial x_{i_k}} \mid i_1 < \ldots < i_k \in \{1, \ldots, n\}\}$.

3   Projected Multilinear Derivatives and Proof of Theorems 1 and 2

This section is devoted to the proof of Theorems 1 and 2. Our proof follows the standard two step approach for proving arithmetic circuit lower bounds: First, define a sub-additive measure that is low for every polynomial computed in the model. Second, show that the measure is exponentially larger for a specific polynomial p. Hence allowing us to conclude that any circuit in the model that computes p requires exponential size.

We consider a variant of the space of partial derivatives, viz., the *projected multilinear derivatives* as the complexity measure for polynomials.

The Complexity Measure. Let $f \in \mathbb{F}[x_1, \ldots, x_n]$. For $S \subseteq \{1, \ldots, n\}$, let $\pi_S : \mathbb{F}[x_1, \ldots, x_n] \to \mathbb{F}[x_1, \ldots, x_n]$ be the projection map that sets all variables in S to zero, i.e., for every $f \in \mathbb{F}[x_1, \ldots, x_n]$, $\pi_S(f) = f(x_i = 0 \mid i \in S)$. Let $\pi_{\mathsf{m}}(f)$ denote the projection of f onto its multilinear monomials, i.e., if $f = \sum_{\alpha \in \mathbb{N}^n} c_\alpha \prod_{i=1}^n x_i^{\alpha_i}$ then $\pi_{\mathsf{m}}(f) = \sum_{\alpha \in \{0,1\}^n} c_\alpha \prod_{i=1}^n x_i^{\alpha_i}$.

For $S \subseteq \{1, \ldots, n\}$ and $0 < k \leq n$, the dimension of Projected Multilinear Derivatives (PMD) of a polynomial f is defined as:

$$\mathrm{PMD}_S^k(f) \triangleq \dim(\mathbb{F}\text{-Span}\{\pi_S(\pi_{\mathsf{m}}(\partial_{\mathsf{ML}}^{=k} f))\}).$$

We omit the subscript S when either S is clear from the context or when it refers to an unspecified set S. It is not hard to see that PMD_S^k is sub-additive.

Lemma 7. *For any $S \subseteq \{1 \ldots, n\}$, $k \geq 1$, and polynomials f and g:*

$$\mathrm{PMD}_S^k(f + g) \leq \mathrm{PMD}_S^k(f) + \mathrm{PMD}_S^k(g).$$

Lower Bound for the Measure

We establish a lower bound on the dimension of projected multilinear derivatives of the polynomial $x_1 \cdots x_n$. This follows from a simple argument and is shown below:

Lemma 8. *For any $S \subseteq \{1, \ldots, n\}$ with $|S| = n/2 + 1$ and $k = 3n/4$ we have:*

$$\mathrm{PMD}_S^k(x_1 \cdots x_n) \geq \binom{n/2 - 1}{n/4} \geq 2^{n/2}/n^2.$$

Proof. Let $T \subseteq \{1, \ldots, n\}$ with $|T| = k$. Then $\frac{\partial^k}{\partial T}(x_1 \cdots x_n) = \prod_{i \notin T} x_i$. Note that if $S \cap \overline{T} = \emptyset$ then we have $\pi_S(\pi_{\mathsf{m}}(\frac{\partial^k}{\partial T}(x_1 \cdots x_n))) = \prod_{i \notin T} x_i$ since setting variables in S to zero does not affect the variables in $\overline{T}$. Otherwise, if $S \cap \overline{T} \neq \emptyset$ then $\pi_S(\pi_{\mathsf{m}}(\frac{\partial^k}{\partial T}(x_1 \cdots x_n))) = 0$. Thus, we have:

$$\mathbb{F}\text{-Span}\{\pi_S(\pi_{\mathsf{m}}(\partial_{\mathsf{ML}}^{=k}(x_1 \cdots x_n)))\} \supseteq \mathbb{F}\text{-Span}\left\{\prod_{i \in \overline{T}} x_i \mid \overline{T} \subseteq \overline{S}, |\overline{T}| \leq n/4\right\}.$$

Hence, $\mathrm{PMD}_S^k(x_1 \cdots x_n) \geq \binom{n/2-1}{n/4} \geq 2^{n/2}/n^2$ using Stirling's approximation of binomial coefficients. $\square$

$\Sigma \wedge \Sigma \wedge$ Circuits: The Curse of Homogeneity

Firstly, we observe that homogeneous $\Sigma \wedge \Sigma \wedge$ circuits of polynomial size cannot compute the monomial $x_1 \cdots \cdots x_n$ by eliminating bottom $\wedge$ gates of degree at least 2:

Observation 9. *Let $f = f_1^{\alpha_1} + \cdots + f_s^{\alpha_s}$ where $f_i = \sum_{j=1}^n \beta_{ij} x_j^{d_i} + \beta_{i0}$, $\beta_{ij} \in \mathbb{F}$. If $f = x_1 \cdots x_n$ then $s = 2^{\Omega(n)}$.*

The homogeneity condition for the bottom power gates is necessary due to the following result in [15]. Let $\mathrm{Sym}_{n,d} = \sum_{S \subseteq [n], |S|=d} \prod_{i \in S} x_i$, the elementary symmetric polynomial of degree d.

Proposition 10. *[15, Corollary 17.16] For any $d > 0$, $\mathrm{Sym}_{n,d}$ can be computed by a $\Sigma \wedge \Sigma \wedge$ circuit of size $2^{O(\sqrt{d})} \mathrm{poly}(n)$.[1]*

Is it all about homogeneity at the bottom Σ gates? The answer is no. In fact, Observation 9 can also be generalized to the case of powers of polynomials in the span of the set $\{x_{i_j}^{\alpha_{i_j}} \mid 1 \le i_j \le n, \ \alpha_{i_j} \ge 2\}$:

Lemma 11. *For any $\beta_0, \beta_1, \ldots, \beta_r \in \mathbb{F}$, $\alpha, d \in \mathbb{N}$ and for any $S \subseteq \{1, \ldots, n\}$ with $|S| + k > n$, we have $\mathrm{PMD}_S^k((\sum_{j=1}^r \beta_j x_{i_j}^{d_j} + \beta_0)^\alpha) \le 1$ where $1 \le i_j \le n$ and either $\forall j \ d_j \ge 2$ or $\forall j \ d_j = 1$.*

We get the following generalization of Observation 9:

Corollary 12. *Let $f = f_1^{\alpha_1} + \cdots + f_s^{\alpha_s}$ where for every i, either f_i is a linear form or $f_i = \sum_{j=1}^n \beta_{i,l_j} x_{l_j}^{d_{i_j}} + \beta_{i0}$ for $d_{i_j} \ge 2$ and $\beta_{i,l_j} \in \mathbb{F}$. If $f = x_1 \cdots x_n$ then $s = 2^{\Omega(n)}$. Moreover, $|\{i \mid f_i \text{ is linear}\}| = 2^{\Omega(n)}$.*

Proof. Let $S \subset \{1, \ldots, n\}$ with $|S| = n/2 + 1$ and $k = 3n/4$. From Lemmas 11 and 7 we have $\mathrm{PMD}_S^k(f) \le \sum_{i=1}^s \mathrm{PMD}_S^k(f_i^{\alpha_i}) \le s$. Hence by Lemma 8 we have $s \ge 2^{n/2}/n^2$ as required. Further, $\mathrm{PMD}_S^k(f_i^{\alpha_i})$ is non-zero only if f_i is a linear form, and hence $|\{i \mid f_i \text{ is linear}\}| = 2^{\Omega(n)}$. □

$\Sigma \wedge \Sigma \wedge \Sigma$ Circuits: Middle Σ Fan-In Versus the Bottom Degree

The argument above fails even when the degree of the power symmetric polynomial is two (i.e., $d = 2$). Let $f = \ell_1^2 + \cdots + \ell_n^2 + \beta$ where $\ell_1, \ldots, \ell_n$ are homogeneous linear functions such that each of the ℓ_i have all n variables with non-zero coefficients and $\beta \ne 0$. It is not hard to see that the space $\partial_{\mathsf{ML}}^k f^\alpha$ of the kth order derivatives of f^α is contained in the span of $\{f^{\alpha-k} \prod_{i=1}^n \ell_i^{\gamma_i} \mid \sum_i \gamma_i \le k\}$. Even after applying the projections π_{m} and π_S for any $S \subseteq \{1, \ldots, n\}$, with $|S| = (n/2) + 1$, obtaining a bound on PMD_S^k better than the lower bound in Lemma 8 seems to be difficult. The reason is that every multilinear monomial of degree $|n/2 - 1 - k|$ appears in at least one of the projected multilinear derivatives of f^α.

A natural approach to overcome the above difficulty could be to obtain a basis for the projected multilinear derivatives of f^α consisting of a small set of monomials and a small set of products of powers of the linear forms multiplied by suitable powers of f. Surprisingly, as shown below in Lemma 13, the approach works when the degree $d \ge 21$, although it requires an involved combinatorial argument.

---

[1] In [15], Corollary 17.16, it is mentioned that the resulting $\Sigma \wedge \Sigma \wedge$ circuit is homogeneous. However, a closer look at the construction shows that the application of Fischer's identity produces sum gates that are not homogeneous.

Lemma 13. *Suppose that* $f = (\ell_1^d + \ldots + \ell_n^d + \beta)$ *for some scalar* β, *and* ℓ_j *homogeneous linear forms,* $1 \leq j \leq n$. *Let* $Y = \{\ell_i^{d-j} \mid 1 \leq i \leq n, 1 \leq j \leq d\}$ *and* $\lambda = 1/4 + \varepsilon$ *for some* $0 < \varepsilon < 1/4$. *Then, for* $k = 3n/4$ *and any* $S \subseteq \{1, \ldots, n\}$ *with* $|S| = n/2 + 1$, *we have:*

$$\pi_S(\pi_m(\partial_{ML}^{=k} f^\alpha)) \subseteq \mathbb{F}\text{-}Span\left\{\pi_S(\pi_m(\mathcal{F} \odot \left(\mathcal{X}_{\lambda n}^{n/2-1}(\overline{S}) \cup \mathcal{M}_{\leq (1+\varepsilon)n/d}(Y)\right)))\right\}$$

where $\mathcal{F} = \cup_{i=1}^k f^{\alpha-i}$ *and* $\overline{S} = \{1, \ldots, n\} \setminus S$.

Proof. Let $T \subseteq \{x_1, \ldots, x_n\}$ with $|T| = k$, let $f_T^{(k)}$ denote kth order partial derivative of f with respect to T. Note that $f_T^{(k)} \in \mathbb{F}\text{-}Span\{\ell_j^{d-k} \mid 1 \leq j \leq n\}$. Let L_i denote $\{\ell_j^{d-i} \mid 1 \leq j \leq n\}$ so that $f_T^{(k)} \in \mathbb{F}\text{-}Span\{L_k\}$. Then

$$\frac{\partial^k f^\alpha}{\partial T} \in \mathbb{F}\text{-}Span\left\{f^{\alpha-i} \odot D_i^T(f) \mid 1 \leq i \leq k\right\} \tag{1}$$

where $D_i^T(f) = \left\{\prod_{r=1}^i f_{T_r}^{(t_r)} \mid T_1 \uplus \ldots \uplus T_i = T, \text{ where } t_r = |T_r| > 0, 1 \leq r \leq i\right\}$. Intuitively, the set D_i^T contains one polynomial for each possible partition of T into i many parts. The polynomial corresponding to a particular partition is the product of the derivatives of f with respect to each of the parts. Now, the following claim bounds the span of D_i^T:

Claim. For any $1 \leq i \leq k$, $D_i^T \subseteq \mathbb{F}\text{-}Span\left\{\odot_{r=1}^k L_r^{\odot j_r} \mid \sum_{r=1}^k r \cdot j_r = k\right\}$.

Proof. Let $T_1 \uplus \cdots \uplus T_i = T$ be a partition and let j_r denote the number of parts with cardinality r, i.e., $j_r = |\{j \mid |T_j| = r\}|$. Then

$$\prod_{|T_j|=r} f_{T_j}^{(r)} \in \mathbb{F}\text{-}Span\left\{\bigodot_{|T_j|=r} L_r\right\} = \mathbb{F}\text{-}Span\left\{L_r^{\odot j_r}\right\}.$$

Thus, $\prod_{r=1}^i f_{T_r}^{(t_r)} \in \mathbb{F}\text{-}Span\left\{\odot_{r=1}^k L_r^{\odot j_r}\right\}$. Since, $\sum_{r=1}^k r \cdot j_r = k$ for any partition $T_1 \uplus \cdots \uplus T_i$ of T, the claim follows. $\square$

Continuing from (1), we have:

$$\frac{\partial^k f^\alpha}{\partial T} \in \mathbb{F}\text{-}Span\left\{f^{\alpha-i} \odot D_i^T(f) \mid 1 \leq i \leq k\right\}$$

$$\subseteq \mathbb{F}\text{-}Span\left\{\mathcal{F} \odot \{D_i^T(f) \mid 1 \leq i \leq d\}\right\}$$

$$\subseteq \mathbb{F}\text{-}Span\left\{\mathcal{F} \odot \left\{\bigodot_{r=1}^d L_r^{\odot j_r} \mid 1 \cdot j_1 + \cdots + d \cdot j_d = k\right\}\right\}. \tag{2}$$

It remains to show that the right side of (2) is spanned by a set of polynomials that satisfy the properties stated in the lemma. The next claim completes the proof of Lemma 13.

Claim.

$$\pi_S\left(\pi_{\mathsf{m}}\left(\left\{\bigodot_{r=1}^{d} L_r^{\odot j_r} \;\Big|\; \sum_{i=1}^{d} i \cdot j_i = k\right\}\right)\right) \subseteq \mathbb{F}\text{-Span}\left\{\mathcal{X}_{\lambda n}^{\frac{n}{2}-1}(\bar{S}) \cup \mathcal{M}_{\leq \frac{(1+\epsilon)n}{d}}(Y)\right\}.$$

Proof. Note that the polynomials in L_j are homogeneous non-constant polynomials of degree $d - j$, and hence the set $\bigodot_{r=1}^{d} L_r^{\odot j_r}$ consists of homogeneous polynomials of degree $\sum_{r=1}^{d} j_r(d-r)$.

Let $\deg(\bigodot_{r=1}^{d} L_r^{\odot j_r})$ denote the degree of polynomials in the set $\bigodot_{r=1}^{d} L_r^{\odot j_r}$.

The remaining argument is split into three cases depending on the value of $\deg(\bigodot_{r=1}^{d} L_r^{\odot j_r})$.

Case 1: $\deg(\bigodot_{r=1}^{d} L_r^{\odot j_r}) \geq n/2$ then $\pi_S(\pi_{\mathsf{m}}(\bigodot_{r=1}^{d} L_r^{\odot j_r})) = \{0\}$. Note that here we have crucially used the fact that the ℓ_j are homogeneous.

Case 2: $\lambda n \leq \deg(\bigodot_{r=1}^{d} L_r^{\odot j_r})) < n/2$. In this case $\pi_S(\pi_{\mathsf{m}}(\bigodot_{r=1}^{d} L_r^{\odot j_r}))$ is spanned by the set of all multilinear monomials in the set of variables $\{x_j \mid j \notin S\}$ of degree at least λn and at most $n/2 - 1$. Therefore we have,

$$\pi_S(\pi_{\mathsf{m}}(\bigodot_{r=1}^{d} L_r^{\odot j_r})) \subseteq \mathbb{F}\text{-Span}\left\{\mathcal{X}_{\lambda n}^{n/2-1}(\bar{S})\right\}.$$

Case 3: $\deg(\bigodot_{r=1}^{d} L_r^{\odot j_r})) < \lambda n$. Recall that $\deg(\bigodot_{r=1}^{d} L_r^{\odot j_r})) = \sum_{r=1}^{d} j_r(d-r) \leq \lambda n$. Then,

$$\sum_{r=1}^{d} j_r \cdot d \leq \sum_{r=1}^{d} j_r \cdot r + \lambda n = k + \lambda n \text{ (since } \sum_{r=1}^{d} r \cdot j_r = k.)$$
$$= (\lambda + 3/4)n = (1 + \varepsilon)n.$$

Hence, $\pi_S(\pi_{\mathsf{m}}(\bigodot_{r=1}^{d} L_r^{\odot j_r}))$ is spanned by the set of all products of at most $(1 + \varepsilon)n/d$ polynomials of the form ℓ_i^{d-j}, i.e.,

$$\pi_S(\pi_{\mathsf{m}}(\bigodot_{r=1}^{d} L_r^{\odot j_r})) \subseteq \mathbb{F}\text{-Span}\left\{\mathcal{M}_{\leq(1+\varepsilon)n/d}(Y)\right\}.$$

$\square$

This completes the proof. $\square$

Using Lemma 13 above and choosing suitable parameters k and S we obtain the following upper bound on the dimension of projected multilinear derivatives:

Theorem 14. *Let $f = (\ell_1^d + \ldots + \ell_n^d + \beta)$ where ℓ_j are homogeneous linear forms. For $d \geq 21$ and any $S \subseteq \{1, \ldots, n\}$ where $|S| = n/2 + 1$. Then*

$$\mathrm{PMD}_S^k(f^\alpha) \leq 2^{(0.498+o(1))n}.$$

Proof. By Lemma 13,

$$\pi_S(\pi_{\mathsf{m}}(\partial_{\mathsf{ML}}^{=k} f^\alpha)) \subseteq \mathbb{F}\text{-Span}\left\{\pi_S(\pi_{\mathsf{m}}(\{f^{\alpha-i}\}_{i=1}^{k} \odot \left\{\mathcal{X}_{\lambda n}^{n/2-1}(\bar{S}) \cup \mathcal{M}_{\leq(1+\epsilon)n/d}(Y)\right\}))\right\}.$$

Recall that $\lambda = \frac{1}{4} + \varepsilon$. We choose $\varepsilon = 1/50$ and hence $\lambda = 0.27$. We have:

$$\mathrm{PMD}_S^k(f^\alpha) \leq k \cdot (|\mathcal{X}_{\lambda n}^{n/2-1}(\bar{S})| + |\mathcal{M}_{\leq(1+\epsilon)n/d}(Y)|).$$

Now, since $1/4 < \lambda < 1/2$, we have

$$|\mathcal{X}_{\lambda n}^{n/2-1}(\bar{S})| \leq (n/2 - 1 - \lambda n) \cdot \binom{n/2-1}{\lambda n} \leq c(n/2) \cdot \binom{n/2}{\lambda n}$$

$$\leq (cn/2) \cdot 2^{\frac{n}{2} \cdot \mathcal{H}(2\lambda)} \leq (cn/2) \cdot 2^{0.498n}.$$

where c is an absolute constant. We bound $|\mathcal{M}_{\leq(1+\varepsilon)n/d}(Y)|$ as follows:

$$|\mathcal{M}_{\leq(1+\varepsilon)n/d}(Y)| = \binom{|Y| + (1+\varepsilon)n/d}{(1+\varepsilon)n/d} = \binom{dn + (1+\varepsilon)n/d}{(1+\varepsilon)n/d}$$

$$\leq 2^{(dn+(1+\varepsilon)n/d)\mathcal{H}\left(\frac{(1+\varepsilon)n/d}{dn+(1+\varepsilon)n/d}\right)}$$

$$= 2^{n(d+(1+\varepsilon)/d)\mathcal{H}\left((1+\varepsilon)/(d^2+(1+\varepsilon))\right)} \leq 2^{0.4955n} \quad \text{for } d \geq 21.$$

For the last inequality, note that for fixed n and ε, $(d + (1 + \varepsilon)/d)\mathcal{H}((1 + \varepsilon)/(d^2 + (1 + \epsilon))$ is a monotonically decreasing function of d, with $\lim_{d\to\infty}(d + (1 + \varepsilon)/d)\mathcal{H}((1 + \varepsilon)/(d^2 + (1 + \epsilon)) = 0$. Therefore, the bound holds for $d \geq 21$. This completes the proof. $\qquad\square$

Corollary 15. *Let* $f = (\ell_1^d + \ldots + \ell_N^d + \beta)$ *where* ℓ_j *are homogeneous linear forms. If d is such that $N \leq 2^{(d/1000)}$, $d \leq n$, and $n/d = o(n)$ then for any $\alpha > 0$,*

$$\mathrm{PMD}_S^k(f^\alpha) \leq 2^{(0.498+o(1))n}.$$

Proof of Theorem 1: Let $S = \{1, \ldots, n/2+1\}$ and $k = 3n/4$. Then by Theorem 14 we have $\mathrm{PMD}_S^k(f_i) \leq 2^{0.498n+o(n)}$. By the sub-additivity of PMD_S^k (Lemma 7), we have $\mathrm{PMD}_S^k(\sum_{i=1}^s f_i^{\alpha_i}) \leq s \cdot 2^{0.498n+o(n)}$. Since $\mathrm{PMD}_S^k(x_1 \cdots x_n) \geq 2^{n/2}/n^2$, we conclude $s \geq 2^{0.001n}$, as required. $\qquad\square$

Proof of Theorem 2: Let $S = \{1, \ldots, n/2+1\}$ and $k = 3n/4$. Since $d_i \geq \sqrt{n}$, it holds that $N_i \leq 2^{d/1000}$. Then, by Corollary 15, we have $\mathrm{PMD}_S^k(f_i^{\alpha_i}) \leq 2^{0.498n+o(n)}$. By the sub-additivity of PMD_S^k (Lemma 7), we have $\mathrm{PMD}_S^k(\sum_{i=1}^s f_i^{\alpha_i}) \leq s \cdot 2^{0.498n+o(n)}$. Since $\mathrm{PMD}_S^k(x_1 \cdots x_n) \geq 2^{n/2}/n^2$, we conclude $s \geq 2^{0.001n}$ for large enough n, as required. $\qquad\square$

A separation within $\Sigma \wedge \Sigma \wedge \Sigma$ Circuits: An alert reader might have wondered if the restriction on the fan-in of the middle layer of Σ gates in Theorem 2 is a limitation of the method of projected multilinear derivatives. By a simple application of Fischer's identity [3], we get:

Lemma 16. *Over fields of characteristic zero or characteristic greater than n, the polynomial $x_1 \cdots x_n$ can be computed by a homogeneous $\Sigma \wedge^{[\sqrt{n}]} \Sigma^{[O(2^{\sqrt{n}})]} \wedge^{[\sqrt{n}]} \Sigma$ circuit of size $2^{O(\sqrt{n})}$.*

This immediately leads to the following separation of homogeneous $\Sigma \wedge^{[\sqrt{n}]}$ $\Sigma \wedge^{[\sqrt{n}]} \Sigma$ circuits:

Corollary 17. *The class of polynomials computed by $\Sigma \wedge^{[\sqrt{n}]} \Sigma^{[2^{\sqrt{n}}/1000]} \wedge^{[\sqrt{n}]} \Sigma$ of size $2^{O(\sqrt{n})}$ is strictly contained in the class computed by $\Sigma \wedge^{[\sqrt{n}]} \Sigma^{[2^{\sqrt{n}}]} \wedge^{[\sqrt{n}]} \Sigma$ of size $2^{O(\sqrt{n})}$.*

4   Dimension of Shifted Partial Derivatives

This section is devoted to the study of shifted partial derivatives of polynomials that are computed by restricted $\Sigma \wedge \Sigma \wedge \Sigma$ circuits and proofs of Theorem 3 and Lemma 4.

We begin with a simple upper bound on the dimension of the derivatives of powers of projections of p_d onto low-dimensional sub-spaces:

Lemma 18. *Let $f = p_d(\ell_1, \ldots, \ell_t)$ where $\ell_1, \ldots, \ell_t$ are linear forms. Then for any $k > 0$, we have $\dim \left(\mathbb{F}\text{-Span} \left\{ \partial_{\mathsf{ML}}^{\leq k} f^\alpha \right\} \right) \leq (k+1)(dk)^r$ where r is the dimension of the span of $\{\ell_1, \ldots, \ell_t\}$.*

Proof. Without loss of generality, assume that $\ell_1, \ldots, \ell_r$ is a basis for the space spanned by $\ell_1, \ldots, \ell_t$ $r \leq t$. Observe that:

$$\partial_{\mathsf{ML}}^{\leq k} f^\alpha \subseteq \mathbb{F}\text{-Span} \left\{ f^{\alpha - i} \cdot \ell_1^{\beta_1} \cdots \ell_r^{\beta_r} \mid \sum_{j=1}^{r} \beta_j \leq dk \right\}_{i \in \{1, \ldots, k\}}$$

and therefore, $\dim \left(\mathbb{F}\text{-Span} \left\{ \partial_{\mathsf{ML}}^{\leq k} f^\alpha \right\} \right) \leq (k+1)(dk)^r$ as required. $\square$

Now, we bound the dimension of shifted partial derivatives of powers of the power symmetric polynomial:

Lemma 19. *Let $f = p_d(x_{j_1}, \ldots, x_{j_m})$ for some $j_1, \ldots, j_m \in \{1, \ldots, n\}$. Then for any $\alpha, l, k \geq 1$*

$$\dim \left(\mathbb{F}\text{-Span} \left\{ x^{\leq l} \partial_{\mathsf{ML}}^{=k} f^\alpha \right\} \right) \leq (k+1) \binom{n+m+k+l}{k+l}.$$

Note that the straightforward bound of $\binom{m}{k}\binom{n+l}{l}$ is better than this bound if m is large. However, when m is small (say $m \leq n/40$), the bound shown above is better for suitable values of k and l. Combining Lemmas 18 and 19 with the sum and product rules for partial derivatives, we get:

Lemma 20. *Let $\ell_1, \ldots \ell_t \in \mathbb{F}[x_1, \ldots, x_n]$ be linear forms and let r denote their rank. Let $f = p_d(x_{j_1}, \ldots, x_{j_m}, \ell_1, \ldots, \ell_t)$. Then for any $d > k > 0$, we have*

$$\dim \left(\mathbb{F}\text{-Span} \left\{ x^{\leq l} \partial_{\mathsf{ML}}^{=k} f^\alpha \right\} \right) \leq (\alpha + 1)(k+1)^3 (dk)^r \binom{m+n+k+l}{k+l}.$$

Finally, using sub-additivity of shifted partial derivatives and Lemma 20 we obtain the following upper bound:

Theorem 21. *Let $d > k > 0$ and $g = \sum_{i=1}^{s} f_i^{\alpha_i}$ where each of the polynomials $f_i = p_{d_i}(x_{i_1}, \ldots, x_{i_{m_i}}, \ell_{i_1}, \ldots, \ell_{i_{r_i}})$ and $\ell_{i_1}, \ldots, \ell_{i_{m_i}}$ are linear forms in $x_1, \ldots, x_n$. Then for any $l > 0$ with $k + l > n + m$:*

$$\dim\left(\mathbb{F}\text{-}Span\left\{x^{\leq l}\partial_{\mathsf{ML}}^{=k}g\right\}\right) \leq s(\alpha+1)(k+1)^3(dk)^r \binom{n+m+k+l}{k+l}$$

where $m = \max_i m_i$ and $r = \max_i\{\dim(\mathbb{F}\text{-}Span\{\ell_{i_1}, \ldots, \ell_{i_{r_i}}\})\}$.

Combining the previous theorem with the lower bound from Proposition 6 gives us the required size lower bound.

Theorem 3. *Let $g = \sum_{i=1}^{s} f_i^{\alpha_i}$ where $f_i = p_{d_i}(x_{i_1}, \ldots, x_{i_{m_i}}, \ell_{i_1}, \ldots, \ell_{i_{r_i}})$, $m_i \leq \frac{1}{40}n$, $r_i \leq n^\epsilon$, $d \leq 2^{n^{1-\gamma}}$ and $\alpha_i \leq 2^{n^\delta}$ for all i, for some $0 < \delta, \epsilon, \gamma < 1$, $\varepsilon < \gamma$. If $g = x_1 x_2 \ldots x_n$ then $s = 2^{\Omega(n)}$.*

Proof. Let $d \geq 2$ and $m = \max_i m_i$. Using Proposition 6 and Theorem 21

$$s \geq \frac{\binom{n}{k}\binom{n-k+l}{l}}{(\alpha+1)(k+1)^3(dk)^r\binom{n+m+k+l}{k+l}}$$

where $\alpha = \max_i \alpha_i$. Taking the logarithm and using that $3\log(k+1) \leq 3\log dk$ since $d \geq 2$ gives us

$$\log s \geq \log\binom{n}{k} + \log\binom{n-k+l}{l} - \left(\log(\alpha+1) + \log\binom{n+m+k+l}{k+l} + (r+3)\log dk\right).$$

Note that $(r+3)\log dk \in o(n)$ if $d \leq 2^{n^{1-\gamma}}$. Now, using the approximation of binomial coefficients in Proposition 5 and setting $k = n/10$ and $l = 10n$ we get $\log s \geq 0.0165n$. This proves the required bound. □

References

1. Agrawal, M., Vinay, V.: Arithmetic circuits: a chasm at depth four. In: FOCS, pp. 67–75 (2008)
2. Baur, W., Strassen, V.: The complexity of partial derivatives. Theor. Comput. Sci. **22**, 317–330 (1983)
3. Fischer, I.: Sums of like powers of multivariate linear forms. Math. Mag. **67**(1), 59–61 (1994)
4. Fournier, H., Limaye, N., Malod, G., Srinivasan, S.: Lower bounds for depth 4 formulas computing iterated matrix multiplication. In: STOC, pp. 128–135 (2014)
5. Grigoriev, D., Karpinski, M.: An exponential lower bound for depth 3 arithmetic circuits. In STOC, pp. 577–582 (1998)
6. Gupta, A., Kamath, P., Kayal, N., Saptharishi, R.: Arithmetic circuits: a chasm at depth three. In: FOCS, pp. 578–587 (2013)

7. Gupta, A., Kamath, P., Kayal, N., Saptharishi, R.: Approaching the chasm at depth four. J. ACM **61**(6), 33:1–33:16 (2014)
8. Kayal, N.: An exponential lower bound for the sum of powers of bounded degree polynomials. ECCC 19:81 (2012)
9. Koiran, P.: Shallow circuits with high-powered inputs. In: ICS, pp. 309–320. Tsinghua University Press (2011)
10. Koiran, P.: Arithmetic circuits: the chasm at depth four gets wider. Theor. Comput. Sci. **448**, 56–65 (2012)
11. Koiran, P., Portier, N., Tavenas, S.: A wronskian approach to the real tau-conjecture. J. Symb. Comput. **68**, 195–214 (2015)
12. Koiran, P., Portier, N., Tavenas, S., Thomassé, S.: A tau-conjecture for newton polygons. Found. Comput. Math. **15**(1), 185–197 (2015)
13. MacKay, D.J.C.: Information Theory, Inference and Learning Algorithms. Cambridge University Press, Cambridge (2003)
14. Mulmuley, K., Sohoni, M.A.: Geometric complexity theory I: an approach to the P vs. NP and related problems. SIAM J. Comput. **31**(2), 496–526 (2001)
15. Saptharishi, R.: A survey of lower bounds in arithmetic circuit complexity. Version 3.1.0 (2016). https://github.com/dasarpmar/lowerbounds-survey/releases
16. Shub, M., Smale, S.: On the intractability of hilbert's nullstellensatz and an algebraic version of "NP != P ?". Duke Math. J. **81**(1), 47–54 (1995)
17. Tavenas, S.: Improved bounds for reduction to depth 4 and depth 3. In: Chatterjee, K., Sgall, J. (eds.) MFCS 2013. LNCS, vol. 8087, pp. 813–824. Springer, Heidelberg (2013). doi:10.1007/978-3-642-40313-2_71
18. Valiant, L.G.: Completeness classes in algebra. In: STOC, pp. 249–261 (1979)

Decidable Weighted Expressions
with Presburger Combinators

Emmanuel Filiot, Nicolas Mazzocchi[✉], and Jean-François Raskin

Université libre de Bruxelles, Brussels, Belgium
nicolas.mazzocchi@ulb.ac.be

Abstract. In this paper, we investigate the expressive power and the algorithmic properties of weighted expressions, which define functions from finite words to integers. First, we consider a slight extension of an expression formalism, introduced by Chatterjee et al. in the context of infinite words, by which to combine values given by unambiguous (max,+)-automata, using Presburger arithmetic. We show that important decision problems such as emptiness, universality and comparison are PSPACE-C for these expressions. We then investigate the extension of these expressions with Kleene star. This allows to iterate an expression over smaller fragments of the input word, and to combine the results by taking their iterated sum. The decision problems turn out to be undecidable, but we introduce the decidable and still expressive class of synchronised expressions.

1 Introduction

Quantitative Languages. Quantitative languages (QL), or series, generalise Boolean languages to function from finite words into some semiring. They have recently received a particular attention from the verification community, for their application in modeling system *quality* [3], lifting classical Boolean verification problems to a quantitative setting. In this paper, we consider the case of integer weights and in this context, the comparison problem asks whether two QL $f, g : \Sigma^* \to \mathbb{Z}$ satisfy $f(u) \leq g(u)$ for all $u \in \Sigma^*$. Similarly, the universality ($f \geq \nu$ where ν is a constant) and equivalence problem ($f = g$) can be defined, as well as emptiness (does there exists a word whose value is above some given threshold). We say that a formalism for QL is decidable if all these problems are decidable. A popular formalism to define QL is that of weighted automata

E. Filiot is a research associate of F.R.S.-FNRS. This work has been supported by the following projects: the ARC Project Transform (Federation Wallonie-Brussels), the FNRS CDR project Flare.

N. Mazzocchi is a PhD funded by a FRIA fellowship from the F.R.S.-FNRS.

J.-F. Raskin is supported by an ERC Starting Grant (279499: inVEST), by the ARC project - Non-Zero Sum Game Graphs: Applications to Reactive Synthesis and Beyond - funded by the Fdration Wallonie-Bruxelles, and by a Professeur Francqui de Recherche grant awarded by the Francqui Fondation.

R. Klasing and M. Zeitoun (Eds.): FCT 2017, LNCS 10472, pp. 243–256, 2017.
DOI: 10.1007/978-3-662-55751-8_20

(WA) [6]. However, WA over the semiring $(\mathbb{Z}, \max, +)$, called $(\max, +)$-automata, are undecidable [12], even if they are linearly ambiguous $(\max, +)$-automata [5].

Decidable Formalisms for Quantitative Languages and Objectives. The largest known class of $(\max, +)$-automata enjoying decidability is that of finitely ambiguous $(\max, +)$-automata, which is also expressively equivalent to the class of finite-valued $(\max, +)$-automata (all the accepting executions over the same input run yields a constant number of different values) [8]. Moreover, $(\max, +)$-automata are not closed under simple operations such as min and the difference $-$ [11]. Basic functions such as $u \mapsto \min(\#_a(u), \#_b(u))$ and[1] (as a consequence) $u \mapsto |f(u) - g(u)|$ are not definable by $(\max, +)$-automata, even if f, g are [11]. To cope with the expressivity and undecidability issues, a class of weighted expressions was introduced in [2] in the context of ω-words. Casted to finite words, the idea is to use deterministic $(\max, +)$-automata as atoms, and to combine them using the operations max, min, $+$, and $-$. The decision problems defined before were shown to be PSPACE-C [13] over ω-words. One limitation of this formalism, casted to finite words, if that it is not expressive enough to capture finitely ambiguous $(max, +)$-automata, yielding two incomparable classes of QL. In this paper, our objective is to push the expressiveness of weighted expressions as far as possible while retaining decidability, and to capture both finitely ambiguous $(\max, +)$-automata and the expressions of [2], for finite words.

Monolithic Expressions with Presburger Combinators. We define in Sect. 3 a class of expressions, inspired from [2], that we call monolithic in contrast to another class of expressions defined in a second contribution. The idea is to use unambiguous $(\max, +)$-automata as atoms, and to combine them using n-ary functions definable in Presburger arithmetics (we call them Presburger combinators). Any finitely ambiguous $(\max, +)$-automaton being equivalent to a finite union of unambiguous ones [8], this formalism captures finitely ambiguous $(\max, +)$-automata (using the Presburger combinator max). We show that all the decision problems are PSPACE-C, matching the complexity of [13]. It is important to mention that this complexity result cannot be directly obtained from [13] which is on ω-words with mean-payoff automata as atoms (hence the value of an infinite word is prefix-independent). Moreover, unlike in [13], we can rely on existing results by encoding expressions into reversal-bounded counter machines [10].

Expressions with Iterated Sum. The previous expressions are monolithic in the sense that first, some values are computed by weighted automata applied on the whole input word, and then these values are combined using Presburger combinators. It is not possible to iterate expressions on factors of the input word, and to aggregate all the values computed on these factors, for instance by a sum operation. The basic operator for iteration is that of Kleene star (extended to quantitative languages), which we call more explicitly *iterated sum*. It has

---

[1] $\#_\sigma(u)$ is the number of occurrences of σ in u.

already been defined in [6], and its unambiguous version considered in [1] to obtain an expression formalism equivalent to unambiguous $(\max, +)$-automata. Inspired by [1], we investigate in Sect. 4 the extension of monolithic expressions with unambiguous iterated sum, which we just call iterated sum in the paper. The idea is as follows: given an expression E which applies on a domain D, the expression $E^{\circledast}$ is defined only on words u that can be uniquely decomposed (hence the name unambiguous) into factors $u_1 u_2 \ldots u_n = u$ such that $u_i \in D$, and the value of u is then $\sum_{i=1}^{n} E(u)$. Unfortunately, we show that such an extension yields undecidability (if 2 or more iterated sum operations occur in the expression). The undecidability is caused by the fact that subexpressions $E^{\circledast}$ may decompose the input word in different ways. We therefore define the class of so called *synchronised* expressions with iterated sum, which forbids this behaviour. We show that while being expressive (for instance, they can define QL beyond finitely ambiguous $(\max, +)$-automata), decidability is recovered. The proof goes via a new weighted automata model (Sect. 5), called *weighted chop automata*, that slice the input word into smaller factors, recursively apply smaller chop automata on the factors to compute their values, which are then aggregated by taking their sum. In their synchronised version, we show decidability for chop automata. We finally discuss some extensions in Sect. 6[2].

2   Quantitative Languages

Words, Languages and Quantitative Languages. Let Σ be a finite alphabet and denote by Σ^* the set of finite words over Σ, with ϵ the empty word. Given two words $u, v \in \Sigma^*$, $|u|$ and $|v|$ denote their length, and the distance between u and v is defined as $d(u, v) = |u| + |v| - 2|\sqcap(u, v)|$, where $\sqcap(u, v)$ denotes the longest common prefix of u and v. A *quantitative language* (QL)[3] is a partial function $f : \Sigma^* \to \mathbb{Z}$, whose domain is denoted by $\mathrm{dom}(f)$. E.g., consider the function mapping any word $w \in \Sigma^*$ to the number of occurrences $\#_\sigma(w)$ of some symbol $\sigma \in \Sigma$ in w. A QL f is *Lipschitz-continuous* if there exists $K \in \mathbb{N}$ such that for all words $u, v \in \Sigma^*$, $|f(u) - f(v)| \leq K \cdot d(u, v)$.

Combinators for Quantitative Languages. Any binary operation $\boxplus : \mathbb{Z}^2 \to \mathbb{Z}$ is extended to quantitative languages by $f_1 \boxplus f_2(w) = f_1(w) \boxplus f_2(w)$ if $w \in \mathrm{dom}(f_1) \cap \mathrm{dom}(f_2)$, otherwise it is undefined. We will consider operations defined in *existential Presburger logic*. An existential Presburger formula (simply called Presburger formula in the sequel) is built over terms t on the signature $\{0, 1, +\} \cup X$, where X is a set of variables, as follows: $\phi ::= t = t \mid t > t \mid \phi \vee \phi \mid \phi \wedge \phi \mid \exists x. \phi$. If a formula ϕ has $n + 1$ free variables $x_1, \ldots, x_{n+1}$, for all $v_1, \ldots, v_{n+1} \in \mathbb{Z}$, we write $\phi(v_1, \ldots, v_{n+1})$ if ϕ holds for the valuation mapping x_i to v_i. When $n \geq 1$, we say that ϕ is *functional* if for all $v_1, \ldots, v_n \in \mathbb{Z}$, there exists a unique $v_{n+1} \in \mathbb{Z}$ such that $\phi(v_1, \ldots, v_{n+1})$ holds. Hence, ϕ defines a (total) function from $\mathbb{Z}^n$ to $\mathbb{Z}$ that we denote $[\![\phi]\!]$. We call n the arity of ϕ and may write $\phi(x_1, \ldots, x_n)$ to

---

[2] Full proofs are given in the full paper version at http://arxiv.org/abs/1706.08855.

[3] Also called *formal series* in [6].

denote the unique x_{n+1} such that $\phi(x_1, \ldots, x_{n+1})$ holds. We say that a function $f : \mathbb{Z}^n \to \mathbb{Z}$ is Presburger-definable if there exists a functional Presburger-formula ϕ such that $f = [\![\phi]\!]$. E.g., the max of values $x_1, \ldots, x_n$ is definable by $\phi_{\max}(x_1, \ldots, x_n, x) \equiv (\bigwedge_{i=1}^{n} x_i \leq x) \wedge (\bigvee_{i=1}^{n} x_i = x)$.

Semi-linear Sets. Let $k \geq 1$. A set $S \subseteq \mathbb{Z}^k$ is *linear* if there exist $x_1, \ldots, x_n \in \mathbb{Z}^k$, called the period vectors, and $x_0 \in \mathbb{Z}^k$, called the base, such that $S = \{x_0 + \sum_{i=1}^{n} a_i x_i \mid a_1, \ldots, a_n \in \mathbb{N}\}$. S is *semi-linear* if it is a finite a union of linear sets. Note that the set of base and periodic vectors of each linear set of the union provides a finite representation of S. It is a folklore result that a set $S \subseteq \mathbb{Z}^k$ is semi-linear iff it is definable by some existential Presburger formula.

Decision Problems. In this paper, we are interested by fundamental decision problems on (finite representations of) quantitative languages, namely universality, emptiness and comparison. Given finitely represented quantitative languages f, f_1, f_2 and $v \in \mathbb{Z}$,

- the v-emptiness (resp. v-universality) problem asks whether there exists $u \in \mathrm{dom}(f)$ such that $f(u) \succsim v$ (resp. whether all $u \in \mathrm{dom}(f)$ satisfies $f(u) \succsim v$), for $\succsim \in \{>, \geq\}$.
- the $\succsim$-inclusion problem with $\succsim \in \{>, \geq\}$ asks whether $\mathrm{dom}(f_1) \supseteq \mathrm{dom}(f_2)$ and for all $w \in \mathrm{dom}(f_2)$, $f_1(w) \succsim f_2(w)$. We write $f_1 \succsim f_2$.
- the equivalence problem asks whether $f_1 \geq f_2 \wedge f_2 \geq f_1$ denoted by $f_1 \equiv f_2$.

Remark 1. For classes of QL (effectively) closed under regular domain restriction and difference, and with decidable domain inclusion, the v-universality, inclusion and equivalence problems, are reducible to the 0-emptiness problem as follows:

1. to establish $\forall w \in \mathrm{dom}(f) : f(w) \geq v$ (universality), it suffices to check that it is not the case that $\exists w \in \mathrm{dom}(f) : -(f(w) - v) > 0$ (0-emptiness).
2. to establish $\mathrm{dom}(f_2) \subseteq \mathrm{dom}(f_1)$ and for all $w \in \mathrm{dom}(f_2)$, $f_1(w) \geq f_2(w)$, when the first check succeeds, we reduce the second one as follows: construct a new QL g on $\mathrm{dom}(f_2)$ such that $\forall w \in \mathrm{dom}(f_2) : g(w) = f_2(w) - f_1(w)$ and check that $\forall w \in \mathrm{dom}(f_2) : g(w) \geq 0$ (0-emptiness).

The other variants with strict inequalities are treated similarly. Note also with similar arguments, we can show that the 0-emptiness problem can be reduced to the universality and the inclusion problems. The quantitative expression formalisms that we define in this paper have those closure properties (in PTIME) and so, we concentrate, in most of our results, on the 0-emptiness problem.

Weighted Automata. Weighted automata (WA) have been defined as a representation of QL (more generally with values in a semiring). Here, we consider weighted automata over the semiring $(\mathbb{Z} \cup \{-\infty\}, \max, +)$ and just call them *weighted automata*. They are defined as tuples $M = (A, \lambda)$ where $A = (Q, I, F, \Delta)$ is a finite automaton over Σ whose language is denoted by $L(A)$ and $\lambda : \Delta \to \mathbb{Z}$ is a weight function on transitions. Given a word $w \in L(A)$

and an accepting run $r = q_1 a_1 \ldots q_n a_n q_{n+1}$ of A on w, the value $V(r)$ of r is defined by $\sum_{i=1}^{n} \lambda(q_i, a_i, q_{i+1})$ if $n > 1$, and by 0 if[4] $n = 1$. Finally, M defines a quantitative language $[\![M]\!] : L(A) \rightarrow \mathbb{Z}$ such that for all $w \in L(A)$, $[\![M]\!](w) = \max\{V(r) \mid r \text{ is an accepting run of } A \text{ on } w\}$. M is called deterministic if A is deterministic. We say that M is k-ambiguous if A is k-ambiguous, i.e. there are at most k accepting runs on words of $L(A)$. A 1-ambiguous WA is also called *unambiguous*. M is k-valued if for all $w \in L(A)$, the set $\{V(r) \mid r \text{ is an accepting run of } A \text{ on } w\}$ has cardinality at most k. In particular, any k-ambiguous WA is k-valued. The converse also holds, and it is decidable whether a WA is k-valued, for a given k [8]. While emptiness is decidable for WA [9], inclusion and universality are undecidable [12]. However, all these problems are decidable for k-valued WA, for a fixed k [8].

3 Monolithic Expressions

We start our study of weighted expressions by a definition directly inspired by [2] where weighted automata[5] are used as building blocs of quantitative expressions that can be inductively composed with functions such as min, max, addition and difference. The equivalence checking problem for those expressions is decidable in PSPACE. We start here with deterministic $(\max, +)$-automata as building blocs.

Definition 1. *A simple expression (s-expression) is a term E generated by*

$$E ::= D \mid \min(E_1, E_2) \mid \max(E_1, E_2) \mid E_1 + E_2 \mid E_1 - E_2$$

where D is a deterministic WA (we remind that by WA we mean $(\max, +)$-automata).

Semantics. Any s-expression E defines a quantitative language $[\![E]\!] : \Sigma^* \rightarrow \mathbb{Z}$ on a domain $\mathrm{dom}(E)$ inductively as follows: if $E \equiv A$, then $\mathrm{dom}(E) = L(A)$ and for all $u \in L(A)$, $[\![E]\!](u) = [\![A]\!](u)$ (the semantics of WA is defined in Sect. 2); if $E \equiv \min(E_1, E_2)$, then $\mathrm{dom}(E) = \mathrm{dom}(E_1) \cap \mathrm{dom}(E_2)$ and for all $u \in \mathrm{dom}(E)$, $[\![E]\!](u) = \min([\![E_1]\!](u), [\![E_2]\!](u))$, symmetrical works for max, $+$ and $-$. We say that two s-expressions E_1, E_2 are equivalent if $[\![E_1]\!] = [\![E_2]\!]$ (in particular $\mathrm{dom}(E_1) = \mathrm{dom}(E_2)$). To characterise the expressiveness of s-expressions, we note that:

Lemma 1. *Any s-expression defines a Lipschitz continuous quantitative language.*

The unambiguous WA can define non Lipschitz continuous functions, hence not definable by s-expressions. On the contrary, the function $u \mapsto \min(\#_a(u), \#_b(u))$ is definable by an s-expression while it is not definable by a WA [11].

---

[4] Sometimes, initial and final weight functions are considered in the literature [6], so that non-zero values can be assigned to ϵ.

[5] Chatterjee et al. studied quantitative expressions on infinite words and the automata that they consider are deterministic mean-payoff automata.

Proposition 1. *There are quantitative languages that are definable by unambiguous weighted automata and not by s-expressions. There are quantitative languages that are definable by s-expressions but not by a WA.*

To unleash their expressive power, we generalise s-expressions. First, instead deterministic WA, we consider unambiguous WA as atoms. This extends their expressiveness beyond finite valued WA. Second, instead of considering a fixed (and arbitrary) set of composition functions, we consider any function that is (existential) Presburger definable. Third, we consider the addition of Kleene star operator. While the first two extensions maintain decidability in PSPACE, the third extension leads to undecidability and sub-cases need to be studied to recover decidability. We study the two first extensions here and the Kleene star operator in the next section.

Definition 2. *Monolithic expressions (m-expression) are terms E generated by the grammar $E ::= A \mid \phi(E_1, \ldots, E_n)$, where A is an unambiguous WA, and ϕ is a functional Presburger formula of arity n.*

The semantics $[\![E]\!] : \Sigma^* \to \mathbb{Z}$ of an m-expression E is defined inductively, and similarly as s-expression. In particular, for $E = \phi(E_1, \ldots, E_n)$, $\mathrm{dom}(E) = \bigcap_{i=1}^n \mathrm{dom}(E_i)$ and for all $u \in \mathrm{dom}(E)$, $[\![E]\!](u) = [\![\phi]\!]([\![E_1]\!](u), \ldots, [\![E_n]\!](u))$ (the semantics of functional Presburger formulas is defined in Sect. 2).

Example 1. As seen in Sect. 2, max is Presburger-definable by a formula $\phi_{\max}$, it is also the case for $\min(E_1, \ldots, E_n)$, $E_1 + E_2$, $E_1 - E_2$ and the unary operation $-E$. For m-expressions E_1, E_2, the distance $|E_1 - E_2| : w \in \mathrm{dom}(E_1) \cap \mathrm{dom}(E_2) \mapsto |E_1(w) - E_2(w)|$ is defined by the m-expression $\max(E_1 - E_2, E_2 - E_1)$. This function is not definable by a WA even if E_1, E_2 are 2-ambiguous WA, as a consequence of the non-expressibility by WA of $\min(\#_a(.), \#_b(.)) = |0 - \max(-\#_a(.), -\#_b(.))|$ [11].

Lemma 2. *M-expressions are more expressive than finite valued WA. There are functions definable by m-expressions and not by a WA.*

Theorem 1. *For m-expressions, the emptiness, universality and comparison problems are PSPACE-COMPLETE.*

Proof (Sketch). By Remark 1, all the problems reduce in PTIME to the 0-emptiness problem for which we establish PSPACE membership. Clearly, by combining Presburger formulas, any m-expression is equivalent to an m-expression $\phi(A_1, \ldots, A_n)$ where A_i are unambiguous WA. Now, the main idea is to construct a product $A_1 \times \cdots \times A_n$ (valued over $\mathbb{Z}^n$), which maps any word $u \in \bigcap_i \mathrm{dom}(A_i)$ to $(A_1(u), \ldots, A_n(u))$. (Effective) semi-linearity of $\mathrm{range}(A_1 \times \cdots \times A_n)$ is a consequence of Parikh's theorem, which implies semi-linearity of $\mathrm{range}(\phi(A_1, \ldots, A_n))$. Then it suffices to check for the existence of a positive value in this set. To obtain PSPACE complexity, the difficulty is that $A_1 \times \cdots \times A_n$ has exponential size. To overcome this, we encode $\phi(A_1, \ldots, A_n)$ into a counter machine. First, $A_1 \times \cdots \times A_n$ is encoded into a machine M whose counter valuation, after reading u, encodes the tuple $(A_1(u), \ldots, A_n(u))$. Then, M is composed with another counter machine M_ϕ that compute, on reading the word ϵ, the value $\phi((A_1(u), \ldots, A_n(u))$ (stored in an extra counter). Finally, the compositional machine $M \cdot M_\phi$ accepts iff this latter value is positive, hence it suffices to check for its emptiness. We define $M \cdot M_\phi$ in such a way that it is *reversal-bounded* (its counters change from increasing to decreasing mode a constant number of times [10]). Reversal-bounded counter machines have decidable emptiness problem. While M_ϕ can be constructed in PTIME, M has an exponential size in general. However, we can use a small witness property given in [10] to devise a PSPACE algorithm that does not construct M explicitly.

PSPACE-HARDNESS for emptiness is obtained from the emptiness problem of the intersection of n DFAs.   □

4   Expressions with Iterated Sum

Given $f : \Sigma^* \to \mathbb{Z}$ a quantitative language, the iterated sum of f (or unambiguous Kleene star), denoted by $f^\circledast$, is defined by $f^\circledast(\epsilon) = 0$, and for all $u \in \Sigma^+$, if there exists at most one tuple $(u_1, \ldots, u_n) \in (\mathrm{dom}(f) \setminus \{\epsilon\})^n$ such that $u_1 \ldots u_n = u$, then $f^\circledast(u) = \sum_{i=1}^n f(u_i)$. Note that $\epsilon \in \mathrm{dom}(f^\circledast)$ for any f. By extending m-expressions with iterated sum, we obtain iterated-sum expressions (i-expressions).

Definition 3. *An iterated-sum expression E (i-expression for short) is a term generated by the grammar $E ::= A \mid \phi(E, E) \mid E^\circledast$, where A is some unambiguous WA over Σ and ϕ is a functional Presburger formula.*

As for m-expressions, the semantics of any i-expression E is a quantitative language $[\![E]\!] : \Sigma^* \to \mathbb{Z}$ inductively defined on the structure of the expression.

Example 2. Assume that $\Sigma = \{a, b, \$\}$ and consider the QL f defined for all $u \in \Sigma^*$ by $u_1\$u_2\$ \ldots u_n\$ \mapsto \sum_{i=1}^n \max(\#_a(u_i), \#_b(u_i))$ where each u_i belongs to $\{a, b\}^*$, and $\#_\sigma$ counts the number of occurrences of σ in a word. Counting the number of σ in $v\$$ where $v \in \{a, b\}^*$ is realisable by a 2 states deterministic WA A_σ. Then, f is defined by the i-expression $\max(A_a, A_b)^\circledast$.

Proposition 2. *The domain of any i-expression is (effectively) regular.*

Theorem 2. *Emptiness, universality and comparisons for i-expressions are undecidable problems, even if only s-expressions are iterated.*

Proof (Sketch). The proof of this theorem, inspired by the proof of [5] for the undecidability of WA universality, consists of a reduction from the 2-counter machine halting problem to the 0-emptiness problem of i-expressions. This establishes undecidability for the other decision problems by Remark 1. In this reduction, a transition between two successive configurations $\ldots (q_1, (x \mapsto c_1, y \mapsto d_1))\delta(q_2, (x \mapsto c_2, y \mapsto d_2))\ldots$ is coded by a factor of word of the form: $\ldots \vdash q_1 a^{c_1} b^{d_1} \triangleleft \delta \triangleright q_2 a^{c_2} b^{d_2} \dashv\vdash q_2 a^{c_2} b^{d_2} \triangleleft \ldots$

We show that such a word encodes an halting computation if it respects a list of simple requirements that are all are regular but two: one that expresses that increments and decrements of variables are correctly executed, and one that imposes that, from one transition encoding to the next, the current configuration is copied correctly. In our example above, under the hypothesis that x is incremented in δ, this amounts to check that the number of a occurrences before δ is equal to the number of occurrences of a after δ minus one. This property can be verified by s-expression on the factor between the $\vdash$ and $\dashv$ that returns 0 if it is the case and a negative value otherwise. The second property amounts to check that the number of occurrences of a between the first $\triangleright$ and $\dashv$ and the number of a between the second $\vdash$ and second $\triangleleft$ are equal. Again, it is easy to see that this can be done with an s-expression that returns 0 if it is the case and a negative value otherwise. Then, with i-expressions we decompose the word into factors that are between the markers $\vdash$ and $\dashv$, and other factors that are between the markers $\triangleright$ and $\triangleleft$, and we iterate the application of the s-expressions mentioned above. The sum of all the values computed on the factors is equal to 0 if the requirements are met and negative otherwise. $\square$

A close inspection of the proof above, reveals that the undecidability stems from the asynchronicity between parallel star operators, and in the way they decompose the input word (decomposition based on $\vdash \cdots \dashv$ or $\triangleright \cdots \triangleleft$). The two overlapping decompositions are needed. By disallowing this, decidability is recovered: subexpressions $F^{\circledast}$ and $G^{\circledast}$ at the same nested star depth must decompose words in exactly the same way.

Let us formalise the notion of star depth. Given an i-expression E, its syntax tree $T(E)$ is a tree labeled by functional Presburger formulas ϕ, star operators $^{\circledast}$, or unambiguous WA A. Any node p of $T(E)$ defines a subexpression $E|_p$ of E. The *star depth* of node p is the number of star operators occurring above it, i.e. the number of nodes q on the path from the root of $T(E)$ to p (excluded) labeled by a star operator. E.g. in the expression $\phi(A_1^{\circledast}, \phi(A_2^{\circledast}))^{\circledast}$, the subexpression $A_1^{\circledast}$ has star depth 1, A_1 has star depth 2, and the whole expression has star depth 0.

Definition 4. *An i-expression E is* synchronised *if for all nodes p, q of $T(E)$ at the same star depth, if $E|_p = F^{\circledast}$ and $E|_q = G^{\circledast}$, then $dom(F) = dom(G)$.*

By Proposition 2, this property is decidable. Asking that F and G have the same domain enforces that any word u is decomposed in the same way by $F^{\circledast}$ and $G^{\circledast}$. Given a set $S = \{E_1, \ldots, E_n\}$ of i-expressions, we write $\text{Sync}(S)$ the predicate which holds iff $\phi(E_1, \ldots, E_n)$ is synchronised, where ϕ is some functional Presburger formula.

Example 3. An i-expression E is star-chain if for any distincts subexpressions $F^{\circledast}$ and $G^{\circledast}$ of E, $F^{\circledast}$ is a subexpression of G, or $G^{\circledast}$ is a subexpression of F. E.g. $\max(A^{\circledast}, B)^{\circledast}$ is star-chain, while $\max(A^{\circledast}, B^{\circledast})$ is not. The expression of Example 2 is also a star-chain, hence it is synchronised, as well as $\min(\max(A_a, A_b)^{\circledast}, A_c)$. Note that A_c applies on the whole input word, while A_a and A_b apply on factors of it.

Finitely ambiguous WA is the largest class of WA for which emptiness, universality and comparisons are decidable [8]. Already for linearly ambiguous WA, universality and comparison problems are undecidable [5]. Example 2 is realisable by a synchronised i-expression or a WA which non-deterministically guess, for each factor u_i, whether it should count the number of a or b. However, as shown in [11] (Sect. 3.5), it is not realisable by any finitely ambiguous WA. As a consequence:

Proposition 3. *There is a quantitative language f such that f is definable by a synchronised i-expression or a WA, but not by a finitely ambiguous WA.*

As a direct consequence of the definition of i-expressions and synchronisation, synchronised i-expressions are closed under Presburger combinators and unambiguous iterated-sum in the following sense:

Proposition 4. *Let $E_1, \ldots, E_n, E$ be i-expressions and ϕ a functional Presburger formula of arity n. If $Sync(E_1, \ldots, E_n)$, then $\phi(E_1, \ldots, E_n)$ is synchronised, and if E is synchronised, so is $E^{\circledast}$.*

Despite the fact that synchronised i-expressions can express QL that are beyond finitely ambiguous WA, we have decidability (proved in the next section):

Theorem 3. *The emptiness and universality problems are decidable for synchronised i-expressions. The comparisons problems for i-expressions E_1, E_2 such that $Sync\{E_1, E_2\}$ are decidable.*

5 Decidability of Synchronised Iterated Sum Expressions

In this section, we introduce a new weighted automata model, called weighted chop automata (WCA), into which we transform i-expressions. It is simple to see that the proof of undecidability of i-expressions (Theorem 2) can be done the same way using WCA. We introduce the class of synchronised WCA, to which synchronised i-expressions can be compiled, and by which we recover decidability, thus proving Theorem 3. The intuitive behaviour of a WCA is as follows. An unambiguous generalised automaton (whose transitions are not reading single letters but words in some regular language) "chop" the input word into factors, on which expressions of the form $\phi(C_1, \ldots, C_n)$, where C_i are smaller WCA, are applied to obtain intermediate values, which are then summed to obtain the value of the whole input word.

Formally, a *generalised finite automaton* is a tuple $A = (Q, I, F, \Delta)$ where Q is a set of states, I its initial states and F its final states, and Δ maps any pair $(p, q) \in Q^2$ to a regular language $\Delta(p, q) \subseteq \Sigma^*$ (finitely represented by some NFA). A run of A over a word $u = u_1 \ldots u_n$ is a sequence $r = q_0 u_1 \ldots q_{n-1} u_n q_n$ such that $u_i \in \Delta(q_{i-1}, q_i)$ for all $1 \leq i \leq n$. It is accepting if $q_0 \in I$ and $q_n \in F$. We say that A is unambiguous if for all $u \in \Sigma^*$, there is at most one accepting run of A on u (and hence its decomposition $u_1 \ldots u_n$ is unique). This property can be decided in PTIME.

Definition 5. *A 0-weighted chop automaton is an unambiguous WA. Let $n > 0$. An n-weighted chop automaton (n-WCA) is a tuple $C = (A, \lambda)$ where A is an unambiguous generalised finite automaton and λ is a function mapping any pair $(p, q) \in Q^2$ to some expression $E = \phi(C_1, \ldots, C_m)$ where for all i, C_i is an n'-WCA, for some $n' < n$, and ϕ is a functional Presburger formula of arity m. Moreover, it is required that at least one C_i is an $(n-1)$-WCA. A WCA is an n-WCA for some n.*

Semantics. A WCA C defines a quantitative language $[\![C]\!]$ of domain $\mathrm{dom}(C)$ inductively defined as follows. If C is a 0-WCA, then its semantics is that of unambiguous WA. Otherwise $C = (A, \lambda)$, and the set $\mathrm{dom}(C)$ is the set of words $u = u_1 \ldots u_n$ on which there exists one accepting run $r = q_0 u_1 \ldots q_{n-1} u_n q_n$ of A such that for all $1 \leq i \leq n$, if $\lambda(q_{i-1}, q_i)$ is of the form $\phi(C_1, \ldots, C_m)$, then $u_i \in \bigcap_{j=1}^{m} \mathrm{dom}(C_j)$, and in this case we let $v_i = [\![\phi]\!]([\![C_1]\!](u), \ldots, [\![C_m]\!](u))$. The value of r (which also defines the value of u) is then $\sum_{i=1}^{n} v_i$. We denote by $\mathrm{dec}_C(u)$ the (unique) sequence $(u_1, \lambda(q_0, q_1)) \ldots (u_n, \lambda(q_{n-1}, q_n))$.

Example 4. Let $\Sigma = \{a, b, c, d\}$ and $\bullet, \$ \notin \Sigma$, the WCA depicted below realises the function mapping any word of the form $u_1\$ \ldots u_n\$ \bullet v_1\$ \ldots v_m\$$, where $u_i, v_i \in \{a, b, c, d\}^*$, to $\sum_{i=1}^{n} \max(\#_a(u_i), \#_b(u_i)) + \sum_{i=1}^{m} \max(\#_c(v_i), \#_d(v_i))$. The automata A_σ are unambiguous WA counting the number of occurences of σ, and C_i are shortcuts for $\phi_{id}(C_i)$ where ϕ_{id} defines the identify function.

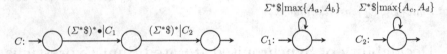

Synchronised WCA. The notion of synchronisation of WCA is inductively defined. Two expressions $\phi_1(C_1, \ldots, C_n)$ and $\phi_2(C'_1, \ldots, C'_m)$ are synchronised if C_i is synchronised with C'_j for all i, j. We say that two WCA C_1, C_2 are synchronised, denoted by $C_1 \| C_2$, if they are either both 0-WCA, or $C_1 = (A_1, \lambda_1)$ and $C_2 = (A_2, \lambda_2)$, and the following holds: for all $u \in L(A_1) \cap L(A_2)$, if $\mathrm{dec}_{C_1}(u) = (u_1, E_1), \ldots, (u_n, E_n)$ and $\mathrm{dec}_{C_2}(u) = (v_1, F_1), \ldots, (v_m, F_m)$, then $n = m$ and for all $1 \leq i \leq n$, we have $u_i = v_i$ and E_i is synchronised with F_i. We write $\mathrm{Sync}(\{C_1, \ldots, C_n\})$ if $C_i \| C_j$ for all $i, j \in \{1, \ldots, n\}$. Now, a WCA C is synchronised if it is an unambiguous WA, or it is of the form (A, λ), and any expression $\phi(C_1, \ldots, C_n)$ in the range of λ satisfies $\mathrm{Sync}(\{C_1, \ldots, C_n\})$. E.g., the WCA of Example 4 is synchronised, and it can be seen that if $C_1 \| C_2$, then both C_1 and C_2 are n-WCA for the same n.

Proposition 5. *Synchronisation is decidable in PTIME for WCA.*

We now investigate the closure properties of WCA. Given two quantitative languages f_1, f_2, let us define their *split sum* $f_1 \odot f_2$ as the function mapping any word u which can be uniquely decomposed into u_1, u_2 such that $u_i \in \mathrm{dom}(f_i)$ for all i, to $f_1(u_1) + f_2(u_2)$ [1]. We also define the *conditional choice* $f_1 \triangleright f_2$ as the mapping of any word $u \in \mathrm{dom}(f_1)$ to $f_1(u)$, and of any word $u \in \mathrm{dom}(f_2) \setminus \mathrm{dom}(f_1)$

to $f_2(u)$ [1]. These operators may be thought of as (unambiguous) concatenation and disjunction in rational expressions. Synchronised WCA are closed under these operations, as well as Presburger combinators and (unambiguous) iterated sum, in the following sense:

Proposition 6. *Let $C_1, \ldots, C_n$ be WCA such that $Sync\{C_1, \ldots, C_n\}$ and C, D two synchronised WCA. Let ϕ be a functional Presburger formula of arity n, and $L \subseteq \Sigma^*$ a regular language. There exists synchronised WCA respectively denoted by $C^{\circledast}$, $C \odot D$, $C \triangleright D$, $C|_L$ and $\phi(C_1, \ldots, C_n)$ such that*

- $[\![C^{\circledast}]\!] = [\![C]\!]^{\circledast}$, $[\![C \odot D]\!] = [\![C]\!] \odot [\![D]\!]$, $[\![C \triangleright D]\!] = [\![C]\!] \triangleright [\![D]\!]$ *and* $[\![C|_L]\!] = [\![C]\!]|_L$
- *for all* $u \in \bigcap_{i=1}^{n} dom(C_i)$, $[\![\phi(C_1, \ldots, C_n)]\!](u) = [\![\phi]\!]([\![C_1]\!](u), \ldots, [\![C_n]\!](u))$ *and* $dom(\phi(C_1, \ldots, C_n)) = \bigcap_{i=1}^{n} dom(C_i)$

The key lemma towards decidability of synchronised WCA is the following:

Lemma 3. *Let C be a synchronised weighted chop automaton. Then $\{[\![C]\!](u) \mid u \in dom(C)\}$ is semi-linear and effectively computable.*

Proof (Sketch). The proof goes by induction on C. If C is an unambiguous WA, then semi-linearity is known (for instance by using Parikh theorem or reversal-bounded counter machine as in the proof of Theorem 1). If $C = (A, \lambda)$ and A has set of states Q, we first assume that for all states $p, q \in Q$, $\lambda(p, q)$ (which is an expression of the form $\phi(C_1, \ldots, C_n)$), has semi-linear range $S_{p,q}$. Consider the morphism μ from the free monoid $(Q \times Q)^*$ to the monoid of semi-linear sets of $\mathbb{Z}$ (with neutral element $\{0\}$ and addition), defined by $\mu((p, q)) = S_{p,q}$. Clearly, for any regular language $L \subseteq (Q \times Q)^*$, $\mu(L)$ is semi-linear, because semi-linear sets are closed under addition, finite union, and Kleene star (see [7] for instance). Then, we can show that range$(C) = \mu(L)$ for L the set of words over $Q \times Q$ of the form $(q_0, q_1)(q_1, q_2) \ldots (q_k, q_{k+1})$ such that q_0 is initial, q_{k+1} final, and for all i, $\Delta(q_i, q_{i+1}) \neq \varnothing$. L is clearly regular, as the $\Delta(q_i, q_{i+1})$ are.

To show that the expressions $\phi(C_1, \ldots, C_n)$ have semi-linear ranges, the key idea is that thanks to synchronisation, we can safely construct a kind of product between the WCA $C_1, \ldots, C_n$. This product is not a proper WCA but a "generalised" WCA with values in $\mathbb{Z}^n$. By induction, we can show that this product has semi-linear range (in fact, our induction is on generalised WCA rather than proper WCA), whose values can be combined into a semilinear set thanks to the Presburger combinator ϕ. □

As a direct consequence of Lemma 3, Remark 1 and Proposition 6:

Theorem 4. *The following problems are decidable: emptiness and universality of synchronised WCA, comparisons of WCA C_1, C_2 such that $Sync\{C_1, C_2\}$.*

We conclude this section by showing that any synchronised i-expression can be converted into a synchronised WCA. This conversion is effective, this entails by Theorem 4 the decidability of synchronised i-expressions (Theorem 3).

Theorem 5. *Any synchronised i-expression E is (effectively) equivalent to some synchronised weighted chop automaton C_E, i.e. $[\![E]\!] = [\![C_E]\!]$.*

Proof (Sketch). Let us illustrate the main idea of this proof on an example. Suppose that $E = \phi(A, B^\circledast)$ for some unambiguous WA A, B, and Presburger formula ϕ. The difficulty with this kind of expression comes from the fact that A is applied on the whole input word, while B is applied iteratively on factors of it. Clearly, A is also a 0-WCA, and B could be inductively converted into some WCA C, in turn used to construct a WCA $C^\circledast$ (as done in Proposition 6). However, A and $C^\circledast$ are not synchronised in general: by definition of synchronisation for WCA, n-WCA are synchronised with n-WCA only. This latter property is crucial to make a product construction of synchronised WCA and to prove semi-linearity of their ranges (Lemma 3).

Hence, the main idea to prove this result is to "chop" A into smaller WA that are synchronised with $\mathrm{dom}(B)$, and to express A as a combination of these smaller automata. More precisely, for all states p, q of A we can define $A_{p,q}$ to be the WA A with initial state p, final state q, whose domain is restricted to $\mathrm{dom}(B)$. Then, all the smaller automata $A_{p,q}$ are combined into a single WCA which simulates successive applications of the automata $A_{p,q}$, by taking care of the fact that the words it accepts must be uniquely decomposable into factors of $\mathrm{dom}(B)$. This resulting WCA, say C', is necessarily synchronised with $C^\circledast$, and we can return the single synchronised WCA $\phi(C', C^\circledast)$, as defined in Proposition 6, which is equivalent to the i-expression $\phi(A, B^\circledast)$. The general case is just a technical generalisation of this main idea.   □

6   Discussion

First, iterating max instead of sum also yields undecidability for i-expressions. Second, the decidability of synchronised i-expressions goes by the model weighted chop automata, which slice the input word into factors on which subautomata are applied. Any synchronised i-expression can be converted into a synchronised chop automaton (Theorem 5). We conjecture that the converse of Theorem 5 is not true, i.e. synchronised WCA are strictly more expressive than synchronised i-expressions. In particular, we conjecture that synchronised i-expressions are not closed under split sum, unlike synchronised WCA (Proposition 6). The quantitative language of Example 4 does not seem to be definable by any synchronised i-expression.

It turns out that extending i-expressions with split sum $\odot$ and conditional choice $\triangleright$, with a suitable notion of synchronisation, gives a formalism equivalent to synchronised WCA. Due to lack of space, and since the notion of synchronisation for such extended expressions is quite technical (and a bit ad-hoc), we decided not to include it.

An expression formalism with unambiguous iterated sum, conditional choice and split sum, whose atoms are constant quantitative languages (any word from a regular language is mapped to a same constant value), was already introduced by Alur et al. [1]. It is shown that this formalism is equivalent to unambiguous WA. Our goal was to go much beyond this expressivity, by having a formalism closed under Presburger combinators. Adding such combinators to the expressions of [1] would immediately yield an undecidable formalism (as a consequence of Theorem 2). This extension would actually correspond exactly to the extension we discussed in the previous paragraph, and one could come up with a notion of synchronisation to recover decidability. We did not do it in this paper, for the reason explained before, but it would be interesting to have an elegant notion of synchronisation for the extension of [1] with Presburger combinators. More generally, our notion of synchronisation is semantical (but decidable). This raises the question of whether another weighted expression formalism with a purely syntactic notion of synchronisation could be defined.

Finally, Chatterjee et al. have introduced a recursive model of WA [4]. They are incomparable to weighted chop automata: they can define QL whose ranges are not semilinear, but the recursion depth is only 1 (a master WA calls slave WA).

Acknowledgements. We are very grateful to Ismaël Jecker and Nathan Lhote for fruitful discussions on this work, and for their help in establishing the undecidability result.

References

1. Alur, R., Freilich, A., Raghothaman, M.: Regular combinators for string transformations. In: CSL, pp. 9:1–9:10 (2014)
2. Chatterjee, K., Doyen, L., Edelsbrunner, H., Henzinger, T.A., Rannou, P.: Mean-payoff automaton expressions. In: Gastin, P., Laroussinie, F. (eds.) CONCUR 2010. LNCS, vol. 6269, pp. 269–283. Springer, Heidelberg (2010). doi:10.1007/978-3-642-15375-4_19
3. Chatterjee, K., Doyen, L., Henzinger, T.A.: Quantitative languages. ACM Trans. Comput. Log. **11**(4) (2010)
4. Chatterjee, K., Henzinger, T.A., Otop, J.: Nested weighted automata. In: LICS (2015)
5. Daviaud, L., Guillon, P., Merlet, G.: Comparison of max-plus automata and joint spectral radius of tropical matrices. CoRR, abs/1612.02647 (2016)
6. Droste, M., Kuich, W., Vogler, H.: Handbook of Weighted Automata. Springer, Heidelberg (2009)
7. Eilenberg, S., Schützenberger, M.P.: Rational sets in commutative monoids. J. Algebra **13**, 173–191 (1969)
8. Filiot, E., Gentilini, R., Raskin, J.-F.: Finite-valued weighted automata. In: FSTTCS, pp. 133–145 (2014)
9. Filiot, E., Gentilini, R., Raskin, J.-F.: Quantitative languages defined by functional automata. LMCS **11**(3) (2015)
10. Gurari, E.M., Ibarra, O.H.: The complexity of decision problems for finite-turn multicounter machines. In: Even, S., Kariv, O. (eds.) ICALP 1981. LNCS, vol. 115, pp. 495–505. Springer, Heidelberg (1981). doi:10.1007/3-540-10843-2_39

11. Klimann, I., Lombardy, S., Mairesse, J., Prieur, C.: Deciding unambiguity and sequentiality from a finitely ambiguous max-plus automaton. TCS **327**(3) (2004)
12. Krob, D.: The equality problem for rational series with multiplicities in the tropical semiring is undecidable. Int. J. Algebra Comput. **4**(3), 405–425 (1994)
13. Velner, Y.: The complexity of mean-payoff automaton expression. In: Czumaj, A., Mehlhorn, K., Pitts, A., Wattenhofer, R. (eds.) ICALP 2012. LNCS, vol. 7392, pp. 390–402. Springer, Heidelberg (2012). doi:10.1007/978-3-642-31585-5_36

The Complexity of Routing with Few Collisions

Till Fluschnik[1]([✉]), Marco Morik[1], and Manuel Sorge[1,2]

[1] Institut für Softwaretechnik und Theoretische Informatik,
TU Berlin, Berlin, Germany
till.fluschnik@tu-berlin.de, marco.t.morik@campus.tu-berlin.de,
sorge@post.bgu.ac.il
[2] Ben Gurion University of the Negev, Beersheba, Israel

Abstract. We study the computational complexity of routing multiple objects through a network in such a way that only few collisions occur: Given a graph G with two distinct terminal vertices and two positive integers p and k, the question is whether one can connect the terminals by at least p routes (e.g. paths) such that at most k edges are time-wise shared among them. We study three types of routes: traverse each vertex at most once (paths), each edge at most once (trails), or no such restrictions (walks). We prove that for paths and trails the problem is NP-complete on undirected and directed graphs even if k is constant or the maximum vertex degree in the input graph is constant. For walks, however, it is solvable in polynomial time on undirected graphs for arbitrary k and on directed graphs if k is constant. We additionally study for all route types a variant of the problem where the maximum length of a route is restricted by some given upper bound. We prove that this length-restricted variant has the same complexity classification with respect to paths and trails, but for walks it becomes NP-complete on undirected graphs.

1 Introduction

We study the computational complexity of determining bottlenecks in networks. Consider a network in which each link has a capacity. We want to send a set of objects from point s to point t in this network, each object moving at a constant rate of one link per time step. We want to determine whether it is possible to send our (predefined number of) objects without congestion and, if not, which links in the network we have to replace by larger-capacity links to make it possible.

Apart from determining bottlenecks, the above-described task arises when securely routing very important persons [15], or packages in a network [2], routing

---

Major parts of this work done while all authors were with TU Berlin. A full version of this paper is available at http://arxiv.org/abs/1705.03673.

T. Fluschnik—Supported by the DFG, project DAMM (NI 369/13-2).

M. Sorge—Supported by the DFG, project DAPA (NI 369/12-2), the People Programme (Marie Curie Actions) of the European Union's Seventh Framework Programme (FP7/2007-2013) under REA grant agreement number 631163.11 and by the Israel Science Foundation (grant no. 551145/14).

R. Klasing and M. Zeitoun (Eds.): FCT 2017, LNCS 10472, pp. 257–270, 2017.
DOI: 10.1007/978-3-662-55751-8_21

container transporting vehicles [17], and generally may give useful insights into the structure and robustness of a network. A further motivation is congestion avoidance in routing fleets of vehicles, a problem treated by recent commercial software products (e.g. http://nunav.net/) and poised to become more important as passenger cars and freight cars become more and more connected. Assume that we have many requests on computing a route for a set of vehicles from a source location to a target location, as it happens in daily commuting traffic. Then the idea is to centrally compute these routes, taking into account the positions in space and time of all other vehicles. To avoid congestion, we try to avoid that on two of the routes the same street appears at the same time.

Formally, we are given an undirected or directed graph with marked source and sink vertex. We ask whether we can construct routes between the source and the sink in such a way that these routes share as few edges as possible. By routes herein we mean either paths, trails, or walks, modeling different restrictions on the routes: A *walk* is a sequence of vertices such that for each consecutive pair of vertices in the sequence there is an edge in the graph. A *trail* is a walk where each edge of the graph appears at most once. A *path* is a trail that contains each vertex at most once. We say that an edge is *shared* by two routes, if the edge appears at the same position in the sequence of the two routes. The sequence of a route can be interpreted as the description of where the object taking this route is at which time. So we arrive at the following core problem:

ROUTING WITH COLLISION AVOIDANCE (RCA)
Input: A graph $G = (V, E)$, two vertices $s, t \in V$, and $p, k \in \mathbb{N}$.
Question: Are there p s-t routes that share at most k edges?

This definition is inspired by the MINIMUM SHARED EDGES (MSE) problem [7,15,19], in which an edge is already shared if it occurs in two routes, regardless of the time of traversal. Finally, note that finding routes from s to t also models the general case of finding routes between a set of sources and a set of sinks.

Considering our introductory motivating scenarios, it is reasonable to restrict the maximal length of the routes. For instance, when routing vehicles in daily commuting traffic while avoiding congestion, the routes should be reasonably short. Motivated by this, we study the following variant of RCA.

FAST ROUTING WITH COLLISION AVOIDANCE (FRCA)
Input: A graph $G = (V, E)$, two vertices $s, t \in V$, and $p, k, \alpha \in \mathbb{N}$.
Question: Are there p s-t routes each of length at most α that share at most k edges?

In the problem variants PATH-RCA, TRAIL-RCA, and WALK-RCA, the routes are restricted to be paths, trails, or walks, respectively (analogously for FRCA).

Our Contributions. We give a full computational complexity classification (see Table 1) of RCA and FRCA (except WALK-FRCA) with respect to the three

Table 1. Overview of our results: DAGs abbreviates directed acyclic graphs; NP-c., W[2]-h., P abbreviate NP-complete, W[2]-hard, and containment in the class P, respectively; Δ denotes the maximum degree; $\Delta_{i/o}$ denotes the maximum over the in- and outdegrees. ^{a} (Theorem 1) ^{b} (Theorem 3) ^{c} (Corollary 2) ^{d} (even on planar graphs)

	Undirected, with k		Directed, with k		DAGs, with k	
	constant	arbitrary	constant	arbitrary	const.	arbitrary
PATH-(F)RCA	NP-c.^{d}, Thm. 4 (Cor. 3)		NP-c.^{d}, Thm. 4 (Cor. 3)		P^a	NP-c.,
	each $k \geq 0$	each $\Delta \geq 4$	each $k \geq 0$	each $\Delta_{i/o} \geq 4$		W[2]-h$^{b/c}$
TRAIL-(F)RCA	NP-c.^{d}, Thm. 5 (Cor. 4)		NP-c.^{d}, Thm. 6 (Cor. 5)		P^a	NP-c.,
	each $k \geq 0$	each $\Delta \geq 5$	each $k \geq 0$	each $\Delta_{i/o} \geq 3$		W[2]-h$^{b/c}$
WALK-RCA	P (Thm. 7)		P (Thm. 9) NP-c.,		P^a	NP-c.,
				W[2]-h.^{b}		W[2]-h.^{b}
WALK-FRCA	open	NP-c., W[2]-h.	P (Thm. 9) NP-c.,		P^a	NP-c.,
		(Thm. 8)		W[2]-h.^{c}		W[2]-h.^{c}

mentioned route types; with respect to undirected, directed, and directed acyclic input graphs; and distinguishing between constant and arbitrary budget.

To our surprise, there is no difference between paths and trails in our classification. Both PATH-RCA (Sect. 4) and TRAIL-RCA (Sect. 5) are NP-complete in all of our cases except on directed acyclic graphs when $k \geq 0$ is constant (Sect. 3). Some of the reductions showcase a strong relationship to the HAMILTON CYCLE problem. We show that PATH-RCA and TRAIL-RCA remain NP-complete on undirected and directed graphs even if $k \geq 0$ is constant or the maximum degree is constant. We note that, in contrast, MSE is solvable in polynomial time when the number of shared edges is constant, highlighting the difference to its time-variant PATH-RCA.

The computational complexity of the length-restricted variant FRCA for paths and trails equals the one of the variant without length restrictions. The variant concerning walks (Sect. 6) however differs from the other two variants as it is tractable in more cases, in particular on undirected graphs. (We note that almost all of our tractability results rely on flow computations in time-expanded networks (see, e.g., Skutella [18]).) Remarkably, the tractability does not transfer to the length-restricted variant WALK-FRCA, as it becomes NP-complete on undirected graphs. This is the only case where RCA and FRCA differ with respect to their computational complexity.

Related Work. As mentioned, MINIMUM SHARED EDGES inspired the definition of RCA. MSE is NP-hard on directed [15] and undirected [6,7] graphs. In contrast to RCA, if the number of shared edges equals zero, then MSE is solvable in polynomial time. Moreover, MSE is W[2]-hard with respect to the number of shared edges and fixed-parameter tractable with respect to the number of paths [7]. MSE is polynomial-time solvable on graphs of bounded treewidth [1,19].

There are various tractability and hardness results for problems related to RCA with $k = 0$ in temporal graphs, in which edges are only available at pre-defined time steps [3,11,13,14]. The goal herein is to find a number of edge or vertex-disjoint time-respecting paths connecting two fixed terminal vertices. Time-respecting means that the time steps of the edges in the paths are nonde-creasing. Apart from the fact that all graphs that we study are static, the crucial difference is in the type of routes: vehicles moving along time-respecting paths may wait an arbitrary number of time steps at each vertex, while we require them to move at least one edge per time step (unless they already arrived at the target vertex).

Our work is related to flows over time, a concept already introduced by Ford and Fulkerson [8] to measure the maximal throughput in a network over a fixed time period. This and similar problems were studied continually, see Skutella [18] and Köhler et al. [12] for surveys. In contrast, our throughput is fixed, our flow may not stand still or go in circles arbitrarily, and we want to augment the network to allow for our throughput.

2   Preliminaries

We use basic notation from parameterized complexity [4].

We define $[n] := \{1, \ldots, n\}$ for every $n \in \mathbb{N}$. Let $G = (V, E)$ be an undirected (directed) graph. Let the sequence $P = (v_1, \ldots, v_\ell)$ of vertices in G be a walk, trail, or path. We call v_1 and v_ℓ the start and end of P. For $i \in [\ell]$, we denote by $P[i]$ the vertex v_i at position i in P. Moreover, for $i, j \in [\ell]$, $i < j$, we denote by $P[i, j]$ the subsequence $(v_i, \ldots, v_j)$ of P. By definition, P has an alternative representation as sequence of edges (arcs) $P = (e_1, \ldots, e_{\ell-1})$ with $e_i := \{v_i, v_{i+1}\}$ ($e_i := (v_i, v_{i+1})$) for $i \in [\ell - 1]$. Using this representation, we say that P contains/uses edge (arc) e at time step i if edge (arc) e appears at the ith position in P represented as sequence of edges (arcs) (analogue for vertices). We call an edge/arc shared if two routes uses the edge/arc at the same time step. Note that in the undirected case, if an edge is traversed in different directions at the same time, we count it as shared. We say that a walk/trail/path Q is an s-t walk/trail/path, if s is the start and t is the end of Q. The length of a walk/trail/path is the number of edges (arcs) contained, where we also count multiple occurrences of an edge (arc) (we refer to a path of length m as an m-chain). (We define the maximum over in- and outdegrees in G by $\Delta_{i/o}(G) := \max_{v \in V(G)}\{\text{outdeg}(v), \text{indeg}(v)\}$.)

We state some preliminary observations on RCA and FRCA. If the termi-nals s and t have distance at most k, then routing any number of paths along the shortest path between them introduces at most k shared edges. Moreover, one can show that RCA and FRCA are contained in NP. Due to space constraints, details of these and other results (marked by $\star$) are deferred to a full version.

3   Everything is Equal on DAGs

Note that on directed acyclic graphs, every walk contains each edge and each vertex at most once. Hence, every walk is a path in DAGs, implying that all three types of routes are equivalent in DAGs.

We prove that RCA is solvable in polynomial time if the number k of shared arcs is constant, but NP-complete if k is part of the input. Moreover, we prove that the same holds for the length-restricted variant FRCA. We start the section with the case of constant $k \geq 0$.

3.1   Constant Number of Shared Arcs

Theorem 1. RCA *and* FRCA *on* n-*vertex* m-*arc DAGs are solvable in* $O(m^{k+1} \cdot n^3)$ *time and* $O(m^{k+1} \cdot \alpha^2 \cdot n)$ *time, respectively.*

We prove Theorem 1 as follows: We first show that RCA and FRCA on DAGs are solvable in polynomial time if $k = 0$ (Theorem 2 below). We then show that an instance of RCA and FRCA on directed graphs is equivalent to deciding, for all k-sized subsets K of arcs, the instance with $k = 0$ and a modified input graph in which each arc in K has been copied p times:

Theorem 2 ($\star$). *If* $k = 0$, RCA *on* n-*vertex* m-*arc DAGs is solvable in* $O(n^3 \cdot m)$ *time.*

We need the notion of time-expanded graphs. Given a directed graph G, we denote a directed graph H the (directed) τ-*time-expanded graph* of G if $V(H) = \{v^i \mid v \in V(G), i = 0, \ldots, \tau\}$ and $A(H) = \{(v^{i-1}, w^i) \mid i \in [\tau], (v, w) \in A(G)\}$. Note that for every directed n-vertex m-arc graph the τ-time-expanded graph can be constructed in $O(\tau \cdot (n+m))$ time. We prove that we can decide RCA and FRCA by flow computation in the time-expanded graph of the input graph:

Lemma 1 ($\star$). *Let* $G = (V, A)$ *be a directed graph with two distinct vertices* $s, t \in V$. *Let* $p \in \mathbb{N}$ *and* $\tau := |V|$. *Let* H *be the* τ-*time-expanded graph of* G *with* p *additional arcs* (t^{i-1}, t^i) *between the copies of* t *for each* $i \in [\tau]$. *Then,* G *allows for at least* p s-t *walks of length at most* τ *not sharing any arc if and only if* H *allows for an* s^0-t^τ *flow of value at least* p.

Lemma 1 is directly applicable to FRCA, by constructing an α-expanded graph.

Corollary 1 *If* $k = 0$, *then* FRCA *on* n-*vertex* m-*arc DAGs is solvable in* $O(\alpha^2 \cdot n \cdot m)$ *time.*

Let $G = (V, A)$ be a directed graph and let $K \subseteq A$ and $x \in \mathbb{N}$. We denote by $G(K, x)$ the graph obtained from G by replacing each arc $(v, w) \in K$ in G by x copies $(v, w)_1, \ldots, (v, w)_x$.

Lemma 2 (⋆). *Let* $(G = (V, A), s, t, p, k)$ *be an instance of* WALK-RCA *with* G *being a directed graph. Then,* (G, s, t, p, k) *is a yes-instance of* WALK-RCA *if and only if there exists a set* $K \subseteq A$ *with* $|K| \leq k$ *such that* $(G(K, p), s, t, p, 0)$ *is a yes-instance of* WALK-RCA. *The same statement holds true for* WALK-FRCA.

Proof (Theorem 1). Let $(G = (V, A), s, t, p, k)$ be an instance of WALK-RCA with G being a directed acyclic graph. For each k-sized subset $K \subseteq A$ of arcs in G, we decide the instance $(G(K, p), s, t, p, 0)$. The statement for RCA then follows from Lemma 2 and Theorem 2. We remark that the value of a maximum flow between two terminals in an n-vertex m-arc graph can be computed in $O(n \cdot m)$ time [16]. The running time of the algorithm is in $O(|A|^k \cdot (|V|^3 \cdot |A|))$. The statement for FRCA follows analogously with Lemma 2 and Corollary 1. □

3.2   Arbitrary Number of Shared Arcs

If the number k of shared arcs is arbitrary, then both RCA and FRCA are hard.

Theorem 3 (⋆). RCA *on DAGs is* NP-*complete and* W[2]-*hard with respect to* k.

The construction in the reduction for Theorem 3 is similar to the one used by Omran et al. [15, Theorem 2]. Herein, we give a (parameterized) many-one reduction from the NP-complete [10] SET COVER problem: given a set $U = \{u_1, \dots, u_n\}$, a set of subsets $\mathcal{F} = \{F_1, \dots, F_m\}$ with $F_i \subseteq U$ for all $i \in [m]$, and an integer $\ell \leq m$, is there a subset $\mathcal{F}' \subseteq \mathcal{F}$ with $|\mathcal{F}'| \leq \ell$ such that $\bigcup_{F \in \mathcal{F}'} F = U$. Note that SET COVER is W[2]-complete with respect to the solution size ℓ in question [5]. In the following Construction 1, given a SET COVER instance, we construct the DAG in an equivalent RCA or FRCA instance.

Construction 1. Let a set $U = \{u_1, \dots, u_n\}$, a set of subsets $\mathcal{F} = \{F_1, \dots, F_m\}$ with $F_i \subseteq U$ for all $i \in [m]$, and an integer $\ell \leq m$ be given. Construct a directed acyclic graph $G = (V, A)$ as follows. Initially, let G be the empty graph. Add the vertex sets $V_U = \{v_1, \dots, v_n\}$ and $V_{\mathcal{F}} = \{w_1, \dots, w_m\}$, corresponding to U and $\mathcal{F}$, respectively. Add the arc (v_i, w_j) to G if and only if $u_i \in F_j$. Next, add the vertex s to G. For each $w \in V_{\mathcal{F}}$, add an $(\ell + 2)$-chain to G connecting s with w, and direct all edges in the chain from s towards w. For each $v \in V_U$, add an $(\ell + 1)$-chain to G connecting s with v, and direct all edges in the chain from s towards v. Finally, add the vertex t to G and add the arcs (w, t) for all $w \in V_{\mathcal{F}}$. □

Lemma 3 (⋆). *Let* $U, \mathcal{F}, \ell,$ *and* G *as in Construction 1. Then there are at most* ℓ *sets in* $\mathcal{F}$ *such that their union is* U *if and only if* G *admits* $n + m$ s-t *walks sharing at most* ℓ *arcs in* G.

Theorem 3 follows then from Construction 1 and Lemma 3. Observe that each s-t walk in the graph obtained from Construction 1 is of length at most $\ell + 3$. Therewith, we obtain the following.

Corollary 2. FRCA *on DAGs is* NP-*complete and* W[2]-*hard with respect to* $k + \alpha$.

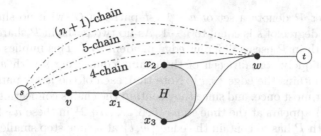

Fig. 1. Graph G' obtained in Construction 2. The gray part represents the graph H. Dashed lines represent chains.

4 Path-RCA

In this section, we prove the following theorem.

Theorem 4 ($\star$). PATH-RCA *both on undirected planar and directed planar graphs is* NP-*complete, even if $k \geq 0$ is constant or $\Delta \geq 4$ is constant.*

In the proof of Theorem 4, we reduce from the following NP-complete [9] problem (a cubic graph is a graph where every vertex has degree exactly three):

PLANAR CUBIC HAMILTONIAN CYCLE (PCHC)
Input: An undirected, planar, cubic graph G.
Question: Is there a cycle in G that visits each vertex exactly once?

Roughly, the instance of PATH-RCA obtained in the reduction consists of a graph G' containing the original graph G connected to the terminals s, t such that t is of degree one (see Fig. 1). We ask for constructing $n-1$ paths connecting the terminals, where n is the number of vertices in the input graph of PCHC. The idea is the following: All but one of these paths must occupy the only edge to t for in total $n - 2$ time steps such that the one remaining path has to visit all the vertices in the original graph G to not introduce a shared edge.

The reduction to prove Theorem 4 uses the following Construction 2.

Construction 2. Let $G = (V, E)$ be an undirected, planar, cubic graph with $n = |V|$. Construct in time polynomial in the size of G an undirected planar graph G' as follows (refer to Fig. 1). Let initially G' be the empty graph. Add a copy of G to G'. Denote the copy of G in G' by H. Next, add the new vertices s, t, v, w to G. Connect s with v, and w with t by an edge. For each $m \in \{4, 5, \ldots, n+1\}$, add an m-chain connecting s with w. Next, consider a fixed plane embedding $\phi(G)$ of G. Let x_1 denote a vertex incident to the outer face in $\phi(G)$. Then, there are two neighbors x_2 and x_3 of x_1 also incident to the outer face in $\phi(G)$. Add the edges $\{v, x_1\}$, $\{x_2, w\}$ and $\{x_3, w\}$ to G' completing the construction of G'. We remark that G' is planar as it allows a plane embedding (see Fig. 1) using ϕ as an embedding of H. $\square$

Lemma 4. *Let G and G' be as in Construction 2. Then G admits a Hamiltonian cycle if and only if G' allows for at least $n - 1$ s-t paths with no shared edge.*

Proof. ($\Leftarrow$) Let $\mathcal{P}$ denote a set of $n-1$ s-t paths in G' with no shared edge. Note that the degree of s is equal to $n-1$. As no two paths in $\mathcal{P}$ share any edge in G', each path in $\mathcal{P}$ uses a different edge incident to s. This implies that $n-2$ paths in $\mathcal{P}$ uniquely contain each of the chains connecting s with w, and one path $P \in \mathcal{P}$ contains the edge $\{s, v\}$. Note that each of the $n-2$ paths contain the vertex w at most once, and since they contain the chains connecting s with w, the edge $\{w, t\}$ appears at the time steps $\{5, 6, \ldots, n+2\}$ in these $n-2$ paths $\mathcal{P}$. Hence, the path P has to contain the edge $\{w, t\}$ at a time step smaller than five or larger than $n+2$. Observe that, by construction, the shortest path between s and w is of length 4 and, thus, P cannot contain the edge $\{w, t\}$ on any time step smaller than five. Hence, P has to contain the edge at time step at least $n+3$. Since the distance between s and x_1 is two, and the distance from x_2, x_3 to w is one, P has to visit each vertex in H exactly once, starting at x_1, and ending at one of the two neighbors x_2 or x_3 of x_1. Hence, P restricted to H describes a Hamiltonian path in H, which can be extended to an Hamiltonian cycle by adding the edge $\{x_1, x_2\}$ in the first or $\{x_1, x_3\}$ in the second case.

($\Rightarrow$) Let G admit a Hamiltonian cycle C. Since C contains every vertex in G exactly once, it contains x_1 and its neighbors x_2 and x_3. Since C forms a cycle in G and G is cubic, at least one of the edges $\{x_1, x_2\}$ or $\{x_1, x_3\}$ appears in C. Let C' denote an ordering of the vertices in C such that x_1 appears first and the neighbor $x \in \{x_2, x_3\}$ of x_1 with $\{x_1, x\}$ contained in C appears last. We construct $n-1$ s-t paths without sharing an edge. First, we construct $n-2$ s-t paths, each containing a different chain connecting s with w and the edge $\{w, t\}$. Observe that since the lengths of each chain is unique, no edge (in particular, not $\{w, t\}$) is shared. Finally, we construct the one remaining s-t path P as follows. We lead P from s to x_1 via v, then following C' in H to x, and then from x to t via w. Observe that P has length $n+3$ and contains the edge $\{w, t\}$ at time step $n+3$. Hence, no edge is shared as the path containing the $(n+1)$-chain contains the edge $\{w, t\}$ at time step $(n+2)$. We constructed $n-1$ s-t paths in G' with no shared edge. □

The remaining proof of Theorem 4 is deferred to a full version. We remark that the statement in Theorem 4 for constant k follows from Lemma 4.

As the length of every s-t path is upper bounded by the number of vertices in the graph, we immediately obtain the following.

Corollary 3. PATH-FRCA *both on undirected planar and directed planar graphs is* NP-*complete, even if* $k \geq 0$ *is constant or* $\Delta \geq 4$ *is constant.*

5   Trail-RCA

We now show that TRAIL-RCA has the same computational complexity fingerprint as PATH-RCA. That is, TRAIL-RCA (TRAIL-FRCA) is NP-complete on undirected and directed planar graphs, even if the number $k \geq 0$ of shared edges (arcs) or the maximum degree $\Delta \geq 5$ ($\Delta_{i/o} \geq 3$) is constant. The reductions are slightly more involved, because it is harder to force trails to take a certain way.

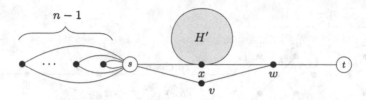

Fig. 2. Graph G' obtained in Construction 3. The gray part represents the graph H'.

5.1   On Undirected Graphs

Theorem 5 ($\star$). TRAIL-RCA *on undirected planar graphs is NP-complete, even if $k \geq 0$ is constant or $\Delta \geq 5$ is constant.*

We provide two constructions supporting the two subresults for constants k, Δ. The reductions are again from PLANAR CUBIC HAMILTONIAN CYCLE (PCHC).

Construction 3. Let $G = (V, E)$ be an undirected planar cubic graph with $n = |V|$. Construct an undirected planar graph G' as follows (refer to Fig. 2). Initially, let G' be the empty graph. Add a copy of G to G' and denote the copy by H. Subdivide each edge in H and denote the resulting graph H'. Note that H' is still planar. Consider a plane embedding $\phi(H')$ of H' and and let $x \in V(H')$ be a vertex incident to the outer face in the embedding. Next, add the vertex set $\{s, v, w, t\}$ to G'. Add the edges $\{s, x\}$, $\{s, v\}$, $\{v, w\}$, and $\{w, t\}$ to G'. Finally, add $n - 1$ vertices $B = \{b_1, \ldots, b_{n-1}\}$ to G' and connect each of them with s by two edges (in the following, we distinguish these edges as $\{s, b_i\}_1$ and $\{s, b_i\}_2$, for each $i \in [n - 1]$). Note that the graph is planar (see Fig. 2 for an embedding, where H' is embedded as $\phi(H')$) but not simple. $\qquad\square$

Lemma 5 ($\star$). *Let G and G' as in Construction 3. Then G admits a Hamiltonian cycle if and only if G' admits $2n$ s-t trails with no shared edge.*

To deal with the parallel edges in graph G' in Construction 3, we now subdivide edges, maintaining an equivalent statement as in Lemma 5.

Lemma 6 ($\star$). *Let G be an undirected graph (not necessarily simple) and $s, t \in V(G)$. Obtain graph G' from G by replacing each edge $\{u, v\} \in E$ in G by a path of length three, identifying its endpoints with u and v. Then G admits $p \in \mathbb{N}$ s-t trails with no shared edge if and only if G' admits p s-t trails with no shared edge.*

We now show how to modify Construction 3 for maximum degree five, giving up, however, a constant upper bound on the number of shared edges.

Construction 4. Let $G = (V, E)$ be an undirected planar cubic graph with $n = |V|$. Construct an undirected planar graph G' as follows (see Fig. 3). Let initially G' be the graph obtained from Construction 3. Subdivide each edge

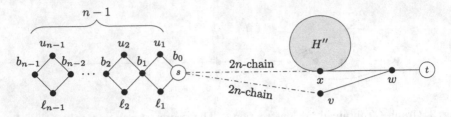

Fig. 3. Graph G' obtained in Construction 4. The gray part represents the graph H'.

in H' and denote the resulting graph by H''. Observe that the distance in H'' between any two vertices in $V(H'') \cap V(H)$ is divisible by four. Next, delete all edges incident with vertex s. Connect s with v via a $2n$-chain, and connect s with x via a $2n$-chain. Connect s with b_1 via two P_2's (2-chains). Denote the two vertices on the P_2's by ℓ_1 and u_1. Finally, for each $i \in [n-2]$, connect b_i with b_{i+1} via two P_2's. For each $i \in [n-2]$, denote the two vertices on the P_2's between b_i and b_{i+1} by ℓ_{i+1} and u_{i+1}. For an easier notation, we denote vertex s also by b_0. □

Lemma 7 $(\star)$. *Let G and G' as in Construction 4. Then G admits a Hamiltonian cycle if and only if G' has $2n$ s-t trails with at most $2n - 4$ shared edges.*

The proof of Theorem 5 then follows from Lemmas 5 and 7. As the length of each s-t trail is upper bounded by the number of edges in the graph, we immediately obtain the following.

Corollary 4. TRAIL-FRCA *on undirected planar graphs is* NP-*complete, even if $k \geq 0$ is constant or $\Delta \geq 5$ is constant.*

5.2   On Directed Graphs

We know that TRAIL-RCA and TRAIL-FRCA are NP-complete on undirected graphs, even if the number of shared edges or the maximum degree is constant. In what follows, we show that this is also the case for TRAIL-RCA and TRAIL-FRCA on directed graphs.

Theorem 6 $(\star)$. TRAIL-RCA *on directed planar graphs is* NP-*complete, even if $k \geq 0$ is constant or $\Delta_{i/o} \geq 3$ is constant.*

As the length of each s-t trail is upper bounded by the number of edges in the graph, we immediately obtain the following.

Corollary 5. TRAIL-FRCA *on directed planar graphs is* NP-*complete, even if $k \geq 0$ is constant or $\Delta_{i/o} \geq 3$ is constant.*

6 Walk-RCA

Regarding their computational complexity fingerprint, PATH-RCA and TRAIL-RCA are equal. In this section, we show that WALK-RCA differs in this aspect. We prove that the problem is solvable in polynomial time on undirected graphs and on directed graphs if $k \geq 0$ is constant.

6.1 On Undirected Graphs

On a high level, the tractability on undirected graphs is due to the fact that a walk can alternate arbitrarily often between two vertices. Hence, we can model a queue on the source vertex s, where at distinct time steps the walks leave s via a shortest path towards t.

Theorem 7 ($\star$). WALK-RCA *on undirected graphs is solvable in linear time.*

The situation changes for the length-restricted variant WALK-FRCA.

Theorem 8 ($\star$). WALK-FRCA *on undirected graphs is NP-complete and* W[2]*-hard with respect to* $k + \alpha$.

We note that the proof is similar to the proof of Theorem 3. It remains open whether WALK-FRCA is NP-complete when k is constant.

6.2 On Directed Graphs

Due to Theorems 2 and 3, we know that WALK-RCA is NP-complete on directed graphs and is solvable in polynomial time on directed acyclic graphs when $k = 0$, respectively. In this section, we prove that if $k \geq 0$ is constant, then WALK-RCA remains tractable on directed graphs (this also holds true for WALK-FRCA). Note that for PATH-RCA and TRAIL-RCA the situation is different, as both become NP-complete on directed graphs, even if $k \geq 0$ is constant.

Theorem 9. WALK-RCA *and* WALK-FRCA *on directed* n-*vertex* m-*arc graphs is solvable in* $O(m^{k+1} \cdot n \cdot (p \cdot n)^2)$ *time and* $O(m^{k+1} \cdot n \cdot \alpha^2)$ *time, respectively.*

Our proof of Theorem 9 follows the same strategy as our proof of Theorem 1. That is, we try to guess the shared arcs, make them infinite capacity in some way, and then solve the problem with zero shared arcs via a network flow formulation in the time-expanded graph. The crucial difference is that here we do not have at first an upper bound on the length of the walks in the solution.

Theorem 10 ($\star$). *If* $k = 0$, *then* WALK-RCA *on directed* n-*vertex* m-*arc graphs is solvable in* $O(n \cdot m \cdot (p \cdot n)^2)$ *time.*

Lemma 8. *Every yes-instance* (G, s, t, p, k) *of* WALK-RCA *on directed graphs admits a solution in which the longest walk is of length at most* $p \cdot d_t$, *where* $d_t = \max_{v \in V : \text{dist}_G(v,t) < \infty} \text{dist}_G(v, t)$.

Observe that d_t is well-defined on every yes-instance of WALK-RCA. Moreover, $d_t \leq |V(G)|$. Below we use the following notation. For two walks $P_1 = (v_1, \ldots, v_\ell)$ and $P_2 = (w_1, \ldots, w_{\ell'})$ with $v_\ell = w_1$, denote by $P_1 \circ P_2$ the walk $(v_1, \ldots, v_\ell, w_2, \ldots, w_{\ell'})$ obtained by the concatenation of the two walks.

Proof (Lemma 8). Let $\mathcal{P}$ be a solution to (G, s, t, p, k) with $|\mathcal{P}| = p$ where the sum of the lengths of the walks in $\mathcal{P}$ is minimum among all solutions to (G, s, t, p, k). Suppose towards a contradiction that the longest walk $P^* \in \mathcal{P}$ is of length $|P^*| > p \cdot d_t$. Then, there is an $i \in [p]$ such that there is no walk in $\mathcal{P}$ of length ℓ with $(i-1) \cdot d_t < \ell \leq i \cdot d_t$.

Let $v = P^*[(i-1) \cdot d_t + 1]$, that is, v is the $((i-1) \cdot d_t + 1)$th vertex on P^*, and let S be a shortest v-t path. Observe that the length of S is at most d_t. Consider the walk $P' := P^*[1, (i-1) \cdot d_t + 1] \circ S$, that is, we concatenate the length-$((i-1) \cdot d_t)$ initial subpath of P^* with S to obtain P'. Observe that $(i-1) \cdot d_t < |P'| \leq i \cdot d_t$. If $\mathcal{P} \backslash P^* \cup P'$ forms a solution to (G, s, t, p, k), then, since $|P'| < |P^*|$, $\mathcal{P} \backslash P^* \cup P'$ is a solution of smaller sum of the lengths of the walks, contradicting the choice of $\mathcal{P}$. Otherwise, P' introduces additional shared arcs and let $A' \subseteq A(G)$ denote the corresponding set. Observe that A' is a subset of the arcs of S. Let $a = (x, y) \in A'$ be the shared arc such that $\text{dist}_S(y, t)$ is minimum among all shared arcs in A', and let $P'[j] = y$. Let $P \in \mathcal{P}$ be a walk sharing the arc a with P'. Note that $|P| > i \cdot d_t$. Thus, $P'' := P[1, j] \circ P'[j, |P'|]$ is a walk of shorter length than P. Moreover, $\mathcal{P} \backslash P \cup P''$ is a solution to (G, s, t, p, k). As $|P''| < |P|$, $\mathcal{P} \backslash P \cup P''$ is a solution of smaller sum of the lengths of the walks, contradicting the choice of $\mathcal{P}$. As either case yields a contradiction, it follows that $|P^*| \leq p \cdot d_t$. $\square$

The proof of Theorem 10 relies on time-expanded graphs. Due to Lemma 8, the time-horizon is bounded polynomially in the input size. The proof is deferred to a full version. Restricting to α-time-expanded graphs yields the following.

Corollary 6. *If $k = 0$, WALK-FRCA on directed n-vertex m-arc graphs is solvable in $O(n \cdot m \cdot \alpha^2)$ time.*

Proof (Theorem 9). Let $(G = (V, E), s, t, p, k)$ be an instance of WALK-RCA with G being a directed graph. For each k-sized subset $K \subseteq A$ of arcs in G, we decide the instance $(G(K, p), s, t, p, 0)$. The statement for WALK-RCA then follows from Lemma 2 and Theorem 10. The running time of the algorithm is in $O(|A|^k \cdot p^2 \cdot (|V|^3 \cdot |A|))$. The statement for WALK-FRCA then follows from Lemma 2 and Corollary 6. $\square$

7   Conclusion and Outlook

Some of our results can be seen as a parameterized complexity study of RCA focusing on the number k of shared edges. It is interesting to study the problem with respect to other parameters. Herein, the first natural parameterization is the number of routes. Recall that the MINIMUM SHARED EDGES (MSE) problem is fixed-parameter tractable with respect to the number of paths [7].

However, most tractability results for MSE rely on the notion of separators—it seems that such a notion has not been sufficiently developed yet for traversals over time. A second parameterization we consider as interesting is the combined parameter maximum degree plus k. In our NP-completeness results for PATH-RCA and TRAIL-RCA it seemed difficult to achieve constant k and constant maximum degree at the same time.

Another research direction is to investigate on which graph classes PATH-RCA and TRAIL-RCA become tractable. We proved that both problems remain NP-complete even on planar graphs. Do PATH-RCA and TRAIL-RCA, like MSE [1,19], become polynomial-time solvable on graphs of bounded treewidth?

Finally, we proved that on undirected graphs, WALK-RCA is solvable in polynomial time while WALK-FRCA is NP-complete. However, we left open whether WALK-FRCA on undirected graphs is NP-complete or polynomial-time solvable when k is constant.

References

1. Aoki, Y., Halldórsson, B.V., Halldórsson, M.M., Ito, T., Konrad, C., Zhou, X.: The minimum vulnerability problem on graphs. In: Zhang, Z., Wu, L., Xu, W., Du, D.-Z. (eds.) COCOA 2014. LNCS, vol. 8881, pp. 299–313. Springer, Cham (2014). doi:10.1007/978-3-319-12691-3_23

2. Assadi, S., Emamjomeh-Zadeh, E., Norouzi-Fard, A., Yazdanbod, S., Zarrabi-Zadeh, H.: The minimum vulnerability problem. Algorithmica 70(4), 718–731 (2014)

3. Berman, K.A.: Vulnerability of scheduled networks and a generalization of Menger's theorem. Networks 28(3), 125–134 (1996)

4. Downey, R.G., Fellows, M.R.: Parameterized Complexity. Monographs in Computer Science. Springer, Heidelberg (1999)

5. Downey, R.G., Fellows, M.R.: Fundamentals of Parameterized Complexity. Texts in Computer Science. Springer, Heidelberg (2013)

6. Fluschnik, T.: The Parameterized Complexity of Finding Paths with Shared Edges. Master thesis, Institut für Softwaretechnik und Theoretische Informatik, TU Berlin (2015). http://fpt.akt.tu-berlin.de/publications/theses/MA-till-fluschnik.pdf

7. Fluschnik, T., Kratsch, S., Niedermeier, R., Sorge, M.: The parameterized complexity of the minimum shared edges problem. In: Proceedings of the 35th IARCS Annual Conference on Foundation of Software Technology and Theoretical Computer Science (FSTTCS 2015), LIPIcs, vol. 45, pp. 448–462. Schloss Dagstuhl - Leibniz-Zentrum fuer Informatik (2015)

8. Ford, L.R., Fulkerson, D.R.: Flows in Networks. Princeton University Press, Princeton (1962)

9. Garey, M.R., Johnson, D.S., Tarjan, R.E.: The planar Hamiltonian circuit problem is NP-complete. SIAM J. Comput. 5(4), 704–714 (1976)

10. Karp, R.M.: Reducibility among combinatorial problems. In: Miller, R.E., Thatcher, J.W., Bohlinger, J.D. (eds.) Complexity of Computer Computations. IRSS, pp. 85–103. Springer, Boston (1972). doi:10.1007/978-1-4684-2001-2_9

11. Kempe, D., Kleinberg, J.M., Kumar, A.: Connectivity and inference problems for temporal networks. J. Comput. Syst. Sci. 64(4), 820–842 (2002)

12. Köhler, E., Möhring, R.H., Skutella, M.: Traffic networks and flows over time. In: Lerner, J., Wagner, D., Zweig, K.A. (eds.) Algorithmics of Large and Complex Networks. LNCS, vol. 5515, pp. 166–196. Springer, Heidelberg (2009). doi:10.1007/978-3-642-02094-0_9

13. Mertzios, G.B., Michail, O., Chatzigiannakis, I., Spirakis, P.G.: Temporal network optimization subject to connectivity constraints. In: Fomin, F.V., Freivalds, R., Kwiatkowska, M., Peleg, D. (eds.) ICALP 2013. LNCS, vol. 7966, pp. 657–668. Springer, Heidelberg (2013). doi:10.1007/978-3-642-39212-2_57

14. Michail, O.: An introduction to temporal graphs: an algorithmic perspective. Internet Math. **12**(4), 239–280 (2016)

15. Omran, M.T., Sack, J., Zarrabi-Zadeh, H.: Finding paths with minimum shared edges. J. Comb. Optim. **26**(4), 709–722 (2013)

16. Orlin, J.B.: Max flows in $O(nm)$ time, or better. In: Proceedings of the 45th ACM Symposium on Theory of Computing (STOC 2013), pp. 765–774. ACM (2013)

17. Rashidi, H., Tsang, E.: Vehicle Scheduling in Port Automation: Advanced Algorithms for Minimum Cost Flow Problems, 2nd edn. CRC Press, Boca Raton (2015)

18. Skutella, M.: An introduction to network flows over time. In: Cook, W., Lovász, L., Vygen, J. (eds.) Research Trends in Combinatorial Optimization, pp. 451–482. Springer, Heidelberg (2009). doi:10.1007/978-3-540-76796-1_21

19. Ye, Z.Q., Li, Y.M., Lu, H.Q., Zhou, X.: Finding paths with minimum shared edges in graphs with bounded treewidths. In: Proceedings of the International Conference on Frontiers of Computer Science (FCS 2013), pp. 40–46 (2013)

Parikh Image of Pushdown Automata

Pierre Ganty[1] and Elena Gutiérrez[1,2(✉)]

[1] IMDEA Software Institute, Madrid, Spain
{pierre.ganty,elena.gutierrez}@imdea.org
[2] Universidad Politécnica de Madrid, Madrid, Spain

Abstract. We compare pushdown automata (PDAs for short) against other representations. First, we show that there is a family of PDAs over a unary alphabet with n states and $p \geq 2n + 4$ stack symbols that accepts one single long word for which every equivalent context-free grammar needs $\Omega(n^2(p - 2n - 4))$ variables. This family shows that the classical algorithm for converting a PDA into an equivalent context-free grammar is optimal even when the alphabet is unary. Moreover, we observe that language equivalence and Parikh equivalence, which ignores the ordering between symbols, coincide for this family. We conclude that, when assuming this weaker equivalence, the conversion algorithm is also optimal. Second, Parikh's theorem motivates the comparison of PDAs against finite state automata. In particular, the same family of unary PDAs gives a lower bound on the number of states of every Parikh-equivalent finite state automaton. Finally, we look into the case of unary deterministic PDAs. We show a new construction converting a unary deterministic PDA into an equivalent context-free grammar that achieves best known bounds.

1 Introduction

Given a context-free language which representation, pushdown automata or context-free grammars, is more concise? This was the main question studied by Goldstine et al. [8] in a paper where they introduced an infinite family of context-free languages whose representation by a pushdown automaton is more concise than by context-free grammars. In particular, they showed that each language of the family is accepted by a pushdown automaton with n states and p stack symbols, but every context-free grammar needs at least $n^2p + 1$ variables if $n > 1$ (p if $n = 1$). Incidentally, the family shows that the translation of a pushdown automaton into an equivalent context-free grammar used in textbooks [9],

P. Ganty—has been supported by the Madrid Regional Government project S2013/ICE-2731, *N-Greens Software - Next-GeneRation Energy-EfficieNt Secure Software*, and the Spanish Ministry of Economy and Competitiveness project No. TIN2015-71819-P, *RISCO - RIgorous analysis of Sophisticated COncurrent and distributed systems*.

E. Gutiérrez—is partially supported by BES-2016-077136 grant from the Spanish Ministry of Economy, Industry and Competitiveness.

R. Klasing and M. Zeitoun (Eds.): FCT 2017, LNCS 10472, pp. 271–283, 2017.
DOI: 10.1007/978-3-662-55751-8_22

which uses the same large number of $n^2p + 1$ variables if $n > 1$ (p if $n = 1$), is optimal in the sense that there is no other algorithm that always produces fewer grammar variables.

In this paper, we revisit these questions but this time we turn our attention to the unary case. We define an infinite family of context-free languages as Goldstine et al. did but our family differs drastically from theirs. Given $n \geq 1$ and $k \geq 1$, each member of our family is given by a PDA with n states, $p = k + 2n + 4$ stack symbols and a *single input symbol*.[1] We show that, for each PDA of the family, every equivalent context-free grammar has $\Omega(n^2(p - 2n - 4))$ variables. Therefore, this family shows that the textbook translation of a PDA into a language-equivalent context-free grammar is *optimal*[2] even when the alphabet is unary. Note that if the alphabet is a singleton, equality over words (two words are equal if the same symbols appear at the same positions) coincides with Parikh equivalence (two words are Parikh-equivalent if each symbol occurs equally often in both words[3]). Thus, we conclude that the conversion algorithm is also optimal for Parikh equivalence. We also investigate the special case of deterministic PDAs over a singleton alphabet for which equivalent context-free grammar representations of small size had been defined [3,10]. We give a new definition for an equivalent context-free grammar given a unary deterministic PDA. Our definition is constructive (as far as we could tell the result of Pighizzini [10] is not) and achieves the best known bounds [3] by combining two known constructions.

Parikh's theorem [11] states that every context-free language has the same Parikh image as some regular language. This allows us to compare PDAs against finite state automata (FSAs for short) for Parikh-equivalent languages. First, we use the same family of PDAs to derive a lower bound on the number of states of every Parikh-equivalent FSA. The comparison becomes simple as its alphabet is unary and it accepts one single word. Second, using this lower bound we show that the 2-step procedure chaining existing constructions: (i) translate the PDA into a language-equivalent context-free grammar [9]; and (ii) translate the context-free grammar into a Parikh-equivalent FSA [4] yields *optimal*[4] results in the number of states of the resulting FSA.

As a side contribution, we introduce a semantics of PDA runs as trees that we call *actrees*. The richer tree structure (compared to a sequence) makes simpler to compare each PDA of the family with its smallest grammar representation.

Due to the lack of space, some proofs and examples are deferred to a long version [7].

Structure of the Paper. After preliminaries in Sect. 2 we introduce the tree-based semantics in 3. In Sect. 4 we compare PDAs and context-free grammars when they represent Parikh-equivalent languages. We will define the infinite family of PDAs and establish their main properties. We dedicate Sect. 4.2 to the

---

[1] Their family has an alphabet of non-constant size.

[2] In a sense that we will precise in Sect. 4 (Remark 10).

[3] But not necessarily at the same positions, e.g. ab and ba are Parikh-equivalent.

[4] In a sense that we will precise in Sect. 5 (Remark 16).

special case of deterministic PDAs over a unary alphabet. Finally, Sect. 5 focuses on the comparison of PDAs against finite state automata for Parikh-equivalent languages.

2   Preliminaries

A *pushdown automaton* (or PDA) is a 6-tuple $(Q, \Sigma, \Gamma, \delta, q_0, Z_0)$ where Q is a finite nonempty set of *states* including q_0, the *initial* state; Σ is the *input alphabet*; Γ is the *stack alphabet* including Z_0, the *initial* stack symbol; and δ is a finite subset of $Q \times \Gamma \times (\Sigma \cup \{\varepsilon\}) \times Q \times \Gamma^*$ called the *actions*. We write $(q, X) \hookrightarrow_b (q', \beta)$ to denote an action $(q, X, b, q', \beta) \in \delta$. We sometimes omit the subscript to the arrow.

An *instantaneous description* (or ID) of a PDA is a pair (q, β) where $q \in Q$ and $\beta \in \Gamma^*$. We call the first component of an ID the *state* and the second the *stack content*. The *initial ID* consists of the initial state and the initial stack symbol for the stack content. When reasoning formally, we use the functions *state* and *stack* which, given an ID, returns its state and stack content, respectively.

An action $(q, X) \hookrightarrow_b (q', \beta)$ is *enabled* at ID I if $state(I) = q$ and $(stack(I))_1 = X$.[5] Given an ID $(q, X\gamma)$ enabling $(q, X) \hookrightarrow_b (q', \beta)$, define the *successor ID* to be $(q', \beta\gamma)$. We denote this fact as $(q, X\gamma) \vdash_b (q', \beta\gamma)$, and call it a *move* that *consumes* b from the input.[6] We sometimes omit the subscript of $\vdash$ when the input consumed (if any) is not important. Given $n \geq 0$, a *move sequence*, denoted $I_0 \vdash_{b_1} \cdots \vdash_{b_n} I_n$, is a finite sequence of IDs $I_0 I_1 \ldots I_n$ such that $I_i \vdash_{b_i} I_{i+1}$ for all i. The move sequence *consumes* w (*from the input*) when $b_1 \cdots b_n = w$. We concisely denote this fact as $I_0 \vdash .\overset{w}{.} \vdash I_n$. A move sequence $I \vdash \cdots \vdash I'$ is a *quasi-run* when $|stack(I)| = 1$ and $|stack(I')| = 0$; and a *run* when, furthermore, I is the initial ID. Define the *language* of a PDA P as $L(P) = \{w \in \Sigma^* \mid P \text{ has a run consuming } w\}$.

The *Parikh image* of a word w over an alphabet $\{b_1, \ldots, b_n\}$, denoted by $\wr w \wr$, is the vector $(x_1, \ldots, x_n) \in \mathbb{N}^n$ such that x_i is the number of occurrences of b_i in w. The *Parikh image of a language* L, denoted by $\wr L \wr$, is the set of Parikh images of its words. When $\wr L_1 \wr = \wr L_2 \wr$, we say L_1 and L_2 are *Parikh-equivalent*.

We assume the reader is familiar with the basics of *finite state automata* (or FSAs for short) and *context-free grammars* (or CFGs for short). Nevertheless we fix their notation as follows. We denote a FSA as a tuple $(Q, \Sigma, \delta, q_0, F)$ where Q is a finite set of *states* including the *initial* state q_0 and the *final* states F; Σ is the *input alphabet* and $\delta \subseteq Q \times (\Sigma \cup \{\varepsilon\}) \times Q$ is the *set of transitions*. We denote a CFG as a tuple (V, Σ, S, R) where V is a finite set of *variables* including S the *start* variable, Σ is the *alphabet* or set of terminals and $R \subseteq V \times (V \cup \Sigma)^*$ is a finite set of *rules*. Rules are conveniently denoted $X \rightarrow \alpha$. Given a FSA A and a CFG G we denote their *languages* as $L(A)$ and $L(G)$, respectively.

Finally, let us recall the translation of a PDA into an equivalent CFG. Given a PDA $P = (Q, \Sigma, \Gamma, \delta, q_0, Z_0)$, define the CFG $G = (V, \Sigma, R, S)$ where

---

[5] $(w)_i$ is the i-th symbol of w if $1 \leq i \leq |w|$; else $(w)_i = \varepsilon$. $|w|$ is the length of w.
[6] When $b = \varepsilon$ the move does not consume input.

– The set V of variables — often called the *triples* — is given by

$$\{[qXq'] \mid q, q' \in Q, X \in \Gamma\} \cup \{S\}. \tag{1}$$

– The set R of production rules is given by

$$\{S \to [q_0 Z_0 q] \mid q \in Q\}$$
$$\cup \{[qXr_d] \to b[q'(\beta)_1 r_1] \dots [r_{d-1}(\beta)_d r_d] \tag{2}$$
$$\mid (q, X) \hookrightarrow_b (q', \beta), d = |\beta|, r_1, \dots, r_d \in Q\}$$

For a proof of correctness, see the textbook of Ullman et al. [9]. The previous definition easily translates into a *conversion algorithm*. Observe that the runtime of such algorithm depends polynomially on $|Q|$ and $|\Gamma|$, but exponentially on $|\beta|$.

3 A Tree-Based Semantics for Pushdown Automata

In this section we introduce a tree-based semantics for PDA. Using trees instead of sequences sheds the light on key properties needed to present our main results.

Given an action a denoted by $(q, X) \hookrightarrow_b (q', \beta)$, q is the *source* state of a, q' the *target* state of a, X the symbol a *pops* and β the (possibly empty) sequence of symbols a *pushes*.

A *labeled tree* $c(t_1, \dots, t_k)$ $(k \geq 0)$ is a finite tree whose nodes are labeled, where c is the label of the root and $t_1, \dots, t_k$ are labeled trees, the children of the root. When $k = 0$ we prefer to write c instead of $c()$. Each labeled tree t defines a sequence, denoted $\bar{t}$, obtained by removing the symbols '(', ')' or ',' when interpreting t as a string, e.g. $\overline{c(c_1, c_2(c_{21}))} = c\, c_1 c_2 c_{21}$. The *size* of a labeled tree t, denoted $|t|$, is given by $|\bar{t}|$. It coincides with the number of nodes in t.

Definition 1. *Given a PDA P, an action-tree (or actree for short) is a labeled tree $a(a_1(\dots), \dots, a_d(\dots))$ where a is an action of P pushing β with $|\beta| = d$ and each children $a_i(\dots)$ is an actree such that a_i pops $(\beta)_i$ for all i. Furthermore, an actree t must satisfy that the source state of $(\bar{t})_{i+1}$ and the target state of $(\bar{t})_i$ coincide for every i.*

An actree t consumes an input resulting from replacing each action in the sequence $\bar{t}$ by the symbol it consumes (or ε, if the action does not consume any). An actree $a(\dots)$ is accepting if the initial ID enables a.

Example 2. Consider a PDA P with actions a_1 to a_5 respectively given by $(q_0, X_1) \hookrightarrow_\varepsilon (q_0, X_0 X_0)$, $(q_0, X_0) \hookrightarrow_\varepsilon (q_1, X_1 \star)$, $(q_1, X_1) \hookrightarrow_\varepsilon (q_1, X_0 X_0)$, $(q_1, X_0) \hookrightarrow_b (q_1, \varepsilon)$ and $(q_1, \star) \hookrightarrow_\varepsilon (q_0, \varepsilon)$. The reader can check that the actree $t = a_1(a_2(a_3(a_4, a_4), a_5), a_2(a_3(a_4, a_4), a_5))$, depicted in Fig. 1, satisfies the conditions of Definition 1 where $\bar{t} = a_1 a_2 a_3 a_4 a_4 a_5 a_2 a_3 a_4 a_4 a_5$, $|t| = 11$ and the input consumed is b^4.

We recall the notion of *dimension* of a labeled tree [5] and we relate dimension and size of labeled trees in Lemma 5 (proof omitted).

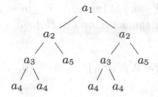

Fig. 1. Depiction of the tree $a_1(a_2(a_3(a_4, a_4), a_5), a_2(a_3(a_4, a_4), a_5))$

Definition 3. *The* dimension *of a labeled tree t, denoted as $d(t)$, is inductively defined as follows. $d(t) = 0$ if $t = c$, otherwise we have $t = c(t_1, \ldots, t_k)$ for some $k > 0$ and*

$$d(t) = \begin{cases} max_{i\in\{1,\ldots,k\}} d(t_i) & \text{if there is a unique maximum,} \\ max_{i\in\{1,\ldots,k\}} d(t_i) + 1 & \text{otherwise.} \end{cases}$$

Example 4. The annotation $\overset{d(t)}{t}(\ldots)$ shows that the actree of Example 2 has dimension 2

$$\overset{2}{a_1}(\overset{1}{a_2}(\overset{1}{a_3}(\overset{0}{a_4}, \overset{0}{a_4}), \overset{0}{a_5}), \overset{1}{a_2}(\overset{1}{a_3}(\overset{0}{a_4}, \overset{0}{a_4}), \overset{0}{a_5})).$$

Lemma 5. $|t| \geq 2^{d(t)}$ *for every labeled tree t.*

The actrees and the quasi-runs of a PDA are in one-to-one correspondence as reflected in Theorem 6 (proof omitted).

Theorem 6. *Given a PDA, its actrees and quasi-runs are in a one-to-one correspondence.*

4   Parikh-Equivalent Context-Free Grammars

In this section we compare PDAs against CFGs when they describe Parikh-equivalent languages. We first study the general class of (nondeterministic) PDAs and, in Sect. 4.2, we look into the special case of unary deterministic PDAs.

We prove that, for every $n \geq 1$ and $p \geq 2n + 4$, there exists a PDA with n states and p stack symbols for which every Parikh-equivalent CFG has $\Omega(n^2(p - 2n - 4))$ variables. To this aim, we present a family of PDAs $P(n, k)$ where $n \geq 1$ and $k \geq 1$. Each member of the family has n states and $k + 2n + 4$ stack symbols, and accepts one single word over a unary input alphabet.

4.1   The Family $P(n, k)$ of PDAs

Definition 7. *Given natural values $n \geq 1$ and $k \geq 1$, define the PDA $P(n, k)$ with states $Q = \{q_i \mid 0 \leq i \leq n - 1\}$, input alphabet $\Sigma = \{b\}$, stack alphabet*

$\Gamma = \{S, \star, \$\} \cup \{X_i \mid 0 \le i \le k\} \cup \{s_i \mid 0 \le i \le n-1\} \cup \{r_i \mid 0 \le i \le n-1\}$,
initial state q_0, initial stack symbol S and actions δ

$$(q_0, S) \hookrightarrow_b (q_0, X_k\, r_0)$$
$$(q_i, X_j) \hookrightarrow_b (q_i, X_{j-1}\, r_m\, s_i\, X_{j-1}\, r_m) \;\; \forall\, i, m \in \{0, \ldots, n-1\}, \forall\, j \in \{1, \ldots, k\},$$
$$(q_j, s_i) \hookrightarrow_b (q_i, \varepsilon) \qquad\qquad\qquad\qquad\qquad\qquad\qquad \forall\, i, j \in \{0, \ldots, n-1\},$$
$$(q_i, r_i) \hookrightarrow_b (q_i, \varepsilon) \qquad\qquad\qquad\qquad\qquad\qquad\qquad\quad \forall\, i \in \{0, \ldots, n-1\},$$
$$(q_i, X_0) \hookrightarrow_b (q_i, X_k\, \star) \qquad\qquad\qquad\qquad\qquad\qquad\quad \forall\, i \in \{0, \ldots, n-1\},$$
$$(q_i, X_0) \hookrightarrow_b (q_{i+1}, X_k\, \$) \qquad\qquad\qquad\qquad\qquad\qquad\; \forall\, i \in \{0, \ldots, n-2\},$$
$$(q_i, \star) \hookrightarrow_b (q_{i-1}, \varepsilon) \qquad\qquad\qquad\qquad\qquad\qquad\qquad\;\; \forall\, i \in \{1, \ldots, n-1\},$$
$$(q_0, \$) \hookrightarrow_b (q_{n-1}, \varepsilon)$$
$$(q_{n-1}, X_0) \hookrightarrow_b (q_{n-1}, \varepsilon)$$

Lemma 8. *Given $n \ge 1$ and $k \ge 1$, $P(n, k)$ has a single accepting actree consuming input b^N where $N \ge 2^{n^2 k}$.*

Proof. Fix values n and k and refer to the member of the family $P(n, k)$ as P. We show that P has exactly one accepting actree. To this aim, we define a witness labeled tree t inductively on the structure of the tree. Later we will prove that the induction is finite. First, we show how to construct the root and its children subtrees. This corresponds to case 1 below. Then, each non-leaf subtree is defined inductively in cases 2 to 5. Note that each non-leaf subtree of t falls into one (and only one) of the cases. In fact, all cases are disjoint, in particular 2, 4 and 5. The reverse is also true: all cases describe a non-leaf subtree that does occur in t. Finally, we show that each case describes *uniquely* how to build the next layer of children subtrees of a given non-leaf subtree.

1. $t = a(a_1(\ldots), a_2)$ where $a = (q_0, S) \hookrightarrow_b (q_0, X_k\, r_0)$ and $a_1(\ldots)$ and a_2 are of the form:

$$a_2 = (q_0, r_0) \hookrightarrow_b (q_0, \varepsilon) \qquad\qquad \text{only action popping } r_0$$
$$a_1 = (q_0, X_k) \hookrightarrow_b (q_0, X_{k-1}\, r_0\, s_0\, X_{k-1}\, r_0) \qquad \text{only way to enable } a_2.$$

Note that the initial ID (q_0, S) enables a which is the only action of P with this property. Note also that $\overset{d}{a}(\overset{d}{a_1}(\ldots), \overset{0}{a_2})$ holds, where $d > 0$.

2. Each subtree whose root is labeled $a = (q_i, X_j) \hookrightarrow_b (q_i, X_{j-1}\, r_m\, s_i\, X_{j-1}\, r_m)$ with $i, m \in \{0, \ldots, n-1\}$ and $j \in \{2, \ldots, k\}$ has the form $a(a_1(\ldots), a_2, a_3, a_1(\ldots), a_2)$ where

$$a_2 = (q_m, r_m) \hookrightarrow_b (q_m, \varepsilon) \qquad\qquad\qquad \text{only action popping } r_m$$
$$a_3 = (q_m, s_i) \hookrightarrow_b (q_i, \varepsilon) \qquad\qquad\quad \text{only action popping } s_i \text{ from } q_m$$
$$a_1 = (q_i, X_{j-1}) \hookrightarrow_b (q_i, X_{j-2}\, r_m\, s_i\, X_{j-2}\, r_m) \qquad \text{only way to enable } a_2.$$

Assume for now that t is unique. Therefore, as the 1st and 4th child of a share the same label a_1, they also root the same subtree. Thus, it holds ($d > 0$)

$$\overset{d+1}{a}(\overset{d}{a_1}(\ldots), \overset{0}{a_2}, \overset{0}{a_3}, \overset{d}{a_1}(\ldots), \overset{0}{a_2}).$$

3. Each subtree whose root is labeled $a = (q_i, X_0) \hookrightarrow_b (q_{i+1}, X_k \, \$)$ with $i \in \{0, \ldots, n-2\}$ has the form $a(a_1(\ldots), a_2)$ where

$$a_2 = (q_0, \$) \hookrightarrow_b (q_{n-1}, \varepsilon) \qquad\qquad\qquad \text{only action popping } \$$$

$$a_1 = (q_{i+1}, X_k) \hookrightarrow_b (q_{i+1}, X_{k-1} \, r_0 \, s_{i+1} \, X_{k-1} \, r_0) \quad \text{only way to enable } a_2.$$

Note that $\overset{d}{a} (\overset{d}{a_1} (\ldots), \overset{0}{a_2})$ holds, where $d > 0$.

4. Each subtree whose root is labeled $a = (q_i, X_1) \hookrightarrow_b (q_i, X_0 \, r_m \, s_i \, X_0 \, r_m)$ with $i \in \{0, \ldots, n-1\}$ and $m \in \{0, \ldots, n-2\}$ has the form

$$a(a_1(a_{11}(\ldots), a_{12}), a_2, a_3, a_1(a_{11}(\ldots), a_{12}), a_2).$$

where

$$a_2 = (q_m, r_m) \hookrightarrow_b (q_m, \varepsilon) \qquad\qquad\qquad\qquad \text{only action popping } r_m$$

$$a_3 = (q_m, s_i) \hookrightarrow_b (q_i, \varepsilon) \qquad\qquad \text{only action popping } s_i \text{ from } q_m$$

$$a_1 = (q_i, X_0) \hookrightarrow_b (q_i, X_k \star) \qquad\qquad\qquad\qquad \text{assume it for now}$$

$$a_{12} = (q_{m+1}, \star) \hookrightarrow_b (q_m, \varepsilon) \qquad\qquad\qquad \text{only way to enable } a_2$$

$$a_{11} = (q_i, X_k) \hookrightarrow_b (q_i, X_{k-1} \, r_{m+1} \, s_i \, X_{k-1} \, r_{m+1}) \qquad \text{only way to enable } a_12.$$

Assume a_1 is given by the action $(q_i, X_0) \hookrightarrow_b (q_{i+1}, X_k \, \$)$ instead. Then following the action popping $\$$, we would end up in the state q_{n-1}, not enabling a_2 since $m < n - 1$.

Again, assume for now that t is unique. Hence, as the 1st and 4th child of a are both labeled by a_1, they root the same subtree. Thus, it holds $(d > 0)$

$$\overset{d+1}{a} (\overset{d}{a_1} (\overset{d}{a_{11}} (\ldots), \overset{0}{a_{12}}), \overset{0}{a_2}, \overset{0}{a_3}, \overset{d}{a_1} (\overset{d}{a_{11}} (\ldots), \overset{0}{a_{12}}), \overset{0}{a_2}).$$

5. Each subtree whose root is labeled $a = (q_i, X_1) \hookrightarrow_b (q_i, X_0 \, r_{n-1} \, s_i \, X_0 \, r_{n-1})$ with $i \in \{0, \ldots, n-1\}$ has the form $a(a_1(\ldots), a_2, a_3, a_1(\ldots), a_2)$ where

$$a_2 = (q_{n-1}, r_{n-1}) \hookrightarrow_b (q_{n-1}, \varepsilon) \qquad\qquad\qquad \text{only action popping } r_{n-1}$$

$$a_3 = (q_{n-1}, s_i) \hookrightarrow_b (q_i, \varepsilon) \qquad\qquad \text{only action popping } s_i \text{ from } q_{n-1}$$

$$a_1 = \begin{cases} (q_i, X_0) \hookrightarrow_b (q_{i+1}, X_k \, \$) & \text{if } i < n - 1 \\ (q_{n-1}, X_0) \hookrightarrow_b (q_{n-1}, \varepsilon) & \text{otherwise} \end{cases} \qquad \text{Assume it for now.}$$

For both cases ($i < n - 1$ and $i = n - 1$), assume a_1 is given by $(q_i, X_0) \hookrightarrow_b (q_i, X_k \star)$ instead. Then, the action popping $\star$ must end up in the state q_{n-1} in order to enable a_2, i.e., it must be of the form $(q_n, \star) \hookrightarrow_b (q_{n-1}, \varepsilon)$. Hence the action popping X_k must be of the form $(q_i, X_k) \hookrightarrow_b (q_i, X_{k-1} \, r_m \, s_i \, X_{k-1} \, r_m)$ where necessarily $m = n$, a contradiction (the stack symbol r_n is not defined in P).

Assume for now that t is unique. Then, as the 1st and 4th child of a are labeled by a_1, they root the same subtree (possibly a leaf). Thus, it holds $(d \geq 0)$

$$\overset{d+1}{a} (\overset{d}{a_1} (\ldots), \overset{0}{a_2}, \overset{0}{a_3}, \overset{d}{a_1} (\ldots), \overset{0}{a_2}).$$

We now prove that t is finite by contradiction. Suppose t is an infinite tree. König's Lemma shows that t has thus at least one infinite path, say p, from the root. As the set of labels of t is finite then some label must repeat infinitely often along p. Let us define a strict partial order between the labels of the non-leaf subtrees of t. We restrict to the non-leaf subtrees because no infinite path contains a leaf subtree. Let $a_1(\ldots)$ and $a_2(\ldots)$ be two non-leaf subtrees of t. Let q_{i_1} be the source state of a_1 and q_{f_1} be the target state of the last action in the sequence $\overline{a_1(\ldots)}$. Define q_{i_2}, q_{f_2} similarly for $a_2(\ldots)$. Let X_{j_1} be the symbol that a_1 pops and X_{j_2} be the symbol that a_2 pops. Define $a_1 \prec a_2$ iff (a) either $i_1 < i_2$, (b) or $i_1 = i_2$ and $f_1 < f_2$, (c) or $i_1 = i_2, f_1 = f_2$ and $j_1 > j_2$. First, note that the label a of the root of t (case 1) only occurs in the root as there is no action of P pushing S. Second, relying on cases 2 to 5, we observe that every pair of non-leaf subtrees $a_1(\ldots)$ and $a_2(\ldots)$ (excluding the root) such that $a_1(\ldots)$ is the parent node of $a_2(\ldots)$ verifies $a_1(\ldots) \prec a_2(\ldots)$. Using the transitive property of the strict partial order $\prec$, we conclude that everypair of subtrees $a_1(\ldots)$ and $a_2(\ldots)$ in p such that $a_1(\ldots a_2(\ldots) \ldots)$ verifies $a_1(\ldots) \prec a_2(\ldots)$. Therefore, no repeated variable can occur in p (contradiction). We conclude that t is finite.

The reader can observe that $t = a(\ldots)$ verifies all conditions of the definition of actree (Definition 1) and the initial ID enables a, thus it is an accepting actree of P. Since we also showed that no other tree can be defined using the actions of P, t is unique.

Finally, we give a lower bound on the length of the word consumed by t. To this aim, we prove that $d(t) = n^2 k$. Then since all actions consume input symbol b, Lemma 5 shows that the word b^N consumed is such that $N \geq 2^{n^2 k}$.

Note that, if a subtree of t verifies case 1 or 3, its dimension remains the same w.r.t. its children subtrees. Otherwise, the dimension always grows. Recall that all cases from 1 to 5 describe a set of labels that does occur in t. Also, as t is unique, no path from the root to a leaf repeats a label. Thus, to compute the dimension of t is enough to count the number of distinct labels of t that are included in cases 2, 4 and 5, which is equivalent to compute the size of the set

$$D = \{(q_i, X_j) \hookrightarrow (q_i, X_{j-1} r_m s_i X_{j-1} r_m) \mid 1 \leq j \leq k, 0 \leq i, m \leq n-1\}.$$

Clearly $|D| = n^2 k$ from which we conclude that $d(t) = n^2 k$. Hence, $|t| \geq 2^{n^2 k}$ and therefore t consumes a word b^N where $N \geq 2^{n^2 k}$ since each action of t consumes a b. □

Theorem 9. *For each $n \geq 1$ and $p > 2n + 4$, there is a PDA with n states and p stack symbols for which every Parikh-equivalent CFG has $\Omega(n^2(p - 2n - 4))$ variables.*

Proof. Consider the family of PDAs $P(n, k)$ with $n \geq 1$ and $k \geq 1$ described in Definition 7. Fix n and k and refer to the corresponding member of the family as P.

First, Lemma 8 shows that $L(P)$ consists of a single word b^N with $N \geq 2^{n^2 k}$. It follows that a language L is Parikh-equivalent to $L(P)$ iff L is language-equivalent to $L(P)$.

Let G be a CFG such that $L(G) = L(P)$. The smallest CFG that generates exactly one word of length ℓ has size $\Omega(log(\ell))$ [2, Lemma 1], where the size of a grammar is the sum of the length of all the rules. It follows that G is of size $\Omega(log(2^{n^2 k})) = \Omega(n^2 k)$. As $k = p - 2n - 4$, then G has size $\Omega(n^2(p - 2n - 4))$. We conclude that G has $\Omega(n^2 (p - 2n - 4))$ variables. $\qquad\square$

Remark 10. According to the classical conversion algorithm, every CFG that is equivalent to $P(n, k)$ needs at most $n^2(k+2n+4)+1 \in \mathcal{O}(n^2 k+n^3)$ variables. On the other hand, Theorem 9 shows that a lower bound for the number of variables is $\Omega(n^2 k)$. We observe that, as long as $n \leq Ck$ for some positive constant C, the family $P(n, k)$ shows that the conversion algorithm is optimal[7] in the number of variables when assuming both language and Parikh equivalence. Otherwise, the algorithm is not optimal as there exists a gap between the lower bound and the upper bound. For instance, if $n = k^2$ then the upper bound is $\mathcal{O}(k^5+k^6) = \mathcal{O}(k^6)$ while the lower bound is $\Omega(k^5)$.

4.2   The Case of Unary Deterministic Pushdown Automata

We have seen that the classical translation from PDA to CFG is optimal in the number of grammar variables for the family of unary nondeterministic PDAs $P(n, k)$ when n is in linear relation with respect to k (see Remark 10). However, for unary *deterministic* PDAs (UDPDAs for short) the situation is different. Pighizzini [10] shows that for every UDPDA with n states and p stack symbols, there exists an equivalent CFG with at most $2np$ variables. Although he gives a definition of such a grammar, we were not able to extract an algorithm from it. On the other hand, Chistikov and Majumdar [3] give a polynomial time algorithm that transforms a UDPDA into an equivalent CFG going through the construction of a pair of straight-line programs. The size of the resulting CFG is linear in that of the UDPDA.

We propose a new polynomial time algorithm that converts a UDPDA with n states and p stack symbols into an equivalent CFG with $\mathcal{O}(np)$ variables. Our algorithm is based on the observation that the conversion algorithm from PDAs to CFGs need not consider all the triples in (1) (see Sect. 2). We discard unnecessary triples using the *saturation procedure* [1,6] that computes the set of reachable IDs.

For a given PDA P with $q \in Q$ and $X \in \Gamma$, define the set of *reachable IDs* $R_P(q, X)$ as follows:

$$R_P(q, X) = \{(q', \beta) \mid \exists(q, X) \vdash \cdots \vdash (q', \beta)\}.$$

Lemma 11. *If P is a UDPDA then the set $\{I \in R_P(q, X) \mid stack(I) = \varepsilon\}$ has at most one element for every state q and stack symbol X.*

---

[7] Note that if $n \leq Ck$ for some $C > 0$ then the n^3 addend in $\mathcal{O}(n^2 k + n^3)$ becomes negligible compared to $n^2 k$, and the lower and upper bound coincide.

Proof. Let P be a UDPDA with $\Sigma = \{a\}$. Since P is *deterministic* we have that (i) for every $q \in Q, X \in \Gamma$ and $b \in \Sigma \cup \{\varepsilon\}$, $|\delta(q,b,X)| \leq 1$ and, (ii) for every $q \in Q$ and $X \in \Gamma$, if $\delta(q,\varepsilon,X) \neq \emptyset$ then $\delta(q,b,X) = \emptyset$ for every $b \in \Sigma$.

The proof goes by contradiction. Assume that for some state q and stack symbol X, there are two IDs I_1 and I_2 in $R_P(q,X)$ such that $stack(I_1) = stack(I_2) = \varepsilon$ and $state(I_1) \neq state(I_2)$.

Necessarily, there exist three IDs J, J_1 and J_2 with $J_1 \neq J_2$ such that the following holds:

$$(q,X) \vdash \cdots \vdash J \vdash_a J_1 \vdash \cdots \vdash I_1$$
$$(q,X) \vdash \cdots \vdash J \vdash_b J_2 \vdash \cdots \vdash I_2.$$

It is routine to check that if $a = b$ then P is not deterministic, a contradiction. Next, we consider the case $a \neq b$. When a and b are symbols, because P is a unary DPDA, then they are the same, a contradiction. Else if either a or b is ε then P is not deterministic, a contradiction. We conclude from the previous that when $stack(I_1) = stack(I_2) = \varepsilon$, then necessarily $state(I_1) = state(I_2)$ and therefore that the set $\{I \in R_P(q,X) \mid stack(I) = \varepsilon\}$ has at most one element. $\square$

Intuitively, Lemma 11 shows that, when fixing q and X, there is at most one q' such that the triple $[qXq']$ generates a string of terminals. We use this fact to prove the following theorem.

Theorem 12. *For every UDPDA with n states and p stack symbols, there is a polynomial time algorithm that computes an equivalent CFG with at most np variables.*

Proof. The conversion algorithm translating a PDA P to a CFG G computes the set of grammar variables $\{[qXq'] \mid q,q' \in Q, X \in \Gamma\}$. By Lemma 11, for each q and X there is at most one variable $[qXq']$ in the previous set generating a string of terminals. The consequence of the lemma is twofold: (i) For the triples it suffices to compute the subset T of the aforementioned generating variables. Clearly, $|T| \leq np$. (ii) Each action of P now yields a single rule in G. This is because in (2) (see Sect. 2) there is at most one choice for r_1 to r_d, hence we avoid the exponential blowup of the runtime in the conversion algorithm. To compute T given P, we use the polynomial time saturation procedure [1,6] which given (q,X) computes a FSA for the set $R_P(q,X)$. Then we compute from this set the unique state q' (if any) such that $(q',\varepsilon) \in R_P(q,X)$, hence T. From the above we find that, given P, we compute G in polynomial time. $\square$

Up to this point, we have assumed the empty stack as the acceptance condition. For general PDAs, assuming final states or empty stack as acceptance condition induces no loss of generality. The situation is different for deterministic PDAs where accepting by final states is more general than empty stack. For this reason, we contemplate the case where the UDPDA accepts by final states. Theorem 13 shows how our previous construction can be modified to accommodate the acceptance condition by final states.

Theorem 13. *For every UDPDA with n states and p stack symbols that accepts by final states, there is a polynomial time algorithm that computes an equivalent CFG with $\mathcal{O}(np)$ variables.*

Proof. Let P be a UDPDA with n states and p stack symbols that accepts by final states. We first translate $P = (Q, \Sigma, \Gamma, \delta, q_0, Z_0, F)^8$ into a (possibly nondeterministic) unary pushdown automaton $P' = (Q', \Sigma, \Gamma', \delta', q_0', Z_0')$ with an empty stack acceptance condition. In particular, $Q' = Q \cup \{q_0', sink\}$; $\Gamma' = \Gamma \cup Z_0'$; and δ' is given by

$$\delta \cup \{(q_0', Z_0') \hookrightarrow_\varepsilon (q_0, Z_0 Z_0')\}$$
$$\cup \{(q, X) \hookrightarrow_\varepsilon (sink, X) \mid X \in \Gamma', q \in F\}$$
$$\cup \{(sink, X) \hookrightarrow_\varepsilon (sink, \varepsilon) \mid X \in \Gamma'\}.$$

The new stack symbol Z_0' is to prevent P' from incorrectly accepting when P is in a nonfinal state with an empty stack. The state *sink* is to empty the stack upon P entering a final state. Observe that P' need not be deterministic. Also, it is routine to check that $L(P') = L(P)$ and P' is computable in time linear in the size of P. Now let us turn to $R_{P'}(q, X)$. For P' a weaker version of Lemma 11 holds: the set $H = \{I \in R_{P'}(q, X) \mid stack(I) = \varepsilon\}$ has at most two elements for every state $q \in Q'$ and stack symbols $X \in \Gamma'$. This is because if H contains two IDs then necessarily one of them has *sink* for state.

Based on this result, we construct T as in Theorem 12, but this time we have that $|T|$ is $\mathcal{O}(np)$.

Now we turn to the set of production rules as defined in (2) (see Sect. 2). We show that each action $(q, X) \hookrightarrow_b (q', \beta)$ of P' yields at most d production rules in G where $d = |\beta|$. For each state r_i in (2) we have two choices, one of which is *sink*. We also know that once a move sequence enters *sink* it cannot leave it. Therefore, we have that if $r_i = sink$ then $r_{i+1} = \cdots = r_d = sink$. Given an action, it thus yields d production rules one where $r_1 = \cdots = r_d = sink$, another where $r_2 = \cdots = r_d = sink$, ..., etc. Hence, we avoid the exponential blowup of the runtime in the conversion algorithm.

The remainder of the proof follows that of Theorem 12.   □

5   Parikh-Equivalent Finite State Automata

Parikh's theorem [11] shows that every context-free language is Parikh-equivalent to a regular language. Using this result, we can compare PDAs against FSAs under Parikh equivalence. We start by deriving some lower bound using the family $P(n, k)$. Because its alphabet is unary and it accepts a single long word, the comparison becomes straightforward.

Theorem 14. *For each $n \geq 1$ and $p > 2n+4$, there is a PDA with n states and p stack symbols for which every Parikh-equivalent FSA has at least $2^{n^2(p-2n-4)} + 1$ states.*

---

[8] The set of final states is given by $F \subseteq Q$.

Proof. Consider the family of PDAs $P(n,k)$ with $n \geq 1$ and $k \geq 1$ described in Definition 7. Fix n and k and refer to the corresponding member of the family as P. By Lemma 8, $L(P) = \{b^N\}$ with $N \geq 2^{n^2 k}$. Then, the smallest FSA that is Parikh-equivalent to $L(P)$ needs $N + 1$ states. As $k = p - 2n - 4$, we conclude that the smallest Parikh-equivalent FSA has at least $2^{n^2(p-2n-4)} + 1$ states.   □

Let us now turn to upper bounds. We give a 2-step procedure computing, given a PDA, a Parikh-equivalent FSA. The steps are: (*i*) translate the PDA into a language-equivalent context-free grammar [9]; and (*ii*) translate the context-free grammar into a Parikh-equivalent finite state automaton [4]. Let us introduce the following definition. A grammar is in *2-1 normal form* (2-1-NF for short) if each rule $(X, \alpha) \in R$ is such that α consists of at most one terminal and at most two variables. It is worth pointing that, when the grammar is in 2-1-NF, the resulting Parikh-equivalent FSA from step (*ii*) has $\mathcal{O}(4^n)$ states where n is the number of grammar variables [4]. For the sake of simplicity, we will assume that grammars are in 2-1-NF which holds when PDAs are in *reduced form*: every move is of the form $(q, X) \hookrightarrow_b (q', \beta)$ with $|\beta| \leq 2$ and $b \in \Sigma \cup \{\varepsilon\}$.

Theorem 15. *Given a PDA in reduced form with $n \geq 1$ states and $p \geq 1$ stack symbols, there is a Parikh-equivalent FSA with $\mathcal{O}(4^{n^2 p})$ states.*

Proof. The algorithm to convert a PDA with $n \geq 1$ states and $p \geq 1$ stack symbols into a CFG that generates the same language [9] uses at most $n^2 p + 1$ variables if $n > 1$ (or p if $n = 1$). Given a CFG of n variables in 2-1-NF, one can construct a Parikh-equivalent FSA with $\mathcal{O}(4^n)$ states [4].

Given a PDA P with $n \geq 1$ states and $p \geq 1$ stack symbols the conversion algorithm returns a language-equivalent CFG G. Note that if P is in reduced form, then the conversion algorithm returns a CFG in 2-1-NF. Then, apply to G the known construction that builds a Parikh-equivalent FSA [4]. The resulting FSA has $\mathcal{O}(4^{n^2 p})$ states.   □

Remark 16. Theorem 14 shows that a every FSA that is Parikh-equivalent to $P(n,k)$ needs $\Omega(2^{n^2 k})$ states. On the other hand, Theorem 15 shows that the number of states of every Parikh-equivalent FSA is $\mathcal{O}(4^{n^2(k+2n+4)})$. Thus, our construction is *close to optimal*[9] when n is in linear relation with respect to k.

We conclude by discussing the reduced form assumption. Its role is to simplify the exposition and, indeed, it is not needed to prove correctness of the 2-step procedure. The assumption can be relaxed and bounds can be inferred. They will contain an additional parameter related to the length of the longest sequence of symbols pushed on the stack.

Acknowledgement. We thank Pedro Valero for pointing out the reference on smallest grammar problems [2]. We also thank the anonymous referees for their insightful comments and suggestions.

---

[9] As the blow up of our construction is $\mathcal{O}(4^{n^2(k+2n+4)})$ for a lower bound of $2^{n^2 k}$, we say that it is *close to optimal* in the sense that $2n^2(k + 2n + 4) \in \Theta(n^2 k)$, which holds when n is in linear relation with respect to k (see Remark 10).

References

1. Bouajjani, A., Esparza, J., Maler, O.: Reachability analysis of pushdown automata: Application to model-checking. In: Mazurkiewicz, A., Winkowski, J. (eds.) CONCUR 1997. LNCS, vol. 1243, pp. 135–150. Springer, Heidelberg (1997). doi:10.1007/3-540-63141-0_10
2. Charikar, M., Lehman, E., Liu, D., Panigrahy, R., Prabhakaran, M., Sahai, A., Shelat, A.: The smallest grammar problem. IEEE Trans. Inf. Theory **51**(7), 2554–2576 (2005)
3. Chistikov, D., Majumdar, R.: Unary pushdown automata and straight-line programs. In: Esparza, J., Fraigniaud, P., Husfeldt, T., Koutsoupias, E. (eds.) ICALP 2014. LNCS, vol. 8573, pp. 146–157. Springer, Heidelberg (2014). doi:10.1007/978-3-662-43951-7_13
4. Esparza, J., Ganty, P., Kiefer, S., Luttenberger, M.: Parikh's theorem: a simple and direct automaton construction. IPL **111**(12), 614–619 (2011)
5. Esparza, J., Luttenberger, M., Schlund, M.: A brief history of strahler numbers. In: Dediu, A.-H., Martín-Vide, C., Sierra-Rodríguez, J.-L., Truthe, B. (eds.) LATA 2014. LNCS, vol. 8370, pp. 1–13. Springer, Cham (2014). doi:10.1007/978-3-319-04921-2_1
6. Finkel, A., Willems, B., Wolper, P.: A direct symbolic approach to model checking pushdown systems (extended abstract). Electron. Notes Theoret. Comput. Sci. **9**, 27–37 (1997)
7. Ganty, P., Gutiérrez, E.: Parikh image of pushdown automata (long version) (2017). Pre-print arXiv arXiv: 1706.08315
8. Goldstine, J., Price, J.K., Wotschke, D.: A pushdown automaton or a context-free grammar: which is more economical? Theoret. Comput. Sci. **18**, 33–40 (1982)
9. Hopcroft, J.E., Motwani, R., Ullman, J.D.: Introduction to Automata Theory, Languages, and Computation, 3rd edn. Addison-Wesley Longman Publishing Co., Inc., Boston (2006)
10. Pighizzini, G.: Deterministic pushdown automata and unary languages. Int. J. Found. Comput. Sci. **20**(04), 629–645 (2009)
11. Rohit, J.P.: On context-free languages. J. ACM **13**(4), 570–581 (1966)

Tropical Combinatorial Nullstellensatz and Fewnomials Testing

Dima Grigoriev[1] and Vladimir V. Podolskii[2,3(✉)]

[1] CNRS, Mathématiques, Université de Lille, 59655 Villeneuve d'Ascq, France
Dmitry.Grigoryev@math.univ-lille1.fr
[2] Steklov Mathematical Institute, Moscow, Russia
podolskii@mi.ras.ru
[3] National Research University Higher School of Economics, Moscow, Russia

Abstract. Tropical algebra emerges in many fields of mathematics such as algebraic geometry, mathematical physics and combinatorial optimization. In part, its importance is related to the fact that it makes various parameters of mathematical objects computationally accessible. Tropical polynomials play an important role in this, especially for the case of algebraic geometry. On the other hand, many algebraic questions behind tropical polynomials remain open. In this paper we address three basic questions on tropical polynomials closely related to their computational properties:

1. Given a polynomial with a certain support (set of monomials) and a (finite) set of inputs, when is it possible for the polynomial to vanish on all these inputs?
2. A more precise question, given a polynomial with a certain support and a (finite) set of inputs, how many roots can polynomial have on this set of inputs?
3. Given an integer k, for which s there is a set of s inputs such that any non-zero polynomial with at most k monomials has a non-root among these inputs?

In the classical algebra well-known results in the direction of these questions are Combinatorial Nullstellensatz, Schwartz-Zippel Lemma and Universal Testing Set for sparse polynomials respectively. In this paper we extensively study these three questions for tropical polynomials and provide results analogous to the classical results mentioned above.

1 Introduction

A *max-plus* or a *tropical semiring* is defined by a set $\mathbb{K}$, which can be $\mathbb{R}$ or $\mathbb{Q}$ endowed with two operations, *tropical addition* $\oplus$ and *tropical multiplication* $\odot$, defined in the following way:

$$x \oplus y = \max(x, y), \quad x \odot y = x + y.$$

The first author is grateful to the grant RSF 16-11-10075 and to MCCME for inspiring atmosphere. The work of the second author is partially supported by the grant of the President of Russian Federation (MK-7312.2016.1) and by the Russian Academic Excellence Project '5–100'.

R. Klasing and M. Zeitoun (Eds.): FCT 2017, LNCS 10472, pp. 284–297, 2017.
DOI: 10.1007/978-3-662-55751-8_23

Tropical polynomials are a natural analog of classical polynomials. In classical terms a tropical polynomial is an expression of the form $f(\vec{x}) = \max_i M_i(\vec{x})$, where each $M_i(\vec{x})$ is a linear polynomial (a tropical monomial) in variables $\vec{x} = (x_1, \ldots, x_n)$, and all the coefficients of all M_i's are nonnegative integers except for a free coefficient which can be any element of $\mathbb{K}$ (free coefficient corresponds to a coefficient of the tropical monomial and other coefficients correspond to the powers of variables in the tropical monomial).

The degree of a tropical monomial M is the sum of its coefficients (except the free coefficient) and the degree of a tropical polynomial f denoted by $\deg(f)$ is the maximal degree of its monomials. A point $\vec{a} \in \mathbb{K}^n$ is a root of the polynomial f if the maximum $\max_i\{M_i(\vec{a})\}$ is attained on at least two different monomials M_i. We defer more detailed definitions on the basics of max-plus algebra to Preliminaries.

Tropical polynomials have appeared in various areas of mathematics and found many applications (see, for example, [15,16,20,21,23,28,33]). An early source of the tropical approach was the Newton's method for solving algebraic equations in Newton-Puiseux series [28]. An important advantage of tropical algebra is that it makes some properties of classical mathematical objects computationally accessible [16,20,28,29]: on one hand tropical analogs reflect certain properties of classical objects and on the other hand tropical objects have much more simple and discrete structure and thus are more accessible to algorithms. One of the main goals of max-plus mathematics is to build a theory of tropical polynomials which would help to work with them and would possibly lead to new results in related areas. Computational reasons, on the other hand, make it important to keep the theory maximally computationally efficient.

The case studied best so far is the one of tropical linear polynomials and systems of tropical linear polynomials. For them an analog of a large part of the classical theory of linear polynomials was established. This includes studies of tropical analogs of the rank of a matrix and the independence of vectors [1,9,17], an analog of the determinant of a matrix and its properties [1,9,10], an analog of Gauss triangular form [10]. Also the solvability problem for tropical linear systems was studied from the complexity point of view. Interestingly, this problem turns out to be polynomially equivalent to the mean payoff games problem [2,12] which received considerable attention in computational complexity theory.

For tropical polynomials of arbitrary degree less is known. In [26] the radical of a tropical ideal was explicitly described. In [23,27] a tropical version of the Bezout theorem was proved for tropical polynomial systems for the case when the number of polynomials in the system is equal to the number of variables. In [8] the Bezout bound was extended to systems with an arbitrary number of polynomials. In [13] the tropical analog of Hilbert's Nullstellensatz was established. In [5] a bound on the number of nondegenerate roots of a system of sparse tropical polynomials was given. In [29] it was shown that the solvability problem for tropical polynomial systems is NP-complete.

Our Results. In this paper we address several basic questions for tropical polynomials.

The first question we address is given a set S of points in $\mathbb{R}^n$ and a set of monomials of n variables, is there a tropical polynomial with these monomials that has roots in all points of the set. In the classical case a famous result in this direction with numerous applications in Theoretical Computer Science is Combinatorial Nullstellensatz [3]. Very roughly, it states that the set of monomials of the polynomial might be substantially larger than the set of the points and the polynomial will still be non-zero on at least one of the points. In the tropical case we show that this is not the case: if the number of monomials is larger than the number of points, there is always a polynomial with roots in all points. We establish the general criteria for existence of a polynomial on a given set of monomials with roots in all points of a given set. From this criteria we deduce that if the number of points is equal to the number of monomials and the points and monomials are structured in the same way, then there is no polynomial with roots in all points. We note that the last statement for the classical case is an open question [24].

The second question is given a finite set $S \subseteq \mathbb{R}$ how many roots can a tropical polynomial of n variables and degree d have in the set S^n? In the classical case the well-known Schwartz-Zippel lemma [25,34] states that the maximal number of roots is $d|S|^{n-1}$. We show that in the tropical case the maximal possible number of roots is $|S|^n - (|S| - d)^n$. We note that this result can be viewed as a generalization and improvement of isolation lemma [7,19,22]. In particular, we prove a more precise version of a technical result in [19, Lemma 4].

The third question is related to a universal testing set for tropical polynomials of n variables with at most k monomials. A universal testing set is a set of points $S \subseteq \mathbb{K}^n$ such that any nontrivial polynomial with at most k monomials has a non-root in one of the points of S. The problem is to find a minimal size of a universal testing set for given n and k. In the classical case this problem is tightly related to the problem of interpolating a polynomial with a certain number of monomials (with a priori unknown support) given its values on some universal set of inputs. The classical problem was studied in [4,11,14,18] and the minimal size of the universal testing set for the classical case turns out to be equal to k (while for the interpolation problem the size is $2k$). In the tropical case it turns out that the answer depends on which tropical semiring $\mathbb{K}$ is considered: for $\mathbb{K} = \mathbb{R}$ we show that as in the classical case the minimal size of the universal testing set is equal to k. For $\mathbb{K} = \mathbb{Q}$ it turns out that the minimal size of the universal testing set is substantially larger. We show that its size is $\Theta(kn)$ (the constants in Θ do not depend on k and n). For $n = 2$ we find the precise size of the minimal universal testing set $s = 2k - 1$. For greater n the precise minimal size of the universal testing set still remains unclear. Finally, we establish an interesting connection of this problem to the following problem in Discrete Geometry: what is the minimal number of disjoint convex polytopes in n-dimensional space that is enough to cover any set of s points in such a way that all s points are on the boundary of the polytopes.

The rest of the paper is organized as follows. In Sect. 2 we introduce necessary definitions and notations. In Sect. 3 we give the results on the tropical analog of Combinatorial Nullstellensatz. In Sect. 4 we prove the tropical analog of Schwartz-Zippel Lemma. In Sect. 5 we give the results on tropical universal sets. Due to the space constraints many of the proofs are omitted.

2   Preliminaries

A *max-plus* or a *tropical semiring* is defined by a set $\mathbb{K}$, which is either $\mathbb{R}$ or $\mathbb{Q}$ endowed with two operations, *tropical addition* $\oplus$ and *tropical multiplication* $\odot$, defined in the following way:

$$x \oplus y = \max\{x, y\}, \quad x \odot y = x + y.$$

A tropical (or max-plus) monomial in variables $\vec{x} = (x_1, \ldots, x_n)$ is defined as

$$m(\vec{x}) = c \odot x_1^{\odot i_1} \odot \ldots \odot x_n^{\odot i_n}, \tag{1}$$

where c is an element of the semiring $\mathbb{K}$ and $i_1, \ldots, i_n$ are nonnegative integers. In the usual notation the monomial is the linear function

$$m(\vec{x}) = c + i_1 x_1 + \ldots + i_n x_n.$$

For $\vec{x} = (x_1, \ldots, x_n)$ and $I = (i_1, \ldots, i_n)$ we introduce the notation

$$\vec{x}^I = x_1^{\odot i_1} \odot \ldots \odot x_n^{\odot i_n} = i_1 x_1 + \ldots + i_n x_n.$$

The degree of the monomial m is defined as the sum $i_1 + \ldots + i_n$. We denote this sum by $|I|$.

A *tropical polynomial* is the tropical sum of tropical monomials

$$p(\vec{x}) = \bigoplus_i m_i(\vec{x})$$

or in the usual notation $p(\vec{x}) = \max_i m_i(\vec{x})$. The *degree* of the tropical polynomial p denoted by $\deg(p)$ is the maximal degree of its monomials. A point $\vec{a} \in \mathbb{K}^n$ is a *root* of the polynomial p if the maximum $\max_i\{m_i(\vec{a})\}$ is attained on at least two distinct monomials among m_i (see e.g. [23] for the motivation of this definition). A polynomial p *vanishes* on the set $S \subseteq \mathbb{K}^n$ if all points of S are roots of p.

Geometrically, a tropical polynomial $p(\vec{x})$ is a convex piece-wise linear function and the roots of p are non-smoothness points of this function.

By the *product* of two tropical polynomials $p(\vec{x}) = \bigoplus_i m_i(\vec{x})$ and $q(\vec{x}) = \bigoplus_j m'_j(\vec{x})$ we naturally call a tropical polynomial $p \odot q$ that has as monomials tropical products $m_i(\vec{x}) \odot m'_j(\vec{x})$ for all i, j. We will make use of the following simple observation.

Lemma 1. *A point $\vec{a} \in \mathbb{K}^n$ is a root of $p \odot q$ iff it is a root of either $p(\vec{x})$ or $q(\vec{x})$.*

For two vectors $\vec{a}, \vec{b} \in \mathbb{R}^n$ throughout the paper we will denote by $\langle \vec{a}, \vec{b} \rangle$ their inner product.

3 Tropical Combinatorial Nullstellensatz

For a polynomial p denote by $\mathsf{Supp}(p)$ the set of all $J = (j_1, \ldots, j_n)$ such that p has a monomial $\vec{x}^J$ with some coefficient.

Consider two finite sets $S, R \subseteq \mathbb{R}^n$ such that $|S| = |R|$. We call S and R *non-singular* if there is a bijection $f \colon S \to R$ such that $\sum_{x \in S} \langle \vec{x}, f(\vec{x}) \rangle$ is greater than the corresponding sum for all other bijections from S to R. Otherwise we say that R and S are *singular*. Note that the notion of singularity is symmetrical.

First we formulate a general criteria for vanishing polynomials with given support.

Theorem 1. *Consider support $S \subseteq \mathbb{N}^n$ and the set of points $R \subseteq \mathbb{K}^n$. There are three cases.*

  (i) *If $|R| < |S|$, then there is a polynomial p with support in S vanishing on R.*
 (ii) *If $|R| = |S|$, then there is a polynomial p with support in S vanishing on R iff S and R are singular.*
(iii) *If $|R| > |S|$ then there is a polynomial p with support in S vanishing on R iff for any subset $R' \subset R$ such that $|R'| = |S|$ we have that R' and S are singular.*

The proof of this theorem is by the reduction to tropical linear systems and applying known results on their solvability.

Now we will derive corollaries of this general criteria.

Suppose we have a set $S \subseteq \mathbb{N}^n$. Suppose we have a set of reals $\{\alpha_j^i\}$ for $i = 1, \ldots, n$, $j \in \mathbb{N}$ and for each i we have

$$\alpha_0^i < \alpha_1^i < \alpha_2^i < \ldots.$$

For $J = (j_1, \ldots, j_n)$ we introduce the notation $\vec{\alpha}_J = (\alpha_{j_1}^1, \ldots, \alpha_{j_n}^n)$. Consider the set $R_S = \{\vec{\alpha}_J \mid J \in S\}$.

We consider the following question. Suppose we have a polynomial p with support $\mathsf{Supp}(p) \subseteq S$. For which sets S' is it possible that p vanishes on $R_{S'}$? A natural question is the case of $S = S'$. We show the following theorem.

Theorem 2. *For any S and for any non-zero tropical polynomial f such that $\mathsf{Supp}(f) \subseteq S$ there is $\vec{r} \in R_S$ such that r is a non-root of f.*

An interesting case of this theorem is $S = \{0, 1, \ldots, k\}^n$. Then the result states that any non-zero polynomial of individual degree at most k w.r.t. each variable x_i, $i = 1, \ldots, n$, does not vanish on a lattice of size $k + 1$.

Theorems 1(i) and 2 answer some customary cases of our first question. We note that the situation here is quite different from the classical case. The classical analog of Theorem 2 for the case of $S = \prod_{i=1}^{n} \{0, 1, \ldots, k_i\}$ is a simple observation. In the tropical setting it already requires some work. On the other hand, in the classical case it is known that for such S the domain of the polynomial can be substantially larger then S and still the polynomial remains non-vanishing on R_S (see Combinatorial Nullstellensatz [3]). In tropical case, however, if we extend the domain of the polynomial even by one extra monomial, then due to Theorem 1(i) there is a vanishin non-zero polynomial.

In the proof of Theorem 2 we will use the following simple technical lemma.

Lemma 2. *Consider two sequences of reals $v_1 \leqslant v_2 \leqslant \ldots \leqslant v_l$ and $u_1 \leqslant u_2 \leqslant \ldots \leqslant u_l$. Consider any permutation $\sigma \in Sym_l$ on l element set. Then*

$$\sum_i v_i u_i \geq \sum_i v_i u_{\sigma(i)}.$$

Moreover, the inequality is strict iff there are i, j such that $v_i < v_j$, $u_{\sigma(j)} < u_{\sigma(i)}$.

Proof (Proof Sketch of Theorem 2). By Theorem 1 it is enough for us to show that S and R_S are non-singular.

Consider the bijection $f \colon S \to R_S$ given by $f(J) = \vec{\alpha}_J$. We claim that the maximum over all possible bijections g of the sum $\sum_{J \in S} \langle J, g(J) \rangle$ is attained on the bijection f and only on it.

Consider an arbitrary bijection $g \colon S \to R_S$. Since $R_S \subseteq \mathbb{R}^n$ it is convenient to denote $g(J) = (g_1(J), \ldots, g_n(J))$ and $f(J) = (f_1(J), \ldots, f_n(J))$. Consider the sum

$$\sum_{J \in S} \langle J, g(J) \rangle = \sum_{J \in S} \sum_{i=1}^{n} j_i g_i(J) = \sum_{i=1}^{n} \sum_{J \in S} j_i g_i(J).$$

Applying Lemma 2 it can be shown that for each i

$$\sum_{J \in S} j_i g_i(J) \leqslant \sum_{J \in S} j_i f_i(J) \tag{2}$$

and for at least one i

$$\sum_{J \in S} j_i g_i(J) < \sum_{J \in S} j_i f_i(J). \tag{3}$$

It is not hard to deduce the theorem from these inequalities. $\square$

4   Tropical Analog of Schwartz-Zippel Lemma

Using the results of the previous section we can prove an analog of Schwartz-Zippel Lemma for tropical polynomials.

Theorem 3. *Let $S_1, S_2, \ldots, S_n \subseteq \mathbb{K}$, denote $|S_i| = k_i$. Then for any $d \leqslant \min_i k_i$ the maximal number of roots a non-vanishing tropical polynomial p of degree d can have in $S_1 \times \ldots \times S_n$ is equal to*

$$\prod_{i=1}^{n} k_i - \prod_{i=1}^{n} (k_i - d).$$

Exactly the same statement is true for the polynomials with individual degree of each variable at most d.

In particular, we have the following corollary.

Corollary 1. *Let $S \subseteq \mathbb{K}$ be a set of size k. Then for any $d \leqslant k$ the maximal number of roots a non-vanishing tropical polynomial p of degree d can have in S^n is equal to*

$$k^n - (k - d)^n.$$

Exactly the same statement is true for the polynomials with individual degree of each variable at most d.

Proof (Proof of Theorem 3).
The upper bound is achieved on the product of d degree-1 polynomials. Indeed, denote $S_i = \{s_{i,1}, s_{i,2}, \ldots, s_{i,k_i}\}$, where $s_{i,1} > s_{i,2} > \ldots > s_{i,k_i}$. For $j = 1, \ldots, d$ denote by p_j the following degree-1 polynomial:

$$p_j(\vec{x}) = (-s_{1,j} \odot x_1) \oplus \ldots \oplus (-s_{i,j} \odot x_i) \oplus \ldots \oplus (-s_{n,j} \odot x_n) \oplus 0.$$

Observe that $\vec{a} \in S_1 \times \ldots \times S_n$ is a root of p_j if for some i $a_i = s_{i,j}$ and for the rest of i we have $a_i \leqslant s_{i,j}$.

Consider a degree d polynomial $p(\vec{x}) = \bigodot_{j=1}^{d} p_j(\vec{x})$. Then from Lemma 1 we have that $\vec{a} \in S_1 \times \ldots \times S_n$ is a non-root of p iff for all i $a_i < s_{i,d}$. Thus the number of non-roots of p is $\prod_{i=1}^{n} (|S_i| - d)$. This proves the upper bound.

For the lower bound, suppose there is a polynomial p with individual degrees d that has more than $\prod_{i=1}^{n} k_i - \prod_{i=1}^{n} (k_i - d)$ roots. Then the number of its non-roots is at most $\prod_{i=1}^{n} (k_i - d) - 1$. Denote the set of all non-roots by R.

Consider a family of all the polynomials of individual degree at most $k_i - d - 1$ in variable x_i for all i. Then their support is of size $\prod_{i=1}^{n} (k_i - d)$. Since the size of the support is greater than R, by Theorem 1 there is a polynomial q with this support that vanishes on R.

Then, by Lemma 1 the non-zero polynomial $p \odot q$ vanishes on $S_1 \times \ldots \times S_n$ and on the other hand has support $\{0, \ldots, k_1 - 1\} \times \ldots \times \{0, \ldots, k_n - 1\}$. This contradicts Theorem 2. Thus there is no such polynomial p and the theorem follows.

5   Tropical Universal Testing Set

In this section we study the minimal size of the universal testing set for sparse tropical polynomials. It turns out that in the tropical case there is a big difference between testing sets over $\mathbb{R}$ and $\mathbb{Q}$. We thus consider these two cases separately below.

Throughout this section we denote by n the number of variables in the polynomials, by k the number of monomials in them and by s the number of points in the universal testing set.

5.1 Testing Sets over $\mathbb{R}$

In this section we will show that the minimal size s of the universal testing set over $\mathbb{R}$ is equal to k.

Theorem 4. *For polynomials over $\mathbb{R}$ the minimal size s of the universal testing set for tropical polynomials with at most k monomials is equal to k.*

Proof (Proof sketch). First of all, it follows from Theorem 1(i) that for any set of s points there is a polynomial with $k = s + 1$ monomials that has roots in all s points. Thus the universal testing set has to contain at least as many points as there are monomials and we have the inequality $s \geq k$.

Next we show that $s \leqslant k$. Consider the set of s points $S = \{\vec{a}_1, \ldots, \vec{a}_s\} \in \mathbb{R}^n$ that has linearly independent over $\mathbb{Q}$ coordinates. Suppose we have a polynomial p with k monomials that has roots in all points $\vec{a}_1, \ldots, \vec{a}_s$. We will show that $k \geq s + 1$. Thus we will establish that S is a universal set for $k = s$ monomials.

Suppose the monomials of p are $m_1, \ldots, m_k$, where $m_i(\vec{x}) = c_i \odot \vec{x}^{J_i}$. Introduce the notation $p(\vec{a}_j) = \max_i(m_i(\vec{a}_j)) = p_j$. Since a_j is a root, the value p_j is achieved on at least two monomials.

Note the monomial m_i has the value p_j in the point $\vec{a}_j$ iff $\langle \vec{a}_j, J_i \rangle + c_i = p_j$.

Now, consider a bipartite undirected graph G. The vertices in the left part correspond to monomials of p (k vertices). The vertices in the right part correspond to the points in S (s vertices). We connect vertex m_i on the left side to the vertex $\vec{a}_j$ on the right side iff $m_i(\vec{a}_j) = p_j$.

Observe, that the degree of vertices on the right hand side is at least 2 (this means exactly that they are roots of p).

Now, we will show that there are no cycles in G. Indeed, suppose there is a cycle. For the sake of convenience of notation assume the sequence of the vertices of the cycle is $m_1, \vec{a}_1, m_2, \vec{a}_2, \ldots, m_l, \vec{a}_l$. Note that since the graph is bipartite, the cycle is of even length. In particular, for all $i = 1, \ldots, l$ we have $m_i(\vec{a}_i) = p_i$, that is

$$\langle \vec{a}_i, J_i \rangle + c_i = p_i. \tag{4}$$

Also for all $i = 1, \ldots, l$ we have $m_{i+1}(\vec{a}_i) = p_i$ (for convenience of notation assume here $m_{l+1} = m_1$), that is

$$\langle \vec{a}_i, J_{i+1} \rangle + c_{i+1} = p_i. \tag{5}$$

Let us sum up all equations in (4) for all $i = 1, \ldots, l$ and subtract from the result all the equations in (5). It is easy to see that all c_i's and p_i's will cancel out and thus we will have

$$\langle \vec{a}_1, J_1 \rangle - \langle \vec{a}_1, J_2 \rangle + \langle \vec{a}_2, J_2 \rangle - \langle \vec{a}_2, J_3 \rangle + \ldots + \langle \vec{a}_l, J_l \rangle - \langle \vec{a}_l, J_1 \rangle = 0.$$

Since $J_1 \neq J_2$, we have a nontrivial linear combination with integer coefficients of the coordinates of vectors $\vec{a}_1, \ldots, \vec{a}_l$. Since the coordinates of these vectors are linearly independent over $\mathbb{Q}$, this is a contradiction. Thus we have shown that there are no cycles in G.

From this it is not hard to deduce that the number of vertices in the left part of G is greater than the number of vertices in the right part, and from this the theorem follows.

5.2   Testing Sets Over $\mathbb{Q}$

The main difference of the problem over the semiring $\mathbb{Q}$ compared to the semiring $\mathbb{R}$ is that now the points of the universal set have to be rational.

Before we proceed with this section we observe that we can assume that tropical polynomials can contain rational (possibly negative) powers of variables: for each such polynomial there is another polynomial with natural exponents with the same set of roots and the same number of monomials. Indeed, suppose p is a polynomial with rational exponents. Recall that

$$p(\vec{x}) = \max(m_1(\vec{x}), \ldots, m_k(\vec{x})), \tag{6}$$

where $m_1, \ldots, m_k$ are monomials. Recall that each monomial is a linear function over $\vec{x}$. Note that if we multiply the whole expression (6) by some positive constant and add the same linear form $m(\vec{x})$ to all monomials, the resulting polynomial will have the same set of roots. Thus, we can get rid of rational degrees in p by multiplying p by large enough integer, and then we can get rid of negative degrees by adding to p a linear form m with large enough coefficients.

Thus, throughout this section we will assume that all polynomials have rational exponents.

It will be convenient to state the results of this section using the following notation. Let $k(s, n)$ be the minimal number such that for any set S of s points in $\mathbb{Q}^n$ there is a tropical polynomial on n variables with at most $k(s, n)$ monomials having roots in all points of S. Note that there is a universal testing set of size s for polynomials with k monomials iff $k < k(s, n)$. Thus, we can easily obtain bounds on the size of the minimal universal testing set from the bounds on $k(s, n)$.

We start with the following upper bound on $k(s, n)$.

Theorem 5. *We have* $k(s, n) \leq \left\lceil \frac{2s}{(n+1)} \right\rceil + 1$.
For the size of the minimal universal testing set the following inequality holds:
$s \geq (k-1)(n+1)/2 + 1$.

We note that this theorem already shows the difference between universal testing sets over $\mathbb{R}$ and $\mathbb{Q}$ semirings.

Proof. We will show that for any set $S = \{\vec{a}_1, \ldots, \vec{a}_s\} \subseteq \mathbb{Q}^n$ of size s there is a nontrivial polynomial with at most $k = \left\lceil \frac{2s}{(n+1)} \right\rceil + 1$ monomials that has roots in all of the points in S. From this the inequalities in the theorem follow.

Throughout this proof we will use the following standard facts about (classical) affine functions on $\mathbb{Q}^n$.

Claim. Suppose π is an $(n-1)$ dimensional hyperplane in $\mathbb{Q}^n$. Let P_1 be a finite set of points in one of the (open) halfspaces w.r.t. π and P_2 be a finite set of points in the other (open) halfspace. Let C_1 and C_2 be some constants. Then the following is true.

1. If $\vec{a}_1, \ldots, \vec{a}_n \in \pi$ are points in the general position in π and $p_1, \ldots, p_n$ are some constants in $\mathbb{Q}$, then there is an affine function f on $\mathbb{Q}^n$ such that $f(\vec{a}_i) = p_i$ for all i, $f(\vec{x}) > C_1$ for all $\vec{x} \in P_1$ and $f(\vec{x}) < C_2$ for all $\vec{x} \in P_2$.

2. If g is an affine function on $\mathbb{Q}^n$ then there is another affine function f on $\mathbb{Q}^n$ such that $f(\vec{x}) = g(\vec{x})$ for all $\vec{x} \in \pi$, $f(\vec{x}) > C_1$ for all $\vec{x} \in P_1$ and $f(\vec{x}) < C_2$ for all $\vec{x} \in P_2$.

The proof of the theorem is by induction on s. The base is $s = 0$. In this case one monomial is enough (and is needed since we require polynomial to be nontrivial).

Consider the convex hull of points of S. Consider a face P of this convex hull. For simplicity of notation assume that the points from S belonging to P are $\vec{a}_1, \ldots, \vec{a}_l$. Consider a $((n-1)$-dimensional) hyperplane π passing through $\vec{a}_1, \ldots, \vec{a}_l$. Since P is a face of the convex hull of S we can pick π in such a way that all points in $S' = \{\vec{a}_{l+1}, \ldots, \vec{a}_s\}$ lie in one halfspace w.r.t. π (the choice of π might be not unique since P might be of the dimension less than $n-1$).

Applying the induction hypothesis we obtain a polynomial $p'(\vec{x}) = \max_i m_i'(\vec{x})$ that has roots in all points of S'. For $j = 1, \ldots, l$ introduce the notation $p_j = p'(\vec{a}_j) = \max_i m_i(\vec{a}_j)$.

We consider three cases: P contains all points of S; P contains not all points of S and $l \leqslant n$; P contains not all points of S and $l > n$.

If P contains all points of S, then the polynomial p' is obtained from the base of induction and consists of one monomial m_1'. Recall, that a monomial is just an affine function on $\mathbb{Q}^n$. Consider a new monomial $m(\vec{x})$ such that $m(\vec{x}) = m_1'(\vec{x})$ on the hyperplane π, but $m(\vec{b}) \neq m_1'(\vec{b})$ for some $\vec{b} \notin \pi$. Then the polynomial $p = p' \oplus m$ has roots in all points of the hyperplane π and thus in all points of S. This polynomial has $2 \leqslant \left\lceil \frac{2s}{(n+1)} \right\rceil + 1$ monomials.

If P contains not all points of S, then the dimension of P is $n-1$ (indeed, otherwise P is not a face).

If additionally $l \leqslant n$, it follows that $l = n$. Thus $\vec{a}_1, \ldots, \vec{a}_n$ are points in the general position in π. Thus due to the claim above we can pick a new monomial m such that $m(\vec{a}_j) = p_j$ for all $j = 1, \ldots, l$ and $m(\vec{a}_j) < p'(\vec{a}_j)$ for all $j > l$. Then the polynomial $p = p' \oplus m$ has roots in all points of S. This polynomial has $1 + \left\lceil \frac{2(s-n)}{(n+1)} \right\rceil + 1 \leqslant \left\lceil \frac{2s}{(n+1)} \right\rceil + 1$ monomials.

Now, if $l \geq n+1$ let $p_0 = \max_{j \leqslant l} p_j$ Applying the claim above take a pair of new distinct monomials m_1 and m_2 such that $m_1(\vec{x}) = m_2(\vec{x}) = p_0$ for all $\vec{x} \in \pi$ and $m_1(\vec{a}_j), m_2(\vec{a}_j) < p'(\vec{a}_j)$ for all $j > l$. Then the polynomial $p = p' \oplus m_1 \oplus m_2$ has roots in all points of S. This polynomial has $2 + \left\lceil \frac{2(s-n-1)}{(n+1)} \right\rceil + 1 = \left\lceil \frac{2s}{(n+1)} \right\rceil + 1$ monomials.

In all three cases we constructed a polynomial with the desired number of monomials.

The construction above leaves the room for improvement. For example, for the case of $n = 2$ we can show the following.

Theorem 6. *For $n = 2$ we have $k(s, 2) \leqslant \lceil \frac{s}{2} \rceil + 1$. For the size of the minimal universal set for polynomials in 2 variables the following inequality holds: $s \geq 2(k - 1) + 1$.*

We now proceed to lower bounds on $k(s, n)$. We start with the following non-constructive lower bound.

Theorem 7. *We have $k(s, n) \geq \left\lceil \frac{s}{n+1} \right\rceil$.*
Then for the minimal size of the universal testing set over $\mathbb{Q}$ we have $s \leqslant k(n + 1) + 1$.

This theorem is proven by a careful counting of dimensions of semialgebraic sets of universal sets and sets of roots of tropical polynomials.

The lower bound on $k(s, n)$ in Theorem 7 is not constructive. In the next section we present some constructive lower bounds. For this we establish a connection of our problem to certain questions in discrete geometry.

5.3   Constructive Lower Bounds

Suppose for some set of points $S = \{\vec{a}_1, \ldots, \vec{a}_s\} \subseteq \mathbb{Q}^n$ there is a polynomial p with monomials $m_1, \ldots, m_k$ that has roots in all points of S.

Recall that the graph of p in $(n + 1)$-dimensional space is a piece-wise linear convex function. Each linear piece corresponds to a monomial and roots of the polynomial are the points of non-smoothness of this function.

Consider the set of all roots of p in $\mathbb{Q}^n$. They partition the space $\mathbb{Q}^n$ into at most k convex (possibly infinite) polytopes. Each polytope corresponds to one of the monomials.

Consider the polytope corresponding to the monomial m_i. Consider all points in S that lie on its boundary and consider their convex hull. We obtain a smaller (finite) convex polytope that we will denote by P_i.

Thus starting from p we arrive at the set of non-intersecting polytopes $P_1, \ldots, P_k$ with vertices in S not containing any points of S in the interior. The fact that p has roots in all points of S means that each point in S belong to at least two of the polytopes $P_1, \ldots, P_k$. We call this structure by a *double covering* of points of S by convex polytopes. The *size* of the covering is the number k of the polytopes in it.

Thus, if we will construct a set S of points that does not have a double covering of size k it will follow that S is a universal set for k monomials. The similar question of single covering has been studied in the literature [6] (in the single cover polytopes cannot intersect even by vertices.).

Denote by $k_1(s, n)$ the minimal number of polytopes that is enough to single cover any s points in n dimensional space. Denote by $k_2(s, n)$ the minimal number of polytopes that is enough to double cover any s points in n dimensional space. The above analysis results in the following theorem.

Theorem 8. $k(s, n) \geq k_2(s, n)$.

For single coverings the following results are known. Let $f(n)$ be the maximal number such that any large enough n-dimensional set of points S contains a convex set of $f(n)$ points that are the vertices of a convex polytope and on the other hand do not contain any other points in S. The function $f(n)$ was studied but is not well understood yet. It is known [32] that the function is at most exponential in n. We can however observe the following.

Lemma 3. *For large enough s we have that $k_1(s, n) \geq s/f(n)$.*

It is known [32] that $f(3) \geq 22$. Thus we get that $k_1(s, 3) \geq s/22$ for large enough s.

It is also known [31] that $\lceil s/2(\log_2 s + 1) \rceil \leqslant k_1(s, 3) \leqslant \lceil 2s/9 \rceil$. For $n = 2$ there are linear upper and lower bounds known [30]. For an arbitrary n in [31] an upper bound $k_1(s, n) \leqslant 2s/(2n+3)$ is shown and $k_1(s, n) = s/2n$ is conjectured.

We establish the following connection between $k_1(s, n)$ and $k_2(s, n)$.

Lemma 4. $k_2(s, n) \geq k_1(s, n)$. *Thus for large enough s we have that $k(s, n) \geq s/f(n)$.*

Lemma 5. $k_1((n + 2)s, n) \geq k_2(s, n)$.

Overall, we have a sequence of inequalities $k(s, n) \geq k_2(s, n) \geq k_1(s, n) \geq k_2(\frac{s}{n+2}, n)$. We do not know how large $k(s, n)$ can be compared to $k_1(s, n)$ and $k_2(s, n)$.

However this connection helps us to show that the lower bound on the size of universal testing set we have established before for the case of $n = 2$ is tight.

Theorem 9. *We have $k(s, 2) \geq k_2(s, 2) \geq \lceil \frac{s}{2} \rceil + 1$.*
For $n = 2$ the size of the minimal universal testing set is equal to $s = 2k - 1$.

We omit the proof and only observe here that the second part of the theorem follows from the first part and Theorem 6 immediately. As a universal set with s points in $\mathbb{Q}^2$ one can pick the set of vertices of a convex polygon M.

References

1. Akian, M., Gaubert, S., Guterman, A.: Linear independence over tropical semirings and beyond. Contemp. Math. **495**, 1–33 (2009)
2. Akian, M., Gaubert, S., Guterman, A.: Tropical polyhedra are equivalent to mean payoff games. Int. J. Algebra Comput. **22**(1), 1250001 (2012)
3. Alon, N.: Combinatorial Nullstellensatz. Comb. Probab. Comput. **8**(1–2), 7–29 (1999)
4. Ben-Or, M., Tiwari, P.: A deterministic algorithm for sparse multivariate polynomial interpolation. In: Proceedings of the Twentieth Annual ACM Symposium on Theory of Computing, STOC 1988, pp. 301–309. ACM, New York (1988)
5. Bihan, F.: Irrational mixed decomposition and sharp fewnomial bounds for tropical polynomial systems. Discrete Comput. Geom. **55**(4), 907–933 (2016)

6. Brass, P., Moser, W.O.J., Pach, J.: Research Problems in Discrete Geometry. Springer, Heidelberg (2005)
7. Chari, S., Rohatgi, P., Srinivasan, A.: Randomness-optimal unique element isolation with applications to perfect matching and related problems. SIAM J. Comput. **24**(5), 1036–1050 (1995)
8. Davydow, A., Grigoriev, D.: Bounds on the number of connected components for tropical prevarieties. Discrete Comput. Geom. **57**(2), 470–493 (2017)
9. Develin, M., Santos, F., Sturmfels, B.: On the rank of a tropical matrix. Comb. Comput. Geom. **52**, 213–242 (2005)
10. Grigoriev, D.: Complexity of solving tropical linear systems. Comput. Complex. **22**(1), 71–88 (2013)
11. Grigoriev, D., Karpinski, M.: The matching problem for bipartite graphs with polynomially bounded permanents is in NC (extended abstract). In: 28th Annual Symposium on Foundations of Computer Science, Los Angeles, California, USA, 27–29, pp. 166–172, October 1987
12. Grigoriev, D., Podolskii, V.V.: Complexity of tropical and min-plus linear prevarieties. Comput. Complex. **24**(1), 31–64 (2015)
13. Grigoriev, D., Podolskii, V.V.: Tropical effective primary and dual Nullstellensätze. In: 32nd International Symposium on Theoretical Aspects of Computer Science, STACS 4–7 2015, Garching, Germany, pp. 379–391, March 2015
14. Grigoriev, D.Y., Karpinski, M., Singer, M.F.: The interpolation problem for k-sparse sums of eigenfunctions of operators. Adv. Appl. Math. **12**(1), 76–81 (1991)
15. Huber, B., Sturmfels, B.: A polyhedral method for solving sparse polynomial systems. Math. Comput. **64**, 1541–1555 (1995)
16. Itenberg, I., Mikhalkin, G., Shustin, E.: Tropical Algebraic Geometry. Oberwolfach Seminars. Birkhäuser, Boston (2009)
17. Izhakian, Z., Rowen, L.: The tropical rank of a tropical matrix. Commun. Algebra **37**(11), 3912–3927 (2009)
18. Kaltofen, E., Yagati, L.: Improved sparse multivariate polynomial interpolation algorithms. In: Gianni, P. (ed.) ISSAC 1988. LNCS, vol. 358, pp. 467–474. Springer, Heidelberg (1989). doi:10.1007/3-540-51084-2_44
19. Klivans, A.R., Spielman, D.: Randomness efficient identity testing of multivariate polynomials. In: Proceedings of the Thirty-Third Annual ACM Symposium on Theory of Computing, STOC 2001, pp. 216–223. ACM, New York (2001)
20. Maclagan, D., Sturmfels, B.: Introduction to Tropical Geometry: Graduate Studies in Mathematics. American Mathematical Society, Providence (2015)
21. Mikhalkin, G.: Amoebas of algebraic varieties and tropical geometry. In: Donaldson, S., Eliashberg, Y., Gromov, M. (eds.) Different Faces of Geometry. International Mathematical Series, vol. 3, pp. 257–300. Springer, US (2004)
22. Mulmuley, K., Vazirani, U.V., Vazirani, V.V.: Matching is as easy as matrix inversion. Combinatorica **7**(1), 105–113 (1987)
23. Richter-Gebert, J., Sturmfels, B., Theobald, T.: First steps in tropical geometry. Idempotent Math. Math. Phys. Contemp. Math. **377**, 289–317 (2003)
24. Risler, J.-J., Ronga, F.: Testing polynomials. J. Symbolic Comput. **10**(1), 1–5 (1990)
25. Schwartz, J.T.: Fast probabilistic algorithms for verification of polynomial identities. J. ACM **27**(4), 701–717 (1980)
26. Shustin, E., Izhakian, Z.: A tropical Nullstellensatz. Proc. Am. Math. Soc. **135**(12), 3815–3821 (2007)
27. Steffens, R., Theobald, T.: Combinatorics and genus of tropical intersections and Ehrhart theory. SIAM J. Discrete Math. **24**(1), 17–32 (2010)

28. Sturmfels, B.: Solving Systems of Polynomial Equations. CBMS Regional Conference in Math, vol. 97. American Mathematical Society, Providence (2002)
29. Theobald, T.: On the frontiers of polynomial computations in tropical geometry. J. Symb. Comput. **41**(12), 1360–1375 (2006)
30. Urabe, M.: On a partition into convex polygons. Discrete Appl. Math. **64**(2), 179–191 (1996)
31. Urabe, M.: Partitioning point sets in space into disjoint convex polytopes. Comput. Geom. **13**(3), 173–178 (1999)
32. Valtr, P.: Sets in $\mathbb{R}^d$ with no large empty convex subsets. Discrete Math. **108**(1), 115–124 (1992)
33. Vorobyev, N.: Extremal algebra of positive matrices. Elektron. Informationsverarbeitung und Kybernetik **3**, 39–71 (1967)
34. Zippel, R.: Probabilistic algorithms for sparse polynomials. In: Ng, E.W. (ed.) Symbolic and Algebraic Computation. LNCS, vol. 72, pp. 216–226. Springer, Heidelberg (1979). doi:10.1007/3-540-09519-5_73

On Weak-Space Complexity
over Complex Numbers

Pushkar S. Joglekar[1]([✉]), B.V. Raghavendra Rao[2], and Siddhartha Sivakumar[2]

[1] Vishwakarma Institute of Technology, Pune, India
joglekar.pushkar@gmail.com
[2] Indian Institute of Technology Madras, Chennai, India
bvrr@cse.iitm.ac.in, sith1992@gmail.com

Abstract. Defining a feasible notion of space over the Blum-Shub-Smale (BSS) model of algebraic computation is a long standing open problem. In an attempt to define a right notion of space complexity for the BSS model, Naurois [CiE 2007] introduced the notion of weak-space. We investigate the weak-space bounded computations and their plausible relationship with the classical space bounded computations. For weak-space bounded, division-free computations over BSS machines over complex numbers with $\overset{?}{=} 0$ tests, we show the following:

1. The Boolean part of the weak log-space class is contained in deterministic log-space, i.e., $\mathsf{BP}(\mathsf{LOGSPACE_W}) \subseteq \mathsf{DLOG}$;
2. There is a set $L \in \mathsf{NC}^1_{\mathbb{C}}$ that cannot be decided by any deterministic BSS machine whose weak-space is bounded above by a polynomial in the input length, i.e., $\mathsf{NC}^1_{\mathbb{C}} \not\subseteq \mathsf{PSPACE_W}$.

The second result above resolves the first part of Conjecture 1 stated in [6] over complex numbers and exhibits a limitation of weak-space. The proof is based on the structural properties of the semi-algebraic sets contained in $\mathsf{PSPACE_W}$ and the result that any polynomial divisible by a degree-$\omega(1)$ elementary symmetric polynomial cannot be sparse. The lower bound on the sparsity is proved via an argument involving Newton polytopes of polynomials and bounds on number of vertices of these polytopes, which might be of an independent interest.

1 Introduction

The theory of algebraic computation aims at classifying algebraic computational problems in terms of their intrinsic algebraic complexity. Valiant [28] developed a non-uniform notion of complexity for polynomial evaluations based on arithmetic circuits as a model of computation. Valiant's work led to intensive research efforts towards classifying polynomials based on their complexity. (See [26] for a survey). Valiant's model is non-uniform and it does not allow comparison operation on the values computed. This led to the seminal work by Blum et al. [3] where a real and complex number counterpart of Turing machines, now known as BSS machines has been proposed.

© Springer-Verlag GmbH Germany 2017
R. Klasing and M. Zeitoun (Eds.): FCT 2017, LNCS 10472, pp. 298–311, 2017.
DOI: 10.1007/978-3-662-55751-8_24

Blum et al. [2] defined the complexity classes such as $P_{\mathbb{R}}$ and $NP_{\mathbb{R}}$ in analogy to the classical complexity classes P and NP and proposed the conjecture: $P_{\mathbb{R}} \neq NP_{\mathbb{R}}$. Several natural problems such as Hilbert's Nullstellensatz, Feasibility of quadratic equations are complete for the class $NP_{\mathbb{R}}$ [2]. Further, there has been a significant amount of work on the structural aspects of real computation with various restrictions placed on the computational model. See [20] for a survey of these results.

One of the fundamental objectives of algebraic complexity theory is to obtain transfer theorems, i.e., to translate separations of algebraic complexity classes to either the Boolean world or other models of algebraic computation. Though establishing a relation between the BSS model of computation and the classical Turing machine is a hard task, Fournier and Koiran [9] showed that proving super polynomial time lower bounds against the BSS model would imply separation of classical complexity classes. Also, there has been a study of *algebraic* circuits leading to the definition of parallel complexity classes $NC_{\mathbb{R}}$. In contrast to the Boolean counterparts, Cucker [4] showed that there are sets in $P_{\mathbb{R}}$ that cannot have efficient parallel algorithms, i.e., $P_{\mathbb{R}} \neq NC_{\mathbb{R}}$.

One of the pre-requisites for transfer theorems would be a comparison with the complexity classes in the Boolean world. One approach towards this is restricting the BSS machines over Boolean inputs. A restriction of a real complexity class to Boolean inputs is called Boolean part and is denoted using the prefix BP, e.g., $BP(P_{\mathbb{R}})$ denotes the class of all languages over $\{0,1\}^*$ that can be decided by polynomial time bounded BSS machines [2,13]. Koiran [13] did an extensive study of Boolean parts of real complexity classes. Cucker and Grigoriev [5] showed that $BP(P_{\mathbb{R}}) \subseteq PSPACE/poly$. Further, Allender et al. [1] studied computational tasks arising from numerical computation and showed that the task of testing positivity of an integer represented as an arithmetic circuit is complete for the class $BP(P_{\mathbb{R}})$.

Though the notion of time complexity has been well understood in the real model of computation, it turned out that, setting up a notion of space is difficult. Michaux [21] showed that any computation over the real numbers in the BSS model can be done with only a constant number of cells. This rules out the possibility of using the number of cells used in the computation as a measure of space. Despite the fact that there has been study of parallel complexity classes, a natural measure of space that leads to interesting space complexity classes in analogy with the classical world is still missing.

Naurois [6] proposed the notion of weak-space for computation over real numbers in the BSS model. This is motivated by the weak BSS model of computation proposed by Koiran [15]. The notion of weak-space takes into account the number of bits needed to represent the polynomials representing each cell of a configuration. (See Sect. 2 or [6] for a formal definition.) Based on this notion of space Naurois [6] introduced weak-space classes $LOGSPACE_W$ and $PSPACE_W$ as analogues of the classical space complexity classes DLOG and PSPACE and showed that $LOGSPACE_W$ is contained in $P_W \cap NC_{\mathbb{R}}^2$, where P_W is the class of sets decidable in weak polynomial time [15]. The notion of weak-space enables space bounded computations to have a finite number of configurations, and hence opening the scope for possible analogy with the classical counterparts. However, [6] left several intriguing questions open. Among them; a real analogue of NC^1 versus DLOG, and an upper bound for the Boolean parts of weak space classes.

In this paper, we continue the study of weak-space classes initiated by Naurois [6] and investigate weak-space bounded division free computations where equality is the only test operation allowed. In particular, we address some of the questions left open in [6].

Our Results: We begin with the study of Boolean parts of weak space complexity classes. We show that the Boolean part of $LOGSPACE_W$ is contained in DLOG. (See Theorem 2.) Our proof involves a careful adaptation of the constant elimination technique used by Koiran [14] to weak space bounded computation.

We show that there is a set $L \in NC_{\mathbb{F}}^1$ that cannot be accepted by any polynomial weak-space bounded BSS machine, i.e., $NC_{\mathbb{F}}^1 \not\subset PSPACE_W$ (Theorem 3 and Corollary 1) where $\mathbb{F} \in \{\mathbb{C}, \mathbb{R}\}$. This resolves the first part of the Conjecture 1 in [6] where the computation is division free and only equality tests are allowed. Also, this result is in stark contrast to the Boolean case, where $NC^1 \subseteq DLOG$.

Our Techniques: For the proof of Theorem 3, we consider the restriction $L_n = L \cap \mathbb{F}^n$ for a set $L \in LOGSPACE_W$ and obtain a characterization for the defining polynomials of L_n as a semi-algebraic set in $\mathbb{F}^n$. Then using properties of the Zarisky topology, we observe that if L_n is an irreducible algebraic set, then the defining polynomial for L_n has small weak size. With this, it suffices to obtain a set $L \in NC_{\mathbb{F}}^1$ such that each slice L_n is a hyper-surface such that any nontrivial hyper-surface containing it cannot have sparse polynomial as its defining equations. We achieve this by considering the elementary symmetric polynomial of degree $n/2$ as the defining equation for L_n. For every polynomial multiple of the elementary symmetric polynomial, we prove a lower bound on its sparsity by appealing to the structure of Newton polytopes of these polynomials. (See Theorem 4 for a precise statement.)

Related Results: Koiran and Perifel [16,17] have studied the notion of polynomial space in Valiant's algebraic model and obtained transfer theorems over the real and complex numbers. Mahajan and Rao [19] obtained small space complexity classes in Valiant's algebraic model. To the best of our knowledge, apart from these, and the results by Michaux [21] and Naurois [6], there have been no significant study of space complexity classes in the broad area of algebraic complexity theory.

Organization of the Paper: In Sect. 2, we briefly review the BSS model of computation, and provide all necessary but non-standard definitions used in the paper. In Sect. 3 we look at the Boolean part of $LOGSPACE_W$. In Sect. 4 we prove the main theorem (Theorem 3) of the paper. Section 5 proves Corollary 2 which is an important component in the proof of Theorem 3. Due to lack of space some of the proofs are skipped, all the proofs are included in the full version of the paper [12].

2   Preliminaries

Throughout the paper, $\mathbb{F}$ denotes a field that is either $\mathbb{C}$ or $\mathbb{R}$. Let $\mathbb{F}^* = \cup_{k \geq 0} \mathbb{F}^k$. We give a brief description of a Blum-Shub-Smale (BSS) machine over $\mathbb{F}$. For details, the reader is referred to [3].

Definition 1. *A Blum-Shub-Smale (BSS) machine M over $\mathbb{F}$ with parameters $\alpha_1, \ldots, \alpha_k \in \mathbb{F}$ with $k \geq 0$ and an admissible input $Y \subseteq \mathbb{F}^*$ is a Random Access Machine with a countable number of registers (or cells) each capable of a storing a value from $\mathbb{F}$. The machine is permitted to perform three kinds of operations:*

Computation: Perform $c_l = c_i$ op c_j, where c_i, c_j and c_l are either cells of M or among the parameters and op $\in \{+, \times, -\}$ and move to the next state.

Branch (test): Perform the test $c \overset{?}{=} 0$ for some cell c and move to the next state depending on the result, i.e., branch as per the outcome of the test.

Copy: $c_i = c_j$, copy the value of the cell c_j into c_i. Here c_j can also be one of the parameters $\alpha_1 \ldots, \alpha_k$ of M.

It should be noted that in the definition of a real BSS machine the test instruction is usually $\geq$?0 rather than equality. Throughout the paper, we restrict ourselves to BSS machines where the test operation is =?0. Also, in general, BSS machines allow the division operation, however, we restrict to BSS machines where division is not allowed.

Notion of acceptance and rejection of an input, configurations and time complexity of computation can be defined similar to the case of classical Turing Machines, see [2] for details.

For a BSS machine that halts on all admissible inputs, the set accepted by M is denoted by $L(M)$. For an input $x \in \mathbb{F}^n$, the size of the input x is n.

Definition 2 (Complexity Class $\mathsf{P}_\mathbb{F}$) [2]. *Let $\mathbb{F}$ be a field of real or complex numbers then the complexity class $\mathsf{P}_\mathbb{F}$ is defined as the set of all languages $L \subseteq \mathbb{F}^*$ such that, there is a polynomial time BSS machine accepting L.*

The class $\mathsf{EXP}_\mathbb{F}$ is defined analogously. For a definition of $\mathsf{NP}_\mathbb{F}$ reader is referred to [2].

An *algebraic circuit* is an arithmetic circuit where in addition to the $\times$ and $+$ gates a test gate =?0 is allowed. A test gate has a single input and outputs either 0 or 1 depending on the outcome of the test. Size and depth of algebraic circuits are defined analogously. For the purpose of comparison with BSS complexity classes, we assume that algebraic circuits have a single output gate which is a =?0 gate. The following complexity classes are defined based on algebraic circuits.

Definition 3 [2]. *Let $\mathbb{F}$ be a field of real or complex numbers then the complexity class $NC_{\mathbb{F}}^i$ is defined as, the set of all languages $L \subseteq \mathbb{F}^*$, for which there is an algebraic circuit family $(C_n)_{n \geq 0}$, size of C_n is polynomial in n and depth of C_n is $O((\log n)^i)$ such that for all $n \geq 0$ and $x \in \mathbb{F}^n$, x is in L iff $C_n(x) = 1$.*

Weak Space. Following the notion of weak time defined by Koiran [15], Naurois [6], introduced the notion of weak space for BSS machines. To begin with, we need a measure of weak size of polynomials with integer coefficients. Let $g \in \mathbb{Z}[x_1, \ldots, x_n]$ be a polynomial of degree d. The binary encoding $\phi(m)$ corresponding to a monomial $m = x_{i_1}^{a_1} x_{i_2}^{a_2} \ldots x_{i_k}^{a_k}$ is simply concatenation of $\lceil \log n \rceil$ bit binary encoding of index i_j and $\lceil \log d \rceil$ bit binary encoding of exponent a_j for $j \in [k]$, *i.e.*, $\phi(m) = \langle i_1 \rangle \langle a_1 \rangle \cdot \langle i_2 \rangle \langle a_2 \rangle \cdots \langle i_k \rangle \langle a_k \rangle$, where $\langle i_j \rangle, \langle a_j \rangle$ denotes binary encoding of integers i_j and a_j respectively. Let $g = \sum_{m \in M} g_m m$ where $g_m \neq 0$ is the coefficient of monomial m in g and $M = \{m_1, m_2, \ldots, m_s\}$ be the set of monomials of g with non-zero coefficients. Then the binary encoding of g is $\phi(g) = b_1 \langle g_{m_1} \rangle \phi(m_1) \cdot b_2 \langle g_{m_2} \rangle \phi(m_2) \cdots b_s \langle g_{m_s} \rangle \phi(m_s)$ where $b_i = 1$ if $g_{m_i} \geq 0$ else $b_i = 0$ and $\langle g_{m_i} \rangle$ denotes $\lceil \log C \rceil$-bit binary encoding of g_{m_i} for $i \in [s]$ where $C = max_i |g_{m_i}|$. We denote length of encoding $\phi(g)$ by $S_{weak}(g)$ and call it weak size of polynomial g. It is easy to see that $S_{weak}(g) \leq s(n(\lceil \log n \rceil + \lceil \log d \rceil) + 1 + \lceil \log C \rceil)$.

Definition 4 (Weak-space complexity). *Let M be a BSS machine with parameters $\beta_1, \beta_2, \ldots, \beta_m \in \mathbb{F}$, and an input $x = (x_1, x_2, \ldots, x_n)$. Let $\mathcal{C}_M(x)$ denote the set of all configurations of M on x reachable from the initial configuration. For a configuration $c \in \mathcal{C}_M(x)$, let $f_1^{(c)}, f_2^{(c)}, \ldots, f_r^{(c)}$ be the formal polynomials representing the non-empty cells in the configuration such that*

$$f_i^{(c)}(x_1, x_2, \ldots, x_n) = g_i^{(c)}(x_1, x_2, \ldots, x_n, \beta_1, \beta_2, \ldots, \beta_m)$$

where $g_i^{(c)} \in \mathbb{Z}[x_1, x_2, \ldots, x_n, y_1, y_2, \ldots, y_m]$ for $i \in [m]$. Define the weak size of a configuration c as $S_{weak}(c) = \sum_{j=1}^{r} S_{weak}(g_j^{(c)})$ Then the weak-space complexity of M is defined as $WSpace_M(n) = \max_{x \in \mathbb{F}^n} \max_{c \in \mathcal{C}_M(x)} S_{weak}(c)$.

Remark 1. Note that in the above polynomials $g_1^{(c)}, \ldots, g_r^{(c)}$ depend only on the number of machine parameters and the input size and not on the actual values $\beta_1, \ldots, \beta_m$ for the parameters, and are uniquely defined for a given configuration c. We can think of polynomials $g_1^{(c)}, \ldots, g_r^{(c)}$ are obtained by unfolding the computation of the machine on input $x_1, \ldots, x_n$ treating constants $\beta_1, \beta_2, \ldots, \beta_m$ as indeterminates.

A BSS machine M is said do be s weak-space bounded if $WSpace_M(n) \leq s(n)$. The following concrete weak space classes have been defined in [6].

Definition 5 (Complexity class $SPACE_W(s)$). *For a non-decreasing space constructible function s, $SPACE_W(s)$ is the set of all languages $L \subseteq \mathbb{F}^*$, for which there is a BSS machine M over $\mathbb{F}$ such that $L(M) = L$ and $WSpace_M(n) = O(s(n))$.*

Note that we have omitted the subscript $\mathbb{F}$ in the above definition, this is not an issue since the field will always be clear from the context. The following inclusions are known from [6].

Proposition 1 [6]. $\text{LOGSPACE}_W \subseteq P_w \cap NC_{\mathbb{R}}^2;$ *and* $\text{PSPACE}_W \subset \text{PAR}_{\mathbb{R}}.$

For definition of an algebraic variety and the Zariski topology, the reader is referred to [25]. The elementary symmetric polynomial of degree d is defined as: $\text{sym}_{n,d}(x_1, \ldots, x_n) = \sum_{S \subseteq [n], |S|=d} \prod_{i \in S} x_i$, where $[n] = \{1, \ldots, n\}$.

Convex Polytopes. For the proof of our lowebound result in Sect. 5 we need to review some basic concepts about convex polytopes. For a detailed exposition on convex polytopes, see e.g. [11,29].

A point set $K \subseteq \mathbb{R}^d$ is *convex* if for any two points $x, y \in K$, the point $\lambda x + (1 - \lambda)y$ is in K for any λ, $0 \leq \lambda \leq 1$. The intersection of convex sets is convex. For any $K \subseteq \mathbb{R}^d$, the intersection of all convex sets containing K is called as *convex-hull* of K, $conv(K) = \bigcap \{T \subseteq \mathbb{R}^d | K \subseteq T, T \text{ is convex}\}$.

From the above definition and a simple inductive argument it follows that

Lemma 1. *If* $K \subseteq \mathbb{R}^d$ *and* $x_1, x_2, \ldots, x_n \in K$ *then* $\sum_{i=1}^n \lambda_i x_i \in conv(K)$ *where* $\lambda_i \geq 0$ *and* $\sum_{i=1}^n \lambda_i = 1$ *and if* $K = \{x_1, \ldots, x_n\}$ *is a finite set of points then* $conv(K) = \{\sum_{i=1}^n \lambda_i x_i | \lambda_i \geq 0 \text{ and } \sum_{i=1}^n \lambda_i = 1\}$.

Definition 6 *(Convex Polytope). A convex-hull of a finite set of points in* $\mathbb{R}^d$ *is called as convex polytope.*

Let $P = conv(\{x_1, \ldots, x_n\}) \subset \mathbb{R}^d$ be a convex polytope. Then the dimension of P (denoted as $dim(P)$) is the dimension of the affine space $\{\sum_i \lambda_i x_i | \lambda_i \in \mathbb{R}, \sum_i \lambda_i = 1\}$. Clearly if $P \subset \mathbb{R}^d$ then $dim(P) \leq d$.

We can equivalently think of convex polytopes as bounded sets which are intersections of finitely many closed half spaces in some $\mathbb{R}^d$. More precisely,

Theorem 1 *(Chap. 1, [29]). P is the convex-hull of a finite set of points in* $\mathbb{R}^d$ *iff there exists* $A \in \mathbb{R}^{m \times d}$ *and* $z \in \mathbb{R}^m$ *such that the set* $\{x \in \mathbb{R}^d | Ax \leq z\}$ *is bounded and* $P = \{x \in \mathbb{R}^d | Ax \leq z\}$.

Definition 7 *(Face of Polytope). Let P is a convex polytope in* $\mathbb{R}^d$. *For* $a = (a_1, a_2, \ldots, a_d) \in \mathbb{R}^d$ *and* $b \in \mathbb{R}$ *we say the linear inequality* $\langle a, x \rangle \leq b$ *(where* $\langle a, x \rangle = \sum_{i=1}^d a_i x_i$) *is valid for P if every point* $x = (x_1, \ldots, x_d) \in P$ *satisfies it. A face of P is any set of points in* $\mathbb{R}^d$ *of the form* $P \cap \{x \in \mathbb{R}^d | \langle a, x \rangle = b\}$ *for some* $a \in \mathbb{R}^d$ *and* $b \in \mathbb{R}$ *such that* $\langle a, x \rangle \leq b$ *is a valid linear inequality for* P.

From the above definition and Theorem 1 it is clear that every face of a convex polytope is also a convex polytope. So we can use notion of dimension of convex polytope to talk about dimension of a face of a convex polytope. The faces of dimension 0 are called as the *vertices* of the polytope. Following proposition gives useful criteria for a point $v \in P$ to be a vertex of P. For the proof of following standard propositions refer to Chap. 1 and 2 of [29].

Proposition 2. *For a convex polytope P, a point $v \in P$ is vertex of P iff for any $n \geq 1$, and any $x_1, \ldots, x_n \in P$, $v \neq \sum_{i=1}^{n} \lambda_i x_i$ for $0 \leq \lambda_i < 1, \sum_i \lambda_i = 1$.*

Proposition 3. *Every convex polytope P is convex-hull of set of its vertices, $P = conv(ver(P))$ and if $P = conv(S)$ for finite S then $ver(P) \subseteq S$, where $ver(P)$ denotes the set of vertices of a polytope P.*

3   Boolean Parts of Weak Space Classes

Though the BSS model is intended to capture the intrinsic complexity of computations over real and complex numbers, it is natural to study the power of such computations restricted to the Boolean input. The Boolean parts of real/complex complexity classes have been well studied in the literature [1]. We consider Boolean parts of the weak- space classes introduced by Naurois [6]

Definition 8. *Let C be a complexity class in the BSS model of computation, then the Boolean part of C denoted by $BP(C)$ is the set $BP(C) = \{L \cap \{0,1\}^*$ $\mid L \in C\}$*

We observe that the Boolean part of $LOGSPACE_W$ is contained in $DLOG$, i.e. the class of languages accepted deterministic logarithmic space bounded Turing Machines.

Theorem 2. *For $\mathbb{F} \in \{\mathbb{C}, \mathbb{R}\}$, $BP(LOGSPACE_W) \subseteq DLOG$.*

Proof. Let $L \in LOGSPACE_W$ and M be a BSS machine over $\mathbb{F}$ with $WSpace_M(n)$ $= s(n) = c \log n$ for some $c > 0$ and such that for all $x \in \mathbb{F}^*$, $x \in L \iff M$ accepts x. Our proof is a careful analysis of the constant elimination procedure developed by Koiran [14]. The argument is divided into three cases:

Case 1: Suppose that M does not use any constants from $\mathbb{F}$. Let $x_1, \ldots, x_n \in \{0,1\}$ be an input. Construct a Turing Machine M' that on input $x_1, \ldots, x_n \in \{0,1\}$ simulates M as follows. M' stores content of each cell of M explicitly as a polynomial. For each step of M:

1. If the step is an arithmetic operation, then M' explicitly computes the resulting polynomial and stores it in the target cell and proceeds.
2. If the step is a comparison operation, then M' evaluates the corresponding polynomial and proceeds to the next step of M.

Since the total number of bits required to store all of the polynomials in any given configuration is bounded by $c \log n$ and the arithmetic operations on log-bit representable polynomials can be done in deterministic log-space, it is not difficult to see that the resulting Turing Machine M is log-space bounded.

Case 2: M uses algebraic constants. Suppose $\beta_1, \ldots, \beta_k \in \mathbb{F}$ are the algebraic constants used in M. We begin with the special case when $k = 1$. Let $p_1(x)$ be the minimal polynomial of β_1 with coefficients in $\mathbb{Z}$. Let d be the degree of p_1. We show that Koiran's [14] technique for elimination of algebraic constants can indeed be implemented in weak log-space. We view the content of each cell of M on a given input $x_1, \ldots, x_n \in \{0, 1\}$ as a polynomial in $x_1, \ldots, x_n$ and a new variable y_1. For any polynomial $q(x_1, \ldots, x_n, y_1)$ with integer coefficients, $q(x_1, \ldots, x_n, \beta_1) = 0$ if and only if $q(x_1, \ldots, x_n, y_1) = 0 \mod p_1$. Consider the Turing machine M' that simulates M as follows. M' stores contents of each cell of M as polynomial $p(x_1, \ldots, x_n, y_1) \mod p_1$. Note that every such polynomial has degree d in the variable y_1. For each step of the machine M, the new Turing machine M' does the following:

1. If the step is an arithmetic (add or multiply) operation, then perform the same arithmetic operation on the corresponding polynomials modulo p_1 and store the resulting polynomial in the polynomial corresponding to the cell where result was designated to be stored in M.

2. If the step is an $\overset{?}{=} 0$ test, then evaluate the polynomial corresponding to the cell whose value is to be tested at the given input $x_1, \ldots, x_n \in \{0, 1\}$ modulo p_1. If the result is zero treat the test as affirmative, else in the negative.

We analyse the space of M' on a given input $x_1, \ldots, x_n \in \{0, 1\}$. Consider a cell c of M. Let $g_c = g_c(x_1, \ldots, x_n, y_1)$ be the polynomial representing the value stored at cell c at a fixed point of time in the computation. Note that degree of y_1 in g_c at most $deg(p_1) - 1 = d - 1$. Suppose $g_c = f_0 + f_1 y_1 + f_2 y_1^2 + \cdots + f_{d-1} y_1^{d-1}$. We have $S_{\mathsf{weak}}(f_i) \leq S_{\mathsf{weak}}(g_c)$ for $0 \leq i \leq d - 1$. The overall work space requirement of M' is bounded by $d \cdot WSpace_M(n) = ds(n) = O(\log n)$.

For the case when $k > 1$, Let $\mathbb{G} = \mathbb{Q}(\beta_1, \ldots, \beta_k)$ be the extension field of $\mathbb{Q}$ obtained by adding $\beta_1, \ldots, \beta_k$. Clearly $\mathbb{G}$ is a finite extension of $\mathbb{Q}$. By the primitive element theorem [22], there is a $\beta \in \mathbb{F}$ such that $\mathbb{Q}(\beta) = \mathbb{G}$. Let p be the minimal polynomial for β of degree σ with coefficients from $\mathbb{Q}$. Let $p_1(y), \ldots, p_k(y)$ be univariate polynomials of minimum degree such that $p_i(\beta) = \beta_i$, and let Δ be the maximum of degrees of p_is.

Consider an input $x_1, \ldots, x_n \in \{0, 1\}$ and a cell c of M. Suppose $g_c = g_c(x_1, \ldots, x_n, y_1, y_2, \ldots, y_k)$ is the polynomial representing the value stored at the cell c at any fixed point of time in the computation. Let D be the degree of g_c and $g_c = \sum_{\delta \in \mathbb{N}^k} f_\delta \prod_{j=1}^k y_j^{\delta_j}$, where f_δ is a polynomial of degree at most $D - \sum_i \delta_i$ in $x_1, \ldots, x_n$. Let $g'_c = g_c(x_1, \ldots, x_n, p_1(y), \ldots, p_k(y)) = \sum_{\delta \in \mathbb{N}^k} f_\delta \prod_{j=1}^k p_j(y)^{\delta_j}$, and the coefficient of y^i in g'_c be $g'_i(x_1, \ldots, x_n)$. Note that,

$$S_{\mathsf{weak}}(g_c) = \sum_{\delta \in \mathbb{N}^k} S_{\mathsf{weak}}(f_\delta)(\sum_i \log \delta_i) \tag{1}$$

We first bound $S_{\mathsf{weak}}(g'_c \mod p)$. For $\delta \in \mathbb{N}^k$ with $\sum_i \delta_i \leq D$, let $q_\delta = \prod_{j=1}^k p_j(y)^{\delta_i}$. q_δ is a polynomial of degree at most $D\Delta$. Then $g'_i = \sum_{\delta:\mathsf{coeff}_{q_\delta}(y^i) \neq 0} f_\delta$, thus the number of bits required to store g'_i is bounded by $\sum_{\delta:\mathsf{coeff}_{q_\delta}(y^i) \neq 0} S_{\mathsf{weak}}(f_\delta)$. Since q_δ is of degree at most $d\Delta$ and hence $S_{\mathsf{weak}}(g'_i)$ can be dependent on d. However, $q_\delta \mod p$ is a polynomial of degree at most $\sigma - 1$ and hence any given f_δ will be a summand for at most σ many g'_is. Therefore, $S_{\mathsf{weak}}(g'_c \mod p)$ is at most $\sigma \cdot S_{\mathsf{weak}}(g_c)$.

To conclude the argument for Case 2, we describe the simulation of the machine M': M' simulates M as in the case when $k = 1$ by storing the polynomials g'_c mod p explicitly, i.e., it stores the polynomials g'_i mod p. The number of bits required to store g'_c is bounded by $S_{\text{weak}}(g'_c)$ which in turn is bounded by $(\sigma + 1)S_{\text{weak}}(g_c)$. Now the simulation is done as in the case $k = 1$.

Case 3: M uses transcendental constants. Let γ be a transcendental number. Then for any polynomial p with integer coefficients, we have $p(\gamma) \neq 0$. Thus, for any cell c of M and for any $x_1, \ldots, x_n \in \{0, 1\}$, $g_c(x_1, \ldots, x_n, \gamma) = 0$ if and only if $g_c(x_1, \ldots, x_n, y) \equiv 0$. The simulation of M by M' can be done the same fashion as in Case 2, except that the polynomials g_c are stored as they are. Suppose $g_c(x_1, \ldots, x_n, y) = \sum_{i=0}^{d} f_i y^i$, then $S_{\text{weak}}(g_c) = \sum_i S_{\text{weak}}(f_i) \log i$, therefore the space required to store g_c by storing f_i's explicitly is bounded by $S_{\text{weak}}(g_c)$, M' requires space at most $O(s(n)) = O(\log n)$. Now, consider the case when M uses more than one transcendental constants, and let $\gamma_1, \ldots, \gamma_k$ be the constants used by M that are transcendental. Suppose $t \leq k$ is such that γ_i is transcendental in $\mathbb{Q}(\gamma_1) \cdots (\gamma_{i-1})$ (where $\mathbb{Q}(\gamma_1)$ is the field extension of $\mathbb{Q}$ that contains γ_1) for $i \leq t$ and γ_j is algebraic over $\mathbb{G} = ((\mathbb{Q}(\gamma_1)) \cdots)(\gamma_t)$ for $j \geq t+1$. By the primitive element theorem, let γ be such that $\mathbb{G}(\gamma) = \mathbb{G}(\gamma_{t+1}, \ldots, \gamma_k)$. Let $p_i(y)$ be a polynomial over $\mathbb{G}$ of minimal degree such that $\gamma_i = p_i(\gamma)$ for $t+1 \leq i \leq k$. Now the simulation of M by M' can be done as in Case 2, however, the only difference is polynomials p_i can have rational functions over $\gamma_1, \ldots, \gamma_t$ as coefficients. However, any coefficient of p_i can be written as an evaluation of fraction of polynomials of constant degree over t variables, hence contributing a constant factor in the overall space requirement. Thus, for any cell c of M at any point of computation on a given input $x_1, \ldots, x_n \in \{0, 1\}$ can be represented as a polynomial $g_c(x_1, \ldots, x_n, y)$ mod p over $\mathbb{G}$. By the observations in Case 2, and the fact that any fixed element in $\mathbb{G}$ can be represented in constant space, the overall space required by M' to simulate M is bounded by $O(\Gamma \cdot s(n)) = O(\log(n))$ where Γ is a constant that depends on k, the maximum degree of the polynomials $p_{t+1}, \ldots, p_k$ and the number bits required to represent the coefficients of these polynomials as rational functions over $\mathbb{Q}$ in $\gamma_1, \ldots, \gamma_t$. $\square$

However, we are unable to show the converse of the above theorem, i.e., the question DLOG $\subseteq$ LOGSPACE$_\text{W}$? remains open. The main difficulty is, we can easily construct deterministic log-space bounded machines that evaluate non-sparse polynomials such as the elementary symmetric polynomials over a Boolean input.

4   Weak Space Lower Bounds

In this section we exhibit languages in $\mathbb{F}^*$ that are not in LOGSPACE$_\text{W}$. We begin with a simple structural observations on the languages in SPACE$_\text{W}(s)$ for any non-decreasing function s.

Lemma 2. *Let $L \in \mathsf{SPACE_W}(s)$, then for every $n > 0$, there exist $t \geq 1$ and polynomials $f_{i,j}$, $1 \leq i \leq t$, $1 \leq j \leq m_i$, $g_{i,j}$ and $1 \leq i \leq t$, $1 \leq j \leq m_i$ in $\mathbb{Z}[x_1, \ldots, x_n]$ such that:*

1. $S_{\mathsf{weak}}(f_{i,j}) \leq s(n)$, *for every* $1 \leq i \leq t_1$, $1 \leq j \leq m_i$; *and*
2. $S_{\mathsf{weak}}(g_{i,j}) \leq s(n)$, *for every* $1 \leq i \leq t_2$, $1 \leq j \leq m_i$; *and*
3. $L \cap \mathbb{F}^n = \bigcup_{i=1}^{t} \bigcap_{j=1}^{m_i} [f_{i,j} = 0] \cap \bigcap_{j=1}^{m_i} [g_{i,j} \neq 0]$.

Definition 9. *For $n \geq 0, d \leq n$, let*

$$S_{n,d} \overset{\text{def}}{=} \{(a_1, \ldots, a_n) \in \mathbb{F}^n \mid \mathsf{sym}_{n,d}(a_1, \ldots, a_n) = 0\}$$

i.e., the hyper surface defined by the n-variate elementary symmetric polynomial of degree d. For $d = d(n) \leq n$ define the language: $L^{(d)} \overset{\text{def}}{=} \bigcup_{n \geq 0} S_{n,d(n)}$.

Theorem 3. *For any constant $c > 0$ $L^{(n/2)} \notin \mathsf{SPACE_W}(n^c)$.*

Proof. We argue for the case $\mathbb{F} = \mathbb{C}$. An exactly similar argument is applicable to the case when $\mathbb{F} = \mathbb{R}$. For any $c > 0$ consider an arbitrary language $L' \in \mathsf{SPACE_W}(n^c)$. Then, for every $n \geq 1$, there are n-variate polynomials $f_{i,j}$, $1 \leq i \leq t$, $1 \leq j \leq m_i$, $g_{i,j}$, $1 \leq i \leq t$, $1 \leq j \leq m_i$ in $\mathbb{Z}[x_1, \ldots, x_n]$ as promised by Lemma 2. Let

$$V_i \overset{\text{def}}{=} \bigcap_{j=1}^{m_i} [f_{i,j} = 0]; \; W_i \overset{\text{def}}{=} \bigcap_{j=1}^{m_i} [g_{i,j} \neq 0]; \; \text{and} \; T_i \overset{\text{def}}{=} V_i \cap W_i.$$

Then we have $L' \cap \mathbb{C}^n = \bigcup_{i=1}^{t} T_i$. We argue that for large enough n, $\bigcup_{i=1}^{t} T_i \neq S_{n,n/2}$ and hence conclude $L' \neq L^{(n/2)}$. Let $\widehat{T_i}$ denote the Zariski closure of the set T_i in $\mathbb{C}^n$, i.e., the smallest algebraic variety containing T_i. Proof is by contradiction. Suppose that $\bigcup_{i=1}^{t} T_i = S_{n,n/2}$. As $S_{n,n/2}$ is a closed set in the Zariski topology over $\mathbb{C}^n$, we have $T_i \subseteq \widehat{T_i} \subseteq S_{n,n/2}$ and hence $\bigcup_{i=1}^{t} \widehat{T_i} = S_{n,n/2}$. Then, there should be an i such that $\widehat{T_i} = S_{n,n/2}$, for, $S_{n,n/2}$ is an irreducible algebraic variety. Now there are two cases:

Case 1: $V_i = \mathbb{C}^n$. In this case, $T_i = W_i$ i.e., an open set in the Zariski topology. Since $\mathbb{C}^n$ is dense in the Zariski topology, closure of any open set is in fact $\mathbb{C}^n$ itself. Therefore, $\widehat{T_i} = \mathbb{C}^n \neq S_{n,n/2}$, hence a contradiction.

Case 2: $V_i \neq \mathbb{C}^n$. Then we have $T_i = V_i \cap W_i \subseteq V_i$, therefore $S_{n,n/2} = \widehat{T_i} \subseteq V_i = \bigcap_{j=1}^{m_i} [f_{ij} = 0]$. It is enough to argue that $S_{n,n/2}$ is not contained in any of the varieties $[f_{i,j} = 0]$. Suppose $S_{n,n/2} \subseteq [f_{i,j} = 0]$ for some $1 \leq j \leq m_i$. Since $\mathsf{sym}_{n,n/2}$ is an irreducible polynomial, we have $\mathsf{sym}_{n,n/2} | f_{i,j}$. By Corollary 2, the number of monomials in $f_{i,j}$ is $n^{\omega(1)}$. However, by Lemma 2, the number of monomials in f_{ij} is at most $O(n^c)$, obtaining a contradiction for large enough n. Thus $S_{n,n/2} \nsubseteq [f_{i,j} = 0]$ for any $1 \leq j \leq m_i$ which in turn implies $S_{n,n/2} \nsubseteq V_i$ and hence $S_{n,n/2} \nsubseteq \widehat{T_i}$. Thus in both of the cases above we obtain a contradiction, as a result we have $S_{n,n/2} \neq \bigcup_{i=1}^{t} T_i$. Thus $L' \neq L^{(n/2)}$ as required.

As an immediate corollary we have:

Corollary 1. $NC_\mathbb{F}^1 \not\subseteq PSPACE_W$.

Proof. It is known that $sym_{n,d}$ is computable by polynomial size arithmetic circuits of logarithmic depth [27] and hence $L^{(d)} \in NC_\mathbb{F}^1$. The result follows.

Now, to complete the proof of Theorem 3, we need to prove Corollary 2. This is done in the next section using the properties of Newton's polytope of elementary symmetric polynomials.

5   Polynomials Divisible by Elementary Symmetric Polynomials

Let g be a polynomial in $\mathbb{F}[x_1, \ldots, x_n]$. In this section we prove that, for any polynomial f which is a polynomial multiple of g, the number of monomials of f is lower bounded by the number of vertices of Newton polytope of g. As an implication, we get an exponential lower bound on number of monomials of any polynomial multiple of $sym_{n,d}$. The key step in the proof is a simple Lemma which lower bounds number of vertices of convex polytope R in terms of number of vertices of convex polytopes P and Q when R is Minkowski sum of P and Q. We begin with definition of Minkowski sum.

Definition 10 (Minkowski sum). *For $A, B \subseteq \mathbb{F}^d$, Minkowski sum of A and B (denoted by $A \oplus B$) is defined as $A \oplus B = \{a + b | a \in A, b \in B\}$.*

Minkowski sums of convex sets have been extensively studied in mathematics literature, and has interesting applications in complexity theory, see for example [10,18,24]. The next proposition shows that the Minkowski sum of two convex polytopes is a convex polytope and every vertex of resulting polytope can be uniquely expressed as sum of vertices of the two polytopes. In fact, a more general statement about unique decomposition of a face (of any dimension) of Minkowski sum of convex polytopes into faces of individual polytopes holds true, see for example [11,24].

Proposition 4. *If $P, Q \subseteq \mathbb{F}^d$ are convex polytopes then the Minkowski sum of P and Q is a convex polytope $P \oplus Q = conv(\{p + q | p \in ver(P), q \in ver(Q)\})$ and for every vertex $r \in ver(P \oplus Q)$ there exist unique $p \in P, q \in Q$ such that $r = p + q$, moreover $p \in ver(P), q \in ver(Q)$.*

Lemma 3. *For convex polytopes $P, Q \subseteq \mathbb{F}^d$,*

$$|ver(P \oplus Q)| \geq max(|ver(P)|, |ver(Q)|)$$

Proof. Let $ver(P) = \{p_1, p_2, \ldots, p_m\}$, $ver(Q) = \{q_1, q_2, \ldots, q_n\}$ and $m \geq n$. To the contrary assume that $|ver(P \oplus Q)| < m$ and let $R = P \oplus Q$ and $ver(R) = \{r_1, r_2, \ldots, r_t\}$, where $t < m$. From Proposition 4, for $\ell \in [t]$ every vertex $r_\ell \in$

$ver(R)$ can be *uniquely* expressed as $r_\ell = p_{i_\ell} + q_{j_\ell}$ where $p_{i_\ell} \in ver(P)$ and $q_{j_\ell} \in ver(Q)$. But as $t = |ver(R)| < m = |ver(P)|$, there must be a vertex $p' \in ver(P)$ which plays no role in determining any vertex of $P \oplus Q$, that is, every $r_\ell \in ver(P \oplus Q)$ can be expressed as $r_\ell = p_{i_\ell} + q_{j_\ell}$ where $p_{i_\ell} \in ver(P) \setminus \{p'\}$ and $q_{j_\ell} \in ver(Q)$. Without loss of generality assume that $p' = p_1$. Since p_1 is a vertex of P, there exist a valid linear inequality $\langle v, p_1 \rangle \leq k$, $k \in \mathbb{R}, v \in \mathbb{R}^d$ such that $\langle v, p_1 \rangle = k$ and for any $x \in P \setminus \{p_1\}$, $\langle v, x \rangle < k$. Let $q \in Q$ such that $\langle v, y \rangle \leq \langle v, q \rangle = k'$, $k' \in \mathbb{R}$ for any $y \in Q$. Let $z = p_1 + q \in P \oplus Q$.

From Proposition 3 we know that $R = P \oplus Q = conv(ver(P \oplus Q))$. So the point $z \in P \oplus Q$ can be expressed as $z = \sum_{\ell=1}^{t} \lambda_\ell (p_{i_\ell} + q_{j_\ell})$ where $\lambda_\ell \geq 0, \sum_\ell \lambda_\ell = 1$ where $p_{i_\ell} \in ver(P) \setminus \{p_1\}$ and $q_{j_\ell} \in ver(Q)$. Let $z_P = \sum_{\ell=1}^{t} \lambda_\ell p_{i_\ell} \in P$ and $z_Q = \sum_{\ell=1}^{t} \lambda_\ell q_{j_\ell} \in Q$. So we get $z = z_P + z_Q = p_1 + q$. First we argue that $p_1 \neq z_P$. Assume $p_1 = z_P = \sum_{\ell=1}^{t} \lambda_\ell p_{i_\ell}$, where $p_{i_\ell} \in ver(P) \setminus \{p_1\}$. Clearly if $\lambda_\ell = 1$ for some $\ell \in [t]$ then $\lambda_i = 0$ for $i \in [t] \setminus \{\ell\}$ and we get $p_1 = p_{i_\ell}$ but that is not possible as $p_{i_\ell} \in ver(P) \setminus \{p_1\}$. So we can express a vertex p_1 of P as a nontrivial convex combination of $p_{i_1}, p_{i_2}, \ldots p_{i_t} \in ver(P) \setminus \{p_1\}$. A contradiction to Proposition 2. So $p_1 \neq z_P$.

We know that $\langle v, p_1 \rangle = k$ and for any $x \in P \setminus \{p_1\}, \langle v, p_1 \rangle < k$. In particular, $\langle v, z_P \rangle < k$. Also, by choice of q we have $\langle v, y \rangle \leq \langle v, q \rangle$ for $y \in Q$. As a result we get $\langle v, z_P \rangle + \langle v, z_Q \rangle < \langle v, p_1 \rangle + \langle v, q \rangle$. A contradiction, since $z = z_P + z_Q = p_1 + q$.

Now we recall the notion of Newton's polytope of polynomial in $\mathbb{F}[x_1, \ldots, x_n]$. Let f be a polynomial in $\mathbb{F}[x_1, \ldots, x_n]$. Let $f_{(\alpha_1, \alpha_2, \ldots, \alpha_n)}$ denotes coefficient of the monomial $x_1^{\alpha_1} x_2^{\alpha_2} \ldots x_n^{\alpha_n}$ in f, $f = \sum f_{(\alpha_1, \alpha_2, \ldots, \alpha_n)} x_1^{\alpha_1} x_2^{\alpha_2} \ldots x_n^{\alpha_n}$. A vector $(\alpha_1, \alpha_2, \ldots, \alpha_n) \in \mathbb{R}^n$ is called as an exponent vector of the monomial $x_1^{\alpha_1} x_2^{\alpha_2} \ldots x_n^{\alpha_n}$ of f. The Newton polytope of f is defined as the convex-hull of set of exponent vectors $(\alpha_1, \alpha_2, \ldots, \alpha_n)$ in $\mathbb{R}^n$ for which $f_{(\alpha_1, \alpha_2, \ldots, \alpha_n)} \neq 0$. The Newton polytope of f is denoted by P_f.

For a polynomial f, let $mon(f)$ denote the set of monomials with non-zero coefficient in f. Following Lemma is from [10]. As per [10] a more general version of Lemma 4 appears in [23].

Lemma 4 [23]. *Let $f, g, h \in \mathbb{F}[x_1, \ldots, x_n]$ with $f = gh$ then $P_f = P_g \oplus P_h$.*

Theorem 4. *Let f, g, h be nonzero polynomials in $\mathbb{F}[x_1, \ldots, x_n]$ with $f = gh$ then $|mon(f)| \geq max(|ver(P_g)|, |ver(P_h)|)$.*

Corollary 2. *For any nonzero polynomial $g \in \mathbb{F}[x_1, \ldots, x_n]$ let $f = g \cdot sym_{n, \frac{n}{2}}$ then $|mon(f)| \in 2^{\Omega(n)}$.*

Remark 2. The Corollary 2 as stated above can be proved more easily by exploiting the fact that elementary symmetric polynomials are multilinear (Sect. 6.4 in [7,8]). Our proof method is more general and applicable for non multilinear polynomials as well, and might be of an independent interest. Forbes [7] attributes the idea of using Newton polytopes for lower bounding sparsity of polynomials to Rafael Oliviera's unpublished personal communication. To the best of our knowledge the proof technique we used doesn't appear in any prior published work.

6   Conclusions and Future Directions

Our study reveals that obtaining a good notion of space for the BSS model of algebraic computation still remains a challenging task. We showed that the Boolean part of $\mathsf{LOGSPACE}_w$ is contained in DLOG, however the converse containment is unlikely and it remains open to show that $\mathsf{DLOG} \not\subset \mathsf{LOGSPACE}_w$.

Acknowledgements. We thank the anonymous reviewers for this and an earlier version of the paper for suggestions that helped to improve the presentation of proofs.

References

1. Allender, E., Bürgisser, P., Kjeldgaard-Pedersen, J., Miltersen, P.B.: On the complexity of numerical analysis. SIAM J. Comput. **38**(5), 1987–2006 (2009)
2. Blum, L., Cucker, F., Shub, M., Smale, S.: Complexity and Real Computation. Springer, New York (1997). doi:10.1007/978-1-4612-0701-6
3. Blum, L., Shub, M., Smale, S.: On a theory of computation and complexity over the real numbers: NP-completeness, recursive functions and universal machines. Bull. (New Ser.) Am. Math. Soc. **21**(1), 1–46 (1989)
4. Cucker, F.: $P_R \mathrel{!=} NC_R$. J. Complex. **8**(3), 230–238 (1992)
5. Cucker, F., Grigoriev, D.: On the power of real turing machines over binary inputs. SIAM J. Comput. **26**(1), 243–254 (1997)
6. de Narois, P.J.: A measure of space for computing over the reals. In: Beckmann, A., Berger, U., Löwe, B., Tucker, J.V. (eds.) CiE 2006. LNCS, vol. 3988, pp. 231–240. Springer, Heidelberg (2006). doi:10.1007/11780342_25
7. Forbes, M.A.: Personal communication
8. Forbes, M.A., Shpilka, A., Tzameret, I., Wigderson, A.: Proof complexity lower bounds from algebraic circuit complexity. CoRR, abs/1606.05050 (2016)
9. Fournier, H., Koiran, P.: Are lower bounds easier over the reals? In: Proceedings of 30th Annual ACM Symposium on Theory of Computing, STOC 1998, New York, NY, USA, pp. 507–513. ACM (1998)
10. Gao, S.: Absolute irreducibility of polynomials via Newton polytopes. J. Algebra **237**(1), 501–520 (1997)
11. Gruenbaum, B.: Convex Polytopes. Interscience Publisher, New York (1967)
12. Joglekar, P., Raghavendra Rao, B.V., Sivakumar, S.: On weak-space complexity over complex numbers. In: Electronic Colloquium on Computational Complexity (ECCC), vol. 24, p. 87 (2017)
13. Koiran, P.: Computing over the reals with addition and order. Theoret. Comput. Sci. **133**(1), 35–47 (1994)
14. Koiran, P.: Elimination of constants from machines over algebraically closed fields. J. Complex. **13**(1), 65–82 (1997)
15. Koiran, P.: A weak version of the Blum, Shub, and Smale model. J. Comput. Syst. Sci. **54**(1), 177–189 (1997)
16. Koiran, P., Perifel, S.: VPSPACE and a transfer theorem over the complex field. Theor. Comput. Sci. **410**(50), 5244–5251 (2009)
17. Koiran, P., Perifel, S.: VPSPACE and a transfer theorem over the reals. Comput. Complex. **18**(4), 551–575 (2009)
18. Koiran, P., Portier, N., Tavenas, S., Thomassé, S.: A tau-conjecture for Newton polygons. Found. Comput. Math. **15**(1), 185–197 (2015)

19. Mahajan, M., Raghavendra Rao, B.V.: Small space analogues of valiant's classes and the limitations of skew formulas. Comput. Complex. **22**(1), 1–38 (2013)
20. Meer, K., Michaux, C.: A survey on real structural complexity theory. Bull. Belg. Math. Soc. Simon Stevin **4**(1), 113–148 (1997)
21. Michaux, C.: Une remarque à propos des machines sur $\mathbb{R}$ introduites par Blum, Shub et Smale. Comptes Rendus de l'Académie des Sciences de Paris **309**(7), 435–437 (1989)
22. Morandi, P.: Field and Galois Theory. Graduate Texts in Mathematics. Springer, Cham (1996). doi:10.1007/978-1-4612-4040-2
23. Ostrowski, A.M.: On multiplication and factorization of polynomials, i. lexicographic ordering and extreme aggregates of terms. Aequationes Mathematicae **13**, 201–228 (1975)
24. Schneider, R.: Convex Bodies: The Brunn-Minkowski Theory. Cambridge University Press, Cambridge (1993)
25. Shafarevich, I.R.: Basic Algebraic Geometry, 3rd edn. Springer, Berlin (2013). doi:10.1007/978-3-642-96200-4
26. Shpilka, A., Yehudayoff, A.: Arithmetic circuits: a survey of recent results and open questions. Found. Trends® Theoret. Comput. Sci. **5**(3–4), 207–388 (2010)
27. Tzamaret, I.: Studies in algebraic and propositional proof complexity. Ph.D. thesis, Tel Aviv University (2008)
28. Valiant, L.G.: The complexity of computing the permanent. Theor. Comput. Sci. **8**, 189–201 (1979)
29. Ziegler, G.M.: Lectures on Polytopes. Springer, New York (1995). doi:10.1007/978-1-4613-8431-1

Deterministic Oblivious Local Broadcast in the SINR Model

Tomasz Jurdziński$^{(\boxtimes)}$ and Michał Różański

Institute of Computer Science, University of Wrocław, Wrocław, Poland
`tju@cs.uni.wroc.pl`

Abstract. Local Broadcast is one of the most fundamental communication problems in wireless networks. The task is to allow each node to deliver a message to all its neighbors. In this paper we consider an oblivious and semi-oblivious variants of the problem. The oblivious algorithm is a fixed deterministic schedule of transmissions that tells each station in which rounds it has to transmit. In semi-oblivious variant of the problem we allow a station to quit the execution of the schedule at some point. We present algorithms with complexity of $O(\Delta^{2+2/(\alpha-2)} \log N)$ for the oblivious variant and $O(\Delta \log N)$ for the semi-oblivious case, where $\alpha > 2$ is a path loss parameter, $[1, N]$ is the range of IDs of stations and Δ is the maximal degree in a network. In the latter case we make use of the acknowledgements, which inform a station, after it sent a message, if all its neighbors had received it.

1 Introduction

1.1 The Network Model

We consider a wireless network consisting of nodes located on the 2-dimensional Euclidean plane. We model transmissions in the network with the SINR (*Signal-to-Interference-and-Noise Ratio*) constraints. The model is determined by fixed parameters: path loss $\alpha > 2$, threshold $\beta > 1$, ambient noise $\mathcal{N} > 0$ and uniform transmission power $\mathcal{P}$. The *communication graph* $G = (V, E)$ of a given network consists of all nodes and edges $\{v, u\}$ between nodes that are within distance of at most $1 - \varepsilon$, where $\varepsilon \in (0, 1)$ is a fixed constant (the connectivity parameter). The communication graph, defined as above, is a standard notion in the analysis of ad hoc communication in the SINR model, cf., [4,18]. The value of $SINR(v, u, \mathcal{T})$, for given stations u, v and a set of concurrently transmitting stations $\mathcal{T}$ is defined as follows.

$$SINR(v, u, \mathcal{T}) = \frac{\mathcal{P}/d(v, u)^\alpha}{\mathcal{N} + \sum_{w \in \mathcal{T} \setminus \{v\}} \mathcal{P}/d(w, u)^\alpha} \qquad (1)$$

A node u receives a message from w iff $w \in \mathcal{T}$ and $SINR(w, u, \mathcal{T}) \geq \beta$, where $\mathcal{T}$ is the set of stations transmitting at the same time. *Transmission range* is

This work was supported by the Polish National Science Centre grants DEC-2012/07/B/ST6/01534 and 2014/13/N/ST6/01850.

R. Klasing and M. Zeitoun (Eds.): FCT 2017, LNCS 10472, pp. 312–325, 2017.
DOI: 10.1007/978-3-662-55751-8_25

the maximal distance at which a station can be heard provided there are no other transmitters in the network. Without loss of generality we assume that the transmission range are all equal to 1.

We assume that algorithms work synchronously in rounds. In a single round, a station transmits a message according to the predefined schedule (the algorithm). In the semi-oblivious variant of the problem, a station can quit the execution of the algorithm at some point. This means that nodes do not perform any calculations, nor react to the content of the messages they receive; the only thing a node can do is to transmit its message in rounds defined by the schedule, or – in the semi-oblivious variant – quit the execution of the algorithm after receiving specific feedback.

The *degree* of the network Δ is the maximal number of stations in any ball of radius 1. Observe, that the maximum node degree of the communication graph is $\Theta(\Delta)$.

Each station has a unique identifier from the set $[N]$, where $N = n^{O(1)}$ is the upper bound on the size n of the network. Moreover, the stations know: N, the SINR parameters – $\mathcal{P}, \alpha, \beta, \varepsilon, \mathcal{N}$, and the degree of the network Δ. In the section presenting semi-oblivious algorithm we allow the stations to use the feedback mechanism, which tells the station if its message had been received by all neighbors in given round. It has been widely used in the randomized algorithms for local broadcast problem [3,13].

1.2 Problem Definition and Related Work

The problem of local broadcast is one of the most fundamental communication tasks. We say that an algorithm solves local broadcast problem if, during its execution, each node sends a message that is received by all its neighbors.

In the last years the SINR model was extensively studied. It regards structural properties of so-called SINR-diagrams and reception areas [2,14,20,21] as well as algorithm design for local and global broadcast [3,4,10,11,16,17,19,24, 26], link scheduling [12], and other problems [15,22]. The first work on local broadcast in SINR model by Goussevskaia et al. [10] presented an $O(\Delta \log n)$ randomized algorithm. After that, the problem was studied in various settings. Halldorsson and Mitra presented an $O(\Delta + \log^2 n)$ algorithm in a model with feedback [13]. Recently, for the same setting Barenboim and Peleg presented solution working in time $O(\Delta + \log n \log \log n)$ [3]. For the scenario when degree Δ is not known Yu et al. in [26] improved on the $O(\Delta \log^3 n)$ solution of Goussevskaia et al. to $O(\Delta \log n + \log^2 n)$. However, no *deterministic (oblivious)* algorithm for local broadcast was known in the scenario considered here, i.e., when stations do not know their coordinates.

The local broadcast problem is a generalization of the extensively studied contention resolution problem in multiple-access channels, in which nodes have to send their messages to a shared channel (that corresponds to a neighborhood in our context) [1,5–9,23].

1.3   Our Contribution and Open Problems

To the best knowledge of the authors this paper is the first to investigate non-adaptive (i.e., oblivious or semi-oblivious) deterministic algorithms in the SINR model.

In Sect. 3.1 we present the application of strongly selective families as local broadcast schedules. The complexity of this result is worse than the one of schedules constructed from balanced strongly selective families in Sect. 3.3, but the advantage of this result is that the strongly selective families are constructive (i.e., can be locally computed in polynomial time, c.f., [25]). Also, this result serves as a part of the semi-oblivious schedule in Sect. 4.2.

In Sect. 3.2 we show existence of balanced strongly selective families (bssf) with certain parameters through the probabilistic method. Then, in Sect. 3.3, we prove that bssf can serve as a local broadcast schedule. The length of the schedule is $O(\Delta^{2+2/(\alpha-2)} \log N)$.

In the last section we show existence of fractional balanced selectors through the probabilistic method. The analysis is not standard for such constructions and might be interesting on its own. Then, we apply the result to show existence of the semi-oblivious schedule of length $O(\Delta \log N)$.

Our results indicate that although non-adaptive deterministic local broadcast is a time consuming task, little adaptivity polynomially improves the performance. The issue of efficient construction of balanced strongly selective families and (fractional) balanced selectors remains open.

2   Preliminaries

In this section we present basic definitions and facts to be used in further sections. We use $[n]$ to denote the set of natural numbers $\{1, ..., n\}$. The central object in this paper are families of sets over $[N]$. Any family of subsets over $[N]$ can be regarded as a transmission schedule, assuming that the sets are ordered within a family. We formalize this notion as follows.

A *transmission schedule* of length t is defined by a sequence $\mathcal{S} = (S_1, ..., S_t)$ of subsets of $[N]$, where the ith set determines nodes transmitting in the ith round of the schedule. That is, a node with ID $v \in [N]$ transmits in round i of an execution of $\mathcal{S}$ if and only if $v \in S_i$.

Most of the results in our paper are of probabilistic nature, hence we introduce a suitable notion of random sequences of subsets. We denote by $\mathcal{Q}_{m,p} = (Q_1, ..., Q_m)$ a random sequence subsets of $[N]$ where each $x \in [N]$ is independently put into each Q_i with probability p for each $1 \le i \le m$.

A set $S \subseteq [N]$ *selects* $x \in A$ from $A \subseteq [N]$ when $S \cap A = \{x\}$.

A sequence $\mathcal{S} = (S_1, ..., S_t)$ of sets over $[N]$ is called (N, k)-strongly selective family (or (N, k)-ssf) if for any subset $A \subseteq [N]$ such that $|A| \le k$, and each $x \in A$ there is $i \in [t]$ such that S_i selects x from A.

A sequence $\mathcal{S} = (S_1, ..., S_t)$ of sets over $[N]$ is called $(N, k, \gamma, \boldsymbol{b})$-balanced strongly selective family (or $(N, k, \gamma, \boldsymbol{b})$-bssf), where $\boldsymbol{b} = (b_1, ..., b_l)$, if for each subset $A \subseteq [N]$ such that $|A| \leq k$, and any $B_1, ..., B_l \subseteq [N]$ such that $|B_i| = b_i$, for any $x \in A$ there is $S \in \mathcal{S}$ such that: (i) $S \cap A = \{x\}$, (ii) $|S \cap B_i| \leq \gamma \cdot b_i$ for all i. The sets B_i are called *bounding sets*.

The definition of fractional balanced selector (fbs) is analogous to the definition of bssf, where we require that at least half of the elements of A are selected instead of all of them. We give full definition of fbs in Subsect. 4.1. Note, that we give the probabilistic argument for existence of fbs with specific parameters suited for local broadcast problem in SINR model, however more general statement is possible.

We say that a schedule $\mathcal{S}$ is a *local broadcast schedule* if for any network of degree Δ, for any node $v \in G$ there is a round, when v transmits and is heard by all its neighbors, where G is the communication graph of that network. Formally, for any $v \in G$ there exists $S \in \mathcal{S}$ such that: (i) $v \in S$, (ii) $w \notin S$ for any $w \in N(v)$, (iii) $\mathrm{SINR}(v, w, S) \geq \beta$ for all $w \in N(v)$.

Let $\mathcal{I}_d$ be the maximal value of interference at node v that guarantees that v can receive a message from any station at distance d.

Below we present some basic mathematical facts.

Fact 1. *For all $x \in \mathbb{R}$ we have*

$$e^x \geq 1 + x.$$

Fact 2. *For any natural $k \leq n$ we have*

$$\binom{n}{k} \leq e^{k \ln(n/k) + k}.$$

Fact 3. *(Chernoff Bound) Let $X \sim Bin(n, p)$, where $p \leq 1/2$. Then for any $\delta > 0$ we have*

$$\mathbb{P}(X \geq (1 + \delta)\mathbb{E}[X]) \leq \exp(-\delta^2 \mathbb{E}[X]/(2 + \delta)).$$

Fact 4. *For $p \in (0, 1/2)$ we have*

$$1/4 \leq (1 - p)^{1/p} \leq 1/e.$$

Fact 5. *One can cover a disc of radius $r > 1$ using $8r^2$ discs of radius 1.*

Fact 6. *Let $a > 1, b > 0$, then for all $x \geq 2 \log_a(2b/\ln a)$ we have*

$$\frac{a^x}{x} \geq b.$$

3 Non-adaptive Algorithms

3.1 Application of Strongly Selective Families

In this subsection we present a preliminary result on application of combinatorial structures to solve local broadcast in SINR networks. We explore the parameters of strongly selective families so that the schedules resulting from them are local broadcast schedules.

Proposition 1. *Let T be a set of transmitting stations and let v be a point on the plane. Assume that any ball of radius r contains at most Δ elements of T and $B(v, xr)$ does not contain elements of T for a natural positive number x. Then, the overall strength $I(v) = \sum_{u \in T} P/d(u, v)^\alpha$ of signals from the set T at v is $O\left(r^{-\alpha} x^{-\alpha+2} \Delta\right)$.*

Thanks to the above proposition, in the following corollary we estimate the distance at which there should be no transmitters in order to limit the total power of the signal to a given value.

Corollary 1. *There exists a constant $\tau > 0$ dependent on α, P such that if there are no stations transmitting in $B(v, \tau (\Delta/y)^{1/(\alpha-2)})$ then $I(v) \leq y$, provided that there are at most Δ stations per disc of radius 1.*

The following is a direct consequence of Corollary 1.

Corollary 2. *Let v be the only one transmitting station in $B(v, x)$, assuming that at most Δ stations are transmitting per each unit ball outside of $B(v, x)$. Let $x_\Delta = \tau(\Delta/\mathcal{I}_{1-\varepsilon})^{1/(\alpha-2)} + 1$. If $x \geq x_\Delta$ then every station within distance at most $1 - \varepsilon$ from v can hear the transmission of v.*

The following construction is a natural consequence of the above observations: If a station is a unique transmitter in the set of all stations within distance x_Δ then it is heard by all its neighbors. A ssf-schedule, guaranteeing that each station has a round in which it transmits in such way, would be a local broadcast schedule. In the following Theorem we present such schedules.

Theorem 1. *Let x_Δ be the value defined in Corollary 2, and $\mathbf{S}$ be a $(N, 8\Delta x_\Delta^2)$-ssf schedule. Then $\mathbf{S}$ is a local broadcast schedule for any network of density Δ, and the length of the schedule is $O(\Delta^{2+4/(\alpha-2)} \log(N/(\Delta^{1+2/(\alpha-2)})))$.*

Proof. Let v be any station. By Fact 5, there are at most $8\Delta x_\Delta^2$ stations in $B(v, x_\Delta)$, and by the definition of the schedule $\mathbf{S}$, there is a round in $\mathbf{S}$ when v is the unique transmitter in $B(v, x_\Delta)$. Then, by Corollary 2, the interference in any point of $B(v, 1 - \varepsilon)$ allows to receive the message from v, which concludes the proof.

3.2   Balanced Strongly Selective Families

We say that a sequence $\mathcal{Q}$ of subsets of $[N]$ is a $(N, k, \gamma, \boldsymbol{b})$−balanced ssf, if for any *central set* $A \subseteq [N]$ of size k, any *bounding sets* $B_1, ..., B_l \subseteq [N]$ such that $|B_i| = b_i$, and *limit coefficient* γ, all elements $x \in A$ are *selected* – an element $x \in A$ is said to be selected if there exists a round $Q \in \mathcal{Q}$ such that

(a) $Q \cap A = \{x\}$,
(b) $|Q \cap B_i| \leq \gamma \cdot b_i$.

Note, that the value of l is not defined here. It is not influencing any calculations, however it is important to state that the construction works for a fixed value of l.

Our goal is to show the existence of balanced strongly selective families of minimal size.

Given the bounding sets $B_1, ..., B_l$ we say that the round Q is *quiet* if it satisfies the condition b). In the next proposition we show that, with appropriate choice of p, for fixed bounding sets $B_1, ..., B_l$ the probability that a single round is good is at least $1/2$.

Proposition 2. *Let $B_1, ..., B_l$ be fixed sets such that $|B_1| < |B_2| < ... < |B_l|$, and $\gamma \leq 1$. Then for $Q \in \mathcal{Q}_{m,p}$ we have*

$$\mathbb{P}\left(\bigcap_{i \leq l} |Q \cap B_i| \leq \gamma \cdot b_i\right) \geq 1/2,$$

provided that the value of p satisfies $p \leq \gamma/2$, and $p \cdot b_i \geq 6 + i$.

A *configuration* is a tuple of form $(A, B_1, ..., B_l)$, representing a possible arrangement of elements into central set and bounding sets. Let us define a set of all possible configurations by $\mathcal{F}$, formally

$$\mathcal{F} = \{(A, B_1, ..., B_l) : A, B_1, ..., B_l \subseteq [N], \ |A| = k, \ \forall_{i \neq j} \ |B_i| = b_i,$$
$$A \cap B_i = \varnothing, \ B_j \cap B_i = \varnothing\}.$$

We have the following bound on the size of $\mathcal{F}$.

Proposition 3. *Let $\mathcal{F}$ be the set of all possible configurations. Then*

$$|\mathcal{F}| \leq N^{k+s},$$

where $s = b_1 + ... + b_l$.

Observe, that $\mathcal{Q}$ is not a $(N, k, \gamma, \boldsymbol{b})$-bssf if for some configuration $(A, B_1, ..., B_l)$ there is an element of A that is not selected. In the next propositions we bound the probability that for a fixed choice of $A, B_1, ..., B_l$ a random sequence $\mathcal{Q}_{m,p}$ does not select some element of A.

Proposition 4. *Let $A, B_1, ..., B_l$ be a fixed configuration, and $v \in A$. The probability that v is selected in a single round $Q \in \mathcal{Q}_{m,p}$ is at least $p/4^{pk+1}$, provided that the value of p satisfies $p \leq \gamma/2$, and $p \cdot b_i \geq i + 6$.*

Now we bound the probability that $\mathcal{Q}_{m,p}$ selects all elements of A in quiet rounds for a fixed configuration $A, B_1, ..., B_l$. We say that a family S over $[N]$ is *good* for a configuration $(A, B_1, ..., B_l) \in \mathcal{F}$ if it selects all elements of A.

Proposition 5. *Let $A, B_1, ..., B_l$ be a fixed configuration. Let $p \leq \gamma/2$, and $p \cdot b_i \geq i + 6$. Then we have*

$$\mathbb{P}(\mathcal{Q}_{m,p} \text{ is not good for } A, B_1, ..., B_l) \leq k \cdot \exp(-m \cdot p/4^{pk}).$$

Lemma 1. *Let $p \leq \gamma/2$, and $p \cdot b_i \geq i + 6$. Then we have*

$$\mathbb{P}(\mathcal{Q}_{m,p} \text{ is } (N, k, \gamma, \boldsymbol{b})\text{-}bssf) > 0,$$

provided that $m > 4^{pk+1} \cdot \frac{2k+s}{p} \ln N$, where $s = b_1 + \ldots + b_l$.

Proof. The sequence $\mathcal{Q}_{m,p}$ is a $(N, k, \gamma, \boldsymbol{b})$-balanced selector if it is good for all $(\tilde{A}, \tilde{B}_1, \ldots, \tilde{B}_l) \in \mathcal{F}$. Thus,

$$\mathbb{P}(\mathcal{Q}_{m,p} \text{ is not } (N, k, \gamma, \boldsymbol{b}\text{-bssf})) \leq \sum_{c \in \mathcal{F}} \mathbb{P}(\mathcal{Q}_{m,p} \text{ is not good for } \mathbf{c})$$
$$\leq |\mathcal{F}| \cdot k \exp(-m \cdot p/4^{pk+1})$$
$$\leq N^{k+s} \cdot k \exp(-m \cdot p/4^{pk})$$
$$\leq \exp((2k+s) \ln N - m \cdot p/4^{pk})$$

Thus, in order to guarantee $\mathbb{P}(\mathcal{Q}_{m,p}$ is not $(N, k, \gamma, \boldsymbol{b}\text{-bssf})) < 1$ it is sufficient that m satisfies the following inequality

$$(2k + s) \ln N - m \cdot p/4^{pk} < 0,$$

which is true given the assumptions regarding the value of m.

3.3 Oblivious Local Broadcast with BSSFs

In this section we provide an application of the balanced selective families to Local Broadcast in the SINR model. We proceed analogously to the result in Theorem 1 – we show that for any network of density Δ each node transmits in some round when it is possible for all its neighbors to receive the message. The main difference with the previous result is that we do not demand that a station will be a unique transmitter among its very broad neighborhood, instead we analyze the interference carefully and exploit its nature by applying balanced selective families.

First, we show for a certain choice of k, b_i, γ there exists a bssf of size guaranteed by the Lemma 1 for such parameters. Then, we prove that a schedule constructed from such bssf is feasible local broadcast schedule.

Let c_1 be a constant, whose value will be defined later. We allow for c_1 to depend on the model constants, that is $\alpha, \beta, \mathcal{P}$.

Let $k = c_1 \Delta$, $b_i = c_1 \Delta \cdot 2^{2i}$, and $\gamma = 1/\Delta$. Since the length of the bssf from Lemma 1 depends highly on $s = b_1 + \ldots + b_l$ we have an incentive to choose the value of l to be as small as possible. On the other hand it is important to capture the behavior of the interference in the network through the bounding sets, which implies that the cardinality of all bounding sets needs to be high. We set the value of l to $\lambda = \lceil \log(2 + \tau(2\Delta/\mathcal{I}_{1-\varepsilon})^{1/(\alpha-2)}) + 5 \rceil$. We give more details for this value later.

Definition 1. *A geometric configuration for a point x on a plane is a partition of the network into $l + 1$ sets $A^{(x)}, B_1^{(x)}, ..., B_l^{(x)}$ in the following way. The k stations that are closest to x constitute the set A. The next b_i stations that are closest to x go to $B_1^{(x)}$, and so on. We call the point x the* perspective.

Note that the above definition is ambiguous, because it is not clear which elements to choose if there is more than one possibility. However, this will not be a problem in the following analysis, since the definition is used only to connect a configuration $A, B_1, ..., B_l$ of sets with its location on a plane through the perspective point.

We identify rounds of communication with the subset of transmitting stations $Q \subseteq [N]$. For a fixed geometric configuration we say that the round Q is *quiet* if for all i we have $|Q \cap B_i^{(x)}| \leq \gamma |B_i^{(x)}|$, which corresponds to the condition present in the definition of balanced strongly selective families.

Proposition 6. *For any geometric configuration $A^{(x)}, B_1^{(x)}, ..., B_l^{(x)}$ and $w \in B_i^{(x)}$, we have $d(x, w) \geq \sqrt{c_1} \cdot 2^{i-5}$.*

Consider some point x on a plane. Our idea of handling the interference in the network is to divide it into three groups. The first group consists of first k stations closest to x, namely $A^{(x)}$. The second group consists of $B_1^{(x)} \cup ... \cup B_\lambda^{(x)}$ (recall that $\lambda = \lceil \log(2 + \tau(2\Delta/\mathcal{I}_{1-\varepsilon})^{1/(\alpha-2)}) + 5 \rceil$), and the last group consists of all other stations. We denote the groups by $H_1^{(x)}, H_2^{(x)}$, and $H_3^{(x)}$. The partition of the network into three groups here is crucial to the complexity of the algorithm. The interference from stations within small distance is far more influential than the interference from distant stations, thus we capture the $H_2^{(x)}$ in the bounding sets, and analyze interference coming from there carefully. On the other hand, we bound the interference from $H_3^{(x)}$ less precisely, using Corollary 1. Note, that it would be possible to fit all the stations into bounding sets (that is H_2), but then the total cardinality of bounding sets, that is $s = b_1 + ... + b_l$, would be $\Omega(N)$ and the length of the bssf would be $\Omega(N \log N)$ (see Lemma 1).

Let us recall that $\mathcal{I}_d$ denotes the maximal value of interference in a fixed point v that guarantees that v can receive a message from any station at distance d, and $\mathcal{I}(X, x) = \sum_{v \in X} \mathcal{P}/d(v, x)^\alpha$. In the following two propositions we bound the interference from $H_2^{(x)}$ and $H_3^{(x)}$ for any configuration $(A^{(x)}, B_1^{(x)}, ..., B_\lambda^{(x)})$.

Proposition 7. *Let $c = (A^{(x)}, B_1^{(x)}, ..., B_\lambda^{(x)}) \in \mathcal{F}$ be a fixed geometric configuration. Then in each quiet round $Q \subseteq [N]$, the maximal interference coming from $H_2^{(x)}$ in $B(x, 2)$ is bounded as follows,*

$$\mathcal{I}(Q \cap H_2^{(x)}, v) \leq \mathcal{I}_{1-\varepsilon}/2 \quad \text{for each } v \in B(x, 2),$$

provided that $c_1 \geq \max\{2^{14}, (\frac{2^{6\alpha+1}\mathcal{P}}{\mathcal{I}_{1-\varepsilon}(2^{\alpha-2}-1)})^{2/(\alpha-2)}\}$.

Proof. The good round for x means that at most γb_i nodes are transmitting from $B_i^{(x)}$ for each i. Let d_i denote the minimal distance from the nodes in $B_i^{(x)}$

to $B(x, 2)$. Thanks to Proposition 6 we have $d_i \geq \sqrt{c_1} \cdot 2^{i-5} - 2 \geq \sqrt{c_1} \cdot 2^{i-6}$ (provided that $c_1 \geq 2^{14}$). Denote by $\mathcal{I}$ the maximal interference in $B(x, 2)$ coming from the stations in $\bigcup_{i \leq \lambda} B_i^{(x)}$ in a good round for c. We have

$$\mathcal{I} \leq \sum_{1 \leq i \leq \lambda} \gamma |B_i| \cdot \frac{\mathcal{P}}{d_i^\alpha}$$

$$\leq \sum_{1 \leq i \leq \lambda} c_1^{1-\alpha/2} \cdot 2^{i(2-\alpha)} \cdot 2^{6\alpha} \cdot \mathcal{P}$$

$$\leq \frac{2^{6\alpha}\mathcal{P}}{c_1^{\alpha/2-1}} \sum_{i \geq 1} \left(\frac{1}{2^{\alpha-2}} \right)^i = \frac{2^{6\alpha}\mathcal{P}}{c_1^{\alpha/2-1}(2^{\alpha-2} - 1)}.$$

Thus, for $c_1 \geq (\frac{2^{6\alpha+1}\mathcal{P}}{\mathcal{I}_{1-\varepsilon}(2^{\alpha-2}-1)})^{2/(\alpha-2)}$ we have $\mathcal{I} \leq \mathcal{I}_{1-\varepsilon}/2$.

Proposition 8. *In every round the maximal interference coming from $H_3^{(x)}$ to any station $v \in B(x, 2)$ is at most $\mathcal{I}_{1-\varepsilon}/2$.*

Proof. Let us recall $\lambda = \lceil \log(2 + \tau(2\Delta/\mathcal{I}_{1-\varepsilon})^{1/(\alpha-2)}) + 5 \rceil$. Thanks to Proposition 6 we know that for each station $v \in H_3^{(x)}$ we have

$$d(v, x) \geq \sqrt{c_1} \cdot 2^{\lambda-5} \geq 2 + \tau(2\Delta/\mathcal{I}_{1-\varepsilon})^{1/(\alpha-2)},$$

since all stations in $H_3^{(x)}$ lay farther than stations from $B_\lambda^{(x)}$. Thus, for each $u \in B(x, 2)$, and $v \in H_3^{(x)}$ we have $d(u, v) \geq \tau(2\Delta/\mathcal{I}_{1-\varepsilon})^{1/(\alpha-2)}$ which by Corollary 1 assures that $\mathcal{I}(H_3^{(x)}, u) \leq \mathcal{I}_{1-\varepsilon}/2$ for each $u \in B(x, 2)$. ∎

We have shown a way to capture interference in a network by partitioning it into groups and analyzing the interference in them separately. We crafted the analysis so it fits the construction of balanced strongly selective family. Now, we put all the pieces together in the following theorem.

Theorem 2. *There exists a local broadcast schedule $\mathbf{S}$ feasible for networks of density Δ of length $O(\Delta^{2+2/(\alpha-2)} \log N)$.*

Proof. We show that a schedule constructed from $(N, k, \gamma, \boldsymbol{b})$-bssf with $\boldsymbol{b} = (b_1, ..., b_\lambda)$, where $b_i = c_1 \Delta \cdot 2^{2i}$, $\lambda = \lceil \log(2 + \tau(2\Delta/\mathcal{I}_{1-\varepsilon})^{1/(\alpha-2)}) + 5 \rceil$, $k = c_1 \Delta$, $c_1 = \max\{2^{14}, (\frac{2^{6\alpha+1}\mathcal{P}}{\mathcal{I}_{1-\varepsilon}(2^{\alpha-2}-1)})^{2/(\alpha-2)}\}$ is feasible local broadcast schedule for any network of density Δ.

Let $\mathbf{S}$ be the smallest $(N, k, \gamma, \boldsymbol{b})$-bssf. We need to show that during the execution of $\mathbf{S}$ each station is heard by all its neighbors. Let us fix a station v and a point x on a plane such that $v \in B(x, 1)$. The definition of $\mathbf{S}$ guarantees that there exists a round $Q \in \mathbf{S}$ such that $Q \cap A^{(x)} = \{v\}$, and Q is quiet for $B_1^{(x)}, ..., B_\lambda^{(x)}$. Let us denote $\tilde{Q} = Q \smallsetminus \{v\}$. The total interference on any neighbor w of v is equal to

$$\mathcal{I}(\tilde{Q} \cap V, w) = \mathcal{I}(\tilde{Q} \cap H_1^{(x)}, w) + \mathcal{I}(\tilde{Q} \cap H_2^{(x)}, w) + \mathcal{I}(\tilde{Q} \cap H_3^{(x)}, w) \leq \mathcal{I}_{1-\varepsilon},$$

thanks to Propositions 7 and 8, and the fact that $\mathcal{I}(\tilde{Q} \cap H_1^{(x)}, w) = 0$ (recall that $H_1^{(x)} = A^{(x)}$, and $\tilde{Q} \cap A^{(x)} = \varnothing$). This shows that in this round v transmits and interference in all its neighbors is at most $\mathcal{I}_{1-\varepsilon}$ which allows them to receive the message.

It remains to show that the length of $\mathbf{S}$ is $O(\Delta^{2+2/(\alpha-2)} \log N)$. Now, we use Lemma 1 to show the existence of small bssf with parameters defined at the beginning of this proof.

Let $p = \gamma/2$ and $m = \lceil 2^{c_1+1} \Delta (2\Delta c_1 + s) \ln N \rceil$. In order to use Lemma 1 we need to meet its assumptions, that is to guarantee that $p \cdot b_i \geq i + 6$. It is easy to check, that this true for all $c_1 \geq 3$. By Lemma 1 we know that $\mathcal{Q}_{m,p}$ is a $(N, k, \gamma, \boldsymbol{b})$-bssf with non-zero probability, thus there exists bssf of given parameters of size at most m. In order to estimate the value of m let us bound the sum of sizes of all bounding sets,

$$s = b_1 + \ldots + b_\lambda = \sum_{1 \leq i \leq \lambda} c_1 \Delta \cdot 2^{2i} = c_1 \Delta \frac{4^{\lambda+1} - 1}{3}$$

$$\leq 2^8 c_1 \Delta \left(2 + \tau(\frac{2\Delta}{\mathcal{I}_{1-\varepsilon}})^{1/(\alpha-2)}\right)^2 = O(\Delta^{\alpha/(\alpha-2)}).$$

Thus the size of $\mathbf{S}$ is at most $m = O(\Delta(\Delta + \Delta^{\alpha/(\alpha-2)}) \log N) = O(\Delta^{2+2/(\alpha-2)} \log N)$, which concludes the proof.

4   Local Broadcast with Feedback

In this section we provide another structure enabling to accomplish *partial* local broadcast – meaning that only a fraction of nodes will be heard by all its neighbors. This allows us to substantially reduce the size of the schedule to $O(\Delta \log N)$ in networks of diameter Δ. Such approach also gives us an efficient solution to local broadcast in a scenario of *semi-oblivious* networks, with acknowledgments. In such networks, the only action of a node, apart from executing a given schedule, is to quit the protocol at the point when the node was heard by all its neighbors.

We construct a combinatorial structure similar to the one used in previous section. However, there are some major changes both in application of the structure to the SINR networks, and in the analysis of the structure itself. The analysis from the previous section does not allow to reduce the size of the family to $O(\Delta \log N)$. To enable this, we introduce changes in the definition of bounding sets, allowing for more stations to be selected in quiet rounds. However, this enforces a more careful analysis of the interference in the network later.

4.1   Fractional Balanced Selectors

In this section we give the definition of fractional balanced selectors (FBS), which is suited for the use in the local broadcast algorithm. That definition may be generalized, allowing for another parameters, and perhaps be used in other algorithms. However, in this section whenever we speak of fractional balanced selector, we mean an FBS with specific parameters, that are defined in the next paragraph. We treat the selectors as if they are transmission schedules already. Particularly, if some element $v \in [N]$ is present in $Q \subseteq [N]$ we say that v transmits in round Q.

Our goal here is to show a *fractional balanced selector*, that is Q such that, for any C, A, such that $C \subseteq A \subseteq [N]$, where $|C| = \Delta$ and $|A| = k$ there are rounds $Q \in \mathcal{Q}$ for at least half of all elements $v \in C$ such that: (i) $Q \cap A = \{v\}$, (ii) $|Q \cap B_i| \leq |B_i|\delta^{i+k}/k$ for all i, where $|B_i| = b_i, b_i = k2^{2i}, k = c_1\Delta, c_1 = (\delta^c)^{(2+\alpha)/(2\alpha)}$, and α is the SINR Model constant, and values of δ, and c are to be determined later.

Wherever we say that $C, A, B_1, ..., B_l$ are fixed configuration, we mean that they are disjunct, fixed subsets of $[N]$, such that $|C| = \Delta, |A| = k$, and $|B_i| = k2^{2i}$.

The general idea in this section is to prove that the family $\mathcal{Q}_{m,p}$ for certain values of m and p is FBS with non-zero probability. We do it in two steps. First, we show that with probability greater than $1/2$ it satisfies the following property: for all configurations $(C, A, B_1, ..., B_l)$ there are at least σ rounds Q such that $|Q \cap B_i| \leq |B_i|\delta^{i+k}/k$ for all i (see the second condition in the above definition). Then, we show that with probability greater than $1/2$ *any* subset of σ rounds satisfies the first condition, that is it selects at least half of elements in C. Hence, with non-zero probability there is a subset of σ rounds in which the number of transmitters from B_i's is bounded and elements of C are being selected.

A sequence of sets $\mathcal{Q}$ is *i-bounding* if $\sum_{Q \in \mathcal{Q}} |Q \cap B| \leq mb_i/k$ for all $B \subseteq [N]$ of size b_i. We say that $\mathcal{Q}$ is *bounding* if it is i-bounding for all $i \leq l$. In the following two propositions we show that with probability greater than $1/2$ the family $\mathcal{Q}_{m,p}$ is bounding.

Proposition 9. *Let $C, A, B_1, ..., B_l$ be a fixed configuration, and $p = 1/(2k)$. For $\mathcal{Q}_{m,p}$ we define $X_i = \sum_{Q \in \mathcal{Q}_{m,p}} |Q \cap B_i|$, that is the total number of "transmissions" from the set B_i. We have*

$$\mathbb{P}(X_i \geq m|B_i|/k) \leq \exp(-m|B_i|/6k).$$

Proposition 10. *Let $p = 1/(2k)$, then we have*

$$\mathbb{P}(\mathcal{Q}_{m,p} \text{ is bounding}) > \frac{1}{2},$$

provided that $m > 6k \left(\ln(2l) + \ln N + 1 \right)$.

The following lemma is purely deterministic. We show there that if a family of sets $\mathcal{Q}$ is bounding, then it cannot have too many spoiled rounds.

Lemma 2. *If a family $\mathcal{Q}$ of size m is bounding then for any configuration $A, B_1, ..., B_l$ the number of spoiled rounds in $\mathcal{Q}$ is at most m/κ, where $\kappa = (\delta^c(\delta - 1))$.*

This concludes the first step of the analysis. Now, we know that with probability at least $1/2$ there at most m/κ rounds are spoiled by too many transmitters from bounding sets. It remains to show that for any subset of $m - m/\kappa$ quiet rounds there are at least $\Delta/2$ distinct selections from the set C.

We say that a family of sets $\mathcal{Q}$ is (N, a, b)-*inner-selective* if for any sets $\hat{A} \subseteq \hat{B}$ such that $|\hat{A}| = a, |\hat{B}| = b$ for at least half elements $x \in \hat{A}$ of $\hat{A}$ there exists an $Q \in \mathcal{Q}$ such that $Q \cap B = \{x\}$. We say that a family $\mathcal{Q}$ of subsets of $[N]$ is (N, Δ, k, σ)-*good* if any subset $\mathcal{Q}' \subseteq \mathcal{Q}$ of size σ is (N, Δ, k)-inner-selective.

Now, we are ready to make the second step in the proof. We show, that for certain parameters $\mathcal{Q}_{m,p}$ is (N, Δ, k, σ)-good with probability at least $1/2$.

Lemma 3. *Let* $p = 1/(2k)$, $\kappa = \delta^c(\delta - 1)$, *and* $\sigma = m(1 - \frac{1}{\kappa})$.

$$\mathbb{P}\left(\mathcal{Q}_{m,p} \text{ is } (N, \Delta, k, \sigma) - good\right) \geq \frac{1}{2},$$

Provided that $m \geq \frac{16k}{\Delta(1-1/\kappa)}(k \ln(N/k) + 3k + \ln 2)$, *and* $c \geq 2 \log_a(2b/\ln a)$, *where* $a = \delta^{(\alpha-2)/(2\alpha)}$, $b = \max\{\frac{32(\ln \delta + \ln(\delta-1)+1)}{\delta-1}, 1\}$.

Theorem 3. *Let* $p = 1/(2k)$, $\kappa = \delta^c(\delta - 1)$, $\sigma = m(1 - 1/\kappa)$

$$\mathbb{P}(\mathcal{Q}_{m,p} \text{ is a fractional balanced selector}) > 0,$$

provided that $m \geq \max\{6k(\ln(2l) + \ln N + 1) + 1, \frac{16k}{\Delta(1-1/\kappa)}(k \ln(N/k) + 3k + \ln 2)\}$, *and* $c \geq 2 \log_a(2b/\ln a)$, *where* $a = \delta^{(\alpha-2)/(2\alpha)}$, $b = \max\{\frac{32(\ln \delta + \ln(\delta-1)+1)}{\delta-1}, 1\}$.

Because of the complicated formulas in the theorem above, we state its main consequence more clearly in the following corollary.

Corollary 3. *There exists a fractional balanced selector of size* $O(\Delta \log(N))$.

4.2 Semi-oblivious Algorithm with Acknowledgements

Observe, that if we allow for a node to get an acknowledgement when all its neighbors receive its message, then it can quit from further execution of the protocol. Then, after first execution of FBS for networks of density Δ, the density of the network drops to $\Delta/2$, since all the nodes that transmitted successfully during the first FBS quit the protocol. Then, we run an FBS for networks of density $\Delta/2$, and so on. When the density of the network drops below $\Delta^{1/(2+4/(\alpha-2))}$ we use the result of Theorem 1 to make sure that all nodes transmitted successfully.

Theorem 4. *There exists a semi-oblivious algorithm for local broadcast in networks of density* Δ *that runs in* $O(\Delta \log N)$ *rounds, assuming that nodes are capable of using acknowledgements of successful transmissions.*

Acknowledgments. The authors would like to thank Darek Kowalski for his comments to the paper.

References

1. Anta, A.F., Mosteiro, M.A., Munoz, J.R.: Unbounded contention resolution in multiple-access channels. Algorithmica **67**, 295–314 (2013)
2. Aronov, B., Katz, M.J.: Batched point location in SINR diagrams via algebraic tools. In: Halldórsson, M.M., Iwama, K., Kobayashi, N., Speckmann, B. (eds.) ICALP 2015. LNCS, vol. 9134, pp. 65–77. Springer, Heidelberg (2015). doi:10. 1007/978-3-662-47672-7_6
3. Barenboim, L., Peleg, D.: Nearly optimal local broadcasting in the SINR model with feedback. In: Scheideler, C. (ed.) Structural Information and Communication Complexity. LNCS, vol. 9439, pp. 164–178. Springer, Cham (2015). doi:10.1007/ 978-3-319-25258-2_12
4. Daum, S., Gilbert, S., Kuhn, F., Newport, C.: Broadcast in the ad hoc SINR model. In: Afek, Y. (ed.) DISC 2013. LNCS, vol. 8205, pp. 358–372. Springer, Heidelberg (2013). doi:10.1007/978-3-642-41527-2_25
5. De Marco, G., Kowalski, D.: Fast nonadaptive deterministic algorithm for conflict resolution in a dynamic multiple-access channel. SIAM J. Comput. **44**(3), 868–888 (2015)
6. De Marco, G., Kowalski, D.: Contention resolution in a non-synchronized multiple access channel. In: 27th Annual IEEE International Symposium on Parallel and Distributed Processing (IPDPS 2013), Cambridge, MA, USA (2013)
7. De Marco, G., Kowalski, D.R.: Towards power-sensitive communication on a multiple-access channel. In: 30th International Conference on Distributed Computing Systems (ICDCS 2010), Genoa, Italy, May 2010
8. De Marco, G., Pellegrini, M., Sburlati, G.: Faster deterministic wakeup in multiple access channels. Discrete Appl. Math. **155**(8), 898–903 (2007)
9. De Marco, G., Stachowiak, G.: Asynchronous Shared Channel, PODC 2017. Accepted
10. Goussevskaia, O., Moscibroda, T., Wattenhofer, R.: Local broadcasting in the physical interference model. In: Segal, M., Kesselman, A. (eds.) DIALM-POMC, pp. 35–44. ACM (2008)
11. Halldórsson, M.M., Holzer, S., Lynch, N.A.: A local broadcast layer for the SINR network model. In: Proceedings of the 2015 ACM Symposium on Principles of Distributed Computing, PODC 2015, Donostia-San Sebastián, 21–23 July 2015, Spain, pp. 129–138 (2015)
12. Halldórsson, M.M., Mitra, P.: Nearly optimal bounds for distributed wireless scheduling in the SINR model. In: Aceto, L., Henzinger, M., Sgall, J. (eds.) ICALP 2011. LNCS, vol. 6756, pp. 625–636. Springer, Heidelberg (2011). doi:10.1007/ 978-3-642-22012-8_50
13. Halldórsson, M.M., Mitra, P.: Towards tight bounds for local broadcasting. In: Kuhn, F., Newport, C.C. (eds.) FOMC, p. 2. ACM (2012)
14. Halldórsson, M.M., Tonoyan, T.: How well can graphs represent wireless interference? In: Proceedings of the Forty-Seventh Annual ACM on Symposium on Theory of Computing, STOC 2015, 14–17 June 2015, Portland, OR, USA, pp. 635–644 (2015)
15. Hobbs, N., Wang, Y., Hua, Q.-S., Yu, D., Lau, F.C.M.: Deterministic distributed data aggregation under the SINR model. In: Agrawal, M., Cooper, S.B., Li, A. (eds.) TAMC 2012. LNCS, vol. 7287, pp. 385–399. Springer, Heidelberg (2012). doi:10.1007/978-3-642-29952-0_38

16. Jurdzinski, T., Kowalski, D.R.: Distributed backbone structure for algorithms in the SINR model of wireless networks. In: Aguilera, M.K. (ed.) DISC 2012. LNCS, vol. 7611, pp. 106–120. Springer, Heidelberg (2012). doi:10.1007/978-3-642-33651-5_8
17. Jurdzinski, T., Kowalski, D.R., Rozanski, M., Stachowiak, G.: Distributed randomized broadcasting in wireless networks under the SINR model. In: Afek, Y. (ed.) DISC 2013. LNCS, vol. 8205, pp. 373–387. Springer, Heidelberg (2013). doi:10.1007/978-3-642-41527-2_26
18. Jurdzinski, T., Kowalski, D.R., Rozanski, M., Stachowiak, G.: On the impact of geometry on ad hoc communication in wireless networks. In: Halldórsson, M.M., Dolev, S. (eds.) ACM Symposium on Principles of Distributed Computing, PODC 2014, 15–18 July 2014, Paris, France, pp. 357–366. ACM (2014)
19. Jurdzinski, T., Kowalski, D.R., Stachowiak, G.: Distributed deterministic broadcasting in wireless networks of weak devices. In: Fomin, F.V., Freivalds, R., Kwiatkowska, M., Peleg, D. (eds.) ICALP 2013. LNCS, vol. 7966, pp. 632–644. Springer, Heidelberg (2013). doi:10.1007/978-3-642-39212-2_55
20. Kantor, E., Lotker, Z., Parter, M., Peleg, D.: The minimum principle of SINR: a useful discretization tool for wireless communication. In: IEEE 56th Annual Symposium on Foundations of Computer Science, FOCS 2015, 17–20 October 2015, Berkeley, CA, USA, pp. 330–349 (2015)
21. Kantor, E., Lotker, Z., Parter, M., Peleg, D.: The topology of wireless communication. J. ACM **62**(5), 37:1–37:32 (2015)
22. Kesselheim, T.: A constant-factor approximation for wireless capacity maximization with power control in the SINR model. In: Randall, D. (ed.) SODA, pp. 1549–1559. SIAM (2011)
23. Komlos, J., Greenberg, A.G.: An asymptotically fast nonadaptive algorithm for conflict resolution in multiple-access channels. IEEE Trans. Inf. Theory **31**(2), 302–306 (1985)
24. Moscibroda, T., Wattenhofer, R.: The complexity of connectivity in wireless networks. In: INFOCOM 2006 25th IEEE International Conference on Computer Communications, Joint Conference of the IEEE Computer and Communications Societies, 23–29 April 2006, Barcelona, Catalunya, Spain (2006)
25. Porat, E., Rothschild, A.: Explicit nonadaptive combinatorial group testing schemes. IEEE Trans. Inf. Theory **57**(12), 7982–7989 (2011)
26. Yu, D., Hua, Q., Wang, Y., Lau, F.C.M.: An o(log n) distributed approximation algorithm for local broadcasting in unstructured wireless networks. In: IEEE 8th International Conference on Distributed Computing in Sensor Systems, DCOSS 2012, 16–18 May 2012, Hangzhou, China, pp. 132–139 (2012)

Undecidability of the Lambek Calculus with Subexponential and Bracket Modalities

Max Kanovich[1], Stepan Kuznetsov[1,2]($\boxtimes$), and Andre Scedrov[1,3]

[1] National Research University Higher School of Economics,
Moscow, Russian Federation
mkanovich@hse.ru, sk@mi.ras.ru, scedrov@math.upenn.edu
[2] Steklov Mathematical Institute, RAS, Moscow, Russian Federation
[3] University of Pennsylvania, Philadelphia, USA

Abstract. The Lambek calculus is a well-known logical formalism for modelling natural language syntax. The original calculus covered a substantial number of intricate natural language phenomena, but only those restricted to the context-free setting. In order to address more subtle linguistic issues, the Lambek calculus has been extended in various ways. In particular, Morrill and Valentín (2015) introduce an extension with so-called exponential and bracket modalities. Their extension is based on a non-standard contraction rule for the exponential that interacts with the bracket structure in an intricate way. The standard contraction rule is not admissible in this calculus. In this paper we prove undecidability of the derivability problem in their calculus. We also investigate restricted decidable fragments considered by Morrill and Valentín and we show that these fragments belong to the NP class.

1 Linguistic Introduction

The Lambek calculus [23] is a substructural, non-commutative logical system (a variant of linear logic [15] in its intuitionistic non-commutative version [1]) that serves as the logical base for categorial grammars, a formalism that aims to describe natural language by means of logical derivability (see Buszkowski [9], Carpenter [11], Morrill [30], Moot and Retoré [28], *etc.*). The idea of categorial grammar goes back to works of Ajdukiewicz [2] and Bar-Hillel [3], and afterwards it developed into several closely related frameworks, including combinatory categorial grammars (CCG, Steedman [39]), categorial dependency grammars (CDG, Dikovsky and Dekhtyar [12]), and Lambek categorial grammars. A categorial grammar assigns syntactic categories (types) to words of the language. In the Lambek setting, types are constructed using two division operations, \ and /, and the product, ·. Intuitively, $A \setminus B$ denotes the type of a syntactic object that lacks something of type A on the left side to become an object of type B; $B \mathbin{/} A$ is symmetric; the product stands for concatenation. The Lambek calculus provides a system of rules for reasoning about syntactic types.

© Springer-Verlag GmbH Germany 2017
R. Klasing and M. Zeitoun (Eds.): FCT 2017, LNCS 10472, pp. 326–340, 2017.
DOI: 10.1007/978-3-662-55751-8_26

For a quick example, consider the sentence *"John loves Mary."* Let *"John"* and *"Mary"* be of type N (noun), and *"loves"* receive the type $(N \backslash S)/N$ of the transitive verb: it takes a noun from the left and a noun from the right, yielding a sentence, S. This sentence is judged as a grammatical one, because $N, (N \backslash S)/N, N \rightarrow S$ is a theorem in the Lambek calculus.

The Lambek calculus is capable of handling more complicated situations, including dependent clauses: *"the girl whom John loves"*, parsed as N using the following types: $N/CN, CN, (CN \backslash CN)/(S/N), N, (N \backslash S)/N \rightarrow N$ (here CN stands for "common noun," a noun without an article), and coordination: *"John loves Mary and Pete loves Kate,"* where *"and"* is $(S \backslash S)/S$.

There are, however, even more sophisticated cases for which the pure Lambek calculus is known to be insufficient (see, for example, [28,30]). On the one hand, for a noun phrase like *"the girl whom John met yesterday"* it is problematic to find a correct type for *"whom,"* since the dependent clause *"John met yesterday"* expects the lacking noun (*"John met ... yesterday"*; the *"..."* place is called *gap*) in the middle, and it is neither of type S/N nor of type $N \backslash S$. This phenomenon is called *medial extraction*. On the other hand, the grammar sketched above generates, for example, **"the girl whom John loves Mary and Pete loves."* The asterisk indicates ungrammaticality—but, nevertheless, *"John loves Mary and Pete loves"* is of type S/N. To avoid this, one needs to block extraction from certain syntactic structures (*e.g.,* compound sentences), called *islands* [30,38].

These issues can be addressed by extending the Lambek calculus with extra connectives (that allow us to derive more theorems) and also with a more sophisticated syntactic structure (that allows blocking unwanted derivations). In the next section, we follow Morrill and Valentín [30,33] and define an extension of the Lambek calculus with a subexponential modality (allowing medial and also so-called parasitic extraction) and brackets (for creating islands).

2   Logical Introduction

In order to block ungrammatical extractions, such as discussed above, Morrill [29] and Moortgat [27] introduce an extension of the Lambek calculus with brackets that create islands. For the second issue, medial extraction, Morrill and Valentín [4,33] suggest using a modality which they call "exponential," in the spirit of Girard's exponential in linear logic [15]. We use the term "subexponential," which is due to Nigam and Miller [34], since this modality allows only some of the structural rules (permutation and contraction, but not weakening). The difference from [34], however, is in the non-commutativity of the whole system and the non-standard nature of the contraction rule.

We consider $!_b\mathbf{L}^1$, the Lambek calculus with the unit constant [24], brackets, and a subexponential controlled by rules from [33]. The calculus $!_b\mathbf{L}^1$ is a conservative fragment of the $\mathbf{Db}!_b$ system by Morrill and Valentín [33].

Due to brackets, the syntax of $!_b\mathbf{L}^1$ is more involved than the syntax of a standard sequent calculus. Derivable objects are sequents of the form $\Pi \rightarrow A$. The antecedent Π is a structure called *meta-formula* (or *configuration*). Meta-formulae are built from formulae (types) using two metasyntactic operators:

comma and brackets. The succedent A is a formula. Formulae are built from variables $p_1, p_2, \ldots$ and the unit constant $\mathbf{1}$ using Lambek's binary connectives: $\backslash$, $/$, and $\cdot$, and three unary connectives, $\langle \rangle$, $[]^{-1}$, and $!$. The first two of them operate brackets; the last one is the subexponential for medial extraction.

Meta-formulae are denoted by capital Greek letters; $\Delta(\Gamma)$ stands for Δ with a designated occurrence of a meta-formula (in particular, formula) Γ. Meta-formulae are allowed to be empty; the empty meta-formula is denoted by Λ.

The axioms of $!_\mathbf{b}\mathbf{L}^1$ are $A \to A$ and $\Lambda \to \mathbf{1}$, and the rules are as follows:

$$\frac{\Gamma \to B \quad \Delta(C) \to D}{\Delta(C/B, \Gamma) \to D} \ (/\to) \qquad \frac{\Gamma, B \to C}{\Gamma \to C/B} \ (\to/) \qquad \frac{\Delta(A, B) \to D}{\Delta(A \cdot B) \to D} \ (\cdot \to)$$

$$\frac{\Gamma \to A \quad \Delta(C) \to D}{\Delta(\Gamma, A \backslash C) \to D} \ (\backslash \to) \qquad \frac{A, \Gamma \to C}{\Gamma \to A \backslash C} \ (\to \backslash) \qquad \frac{\Gamma_1 \to A \quad \Gamma_2 \to B}{\Gamma_1, \Gamma_2 \to A \cdot B} \ (\to \cdot) \qquad .$$

$$\frac{\Delta(\Lambda) \to A}{\Delta(\mathbf{1}) \to A} \ (\mathbf{1} \to) \qquad \frac{\Delta([A]) \to C}{\Delta(\langle \rangle A) \to C} \ (\langle \rangle \to) \qquad \frac{\Pi \to A}{[\Pi] \to \langle \rangle A} \ (\to \langle \rangle)$$

$$\frac{\Gamma(A) \to B}{\Gamma(!A) \to B} \ (! \to) \qquad \frac{\Delta(A) \to C}{\Delta([[]^{-1}A]) \to C} \ ([]^{-1} \to) \qquad \frac{[\Pi] \to A}{\Pi \to []^{-1}A} \ (\to []^{-1})$$

$$\frac{!A_1, \ldots, !A_n \to A}{!A_1, \ldots, !A_n \to !A} \ (\to !) \qquad \frac{\Delta(!A_1, \ldots, !A_n, [!A_1, \ldots, !A_n, \Gamma]) \to B}{\Delta(!A_1, \ldots, !A_n, \Gamma) \to B} \ (\text{contr}_\mathbf{b})$$

$$\frac{\Delta(!A, \Gamma) \to B}{\Delta(\Gamma, !A) \to B} \ (\text{perm}_1) \qquad \frac{\Delta(\Gamma, !A) \to B}{\Delta(!A, \Gamma) \to B} \ (\text{perm}_2) \qquad \frac{\Pi \to A \quad \Delta(A) \to C}{\Delta(\Pi) \to C} \ (\text{cut})$$

The analysis of syntactic phenomena using $!_\mathbf{b}\mathbf{L}^1$ below follows Morrill [32].

Permutation rules ($\text{perm}_{1,2}$) for $!$ allow medial extraction. The relative pronoun *"whom"* now receives the type $(CN \backslash CN)/(S/!N)$, and the noun phrase *"the girl whom John met yesterday"* now becomes derivable (the type for *"yesterday"* is $(N \backslash S)\backslash(N \backslash S)$, modifier of verb phrase):

$$\frac{\dfrac{\dfrac{\dfrac{\dfrac{\dfrac{\dfrac{\dfrac{N \to N \quad S \to S}{N \backslash S \to N \backslash S} \quad \dfrac{N \to N \quad S \to S}{N, N \backslash S \to S}}{N, N \backslash S, (N \backslash S)\backslash(N \backslash S) \to S}}{N, (N \backslash S)/N, N, (N \backslash S)\backslash(N \backslash S) \to S}}{N, (N \backslash S)/N, !N, (N \backslash S)\backslash(N \backslash S) \to S}}{N, (N \backslash S)/N, (N \backslash S)\backslash(N \backslash S), !N \to S}}{N, (N \backslash S)/N, (N \backslash S)\backslash(N \backslash S) \to S/!N} \quad \dfrac{\dfrac{CN \to CN \quad CN \to CN}{CN, CN \backslash CN \to CN} \quad N \to N}{N/CN, CN, CN \backslash CN \to N}}{N/CN, CN, (CN \backslash CN)/(S/!N), N, (N \backslash S)/N, (N \backslash S)\backslash(N \backslash S) \to N}$$

The permutation rule puts $!N$ to the correct place (*"John met ... yesterday"*).

For brackets, consider the following ungrammatical example: **"the book which John laughed without reading."* In the original Lambek calculus, it would be generated by the following derivable sequent:

$$N/CN, CN, (CN \backslash CN)/(S/N), N, N \backslash S, ((N \backslash S)\backslash(N \backslash S))/(N \backslash S), (N \backslash S)/N \to N.$$

In the grammar with brackets, however, *"without"* receives the syntactic type $[]^{-1}((N \setminus S) \setminus (N \setminus S))/(N \setminus S)$, making the *without*-clause an island that cannot be penetrated by extraction. Thus, the following sequent is not derivable

$$N/CN, CN, (CN \setminus CN)/(S/N), N, N \setminus S, [[]^{-1}((N \setminus S) \setminus (N \setminus S))/(N \setminus S), (N \setminus S)/N] \to N,$$

and the ungrammatical example gets ruled out.

Finally, the non-standard contraction rule, (contr_b), that governs both ! and brackets, was designed for handling a more rare phenomenon called *parasitic extraction*. It appears in examples like *"the paper that John signed without reading."* Compare with the ungrammatical example considered before: now in the dependent clause there are two gaps, and one of them is inside an island (*"John signed ... [without reading ...]"*); both gaps are filled with the same $!N$:

$$
\begin{array}{c}
\dfrac{N \setminus S \to N \setminus S \quad N, N \setminus S \to S}{N, N \setminus S, (N \setminus S) \setminus (N \setminus S) \to S} \\[4pt]
\dfrac{N \setminus S \to N \setminus S \quad N, N \setminus S, [[]^{-1}((N \setminus S) \setminus (N \setminus S))] \to S}{N \to N \quad N, N \setminus S, [[]^{-1}((N \setminus S) \setminus (N \setminus S))/(N \setminus S), N \setminus S] \to S} \\[4pt]
\dfrac{}{N \to N \quad N, N \setminus S, [[]^{-1}((N \setminus S) \setminus (N \setminus S))/(N \setminus S), (N \setminus S)/N, N] \to S} \\[4pt]
\dfrac{}{N, (N \setminus S)/N, N, [[]^{-1}((N \setminus S) \setminus (N \setminus S))/(N \setminus S), (N \setminus S)/N, N] \to S} \\[4pt]
\dfrac{}{N, (N \setminus S)/N, !N, [[]^{-1}((N \setminus S) \setminus (N \setminus S))/(N \setminus S), (N \setminus S)/N, N] \to S} \\[4pt]
\dfrac{}{N, (N \setminus S)/N, !N, [[]^{-1}((N \setminus S) \setminus (N \setminus S))/(N \setminus S), (N \setminus S)/N, !N] \to S} \\[4pt]
\dfrac{}{N, (N \setminus S)/N, !N, [!N, []^{-1}((N \setminus S) \setminus (N \setminus S))/(N \setminus S), (N \setminus S)/N] \to S} \\[4pt]
\dfrac{}{N, (N \setminus S)/N, !N, []^{-1}((N \setminus S) \setminus (N \setminus S))/(N \setminus S), (N \setminus S)/N \to S} \\[4pt]
\dfrac{}{N, (N \setminus S)/N, []^{-1}((N \setminus S) \setminus (N \setminus S))/(N \setminus S), (N \setminus S)/N, !N \to S} \\[4pt]
\dfrac{N, (N \setminus S)/N, []^{-1}((N \setminus S) \setminus (N \setminus S))/(N \setminus S), (N \setminus S)/N \to S/!N \quad N/CN, CN, CN \setminus CN \to N}{N/CN, CN, (CN \setminus CN)/(S/!N), N, (N \setminus S)/N, []^{-1}((N \setminus S) \setminus (N \setminus S))/(N \setminus S), (N \setminus S)/N \to N}
\end{array}
$$

This construction allows potentially infinite recursion, nesting islands with parasitic extraction. On the other hand, ungrammatical examples, like **"the book that John gave to"* with two gaps outside islands (*"John gave ... to ..."*) are not derived with (contr_b), but can be derived using the contraction rule in the standard, not bracket-aware form: $\dfrac{\Delta(!A,!A) \to C}{\Delta(!A) \to C}$ (contr).

The system with (contr) instead of (contr_b) is a conservative extension of its fragment without brackets. In an earlier paper [19] we show that the latter is undecidable. For (contr_b), however, in the bracket-free fragment there are only permutation rules for !, and this fragment is decidable (in fact, it belongs to NP). Therefore, in contrast to [19], the undecidability proof in this paper (Sect. 5) crucially depends on brackets. On the other hand, in [19] we've also proved decidability of a fragment of a calculus with !, but without brackets. In the calculus considered in this paper, $!_b L^1$, brackets control the number of (contr_b) applications, whence we are now able to show membership in NP for a different, broad fragment of $!_b L^1$ (Sect. 6), which includes brackets.

It can be easily seen that the calculus with bracket modalities but without ! also belongs to the NP class. Moreover, as shown in [21], there exists even a polynomial algorithm for deriving formulae of bounded order (connective alternation and bracket nesting depth) in the calculus with brackets but without !.

This algorithm uses proof nets, following the ideas of Pentus [36]. As opposed to [21], as we show here, in the presence of ! the derivability problem is undecidable.

In short, [19] is about the calculus with !, but without brackets; [21] is about the calculus with brackets, but without !. This paper is about the calculus with both ! and brackets, interacting with each other, governed by (contr$_b$).

The rest of this paper is organised as follows. In Sect. 3 we state cut elimination for $!_bL^1$ and sketch the proof strategy. In Sect. 4 we define two intermediate calculi used in our undecidability proof. In Sect. 5 we prove the main result—the fact that $!_bL^1$ is undecidable. This solves an open question posed by Morrill and Valentín [33] (the other open question from [33], undecidability for the case without brackets, is solved in our previous paper [19]). In Sect. 6 we consider a practically interesting fragment of $!_bL^1$ for which Morrill and Valentín [33] give an exponential time algorithm, and strengthen their result by proving an NP upper bound for the derivability problem in this fragment. Section 7 is for conclusion and future research.

3   Cut Elimination in $!_bL^1$

Cut elimination is a natural property that one expects a decent logical system to have. For example, cut elimination entails the *subformula property:* each formula that appears somewhere in the cut-free derivation is a subformula of the goal sequent. (Note that for meta-formulae this doesn't hold, since brackets get removed by applications of some rules, namely, $(\langle\rangle \rightarrow)$, $(\rightarrow []^{-1})$, and (contr$_b$).)

Theorem 1 is claimed in [33], but without a detailed proof. In this section we give a sketch of the proof strategy; the complete proof can be found in the extended version of this paper on arXiv [20].

For the original Lambek calculus cut elimination was shown by Lambek [23] and goes straightforwardly by induction; Moortgat [27] extended Lambek's proof to the Lambek calculus with brackets (but without !). It is well-known, however, that in the presence of a contraction rule direct induction doesn't work. Therefore, one needs to use more sophisticated cut elimination strategies.

The standard strategy, going back to Gentzen's *Hauptsatz* [14], replaces the cut *(Schnitt)* rule with a more general rule called mix *(Mischung)*. Mix is a combination of cut and contraction, and this more general rule can be eliminated by straightforward induction. For linear logic with the exponential obeying standard rules, cut elimination is due to Girard [15]; a detailed exposition of the cut elimination procedure using mix is presented in [25, Appendix A].

For $!_bL^1$, however, due to the subtle nature of the contraction rule, (contr$_b$), formulating the mix rule is problematic. That's why we follow another strategy, "deep cut elimination" by Braüner and de Paiva [5,6]; similar ideas are also used in [7,13]. As usual, we eliminate one cut, and then proceed by induction.

Lemma 1. *Let $\Delta(\Pi) \rightarrow C$ be derived from $\Pi \rightarrow A$ and $\Delta(A) \rightarrow C$ using the cut rule, and $\Pi \rightarrow A$ and $\Delta(A) \rightarrow C$ have cut-free derivations $\mathscr{D}_{\text{left}}$ and $\mathscr{D}_{\text{right}}$. Then $\Delta(\Pi) \rightarrow C$ also has a cut-free derivation.*

We proceed by nested induction on two parameters: (1) the complexity κ of the formula A being cut; (2) the total number σ of rule applications in $\mathscr{D}_{\text{left}}$ and $\mathscr{D}_{\text{right}}$. Induction goes in the same way as for systems without contraction (see Lambek's original paper [23]) for all cases, except $(\to !)$ vs. $(\text{contr}_{\text{b}})$:

$$\cfrac{\cfrac{!\Pi \to A}{!\Pi \to !A}\ (\to !) \quad \cfrac{\Delta(!\Phi_1, !A, !\Phi_2, [!\Phi_1, !A, !\Phi_2, \Gamma]) \to C}{\Delta(!\Phi_1, !A, !\Phi_2, \Gamma) \to C}\ (\text{contr}_{\text{b}})}{\Delta(!\Phi_1, !\Pi, !\Phi_2, \Gamma) \to C}\ (\text{cut})$$

(Here $!\Phi$ stands for $!F_1, \ldots, !F_m$, if $\Phi = F_1, \ldots, F_m$.) The naïve attempt,

$$\cfrac{!\Pi \to !A \quad \cfrac{!\Pi \to !A \quad \Delta(!\Phi_1, !A, !\Phi_2, [!\Phi_1, !A, !\Phi_2, \Gamma]) \to C}{\Delta(!\Phi_1, !A, !\Phi_2, [!\Phi_1, !A, !\Phi_2, \Gamma]) \to C}\ (\text{cut})}{\cfrac{\Delta(!\Phi_1, !\Pi, !\Phi_2, [!\Phi_1, !\Pi, !\Phi_2, \Gamma]) \to C}{\Delta(!\Phi_1, !\Pi, !\Phi_2, \Gamma) \to C}\ (\text{contr}_{\text{b}})}\ (\text{cut})$$

fails, since for the lower (cut) the κ parameter is the same, and σ is unbound. Instead of that, we go inside $\mathscr{D}_{\text{right}}$ and trace the active $!A$ occurrences up to the applications of $(! \to)$ which introduced them. We replace these applications with applications of (cut), and replace $!A$ with $!\Pi$ down the traces, as shown below:

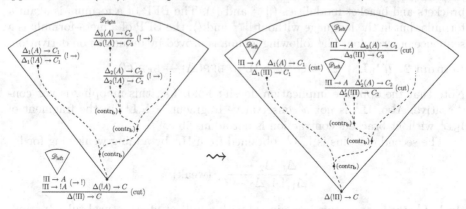

The new (cut) instances have a smaller κ parameter (A is simpler than $!A$) and can be eliminated by induction.

Lemma 1 immediately yields cut elimination for $!_{\text{b}}\mathbf{L}^1$:

Theorem 1. *Every sequent derivable in $!_{\text{b}}\mathbf{L}^1$ has a derivation without* (cut).

4 Calculi Without Brackets: $!\mathbf{L}^1$, $!_{\text{w}}\mathbf{L}^1$, $\mathbf{L}^1$

In this section we consider more traditional versions of the Lambek calculus with ! that don't include bracket modalities. This is needed as a technical step in our undecidability proof (Sect. 5). Types (formulae) of these calculi are built from variables using Lambek's connectives, \, /, and ·, and the subexponential, !. Unlike in $!_b\mathbf{L}^1$, meta-formulae now are merely linearly ordered sequences of formulae (possibly empty), and we can write Δ_1, Π, Δ_2 instead of $\Delta(\Pi)$.

First we define the calculus $!\mathbf{L}^1$. It includes the standard axioms and rules for Lambek connectives and the unit constant—see the rules of $!_b\mathbf{L}^1$ in Sect. 2. For the subexponential modality, !, introduction rules, $(! \rightarrow)$ and $(\rightarrow !)$, and permutation rules are also the same as in $!_b\mathbf{L}^1$, with the natural modification due to a simpler antecedent syntax. The contraction rule, however, is significantly different, since now it is not controlled by brackets:

$$\frac{\Delta_1, !A, !A, \Delta_2 \rightarrow B}{\Delta_1, !A, \Delta_2 \rightarrow B} \text{ (contr)}$$

(A complete list of axioms and rules of $!\mathbf{L}^1$ can be found in the extended version of this paper on arXiv [20].)

This calculus $!\mathbf{L}^1$ is a conservative fragment of $\mathbf{Db!}$, also by Morrill and Valentín [33]. This system could also be used for modelling medial and parasitic extraction, but is not as fine-grained as the bracketed system, being able to derive ungrammatical examples like *"the paper that John sent to" (see Sect. 2).

In order to construct a mapping of $!_b\mathbf{L}^1$ into $!\mathbf{L}^1$, we define the *bracket-forgetting projection (BFP)* of formulae and meta-formulae that removes all brackets and bracket modalities ($[]^{-1}$ and $\langle\rangle$). The BFP of a formula is again a formula, but in the language without $[]^{-1}$ and $\langle\rangle$; the BFP of a meta-formula is a sequence of formulae. The following lemma is proved by induction on derivation.

Lemma 2. *If* $!_b\mathbf{L}^1 \vdash \Delta \rightarrow C$, *then* $!\mathbf{L}^1 \vdash \text{BFP}(\Delta) \rightarrow \text{BFP}(C)$.

Note that the opposite implication doesn't hold, *i.e.*, this mapping is not conservative. Also, $!\mathbf{L}^1$ is not a conservative fragment of $!_b\mathbf{L}^1$: in the fragment of $!_b\mathbf{L}^1$ without brackets contraction is not admissible.

The second calculus is $!_w\mathbf{L}^1$, obtained from $!\mathbf{L}^1$ by adding weakening for !:

$$\frac{\Delta_1, \Delta_2 \rightarrow C}{\Delta_1, !A, \Delta_2 \rightarrow C} \text{ (weak)}$$

In $!_w\mathbf{L}^1$, the ! connective is equipped with a full set of structural rules (permutation, contraction, and weakening), *i.e.*, it is the exponential of linear logic [15].

The cut rule in $!\mathbf{L}^1$ and $!_w\mathbf{L}^1$ can be eliminated by the same "deep" strategy as for $!_b\mathbf{L}^1$. On the other hand, since the contraction rule in these calculi is standard, one can also use the traditional way with mix, like in [25, Appendix A].

Finally, if we remove ! with all its rules, we get the Lambek calculus with the unit constant [24]. We denote it by $\mathbf{L}^1$.

5   Undecidability of $!_b\mathbf{L}^1$

Theorem 2. *The derivability problem for* $!_b\mathbf{L}^1$ *is undecidable.*

As a by-product of our proof we also obtain undecidability of $!\mathbf{L}^1$, which was proved in [19] by a different method. We also obtain undecidability of $!_{\mathbf{w}}\mathbf{L}^1$, which also follows from the results of [25], as shown in [16,17].

We prove Theorem 2 by encoding derivations in generative grammars, or semi-Thue [40] systems. A *generative grammar* is a quadruple $G = \langle N, \Sigma, P, s \rangle$, where N and Σ are disjoint alphabets, $s \in N$ is the *starting symbol*, and P is a finite set of *productions* of the form $\alpha \Rightarrow \beta$, where α and β are words over $N \cup \Sigma$. The production can be *applied* as follows: $\eta \alpha \theta \Rightarrow_G \eta \beta \theta$ for arbitrary (possibly empty) words η and θ over $N \cup \Sigma$. The *language generated by* G is the set $\{\omega \in \Sigma^* \mid s \Rightarrow_G^* \omega\}$, where $\Rightarrow_G^*$ is the reflexive-transitive closure of $\Rightarrow_G$.

We use the following classical result by Markov [26] and Post [37].

Theorem 3. *There exists a generative grammar G that generates an algorithmically undecidable language* [26,37].

In our presentation for every production $(\alpha \Rightarrow \beta) \in P$ we require α and β to be non-empty. This class still includes an undecidable language (cf. [10]).

Further we use two trivial lemmas about derivations in a generative grammar:

Lemma 3. *If $\alpha_1 \Rightarrow_G^* \beta_1$ and $\alpha_2 \Rightarrow_G^* \beta_2$, then $\alpha_1 \alpha_2 \Rightarrow_G^* \beta_1 \beta_2$.*

Lemma 4. *If $\alpha \Rightarrow_G^* \beta$ and $\gamma \Rightarrow_G^* \eta \alpha \theta$, then $\gamma \Rightarrow_G^* \eta \beta \theta$.*

The second ingredient of our undecidability proof is the concept of *theories* over $\mathbf{L}^1$. Let $\mathcal{T}$ be a finite set of sequents in the language of $\mathbf{L}^1$. Then $\mathbf{L}^1 + \mathcal{T}$ is the calculus obtained from $\mathbf{L}^1$ by adding sequents from $\mathcal{T}$ as extra axioms.

In general, the cut rule in $\mathbf{L}^1 + \mathcal{T}$ is not eliminable. However, the standard cut elimination procedure (see [23]) yields the following *cut normalization* lemma:

Lemma 5. *If a sequent is derivable in $\mathbf{L}^1 + \mathcal{T}$, then this sequent has a derivation in which every application of (cut) has a sequent from $\mathcal{T}$ as one of its premises.*

This lemma yields a weak version of the subformula (subconnective) property:

Lemma 6. *If $\mathbf{L}^1 + \mathcal{T} \vdash \Pi \to A$, and both $\Pi \to A$ and $\mathcal{T}$ include no occurrences of $\backslash$, $/$, and $\mathbf{1}$, then there is a derivation of $\Pi \to A$ in $\mathbf{L}^1 + \mathcal{T}$ that includes no occurrences of $\backslash$, $/$, and $\mathbf{1}$.*

The third core element of the construction is the (inst) rule which allows to position a specific formula A into an arbitrary place in the sequent.

Lemma 7. *The rule* $\dfrac{\Delta_1, ![]^{-1}A, \Delta_2, A, \Delta_3 \to C}{\Delta_1, ![]^{-1}A, \Delta_2, \Delta_3 \to C}$(inst) *is admissible in* $!_{\mathbf{b}}\mathbf{L}^1$.

Proof. By consequent application of $([]^{-1} \to)$, $(! \to)$, (perm), (contr$_{\mathbf{b}}$), (perm). $\qquad \square$

Now we are ready to prove Theorem 2. Let $G = \langle N, \Sigma, P, s \rangle$ be the grammar provided by Theorem 3, and the set of variables include $N \cup \Sigma$. We convert productions of G into Lambek formulae in the following natural way:
$$\mathcal{B}_G = \{(u_1 \cdot \ldots \cdot u_k)/(v_1 \cdot \ldots \cdot v_m) \mid (u_1 \ldots u_k \Rightarrow v_1 \ldots v_m) \in P\}.$$

For $\mathcal{B}_G = \{B_1, \ldots, B_n\}$, we define the following sequences of formulae:

$$\Gamma_G = {!B_1, \ldots, !B_n}, \qquad \Phi_G = {!(\mathbf{1}/(!B_1)), \ldots, !(\mathbf{1}/(!B_n))},$$
$$\widetilde{\Gamma}_G = {!\,[]^{-1}B_1, \ldots, !\,[]^{-1}B_n}, \quad \widetilde{\Phi}_G = {!(\mathbf{1}/(!\,[]^{-1}B_1)), \ldots, !(\mathbf{1}/(!\,[]^{-1}B_n))}.$$

(Since in all calculi we have permutation rules for formulae under !, the ordering of $\mathcal{B}_G$ doesn't matter.) We also define a theory $\mathcal{T}_G$ associated with G, as follows:
$$\mathcal{T}_G = \{v_1, \ldots, v_m \to u_1 \cdot \ldots \cdot u_k \mid (u_1 \ldots u_k \Rightarrow v_1 \ldots v_m) \in P\}.$$

Lemma 8. *The following are equivalent:*

1. $s \Rightarrow_G^* a_1 \ldots a_n$ *(i.e., $a_1 \ldots a_n$ belongs to the language defined by G);*
2. $!_{\mathbf{b}}\mathbf{L}^1 \vdash \widetilde{\Phi}_G, \widetilde{\Gamma}_G, a_1, \ldots, a_n \to s$;
3. $!\mathbf{L}^1 \vdash \Phi_G, \Gamma_G, a_1, \ldots, a_n \to s$;
4. $!_{\mathbf{w}}\mathbf{L}^1 \vdash \Gamma_G, a_1, \ldots, a_n \to s$;
5. $\mathbf{L}^1 + \mathcal{T}_G \vdash a_1, \ldots, a_n \to s$.

Proof. $\boxed{1 \Rightarrow 2}$ Proceed by induction on $\Rightarrow_G^*$. The base case is handled as follows:

$$\frac{\dfrac{!\,[]^{-1}B_1 \to !\,[]^{-1}B_1 \quad \ldots \quad !\,[]^{-1}B_n \to !\,[]^{-1}B_n \quad \dfrac{\dfrac{s \to s}{1, \ldots, 1, s \to s}\,(\mathbf{1}\to)^*}{}}{\dfrac{\mathbf{1}/!\,[]^{-1}B_1, !\,[]^{-1}B_1, \ldots, \mathbf{1}/!\,[]^{-1}B_n, !\,[]^{-1}B_n, s \to s}{\dfrac{!(\mathbf{1}/!\,[]^{-1}B_1), !\,[]^{-1}B_1, \ldots, !(\mathbf{1}/!\,[]^{-1}B_n), !\,[]^{-1}B_n, s \to s}{\widetilde{\Phi}_G, \widetilde{\Gamma}_G, s \to s}\,(\mathrm{perm})^*}\,(!\to)^*}\,(/\to)^*}{}$$

For induction, let the last step be $s \Rightarrow_G^* \eta\, u_1 \ldots u_k\, \theta \Rightarrow_G \eta\, v_1 \ldots v_m\, \theta$. Then, since $!\,[]^{-1}((u_1 \cdot \ldots \cdot u_k)/(v_1 \cdot \ldots \cdot v_m))$ is in $\widetilde{\Gamma}_G$, we have

$$\frac{\dfrac{\dfrac{v_1 \to v_1 \quad \ldots \quad v_m \to v_m}{v_1, \ldots, v_m \to v_1 \cdot \ldots \cdot v_m}\,(\to \cdot)^* \quad \dfrac{\widetilde{\Phi}_G, \widetilde{\Gamma}_G, \eta, u_1, \ldots, u_k, \theta \to s}{\widetilde{\Phi}_G, \widetilde{\Gamma}_G, \eta, u_1 \cdot \ldots \cdot u_k, \theta \to s}\,(\cdot \to)^*}{\dfrac{\widetilde{\Phi}_G, \widetilde{\Gamma}_G, \eta, (u_1 \cdot \ldots \cdot u_k)/(v_1 \cdot \ldots \cdot v_m), v_1, \ldots, v_m, \theta \to s}{\widetilde{\Phi}_G, \widetilde{\Gamma}_G, \eta, v_1, \ldots, v_m, \theta \to s}\,(\mathrm{inst})}\,(/\to)}{}$$

Here $\widetilde{\Phi}_G, \widetilde{\Gamma}_G, \eta, u_1, \ldots, u_k, \theta \to s$ is derivable in $!_{\mathbf{b}}\mathbf{L}^1$ by induction hypothesis, and the (inst) rule is admissible due to Lemma 7.

$\boxed{2 \Rightarrow 3}$ Immediate by Lemma 2, since $\Phi_G = \mathrm{BFP}(\widetilde{\Phi}_G)$ and $\Gamma_G = \mathrm{BFP}(\widetilde{\Gamma}_G)$.

$\boxed{3 \Rightarrow 4}$ For each formula $!(\mathbf{1}/!B_i)$ from Φ_G the sequent $\Lambda \to !(\mathbf{1}/!B_i)$ is derivable in $!_{\mathbf{w}}\mathbf{L}^1$ by consequent application of (weak), $(\to /)$, and $(\to !)$ to the $\Lambda \to \mathbf{1}$ axiom. The sequent $\Phi_G, \Gamma_G, a_1, \ldots, a_n \to s$ is derivable in $!\mathbf{L}^1$ and therefore in $!_{\mathbf{w}}\mathbf{L}^1$, and applying (cut) for each formula of Φ_G yields $\Gamma_G, a_1, \ldots, a_n \to s$.

$\boxed{4 \Rightarrow 5}$ In this part of our proof we follow [17,25]. Consider the derivation of $\Gamma_G, a_1, \ldots, a_n \to s$ in $!_{\mathbf{w}}\mathbf{L}^1$ (recall that by default all derivations are cut-free) and remove all the formulae of the form $!B$ from all sequents in this derivation. After this transformation the rules not operating with ! remain valid. Applications of (perm_i), (weak), and (contr) do not alter the sequent. The $(\to !)$ rule is never applied in the original derivation, since our sequents never have formulae of the form $!B$ in their succedents. Finally, an application of $(! \to)$, hiding $(u_1 \cdot \ldots \cdot u_k)/(v_1 \cdot \ldots \cdot v_m)$ under !, is simulated in $\mathbf{L}^1 + \mathcal{T}_G$ as follows:

$$\frac{\dfrac{\dfrac{v_1, \ldots, v_m \to u_1 \cdot \ldots \cdot u_k}{v_1 \cdot \ldots \cdot v_m \to u_1 \cdot \ldots \cdot u_k} \, (\cdot \to)^*}{\varLambda \to (u_1 \cdot \ldots \cdot u_k)/(v_1 \cdot \ldots \cdot v_m)} \, (\to /) \qquad \varDelta_1, (u_1 \cdot \ldots \cdot u_k)/(v_1 \cdot \ldots \cdot v_m), \varDelta_2 \to C}{\varDelta_1, \varDelta_2 \to C} \, \text{(cut)}$$

$\boxed{5 \Rightarrow 1}$ In this part we follow [25]. Let $\mathbf{L}^1 + \mathcal{T}_G \vdash a_1, \ldots, a_n \to s$. By Lemma 6, this sequent has a derivation without occurrences of $\backslash$, $/$, and $\mathbf{1}$. In other words, all formulae in this derivation are built from variables using only the product. Since it is associative, we can omit parenthesis in the formulae; we shall also omit the "$\cdot$"s. The rules used in this derivation can now be written as follows:

$$\frac{\beta_1 \to \alpha_1 \quad \beta_2 \to \alpha_2}{\beta_1 \beta_2 \to \alpha_1 \alpha_2} \, (\to \cdot) \qquad \frac{\beta \to \alpha \quad \eta \alpha \theta \to \gamma}{\eta \beta \theta \to \gamma} \, \text{(cut)}$$

The $(\cdot \to)$ rule is trivial. The axioms are productions of G with the arrows inversed, and $\alpha \to \alpha$. By induction, using Lemmas 3 and 4, we show that if $\beta \to \alpha$ is derivable using these rules and axioms, then $\alpha \Rightarrow_G^* \beta$. Now the derivability of $a_1, \ldots, a_n \to s$ implies $s \Rightarrow_G^* a_1 \ldots a_n$.

Lemma 8 and Theorem 3 conclude the proof of Theorem 2.

6   A Decidable Fragment

The undecidability results from the previous section are somewhat unfortunate, since the new operations added to $\mathbf{L}^1$ have good linguistic motivations [30,33]. As a compensation, in this section we show NP-decidability for a substantial fragment of $!_b \mathbf{L}^1$, introduced by Morrill and Valentín [33] (see Definition 1 below). This complexity upper bound is tight, since the original Lambek calculus is already known to be NP-complete [35]. Notice that Morrill and Valentín present an exponential time algorithm for deciding derivability in this fragment; this algorithm was implemented as part of a parser called CatLog [31].

First we recall the standard notion of *polarity* of occurrences of subformulae in a formula. Every formula occurs positively in itself; subformula polarities get inverted (positive becomes negative and vice versa) when descending into denominators of $\backslash$ and $/$ and also for the left-hand side of the sequent; brackets and unary operations do not change polarity. All inference rules of $!_b \mathbf{L}^1$ respect polarity: a positive (resp., negative) occurrence of a subformula in the premise(s) of the rule translates into a positive (resp., negative) occurrence in the goal.

Definition 1. *An $!_b \mathbf{L}^1$-sequent $\Gamma \to B$ obeys the bracket non-negative condition, if any negative occurrence of a subformula of the form $!A$ in $\Gamma \to B$ includes neither a positive occurrence of a subformula of the form $[]^{-1}C$, nor a negative occurrence of a subformula of the form $\langle\rangle C$.*

Note that sequents used in our undecidability proof are exactly the minimal violations of this bracket non-negative condition.

Theorem 4. *The derivability problem in $!_\mathsf{b}\mathbf{L}^1$ for sequents that obey the bracket non-negative condition belongs to the* NP *class.*

In $!_\mathsf{b}\mathbf{L}^1$, redundant applications of permutation rules could make the proof arbitrarily large without increasing its "real" complexity. In order to get rid of that, we introduce a *generalised form of permutation rule:*

$$\frac{\Delta_0, !A_1, \Delta_1, !A_2, \Delta_2, \dots, \Delta_{k-1}, !A_k, \Delta_k \to C}{\Delta_0', !A_{i_1}, \Delta_1', !A_{i_2}, \Delta_2', \dots, \Delta_{i_{k-1}}', !A_{i_k}, \Delta_{i_k}' \to C} \ (\mathrm{perm})^*$$

where the sequence $\Delta_0', \dots, \Delta_k'$ coincides with $\Delta_0, \dots, \Delta_k$, and $\{i_1, \dots, i_k\} = \{1, \dots, k\}$. Obviously, $(\mathrm{perm})^*$ is admissible in $!_\mathsf{b}\mathbf{L}^1$, and it subsumes $(\mathrm{perm}_{1,2})$, so further we consider a formulation of $!_\mathsf{b}\mathbf{L}^1$ with $(\mathrm{perm})^*$ instead of $(\mathrm{perm}_{1,2})$. Several consecutive applications of $(\mathrm{perm})^*$ can be merged into one. We call a derivation *normal*, if it doesn't contain consecutive applications of $(\mathrm{perm})^*$. If a sequent is derivable in $!_\mathsf{b}\mathbf{L}^1$, then it has a normal cut-free derivation.

Lemma 9. *Every normal cut-free derivation of a sequent that obeys the bracket non-negative restriction is of quadratic size (number of rule applications) w.r.t. the size of the goal sequent.*

Proof. Let us call $(\mathrm{contr}_\mathsf{b})$ and $(\mathrm{perm})^*$ *structural* rules, and all others *logical*.

First, we track all pairs of brackets that occur in this derivation. Pairs of brackets are in one-to-one correspondence with applications of $([\,]^{-1} \to)$ or $(\to \langle\rangle)$ rules that introduce them. Then a pair of brackets either traces down to the goal sequent, or gets destroyed by an application of $(\langle\rangle \to)$, $(\to [\,]^{-1})$, or $(\mathrm{contr}_\mathsf{b})$. Therefore, the number of $(\mathrm{contr}_\mathsf{b})$ applications is less or equal to the number of $([\,]^{-1} \to)$ and $(\to \langle\rangle)$ applications. Each $([\,]^{-1} \to)$ application introduces a negative occurrence of a $[\,]^{-1}C$ formula; each $(\to \langle\rangle)$ occurrence introduces a positive occurrence of a $\langle\rangle C$ formula. Due to the bracket non-negative condition these formulae could not occur in a $!A$ to which $(\mathrm{contr}_\mathsf{b})$ is applied, and therefore they trace down to *distinct* subformula occurrences in the goal sequent. Hence, the total number of $(\mathrm{contr}_\mathsf{b})$ applications is bounded by the number of subformulae of a special kind in the goal sequent, and thus by the size of the sequent.

Second, we bound the number of logical rules applications. Each logical rule introduces exactly one connective occurrence. Such an occurrence traces down either to a connective occurrence in the goal sequent, or to an application of $(\mathrm{contr}_\mathsf{b})$ that merges this occurrence with the corresponding occurrence in the other $!A$. If n is the size of the goal sequent, then the first kind of occurrences is bounded by n; for the second kind, each application of $(\mathrm{contr}_\mathsf{b})$ merges not more than n occurrences (since the size of the formula being contracted, $!A$, is bounded by n due to the subformula property), and the total number of $(\mathrm{contr}_\mathsf{b})$ applications is also bounded by n. Thus, we get a quadratic bound for the number of logical rule applications.

Third, the derivation is a tree with binary branching, so the number of leafs (axioms instances) in this tree is equal to the number of branching points plus one. Each branching point is an application of a logical rule (namely, $(\backslash \rightarrow)$, $(/ \rightarrow)$, or $(\rightarrow \cdot)$). Hence, the number of axiom instances is bounded quadratically.

Finally, the number of (perm)* applications is also quadratically bounded, since each application of (perm)* in a normal proof is preceded by an application of another rule or by an axiom instance.

Proof (of Theorem 4). The normal derivation of a sequent obeying the bracket non-negative condition is an NP-witness for derivability: it is of polynomial size, and correctness is checked in linear time (w.r.t. the size of the derivation).

For the case without brackets, $!\mathbf{L}^1$, considered in our earlier paper [19], the NP-decidable fragment is substantially smaller. Namely, it includes only sequents in which ! can be applied only to variables. Indeed, as soon as we allow formulae of implication nesting depth at least 2 under !, the derivablity problem for $!\mathbf{L}^1$ becomes undecidable [19]. In contrast to $!\mathbf{L}^1$, in $!_{\mathbf{b}}\mathbf{L}^1$, due to the non-standard contraction rule, brackets control the number of (contr$_\mathbf{b}$) applications in the proof, and this allows to construct an effective decision algorithm for derivability of a broad class of sequents, where, for example, any formulae without bracket modalities can be used under !. Essentially, the only problematic situation, that gives rise to undecidability (Theorem 2), is the construction where one forcedly removes the brackets that appear in the (contr$_\mathbf{b}$) rule, *i.e.*, uses constructions like $!\langle\rangle^{-1}B$ (as in our undecidability proof). The idea of the bracket non-negative condition is to rule out such situations while keeping all other constructions allowed, as they don't violate decidability [33].

7   Conclusions and Future Work

In this paper we study an extension of the Lambek calculus with subexponential and bracket modalities. Bracket modalities were introduced by Morrill [29] and Moortgat [27] in order to represent the linguistic phenomenon of islands [38]. The interaction of subexponential and bracket modalities was recently studied by Morrill and Valentín [33] in order to represent correctly the phenomenon of medial and parasitic extraction [4,38]. We prove that the calculus of Morrill and Valentín is undecidable, thus solving a problem left open in [33]. Morrill and Valentín also considered the so-called bracket non-negative fragment of this calculus, for which they presented an exponential time derivability decision procedure. We improve their result by showing that this problem is in NP.

For undecidability, we encode semi-Thue systems by means of sequents that lie just outside the bracket non-negative fragment. More precisely, the formulae used in our encoding are of the form $!\langle\rangle^{-1}A$, where A is a pure Lambek formula of order 2. It remains for further studies whether these formulae could be simplified.

Our undecidability proof could be potentially made stronger by restricting the language. Now we use three Lambek's connectives: $/$, $\cdot$, and $\mathbf{1}$, plus $\langle\rangle^{-1}$ and !. One could get rid of $\mathbf{1}$ using the substitution from [22]. Further, one might also encode a more clever construction by Buszkowski [8] in order to restrict ourselves to the product-free one-division fragment. Finally, one could adopt substitutions from [18] and obtain undecidability for the language with only one variable.

There are also several other linguistically motivated extensions of the Lambek calculus (see, for instance, [28,30,32]) and their algorithmic and logical properties should be investigated.

Acknowledgements. The work of M. Kanovich and A. Scedrov was supported by the Russian Science Foundation under grant 17-11-01294 and performed at National Research University Higher School of Economics, Russia. The work of S. Kuznetsov was supported by grants RFBR 15-01-09218-a and NŠ-9091.2016.1. Sections 1 and 3 were contributed by S. Kuznetsov, Sects. 2 and 4 by M. Kanovich and A. Scedrov, and Sects. 5, 6, and 7 were contributed jointly and equally by all coauthors.

The authors are grateful to V. de Paiva and T. Braüner for helpful comments on the deep cut elimination procedure.

References

1. Abrusci, V.M.: A comparison between Lambek syntactic calculus and intuitionistic linear propositional logic. Zeitschr. für math. Log. Grundl. Math. (Math. Logic Q.) **36**, 11–15 (1990)
2. Ajdukiewicz, K.: Die syntaktische Konnexität. Studia Philos. **1**, 1–27 (1935)
3. Bar-Hillel, Y.: A quasi-arithmetical notation for syntactic description. Language **29**, 47–58 (1953)
4. Barry, G., Hepple, M., Leslie, N., Morrill, G.: Proof figures and structural operators for categorial grammar. In: Proceedings of 5th Conference of the European Chapter of ACL, Berlin (1991)
5. Braüner, T., de Paiva, V.: Cut elimination for full intuitionstic linear logic. BRICS report RS-96-10, April 1996
6. Braüner, T., de Paiva, V.: A formulation of linear logic based on dependency-relations. In: Nielsen, M., Thomas, W. (eds.) CSL 1997. LNCS, vol. 1414, pp. 129–148. Springer, Heidelberg (1998). doi:10.1007/BFb0028011
7. Braüner, T.: A cut-free Gentzen formulation of modal logic S5. Log. J. IGPL **8**(5), 629–643 (2000)
8. Buszkowski, W.: Some decision problems in the theory of syntactic categories. Zeitschr. für math. Logik und Grundl. der Math. (Math. Logic Q.) **28**, 539–548 (1982)
9. Buszkowski, W.: Type logics in grammar. In: Hendricks, V.F., Malinowski, J. (eds.) Trends in Logic: 50 Years of Studia Logica, pp. 337–382. Springer, Dordrecht (2003). doi:10.1007/978-94-017-3598-8_12
10. Buszkowski, W.: Lambek calculus with nonlogical axioms. In: Language and Grammar. CSLI Lecture Notes, vol. 168, pp. 77–93 (2005)
11. Carpenter, B.: Type-Logical Semantics. MIT Press, Cambridge (1997)
12. Dekhtyar, M., Dikovsky, A.: Generalized categorial dependency grammars. In: Avron, A., Dershowitz, N., Rabinovich, A. (eds.) Pillars of Computer Science. LNCS, vol. 4800, pp. 230–255. Springer, Heidelberg (2008). doi:10.1007/978-3-540-78127-1_13

13. Eades III, H., de Paiva, V.: Multiple conclusion linear logic: cut elimination and more. In: Artemov, S., Nerode, A. (eds.) LFCS 2016. LNCS, vol. 9537, pp. 90–105. Springer, Cham (2016). doi:10.1007/978-3-319-27683-0_7
14. Gentzen, G.: Untersuchungen über das logische Schließen I. Math. Z. **39**, 176–210 (1935)
15. Girard, J.-Y.: Linear logic. Theor. Comput. Sci. **50**, 1–102 (1987)
16. de Groote, P.: On the expressive power of the Lambek calculus extended with a structural modality. In: Language and Grammar. CSLI Lecture Notes, vol. 168, pp. 95–111 (2005)
17. Kanazawa, M.: Lambek calculus: recognizing power and complexity. In: Gerbrandy, J., et al. (eds.) JFAK. Essays dedicated to Johan van Benthem on the occasion of his 50th birthday. Vossiuspers, Amsterdam University Press, Amsterdam (1999)
18. Kanovich, M.: The complexity of neutrals in linear logic. In: Proceedings of LICS 1995, pp. 486–495 (1995)
19. Kanovich, M., Kuznetsov, S., Scedrov, A.: Undecidability of the Lambek calculus with a relevant modality. In: Foret, A., Morrill, G., Muskens, R., Osswald, R., Pogodalla, S. (eds.) FG 2015-2016. LNCS, vol. 9804, pp. 240–256. Springer, Heidelberg (2016). doi:10.1007/978-3-662-53042-9_14. arXiv:1601.06303
20. Kanovich, M., Kuznetsov, S., Scedrov, A.: Undecidability of the Lambek calculus with subexponential and bracket modalities (extended technical report). arXiv:1608.04020 (2017)
21. Kanovich, M., Kuznetsov, S., Morrill, G., Scedrov, A.: A polynomial time algorithm for the Lambek calculus with brackets of bounded order. arXiv:1705.00694 (2017). Accepted to FSCD 2017
22. Kuznetsov, S.L.: On the Lambek calculus with a unit and one division. Moscow Univ. Math. Bull. **66**(4), 173–175 (2011)
23. Lambek, J.: The mathematics of sentence structure. Amer. Math. Mon. **65**(3), 154–170 (1958)
24. Lambek, J.: Deductive systems and categories II. Standard constructions and closed categories. In: Hilton, P.J. (ed.) Category Theory, Homology Theory and their Applications I. LNM, vol. 86, pp. 76–122. Springer, Heidelberg (1969). doi:10.1007/BFb0079385
25. Lincoln, P., Mitchell, J., Scedrov, A., Shankar, N.: Decision problems for propositional linear logic. APAL **56**, 239–311 (1992)
26. Markov, A.: On the impossibility of certain algorithms in the theory of associative systems. Doklady Acad. Sci. USSR (N.S.) **55**, 583–586 (1947)
27. Moortgat, M.: Multimodal linguistic inference. J. Log. Lang. Inform. **5**(3–4), 349–385 (1996)
28. Moot, R., Retoré, C.: The Logic of Categorial Grammars: A Deductive Account of Natural Language Syntax and Semantics. Springer, Heidelberg (2012). doi:10.1007/978-3-642-31555-8
29. Morrill, G.: Categorial formalisation of relativisation: pied piping, islands, and extraction sites. Technical report LSI-92-23-R, Universitat Politècnica de Catalunya (1992)
30. Morrill, G.V.: Categorial Grammar: Logical Syntax, Semantics, and Processing. Oxford University Press, Oxford (2011)
31. Morrill, G.: CatLog: a categorial parser/theorem-prover. In: System Demonstration, LACL 2012, Nantes (2012)
32. Morrill, G.: Grammar logicised: relativisation. Linguist. Philos. **40**(2), 119–163 (2017)

33. Morrill, G., Valentín, O.: Computational coverage of TLG: nonlinearity. In: Proceedings of NLCS 2015. EPiC Series, vol. 32, pp. 51–63 (2015)
34. Nigam, V., Miller, D.: Algorithmic specifications in linear logic with subexponentials. In: Proceedings of PPDP 2009, pp. 129–140. ACM (2009)
35. Pentus, M.: Lambek calculus is NP-complete. Theor. Comput. Sci. **357**(1), 186–201 (2006)
36. Pentus, M.: A polynomial time algorithm for Lambek grammars of bounded order. Linguist. Anal. **36**(1–4), 441–471 (2010)
37. Post, E.L.: Recursive unsolvability of a problem of Thue. J. Symb. Log. **12**, 1–11 (1947)
38. Ross, J.R.: Constraints on variables in syntax. Ph.D. thesis, MIT (1967)
39. Steedman, M.: The Syntactic Process. MIT Press, Cambridge, MA (2000)
40. Thue, A.: Probleme über Veränderungen von Zeichenreihen nach gegebener Regeln. Kra. Vidensk. Selsk. Skrifter. **10** (1914). (In: Selected Math. Papers, Univ. Forlaget, Oslo, pp. 493–524 (1977))

Decision Problems for Subclasses of Rational Relations over Finite and Infinite Words

Christof Löding[1] and Christopher Spinrath[2(✉)]

[1] RWTH Aachen University, Aachen, Germany
loeding@informatik.rwth-aachen.de
[2] TU Dortmund University, Dortmund, Germany
christopher.spinrath@tu-dortmund.de

Abstract. We consider decision problems for relations over finite and infinite words defined by finite automata. We prove that the equivalence problem for binary deterministic rational relations over infinite words is undecidable in contrast to the case of finite words, where the problem is decidable. Furthermore, we show that it is decidable in doubly exponential time for an automatic relation over infinite words whether it is a recognizable relation. We also revisit this problem in the context of finite words and improve the complexity of the decision procedure to single exponential time. The procedure is based on a polynomial time regularity test for deterministic visibly pushdown automata, which is a result of independent interest.

Keywords: Rational relations · Automatic relations · ω-automata · Finite transducers · Visibly pushdown automata

1 Introduction

We consider in this paper algorithmic problems for relations over words that are defined by finite automata. Relations over words extend the classical notion of formal languages. However, there are different ways of extending the concept of regular language and finite automaton to the setting of relations. Instead of processing a single input word, an automaton for relations has to read a tuple of input words. The existing finite automaton models differ in the way how the components can interact while being read. In the following, we briefly sketch the four main classes of automaton definable relations, and then describe our contributions.

A (nondeterministic) finite transducer (see, e.g., [4,25]) has a standard finite state control and at each time of a computation, a transition can consume the next input symbol from any of the components without restriction (equivalently, one can label the transitions of a transducer with tuples of finite words). The class of relations that are definable by finite transducers, referred to as the class of rational relations, is not closed under intersection and complement, and many algorithmic problems, like universality, equivalence, intersection emptiness,

© Springer-Verlag GmbH Germany 2017
R. Klasing and M. Zeitoun (Eds.): FCT 2017, LNCS 10472, pp. 341–354, 2017.
DOI: 10.1007/978-3-662-55751-8_27

are undecidable [24]. A deterministic version of finite transducers defines the class of deterministic rational relations (see [25]) with slightly better properties compared to the nondeterministic version, in particular the equivalence problem is decidable [5, 18].

Another important subclass of rational relations are the synchronized rational relations [17] which are defined by automata that synchronously read all components in parallel (using a padding symbol for words of different length). These relations are often referred to as automatic relations, a terminology that we also adopt, and basically have all the good properties of regular languages because synchronous transducers can be viewed as standard finite automata over a product alphabet. These properties lead to applications of automatic relations in algorithmic model theory as a finite way of representing infinite structures with decidable logical theories (so called automatic structures) [6, 19], and in regular model checking, a verification technique for infinite state systems (cf. [1]).

Finally, there is the model of recognizable relations, which can be defined by a tuple of automata, one for each component of the relation, that independently read their components and only synchronize on their terminal states, i.e., the tuple of states at the end determines whether the input tuple is accepted. Equivalently, one can define recognizable relations as finite unions of products of regular languages. Recognizable relations play a role, for example, in [11] where relations over words are used for identifying equivalent plays in incomplete information games. The task is to compute a winning strategy that does not distinguish between equivalent plays. While this problem is undecidable for automatic relations, it is possible to synthesize strategies for recognizable equivalence relations. In view of such results, it is an interesting question whether one can decide for a given relation whether it is recognizable.

All these four concepts of automaton definable relations can directly be adapted to infinite words using the notion of ω-automata (see [29] for background on ω-automata), leading to the classes of (deterministic) ω-rational, ω-automatic, and ω-recognizable relations. Applications like automatic structures and regular model checking have been adapted to relations over infinite words [6, 9], for example for modeling systems with continuous parameters represented by real numbers (which can be encoded as infinite words [8]).

Our contributions are the following, where some background on the individual results is given below. We note that (4) is not a result on relations over words. It is used in the proof of (3) but we state it explicitly because we believe that it is an interesting result on its own.

(1) We show that the equivalence problem for binary deterministic ω-rational relations is undecidable, already for the Büchi acceptance condition (which is weaker than parity or Muller acceptance conditions in the case of deterministic automata).

(2) We show that it is decidable in doubly exponential time for an ω-automatic relation whether it is ω-recognizable.

(3) We reconsider the complexity of deciding for a binary automatic relation whether it is recognizable, and prove that it can be done in exponential time.

(4) We prove that the regularity problem for deterministic visibly pushdown automata [2] is decidable in polynomial time.

The algorithmic theory of deterministic ω-rational relations has not yet been studied in detail. We think, however, that this class is worth studying in order to understand whether it can be used in applications that are studied for ω-automatic relations. One such scenario could be the synthesis of finite state machines from (binary) ω-automatic relations. In this setting, an ω-automatic relation is viewed as a specification that relates input streams to possible output streams. The task is to automatically synthesize a synchronous sequential transducer (producing one output letter for each input letter) that outputs a string for each possible input such that the resulting pair is in the relation (see, e.g., [31] for an overview of this kind of automata theoretic synthesis). It has recently been shown that this synchronous synthesis problem can be lifted to the case of asynchronous automata if the relation is deterministic rational [16]. This shows that the class of deterministic rational relations has some interesting properties, and motivates our study of the corresponding class over infinite words. Our contribution (1) contrasts the decidability of equivalence for deterministic rational relations over finite words [5,18] and thus exhibits a difference between deterministic rational relations over finite and over infinite words. We prove the undecidability by a reduction from the intersection emptiness problem for deterministic rational relations over finite words. The reduction is inspired by a recent construction for proving the undecidability of equivalence for deterministic Büchi one-counter automata [7].

Contributions (2) and (3) are about the effectiveness of the hierarchies formed by the four classes of (ω-)rational, deterministic (ω-)rational, (ω-)automatic, and (ω-)recognizable relations. A systematic overview and study on the effectiveness of this hierarchy for finite words is provided in [14]: For a given rational relation it is undecidable whether it belongs to one of the other classes, for deterministic rational and automatic relations it is decidable whether they are recognizable, and the problem of deciding for a deterministic rational relation whether it is automatic is open.

The question of the effectiveness of the hierarchy for relations over infinite words has already been posed in [30] (where the ω-automatic relations are called Büchi recognizable ω-relations). The undecidability results easily carry over from finite to infinite words. Our result (2) lifts one of the two known decidability results for finite words to infinite words. The algorithm is based on a reduction to a problem over finite words: Using a representation of ω-languages by finite encodings of ultimately periodic words as in [13], we are able to reformulate the recognizability of an ω-automatic relation in terms of *slenderness* of a finite number of languages of finite words. A language of finite words is called slender [22] if there is a bound k such that the language contains for each length at most k words of this length. While slenderness for regular languages is easily seen to be decidable, the complexity of the problem has to the best of our knowledge not been analyzed. We prove that slenderness of a language defined by a nondeterministic finite automaton can be decided in NL.

As mentioned above, the decidability of recognizability of an automatic relation is known from [14]. However, the exponential time complexity claimed in that paper does not follow from the proof presented there. So we revisit the problem and prove the exponential time upper bound for binary relations based on the connection between binary rational relations and pushdown automata: For a relation R over finite words, consider the language L_R consisting of the words $\mathrm{rev}(u)\#v$ for all $(u,v) \in R$, where $\mathrm{rev}(u)$ denotes the reverse of u. It turns out that L_R is linear context-free iff R is rational, L_R is deterministic context-free iff R is deterministic rational, and L_R is regular iff R is recognizable [14]. Since L_R is regular iff R is recognizable, the recognizability test for binary deterministic rational relations reduces to the regularity test for deterministic pushdown automata, which is known to be decidable [27], in doubly exponential time [32].[1] We adapt this technique to automatic relations R and show that L_R can in this case be defined by a visibly pushdown automaton (VPA) [2], in which the stack operation (pop, push, skip) is determined by the input symbol, and no ε-transitions are allowed. The deterministic VPA for L_R is exponential in the size of the automaton for R, and we prove that the regularity test can be done in polynomial time, our contribution (4). We note that the polynomial time regularity test for visibly pushdown processes as presented in [26] does not imply our result. The model in [26] cannot use transitions that cause a pop operation when the stack is empty. For our translation from automatic relations to VPAs we need these kind of pop operations, which makes the model different and the decision procedure more involved (and a reduction to the model of [26] by using new internal symbols to simulate pop operations on the empty stack will not preserve regularity of the language, in general).

The paper is structured as follows. In Sect. 2 we give the definitions of transducers, relations, and visibly pushdown automata. In Sect. 3 we prove the undecidability of the equivalence problem for deterministic ω-rational relations. Section 4 contains the decision procedure for recognizability of ω-automatic relations, and Sect. 5 presents the polynomial time regularity test for deterministic VPAs and its use for the recognizability test of automatic relations. Finally, we conclude in Sect. 6.

2   Preliminaries

We start by briefly introducing transducers and visibly pushdown automata as we need them for our results. For more details we refer to [17,25,29] and [2,3], respectively.

A transducer $\mathcal{A}$ is a tuple $(Q, \Sigma_1, \ldots, \Sigma_k, q_0, \Delta, F)$ where Q is the state set, $\Sigma_i, 1 \leq i \leq k$ are (finite) alphabets, $q_0 \in Q$ is the initial state, $F \subseteq Q$ denotes the

---

[1] Recognizability is decidable for deterministic rational relations of arbitrary arity [14] but we are not aware of a proof preserving the doubly exponential runtime.

accepting states, and $\Delta \subseteq Q \times (\Sigma_1 \cup \{\varepsilon\}) \times \ldots \times (\Sigma_k \cup \{\varepsilon\}) \times Q$ is the transition relation. $\mathcal{A}$ is deterministic if there is a state partition $Q = Q_1 \cup \ldots \cup Q_k$ such that Δ can be interpreted as partial function $\delta : \bigcup_{j=1}^{k}(Q_j \times (\Sigma_j \cup \{\varepsilon\})) \to Q$ with the restriction that if $\delta(q, \varepsilon)$ is defined then no $\delta(q, a)$, $a \neq \varepsilon$ is defined. Note that the state determines which component the transducer processes. $\mathcal{A}$ is *complete* if δ is total (up to the restriction for ε-transitions).

A run of $\mathcal{A}$ on a tuple $u \in \Sigma_1^* \times \ldots \times \Sigma_k^*$ is a sequence $\rho = p_0 \ldots p_n \in Q^*$ such that there is a decomposition $u = (a_{1,1}, \ldots, a_{1,k}) \ldots (a_{n,1}, \ldots, a_{n,k})$ where the $a_{i,j}$ are in $\Sigma_j \cup \{\varepsilon\}$ and for all $i \in \{1, \ldots, n\}$ it holds that $(p_{i-1}, a_{i,1}, \ldots, a_{i,k}, p_i) \in \Delta$. The run of $\mathcal{A}$ on a tuple over infinite words in $\Sigma_1^\omega \times \ldots \times \Sigma_k^\omega$ is an infinite sequence $p_0 p_1 \ldots \in Q^\omega$ defined analogously to the case of finite words. A run on u is called *accepting* if it ends in an accepting state and $\mathcal{A}$ accepts u if there is an accepting run starting in the initial state q_0. Then $\mathcal{A}$ defines the relation $R_*(\mathcal{A}) \subseteq \Sigma_1^* \times \ldots \times \Sigma_k^*$ containing precisely those tuples accepted by $\mathcal{A}$. To enhance the expressive power of deterministic transducers, the relation $R_*(\mathcal{A})$ is defined as the relation of all u such that $\mathcal{A}$ accepts $u(\#, \ldots, \#)$ for some fresh fixed symbol $\# \notin \bigcup_{j=1}^{k} \Sigma_j$. The relations definable by a (deterministic) transducer are called *(deterministic) rational relations*. For tuples over infinite words $u \in \Sigma_1^\omega \times \ldots \times \Sigma_k^\omega$ we utilize the Büchi condition [12]. That is, a run $\rho \in Q^\omega$ is accepting if a state $f \in F$ occurs infinitely often in ρ. Then $\mathcal{A}$ accepts u if there is an accepting run of $\mathcal{A}$ on u starting in q_0 and $R_\omega(\mathcal{A}) \subseteq \Sigma_1^\omega \times \ldots \times \Sigma_k^\omega$ is the relation of all tuples of infinite words accepted by $\mathcal{A}$. We refer to $\mathcal{A}$ as Büchi transducer if we are interested in the relation of infinite words defined by it. The class of ω-rational relations consists of all relations definable by Büchi transducers.

It is well-known that deterministic Büchi automata are not sufficient to capture the ω-regular languages (see [29]) which are the ω-rational relations of arity one. Therefore, we use another kind of transducer to define deterministic ω-rational relations: a deterministic parity transducer is a tuple $\mathcal{A} = (Q, \Sigma_1, \ldots, \Sigma_k, q_0, \delta, \Omega)$ where the first $k+3$ items are the same as for deterministic transducers and $\Omega : Q \to \mathbb{N}$ is the priority function. A run is accepting if the maximal priority occurring infinitely often on the run is even (cf. [23]).

A transducer is synchronous if for each pair $(p, a_1, \ldots, a_k, q), (q, b_1, \ldots, b_k, r)$ of successive transitions it holds that $a_j = \varepsilon$ implies $b_j = \varepsilon$ for all $j \in \{1, \ldots, k\}$. Intuitively, a synchronous transducer is a finite automaton over the vector alphabet $\Sigma_1 \times \ldots \times \Sigma_k$ and, if it operates on tuples $(u_1, \ldots, u_k)$ of finite words, the components u_j may be of different length (i.e. if a u_j has been processed completely, the transducer may use transitions reading ε in the j-th component to process the remaining input in the other components). In fact, synchronous transducers inherit the rich properties of finite automata – e.g., they are closed under all boolean operations and can be determinized. In particular, synchronous (nondeterministic) Büchi transducer and deterministic synchronous parity transducer can be effectively transformed into each other [17,23,25]. Synchronous (Büchi) transducers define the class of (ω-)automatic relations.

Finally, the last class of relations we consider are (ω-)recognizable relations. A relation $R \subseteq \Sigma_1^* \times \ldots \times \Sigma_k^*$ (or $R \subseteq \Sigma_1^\omega \times \ldots \times \Sigma_k^\omega$) is ($\omega$-)*recognizable* if it is the finite union of direct products of (ω-)regular languages—i.e. $R = \bigcup_{i=1}^{\ell} L_{i,1} \times \ldots \times L_{i,k}$ where the $L_{i,j}$ are (ω-)regular languages.

It is well-known that (ω-)recognizable, (ω-)automatic, deterministic (ω-)rational relations, and (ω-)rational relations form a strict hierarchy [25].

In Sect. 5 we use visibly pushdown automata (VPAs) [2], which operate on typed alphabets, called pushdown alphabets below, where the type of an input symbol determines the stack operation. Formally, a *pushdown alphabet* is an alphabet Σ consisting of three disjoint parts—namely, a set Σ_c of *call symbols* enforcing a push operation, a set Σ_r of *return symbols* enforcing a pop operation and internal symbols Σ_{int} which do not permit any stack operation. A VPA is a tuple $\mathcal{P} = (P, \Sigma, \Gamma, p_0, \bot, \Delta, F)$ where P is a finite set of states, $\Sigma = \Sigma_c \mathbin{\dot\cup} \Sigma_r \mathbin{\dot\cup} \Sigma_{\text{int}}$ is a finite pushdown alphabet, Γ is the stack alphabet and $\bot \in \Gamma$ is the stack bottom symbol, $p_0 \in P$ is the initial state, $\Delta \subseteq (P \times \Sigma_c \times P \times (\Gamma \setminus \{\bot\})) \cup (P \times \Sigma_r \times \Gamma \times P) \cup (P \times \Sigma_{\text{int}} \times P)$ is the transition relation, and F is the set of accepting states.

A *configuration* of $\mathcal{P}$ is a pair in $(p, \alpha) \in P \times (\Gamma \setminus \{\bot\})^* \{\bot\}$ where p is the current state of $\mathcal{P}$ and α is the current stack content ($\alpha[0]$ is the top of the stack). Note that the stack bottom symbol $\bot$ occurs precisely at the bottom of the stack. The stack whose only content is $\bot$ is called the *empty stack*. $\mathcal{P}$ can *proceed* from a configuration (p, α) to another configuration (q, β) *via* $a \in \Sigma$ if $a \in \Sigma_c$ and there is a $(p, a, q, \gamma) \in \Delta \cap (P \times \Sigma_c \times P \times (\Gamma \setminus \{\bot\}))$ such that $\beta = \gamma\alpha$ (*push* operation), $a \in \Sigma_r$ and there is a $(p, a, \gamma, q) \in \Delta \cap (P \times \Sigma_r \times \Gamma \times P)$ such that $\alpha = \gamma\beta$ or $\gamma = \alpha = \beta = \bot$—that is, the empty stack may be popped arbitrarily often (*pop* operation), or $a \in \Sigma_{\text{int}}$ and there is a $(p, a, q) \in \Delta \cap (P \times \Sigma_{\text{int}} \times P)$ such that $\alpha = \beta$ (*noop*). A *run* of $\mathcal{P}$ on a word $u = a_1, \ldots, a_n \in \Sigma^*$ is a sequence of configurations $(p_1, \alpha_1) \ldots (p_{n+1}, \alpha_{n+1})$ connected by transitions using the corresponding input letter. A run is accepting if $p_{n+1} \in F$. Furthermore, we say that $\mathcal{P}$ accepts u if (p_1, α_1) is the initial configuration $(p_0, \bot)$ and write $L(p, \alpha)$ for the set of all words accepted from a configuration (p, α). In the case $(p, \alpha) = (p_0, \bot)$ we just say that $\mathcal{P}$ accepts u and write $L(\mathcal{P}) := L(p_0, \bot)$ for the language defined by $\mathcal{P}$. Two configurations $(p, \alpha), (q, \beta)$ are $\mathcal{P}$-equivalent if $L(p, \alpha) = L(q, \beta)$. Lastly, a configuration (p, α) is *reachable* if there is a run from $(p_0, \bot)$ to (p, α).

A *deterministic* VPA (DVPA) $\mathcal{P}$ is a VPA that can proceed to at most one configuration for each given configuration and $a \in \Sigma$.

Viewing the call symbols as opening and the return symbols as closing parenthesis, one obtains a natural notion of a return matching a call, and unmatched call or return symbols.

3   The Equivalence Problem for Deterministic Büchi Transducers

In this section we show that the equivalence for deterministic Büchi transducers is undecidable – in difference to its analogue for relations over finite words

[5,18]. Our proof is derived from a recent construction for proving that the equivalence problem for one-counter Büchi automata is undecidable [7]. We reduce the intersection emptiness problem for relations over finite words to the equivalence problem for deterministic Büchi transducers.

Proposition 1 [4,24]. *The intersection emptiness problem, asking for two binary relations given by deterministic transducers $\mathcal{A}, \mathcal{B}$ whether $R_*(\mathcal{A}) \cap R_*(\mathcal{B}) = \emptyset$ holds, is undecidable.*

Theorem 1. *The equivalence problem for ω-rational relations of arity at least two is undecidable for deterministic Büchi transducers.*

Proof. Let $\mathcal{A}_R$, $\mathcal{A}_S$ be deterministic transducers defining binary relations R and S over finite words, respectively. We construct deterministic Büchi transducers $\mathcal{B}_R$ and $\mathcal{B}_S$ such that each tuple in $R \cap S$ induces a witness for $R_\omega(\mathcal{B}_R) \neq R_\omega(\mathcal{B}_S)$ and vice versa. Recall that $\mathcal{A}_R$, $\mathcal{A}_S$ accept a tuple (u, v) if there is an accepting run on $(u, v)(\#, \#)$ (where $\#$ is an endmarker symbol not contained in any alphabet involved). Then it is easy to see that we can assume that $\mathcal{A}_R$ enters either a unique accepting state q_a^R or unique rejecting state q_r^R after it has read the endmarker symbol in both components. Moreover, q_a^R and q_r^R have no outgoing transitions. Furthermore, let q_0^R be the initial state of $\mathcal{A}_R$ and let q_a^S, q_r^S, q_0^S be the respective states of $\mathcal{A}_S$.

The construction of $\mathcal{B}_R$ and $\mathcal{B}_S$ is illustrated in Fig. 1. Both Büchi transducers are almost the same except for the initial state: both consist of the union of the transition structures of $\mathcal{A}_R$ and $\mathcal{A}_S$ complemented by transitions labeled ε/ε from q_a^X to q_0^X and q_r^X to q_0^Y for $X, Y \in \{R, S\}$, $X \neq Y$. That is, upon reaching a rejecting state of $\mathcal{A}_R$ or $\mathcal{A}_S$ the new transducers will switch to the initial state of the other subtransducer and upon reaching an accepting state they will return to the initial state of the current subtransducer. The new accepting states are q_a^R, q_r^R, q_r^S (note that q_a^S is not accepting introducing an asymmetry). Finally, the initial state of $\mathcal{B}_X$ is q_0^X.

Pick a tuple (u, v) in $R \cap S$. Then the unique run of $\mathcal{A}_X$ on $(u\#, v\#)$ leads to q_a^X and, thus, q_0^X. Hence, the induced unique run of $\mathcal{B}_R$ on $(u\#, v\#)^\omega$ is accepting while the unique run of $\mathcal{B}_S$ is rejecting. On the contrary, assume that there is a pair of infinite words that is rejected by one of the transducers, and accepted by the other. Each such tuple has to contain infinitely often $\#$ in both components, since one of the Büchi transducers accepts. The only way to reject in presence of infinitely many $\#$, is to finally remain in the copy of $\mathcal{A}_S$ visiting q_a^S infinitely often. The accepting run then has to remain in the copy of $\mathcal{A}_R$ because otherwise the runs would meet in the same initial state and thus both accept or both reject. This implies that there is an input $(u\#, v\#)$ that takes $\mathcal{A}_R$ from q_0^R to q_a^R, and $\mathcal{A}_S$ from q_0^S to q_a^S. In other words, $(u, v) \in R \cap S$.

We note that our reduction is rather generic an could be applied to other classes of automata for which the intersection emptiness problem on finite words is undecidable.

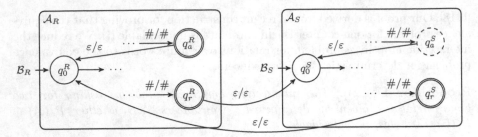

Fig. 1. Illustration of the transducers $\mathcal{B}_R$, $\mathcal{B}_S$. The labels $\#/\#$ are just used for comprehensibility. In the formal construction the $\#$ symbols are read in succession and the transducers may even read other symbols between them (but only in the component where no $\#$ has been read yet).

4 Deciding Recognizability of ω-Automatic Relations

Our aim in this section is to decide ω-recognizability of ω-automatic relations in doubly exponential time. That is, given a deterministic synchronous transducer decide whether it defines an ω-recognizable relation. The proof approach is based on an algorithm for relations over finite words given in [14] which we briefly recall.

Let R be an (ω-)automatic relation of arity k. For each $j \leq k$ we define the equivalence relation

$$E_j := \{((u_1, \ldots, u_j), (v_1, \ldots, v_j)) \mid \forall w_{j+1}, \ldots, w_k :$$
$$(u_1, \ldots, u_j, w_{j+1}, \ldots, w_k) \in R \Leftrightarrow (v_1, \ldots, v_j, w_{j+1}, \ldots, w_k) \in R\}.$$

Then the key to decide (ω-)recognizability is the following result which has been proven in [14] for relations over finite words and is easily extensible to infinite words:

Lemma 1 [14]. *Let R be an (ω-)automatic relation of arity k. Then for all $1 \leq j \leq k$ the equivalence relation E_j has finite index if and only if R is (ω-)recognizable.*

Based on that lemma, the recognizability test presented in [14] proceeds as follows. It is shown that each E_j is an automatic equivalence relation by constructing a synchronous transducer for E_j. It remains to decide for an automatic equivalence relation whether it is of finite index. This can be achieved by constructing a synchronous transducer that accepts a set of representatives of the equivalence classes of E_j (based on a length-lexicographic ordering). Then E_j has finite index if and only if this set of representatives is finite, which can be decided in polynomial time.

It is unclear whether this approach can be used to obtain an exponential time upper bound for the recognizability test.[2] One can construct a family $(R_n)_{n \in \mathbb{N}}$ of

---

[2] The authors of [14] mainly focused on decidability, and they agree that the proof as presented in that paper does not yield an exponential time upper bound.

automatic binary relations R_n defined by a deterministic synchronous transducer of size $\mathcal{O}(n^2)$ such that every synchronous transducer defining E_1 has size (at least) exponential in n. It is unclear whether it is possible to decide in polynomial time for such a transducer whether the equivalence relation it defines is of finite index. For this reason, we revisit the problem for finite words in Sect. 5 and provide an exponential time upper bound for binary relations using a different approach.

We now turn to the case of infinite words. The relation E_j can be shown to be ω-automatic, similarly to the case of finite words. However, it is not possible, in general, for a given ω-automatic relation to define a set of representatives by means of a synchronous transducer, as shown in [20]: There exists a binary ω-automatic equivalence relation such that there is no ω-regular set of representatives of the equivalence classes.

Here is how we proceed instead. The first step is similar to [14]: We construct synchronous transducers for the complements $\overline{E_j}$ of the equivalence relations E_j in polynomial time (starting from a deterministic transducer for R). We then provide a decision procedure to decide for a given transducer for $\overline{E_j}$ whether the index of E_j is finite in doubly exponential time. This procedure is based on an encoding of ultimately periodic words by finite words.

First observe that a tuple in $\Sigma_1^\omega \times \ldots \times \Sigma_j^\omega$ can be seen as an infinite word over $\Sigma = \Sigma_1 \times \ldots \times \Sigma_j$ (this is not the case for tuples over finite words, since the words may be of different length). Hence, we can view each E_j as a binary equivalence relation $E \subseteq \Sigma^\omega \times \Sigma^\omega$. For this reason, we only work with binary relations in the following.

We start by showing that for deciding whether E has finite index it suffices to consider sets of ultimately periodic representatives $u_i v_i^\omega$ such that the periods $|v_i|$ and prefix lengths $|u_i|$ are the same, respectively, for all the representatives (Lemma 2). In the second step E is transformed into an automatic equivalence relation $E_\#$ over finite words using encodings of ultimately periodic words as finite words, where a word uv^ω is encoded by $u\#v$ as in [13] (Definition 2 and Lemma 3). Since $E_\#$ is an automatic relation over finite words, it is possible to obtain a finite automaton for a set of representatives of $E_\#$. Finally, we reduce the decision problem whether E has finite index to deciding slenderness (see Definition 1 below) for polynomially many languages derived from the set of representatives of $E_\#$ (Lemmas 4 and 5). Therefore, by proving that deciding slenderness for (nondeterministic) finite automata is NL-complete (Lemma 6) we obtain our result.

Definition 1 [22]. *A language $L \subset \Sigma^*$ is slender if there exists a $k < \omega$ such that for all $\ell < \omega$ it holds that $|L \cap \Sigma^\ell| < k$.*

We now formalize the ideas sketched above.

Lemma 2. *Let $E \subseteq \Sigma^\omega \times \Sigma^\omega$ be an ω-automatic equivalence relation. Then E has not finite index if and only if for each $k > 0$ there are*

$$u_1, \ldots, u_k,\ v_1, \ldots v_k \in \Sigma^* \text{ with } |u_i| = |u_j| \text{ and } |v_i| = |v_j| \text{ for all } 1 \le i \le j \le k$$

such that $(u_i v_i^\omega, u_j v_j^\omega) \notin E$ for all $1 \le i < j \le k$.

We proceed by transforming E into an automatic equivalence relation $E_\#$ and showing that it is possible to compute in exponential time a synchronous transducer for it, given a synchronous Büchi transducer for $\overline{E}$.

Definition 2. *Let $E \subseteq \Sigma^\omega \times \Sigma^\omega$ be an ω-automatic equivalence relation. Furthermore, let $\Gamma := \Sigma \cup \{\#\}$ for a fresh symbol $\# \notin \Sigma$. Then the relation $E_\# \subseteq \Gamma^* \times \Gamma^*$ is defined by*

$$E_\# := \{(u\#v, x\#y) \mid u, v, x, y \in \Sigma^*, |u| = |x|, |v| = |y|, (uv^\omega, xy^\omega) \in E\}.$$

Lemma 3. *Let $E \subseteq \Sigma^\omega \times \Sigma^\omega$ be an ω-automatic equivalence relation and $\mathcal{A}$ a synchronous Büchi transducer defining the complement $\overline{E}$ of E. Then there is a synchronous transducer $\mathcal{A}_\#$ exponential in the size of $\mathcal{A}$ which defines $E_\#$. In particular, $E_\#$ is an automatic relation.*

With a synchronous transducer for $E_\#$ at hand, we can compute a synchronous transducer defining a set of unique representatives of $E_\#$ similarly to [14]. For convenience, we will denote the set of representatives obtained by this construction by $L_\#(E)$ (although it is not unique in general). We can now readjust Lemma 2 to $E_\#$ (or, more precisely, $L_\#(E)$).

Lemma 4. *Let $E \subseteq \Sigma^\omega \times \Sigma^\omega$ be an ω-automatic equivalence relation. Then E has finite index if and only if there is a $k < \omega$ such that for all $m, n > 0$: $|L_\#(E) \cap \Sigma^n \{\#\} \Sigma^m| \leq k$.*

Note that the condition in Lemma 4 is similar to slenderness but not equivalent to the statement that $L_\#(E)$ is slender. For instance, consider the language L given by the regular expression $a^*\#b^*$. For any $m, n > 0$ we have that $|L \cap \Sigma^n \{\#\} \Sigma^m| = |\{a^n \# b^m\}| \leq 1$. But L is not slender: Let $\ell > 0$. Then $a^{\ell-1-i} \# b^i \in L \cap \Sigma^\ell$ for all $0 \leq i < \ell$. Hence, $|L \cap \Sigma^\ell| \geq \ell$ and, thus, L cannot be slender. However, the next result shows that there is a strong connection between the condition in Lemma 4 and slenderness.

Lemma 5. *Let L be a language of the form $L = \bigcup_{(i,j) \in I} L_i \{\#\} L_j$ where $I \subset \mathbb{N}^2$ is a finite index set and $L_i, L_j \subseteq (\Sigma \setminus \{\#\})^*$ are non-empty regular languages for each pair $(i, j) \in I$. Then there is a $k < \omega$ such that for all $m, n \geq 0$: $|L \cap \Sigma^n \{\#\} \Sigma^m| \leq k$ if and only if for all $(i, j) \in I$ it holds that L_i and L_j are slender.*

The last ingredient we need is the decidability of slenderness in polynomial time. Lemma 6 can be shown analogously to the proof in [28] where it is shown that the finiteness problem for Büchi automata is NL-complete.

Lemma 6. *Deciding slenderness for (nondeterministic) finite automata is NL-complete.*

Finally, we can combine our results to obtain the main result of this section. Firstly, we state our approach to check whether an automatic equivalence has finite index and, afterwards, join it with the approach of [14].

Theorem 2. *Let $E \subseteq \Sigma^\omega \times \Sigma^\omega$ be an ω-automatic equivalence relation and $\mathcal{A}_\#$ be a (nondeterministic) synchronous transducer defining $E_\#$. Then it is decidable in single exponential time whether E has finite index.*

Proof (sketch). Given $E_\#$ it suffices to check the condition of Lemma 4. This can be achieved by checking slenderness for the factors L_i, L_j of a decomposition $L_\#(E) = \bigcup_{(i,j)\in I} L_i\{\#\}L_j$ due to Lemma 5. Given $\mathcal{A}_\#$ we can compute a synchronous transducer $\mathcal{B}$ with initial state q_0 and accepting states F defining a set $L_\#(E)$ of representatives of $E_\#$ (cf. [14]). Then $\mathcal{B}$ induces a decomposition of $L_\#(E)$ of the required shape: each transition $(p, \#, q)$ induces two factors $L_{q_0\{p\}}$ and L_{qF} where L_{yX} is defined by the modification of $\mathcal{B}$ having initial state y, accepting states X, and no $\#$ transition. Hence, there are only polynomially many factors in the size of $\mathcal{B}$ that have to be checked for slenderness. The exponential time claim follows from Lemma 6 and because $\mathcal{B}$ has size exponential in $\mathcal{A}_\#$. $\qquad\blacksquare$

Theorem 3. *Given a complete deterministic synchronous parity transducer $\mathcal{A}$ it is decidable in double exponential time whether $R_\omega(\mathcal{A})$ is ω-recognizable.*

5   Deciding Recognizability of Automatic Relations

At the beginning of Sect. 4 we have sketched the approach presented in [14] for deciding recognizability of an automatic relation. In this section we revisit the problem to obtain an exponential time upper bound for the case of binary relations. The procedure is based on a reduction to the regularity problem for VPAs (Lemma 8). The other main contribution in this section, which is interesting on its own, is a polynomial time algorithm to solve the regularity problem for DVPAs. We start by describing the regularity test. The key to decide regularity in polynomial time is the following result.

Lemma 7. *Let $\mathcal{P}$ be a DVPA with n states. Then $L(\mathcal{P})$ is regular if and only if all pairs $(p, \alpha\beta), (p, \alpha\beta')$ of reachable configurations of $\mathcal{P}$ with $|\alpha| \geq n^3 + 2$ are $\mathcal{P}$-equivalent.*

Theorem 4. *It is decidable in polynomial time whether a given DVPA defines a regular language.*

Proof (sketch). We construct a synchronous transducer $\mathcal{A}$ accepting pairs of configurations violating the right hand side of the equivalence of Lemma 7. Then it suffices to decide the emptiness problem for transducers which boils down to a graph search doable in polynomial time. The transducer has to check whether the length constraint is satisfied, both configurations start in the same state, are reachable and are not $\mathcal{P}$-equivalent (recall that synchronous transducers are closed under intersection). The first two conditions are easy. For the third condition, a finite automaton for the set of reachable configurations of $\mathcal{P}$ can be constructed in polynomial time [10]. The crucial part is to check for non-equivalence. The idea is to guess a separating word and simulate $\mathcal{P}$ in parallel starting in the two configurations given as input to $\mathcal{A}$. One can show that only the return symbols of a separating word that are responsible for popping a symbol from the stacks have to be guessed (instead of the whole separating word).

With Theorem 4 established we turn towards our second objective which is to decide recognizability of binary automatic relations. Recall that for a word u we denote its reversal by $\mathrm{rev}(u)$.

Lemma 8. *Let $R \subseteq \Sigma_1^* \times \Sigma_2^*$ with $\Sigma_1 \cap \Sigma_2 = \emptyset$ be an automatic relation and $\# \notin \Sigma_1 \cup \Sigma_2$ be a fresh symbol. Furthermore, let $\mathcal{A}$ be a (nondeterministic) synchronous transducer defining R. Then $L_R := \{\mathrm{rev}(u)\#v \mid (u,v) \in R\}$ is definable by a DVPA whose size is single exponential in $|\mathcal{A}|$.*

Since L_R is regular if and only if R is a recognizable relation [14], we obtain the second result of this section as corollary of Theorem 4 and Lemma 8.

Corollary 1. *Let $\mathcal{A}$ be a (possibly nondeterministic) synchronous transducer defining a binary relation. Then it is decidable in single exponential time whether $R_*(\mathcal{A})$ is recognizable.*

6   Conclusion

The undecidability of the equivalence problem for deterministic ω-rational relations presented in Sect. 3 exhibits an interesting difference between deterministic transducers on finite and on infinite words. We believe that it is worth to further study the algorithmic theory of this class of relations. For example, the decidability of recognizability for a given deterministic ω-rational relation is an open question. The technique based on the connection between binary rational relations and context-free languages as presented in Sect. 5 that is used in [14] for deciding recognizability of deterministic rational relations cannot be (directly) adapted. First of all, the idea of pushing the first component on the stack and then simulating the transducer while reading the second component fails because this would require an infinite stack. Furthermore, the regularity problem for deterministic ω-pushdown automata is not known to be decidable (only for the subclass of deterministic weak Büchi automata [21]).

It would also be interesting to understand whether the decidability of the synthesis problem (see the introduction) for deterministic rational relations over finite words [16] can be transferred to infinite words.

For the recognizability problem of (ω-)automatic relations we have shown decidability with a doubly exponential time algorithm for infinite words. We also provided a singly exponential time algorithm for the binary case over finite words (improving the complexity of the approach from [14] as explained in Sect. 4). It is open whether there are matching lower complexity bounds.

The connection between automatic relations and VPAs raises the question whether extensions of VPAs studied in the literature (as for example in [15]) can be used to identify interesting subclasses of relations between the (ω-)automatic and deterministic (ω-)rational relations. The problem of identifying such classes for the case of infinite words has already been posed in [30].

References

1. Abdulla, P.A.: Regular model checking. STTT **14**(2), 109–118 (2012). doi:10.1007/s10009-011-0216-8
2. Alur, R., Madhusudan, P.: Visibly pushdown languages. In: Proceedings of the Thirty-sixth Annual ACM Symposium on Theory of Computing, pp. 202–211. ACM (2004)
3. Bárány, V., Löding, C., Serre, O.: Regularity problems for visibly pushdown languages. In: Durand, B., Thomas, W. (eds.) STACS 2006. LNCS, vol. 3884, pp. 420–431. Springer, Heidelberg (2006). doi:10.1007/11672142_34
4. Berstel, J.: Transductions and Context-Free Languages. Springer, Stuttgart (1979)
5. Bird, M.: The equivalence problem for deterministic two-tape automata. J. Comput. Syst. Sci. **7**(2), 218–236 (1973)
6. Blumensath, A., Grädel, E.: Automatic structures. In: Proceedings of the 15th IEEE Symposium on Logic in Computer Science, LICS 2000, pp. 51–62. IEEE Computer Society Press (2000)
7. Böhm, S., Göller, S., Halfon, S., Hofman, P.: On Büchi one-counter automata. In: Vollmer, H., Vallée, B. (eds.) 34th Symposium on Theoretical Aspects of Computer Science (STACS 2017). Leibniz International Proceedings in Informatics (LIPIcs), vol. 66, pp. 14:1–14:13 (2017). doi:10.4230/LIPIcs.STACS.2017.14. ISBN 978-3-95977-028-6
8. Boigelot, B., Jodogne, S., Wolper, P.: An effective decision procedure for linear arithmetic over the integers and reals. ACM Trans. Comput. Log. **6**(3), 614–633 (2005). doi:10.1145/1071596.1071601
9. Boigelot, B., Legay, A., Wolper, P.: Omega-regular model checking. In: Jensen, K., Podelski, A. (eds.) TACAS 2004. LNCS, vol. 2988, pp. 561–575. Springer, Heidelberg (2004). doi:10.1007/978-3-540-24730-2_41
10. Bouajjani, A., Esparza, J., Maler, O.: Reachability analysis of pushdown automata: application to model-checking. In: Mazurkiewicz, A., Winkowski, J. (eds.) CONCUR 1997. LNCS, vol. 1243, pp. 135–150. Springer, Heidelberg (1997). doi:10.1007/3-540-63141-0_10
11. Bozzelli, L., Maubert, B., Pinchinat, S.: Uniform strategies, rational relations and jumping automata. Inf. Comput. **242**, 80–107 (2015). doi:10.1016/j.ic.2015.03.012
12. Büchi, J.R.: On a decision method in restricted second order arithmetic. In: Nagel, E. (ed.) Logic, Methodology, and Philosophy of Science: Proceedings of the 1960 International Congress, pp. 1–11. Stanford University Press, Palo Alto (1962)
13. Calbrix, H., Nivat, M., Podelski, A.: Ultimately periodic words of rational ω-languages. In: Brookes, S., Main, M., Melton, A., Mislove, M., Schmidt, D. (eds.) MFPS 1993. LNCS, vol. 802, pp. 554–566. Springer, Heidelberg (1994). doi:10.1007/3-540-58027-1_27
14. Carton, O., Choffrut, C., Grigorieff, S.: Decision problems among the main subfamilies of rational relations. RAIRO-Theor. Inform. Appl. **40**(02), 255–275 (2006)
15. Caucal, D.: Synchronization of pushdown automata. In: Ibarra, O.H., Dang, Z. (eds.) DLT 2006. LNCS, vol. 4036, pp. 120–132. Springer, Heidelberg (2006). doi:10.1007/11779148_12
16. Filiot, E., Jecker, I., Löding, C., Winter, S.: On equivalence and uniformisation problems for finite transducers. In: 43rd International Colloquium on Automata, Languages, and Programming, ICALP 2016. LIPIcs, vol. 55, p. 125:1–125:14. Schloss Dagstuhl - Leibniz-Zentrum fuer Informatik (2016). doi:10.4230/LIPIcs.ICALP.2016.125

17. Frougny, C., Sakarovitch, J.: Synchronized rational relations of finite and infinite words. Theor. Comput. Sci. **108**(1), 45–82 (1993)
18. Harju, T., Karhumäki, J.: The equivalence problem of multitape finite automata. Theor. Comput. Sci. **78**(2), 347–355 (1991). doi:10.1016/0304-3975(91)90356-7
19. Khoussainov, B., Nerode, A.: Automatic presentations of structures. In: Leivant, D. (ed.) LCC 1994. LNCS, vol. 960, pp. 367–392. Springer, Heidelberg (1995). doi:10.1007/3-540-60178-3_93
20. Kuske, D., Lohrey, M.: First-order and counting theories of ω-automatic structures. In: Aceto, L., Ingólfsdóttir, A. (eds.) FoSSaCS 2006. LNCS, vol. 3921, pp. 322–336. Springer, Heidelberg (2006). doi:10.1007/11690634_22
21. Löding, C., Repke, S.: Regularity problems for weak pushdown ω-automata and games. In: Rovan, B., Sassone, V., Widmayer, P. (eds.) MFCS 2012. LNCS, vol. 7464, pp. 764–776. Springer, Heidelberg (2012). doi:10.1007/978-3-642-32589-2_66
22. Păun, G., Salomaa, A.: Thin and slender languages. Discret. Appl. Math. **61**(3), 257–270 (1995)
23. Piterman, N.: From nondeterministic Büchi and Streett automata to deterministic parity automata. In: 21st Annual IEEE Symposium on Logic in Computer Science (LICS 2006), pp. 255–264. IEEE (2006)
24. Rabin, M.O., Scott, D.: Finite automata and their decision problems. IBM J. Res. Dev. **3**(2), 114–125 (1959)
25. Sakarovitch, J.: Elements of Automata Theory. Cambridge University Press, New York (2009)
26. Srba, J.: Visibly pushdown automata: from language equivalence to simulation and bisimulation. In: Ésik, Z. (ed.) CSL 2006. LNCS, vol. 4207, pp. 89–103. Springer, Heidelberg (2006). doi:10.1007/11874683_6
27. Stearns, R.E.: A regularity test for pushdown machines. Inf. Control **11**(3), 323–340 (1967)
28. Tao, Y.: Infinity problems and countability problems for ω-automata. Inf. Process. Lett. **100**(4), 151–153 (2006)
29. Thomas, W.: Automata on infinite objects. In: Handbook of Theoretical Computer Science, vol. B, pp. 133–191 (1990)
30. Thomas, W.: Infinite trees and automation-definable relations over ω-words. Theor. Comput. Sci. **103**(1), 143–159 (1992). doi:10.1016/0304-3975(92)90090-3
31. Thomas, W.: Facets of synthesis: revisiting Church's problem. In: Alfaro, L. (ed.) FoSSaCS 2009. LNCS, vol. 5504, pp. 1–14. Springer, Heidelberg (2009). doi:10.1007/978-3-642-00596-1_1
32. Valiant, L.G.: Regularity and related problems for deterministic pushdown automata. J. ACM (JACM) **22**(1), 1–10 (1975)

Listing All Fixed-Length Simple Cycles in Sparse Graphs in Optimal Time

George Manoussakis$^{(\boxtimes)}$

LRI, CNRS, Université Paris Sud, Université Paris Saclay, 91405 Orsay, France
george@lri.fr

Abstract. The *degeneracy* of an n-vertex graph G is the smallest number k such that every subgraph of G contains a vertex of degree at most k. We present an algorithm for enumerating all simple cycles of length p in an n-order k-degenerate graph running in time $\mathcal{O}(n^{\lfloor p/2 \rfloor} k^{\lceil p/2 \rceil})$. We then show that this algorithm is worst-case output size optimal by proving a $\Theta(n^{\lfloor p/2 \rfloor} k^{\lceil p/2 \rceil})$ bound on the maximal number of simple p-length cycles in these graphs. Our results also apply to induced (chordless) cycles.

Keywords: Sparse graphs · k-degenerate · Fixed-size cycles listing

1 Introduction

Degeneracy, introduced by Lick and White [15], is a common measure of the sparseness of a graph and is closely related to other sparsity measures such as arboricity and thickness. Degenerate graphs appear often in practice. For instance, the World Wide Web graph, citation networks, and collaboration graphs have low arboricity, and therefore have low degeneracy [12]. Furthermore, planar graphs have degeneracy at most five [15] and the Barabàsi-Albert model of preferential attachment [2], frequently used as a model for social networks, produces graphs with bounded degeneracy. From this point of view it seems pertinent to design fixed-parameter tractable algorithms parametrized by degeneracy. That is, given a n-order k-degenerate graph, to design algorithms of the form $f(k)n^{\mathcal{O}(1)}$ where f may grow exponentially with k but is independent of n.

The question of finding *fixed* length simple induced and non induced cycles in planar and k-degenerate graphs has been extensively studied. Among other contributions, Papadimitriou and Yannakakis [18] presented an algorithm finding C_3's in planar graphs. Chiba and Nishizeki [7] and Chrobak and Eppstein [8] proposed simpler linear time algorithms to find C_3's and the first of these papers also presents an algorithm finding C_4's. Both papers also apply their techniques to k-degenerate graphs. Richards [19] gave an $\mathcal{O}(n \log n)$ algorithm finding C_5's and C_6's. For any fixed length, Alon et al. [1], gave algorithms for both general and k-degenerate graphs. Cai et al. [6] proposed algorithms finding induced cycles of any fixed size. For the problem of finding all occurrences of any p-length simple cycle in planar graphs, assuming some constant bound on p, Eppstein

© Springer-Verlag GmbH Germany 2017
R. Klasing and M. Zeitoun (Eds.): FCT 2017, LNCS 10472, pp. 355–366, 2017.
DOI: 10.1007/978-3-662-55751-8_28

[10] proposes an algorithm running in time $\mathcal{O}(n + occ)$ where occ is the number of simple p-length cycles in the graph. His algorithm works for any subgraph of fixed size p and solves in fact the more general problem of *subgraph isomorphism in planar graphs*. His result has been later improved by Dorn [9], who reduces the time dependence in p. For short cycles of size six or less, Kowalik [14], proposes an algorithm listing all occurrences of these cycles in time $\mathcal{O}(n+occ)$. His algorithm is faster in practice than the one of Eppstein for planar graphs and also works for k-degenerate graphs, with complexity $\mathcal{O}(k^2m + occ)$, for cycles of size up to five. He also proves that the maximal number of simple p-length cycles in a planar graph is $\Theta(n^{\lfloor p/2 \rfloor})$. More recently, Meeks [16] proposed a randomized algorithm, given a general graph G, enumerating any p-sized subgraph H in time $\mathcal{O}(occ * f)$ where f is the time needed to find one occurrence of H in G. This result, together with the one of Alon *et al.* [1] for instance, yields an $\mathcal{O}(occ * n^{\mathcal{O}(1)} k^{\mathcal{O}(1)})$ time algorithm finding all occurrences of a p-length simple cycle in k-degenerate graphs, assuming p constant.

Other contributions have also been made for general graphs. For the problem of finding *all cycles* (any length), Tarjan [20] gives an $\mathcal{O}(nm * occ)$ time algorithm, where occ is the total number of cycles of the graph. This complexity has been improved to $\mathcal{O}((n + m) * occ)$ by Johnson [13]. More recently, Birmelé *et al.* [4] proposed an $\mathcal{O}(c)$ time algorithm where c is the number of edges of all the cycles of the graph. Uno and Satoh [21] proposed an $\mathcal{O}((n + m) * occ)$ algorithm finding all chordless cycles. We are not sure whether these algorithms can be easily adapted to output exactly all p-length simple cycles in k-degenerate or general graphs with similar complexities but where occ would be the number of p-length simple cycles (instead of the number of all cycles). For the problem of counting all cycles of size less than some constant p, Giscard *et al.* [11] propose an algorithm running in time $\mathcal{O}(\Delta|S_p|)$ where Δ is the maximum degree and $|S_p|$ the number of induced subgraphs with p or less vertices in the graph. Alon *et al.* [1] proposed an algorithm counting cycles of size less than 7 in time $\mathcal{O}(n^\omega)$ where ω is the exponent of matrix multiplication. Williams and Williams [22] and Björklund *et al.* [5] also give algorithms which can be used to count fixed-size cycles.

Our main contribution is a simple algorithm listing all p-length simple cycles in an n-order k-degenerate graph in time $\mathcal{O}(n^{\lfloor p/2 \rfloor} k^{\lceil p/2 \rceil} \log k)$, assuming that the graph is stored in an adjacency list data structure. If we have its adjacency matrix the time complexity can be improved to $\mathcal{O}(n^{\lfloor p/2 \rfloor} k^{\lceil p/2 \rceil})$. We then show that this complexity is worst-case output size optimal by proving that the maximal number of p-length simple cycles in an n-order k degenerate graph is $\Theta(n^{\lfloor p/2 \rfloor} k^{\lceil p/2 \rceil})$. These results also hold for induced cycles. To the best of our knowledge, this is the first such algorithm. It differs from the one of Meeks described before since it is deterministic, self-contained and can have better or worst time complexity depending on the number of simple p-length cycles of the input graph. Further improvements are discussed in the conclusion.

Our complexities are given assuming a constant bound on p. The exact dependence in p is described later but is exponential. Our approach for the main algorithm of the paper is the following. We first show that in a k-degenerate graph, any p-length cycle can be decomposed into small special paths, namely t-paths, introduced in Definition 2. We then prove that these t-paths can be computed and combined efficiently to generate candidate cycles. With some more work, we can then output exactly all p-length simple cycles of the graph.

The organization of the document is as follows. In Sect. 2 we introduce notations and definitions. In Sect. 3 we prove preliminary results for paths, t-paths and cycles. Using these results, we describe and prove algorithms in Sect. 4 which is the main section of the paper. The main algorithm is proved in Theorem 4 and the bound for the number of cycles in Theorem 5.

2 Notations and Definitions

In this section, we introduce notations and definitions that we use throughout the paper. By k we will denote the degeneracy of the graph and by p the length of the cycles we seek to list. When not specified, $G = (V, E)$ is a simple graph with $n = |V|$ and $m = |E|$. Given a k-degenerate graph G, we can compute in time $\mathcal{O}(m)$ its degeneracy ordering [3]. A degeneracy ordering is an ordering of the vertices such that each vertex has less than k neighbours with higher ranking in the ordering. This ordering also yields an acyclic orientation of the edges such that every vertex has out-degree at most k. From now on we will consider k-degenerate graph as oriented acyclic graphs with out-degree bounded by k. If (x, y) is an edge of some oriented graph G we will write $x \to y$ and say that x is oriented towards y if edge (x, y) is oriented towards y.

Definition 1. *An oriented path $P : p_1, p_2, ..., p_x$ is increasing (resp. decreasing) with respect to p_1 if $p_1 \to p_2 \to ... \to p_x$ (resp. $p_1 \leftarrow p_2 \leftarrow ... \leftarrow p_x$).*

Definition 2. *Let G be an oriented graph and $i, j \in \mathbb{N}$. A t-path $\mathcal{P}$ of size (i, j) is a path of $i + 1 + j$ vertices, $v_1, v_2, ..v_i, r, u_1, u_2, ...u_j$ such that $\mathcal{P}_l : r, v_i, v_{i-1}, ..., v_1$ and $\mathcal{P}_r : r, u_1, u_2, ..., u_j$ are increasing paths with respect to r. Vertices v_1 and u_j are called the end vertices of $\mathcal{P}$, vertex r its center. If $i, j \in \mathbb{N}^+$, we say that $\mathcal{P}$ is a strict t-path.*

Definition 3. *Let G be an oriented graph and $\mathcal{P}_1$, $\mathcal{P}_2$ two t-paths of G. They are adjacent if they do not have any vertex in common but one or two of their end vertices.*

Definition 4. *Let G be an oriented graph and $x = (i, j) \in \mathbb{N}^2$. We say that we can associate a t-path to x in G if there exists a t-path of $i + 1 + j$ vertices in G.*

Definition 5. *Let x, y, z be three consecutive nodes in C, an oriented simple cycle. Node y is a root of C if $x \leftarrow y \to z$.*

Observation 1. *An acyclic oriented simple cycle has a root.*

Definition 6. *Let* $G = (V, E)$ *be an acyclic oriented* k-*degenerate graph. Assume that* G *is given by the adjacency lists for each vertex. The sorted degenerate adjacency list of a vertex* $x \in V$ *is its adjacency list in which every vertex that is pointing towards* x *has been deleted and which has been sorted afterwards.*

Lemma 1. *The sorted degenerate adjacency lists of a* n-*order* k-*degenerate graph* G *can be computed in time* $\mathcal{O}(nk \log k)$ *and adjacency queries can be done in time* $\mathcal{O}(\log k)$ *using these modified lists.*

Proof. Assume that we have the adjacency lists of G. Let $x \in V$ and let d_x be its degree and d_x^+ its out-degree. In time $\mathcal{O}(d_x)$ remove all vertices from its adjacency lists that are pointing towards it. Then sort the remaining vertices in time $\mathcal{O}(d_x^+ \log d_x^+) = \mathcal{O}(k \log k)$ since, as specified in the beginning of the section, we consider k-degenerate graphs as acyclic oriented graphs with out-degree at most k. Repeat the procedure for all the vertices of the graph. This is done in total time $\mathcal{O}(nk \log k + m) = \mathcal{O}(nk \log k)$. $\square$

3   Basic Results

In this section we prove simple results concerning paths, t-paths and cycles. These lemmas are used to prove the correctness and time complexity of Algorithm 1, presented in Sect. 4.

Lemma 2. *Let* $\mathcal{C}$ *be a simple cycle* $\mathcal{C}$ *of size* p. *For any possible acyclic orientation of* $\mathcal{C}$, *we can compute a list* $\mathcal{L}$ *of strictly positive integer pairs* $(l_1, l_2), ..., (l_{r-1}, l_r)$ *in time* $\mathcal{O}(p)$ *with the following properties:*

(i) *To each pair* (l_i, l_{i+1}) *we can associate a strict* t-*path such that* $\mathcal{C}$ *can be decomposed into* $|\mathcal{L}|$ *strict adjacent* t-*paths: one for each pair of* $\mathcal{L}$.

(ii) *If* $|\mathcal{L}| > 2$, *two strict* t-*paths associated to pairs of* $\mathcal{L}$ *are adjacent if and only if their associated pairs are consecutive in* $\mathcal{L}$ *(modulo* $|\mathcal{L}|$).

(iii) *If* $|\mathcal{L}| > 2$, *two* t-*paths* t_1 *and* t_2 *associated to two consecutive pairs* (l_i, l_{i+1}) *and* (l_{i+2}, l_{i+3}) *(modulo* $|\mathcal{L}|$) *have one common vertex: it is the end vertex of the increasing path* $P_{l_{i+1}}$ *of* t_1 *and the end vertex of the increasing path* $P_{l_{i+2}}$ *of* t_2.

Proof. We proceed as follows. The first step is to find in time $\mathcal{O}(p)$ a root r of $\mathcal{C}$. Let x and y be its two neighbours in $\mathcal{C}$. Find the two longest increasing paths P_1 and P_2 with respect to r in $\mathcal{C}$ going though x and y. Notice that paths P_1 and P_2 are well defined. This can be done in time $\mathcal{O}(|P_1| + |P_2|)$. After that put the corresponding pair in $\mathcal{L}$. If the end vertices of P_1 and P_2 are the same, we are done: the cycle itself is a t-path and $\mathcal{L} = (|P_1|, |P_2|)$. Observe that in that case property (i) is verified.

Otherwise, if the end vertices of P_1 and P_2 are not the same, we proceed with step two. Start from the end vertex v_2 of P_2 and find the longest decreasing path P_3 in $\mathcal{C}$ with respect to v_2. Observe that P_3 exists necessarily, by definition of P_2. Finding the vertices of P_3 is done in $\mathcal{O}(|P_3|)$. The end vertex v_3 of P_3

is a root, by definition of P_3. Observe also that v_3 and v_1 (the end vertex of P_1) are distinct. If that was not the case then P_1 would not have been the longest increasing path starting from r, which is a contradiction by construction of P_1. Now find the longest increasing path with respect to v_3 going in the other direction than P_3, call it P_4. It exists since v_3 and v_1 are distinct. If its end vertex v_4 is equal to v_1, then C can be decomposed into two strict adjacent t-paths and $\mathcal{L} = ((|P_1|, |P_2|), (|P_3|, |P_4|))$. Observe that the three properties for $\mathcal{L}$ are verified in that case.

Otherwise, if v_4 and v_1 are distinct, we proceed exactly as in step 2, but starting this time with the end vertex v_4 of P_4. We proceed in this fashion until we reach vertex v_1.   □

Lemma 3. *Let C be a simple cycle of size p with an acyclic orientation. Let $\mathcal{L} : (l_1, l_2), ..., (l_{r-1}, l_r)$ be the list of integer pairs associated to C, as defined in Lemma 2. List $\mathcal{L}$ is at most of size $\lfloor p/2 \rfloor$.*

Proof. Assume first that p is even. Assume by contradiction that list $\mathcal{L}$ is of size $s > p/2$. By construction of $\mathcal{L}$, each pair (l_j, l_{j+1}) of $\mathcal{L}$ can be associated to a strict t-path of G. By Definition 2 we have that $l_j, l_{j+1} \in \mathbb{N}^*$. Thus each t-path associated to a pair in $\mathcal{L}$ has at least three vertices. This implies that the s many t-paths have $3s$ vertices altogether. But two consecutive t-paths in $\mathcal{L}$ have also a common vertex. Thus, in total, the s many t-paths have at least $3s - s = 2s > p$ vertices which gives the contradiction in that case.

Assume now that p is odd. We first show that there exists at least one strict t-path associated to some pair in $\mathcal{L}$ which has four or more vertices. By definition a strict t-path can not have only one or two vertices. Thus assume by contradiction that all s pairs in $\mathcal{L}$ are associated to strict t-paths of size three. As in the previous case, the total number of vertices of these t-paths is $3s - s = 2s$. By definition, this number is equal to p the size of C which is odd in that case, thus we have a contradiction. This implies that there exists at least one strict t-path associated to some pair in $\mathcal{L}$ which has four or more vertices. Assume now by contradiction that $s > \lfloor p/2 \rfloor = \frac{p-1}{2}$. Since at least one t-path of $\mathcal{L}$ has size four, then the s many t-paths (to which we remove the common vertices) have in total at least $3(s - 1) + 4 - s = 2s + 1$ vertices. To conclude the proof observe that $2s + 1 > 2\frac{p-1}{2} + 1 = p$, which yields the contradiction.   □

Lemma 4. *Let C be a simple cycle of size p with an acyclic orientation. Let $\mathcal{L} : (l_1, l_2), ..., (l_{r-1}, l_r)$ be the list of integer pairs associated to C, as defined in Lemma 2. We have that $\sum_{j=1}^{r} l_j = p$.*

Proof. As defined in Lemma 2, C can be decomposed into $|\mathcal{L}|$ t-paths, one for each pair of $\mathcal{L}$ such that two consecutive pairs (modulo $|\mathcal{L}|$) correspond to adjacent t-paths with one end vertex in common. By Definition 4, a t-path associated to a pair (l_i, l_{i+1}) is of size $l_i + 1 + l_{i+1}$. This implies that the t-paths of $\mathcal{L}$ have total size $(\sum_{j=1}^{r} l_j) + r$. Since cycle C can be decomposed into the t-paths associated to list $\mathcal{L}$, that two consecutive t-paths have one end vertex in common and that there are r such t-paths, we get that $(\sum_{j=1}^{r} l_j) + r = p + r$, which completes the proof. $\square$

Lemma 5. *Let $G = (V, E)$ be a k-degenerate graph. Assume that we have the sorted degenerate adjacency lists of G. Given a vertex $x \in V$ and $i \in \mathbb{N}$, we can compute all increasing paths of size i, starting with x in graph G in time $\mathcal{O}(k^i)$. There are at most $\mathcal{O}(k^i)$ such paths.*

Proof. Start with the sorted degenerate adjacency list of vertex x. By definition it is of size at most k. For every vertex y in this list, construct a candidate path of size one containing x and y. Notice that there are at most k such candidate paths. For each such path, generate all k candidates paths of size two where the third vertex is a vertex of the degenerate adjacency list of the second vertex of the path. There are k^2 such candidates paths of size two in total. Go on in this fashion until all paths of size i have been generated. This procedure takes time $\mathcal{O}(k^i)$: at step $h < i$ we must consider at most k vertices from each of the previously computed k^{h-1} degenerate adjacency lists. $\square$

Corollary 1. *Let $G = (V, E)$ be a k-degenerate graph. Assume that we have the sorted degenerate adjacency lists of G. Given a vertex $x \in V$ and $i, j \in \mathbb{N}$, we can compute all t-paths of size (i, j) with center x, in G, in time $\mathcal{O}(k^{i+j})$. There are at most $\mathcal{O}(k^{i+j})$ such t-paths.*

Corollary 2. *Let $G = (V, E)$ be a k-degenerate graph. Assume that we have the sorted degenerate adjacency lists of G. Given $i, j \in \mathbb{N}$, we can compute all t-paths of size (i, j) in G, in time $\mathcal{O}(nk^{i+j})$. There are at most $\mathcal{O}(nk^{i+j})$ such t-paths.*

4 Algorithm

In this section we prove Theorem 4 which is the main result of the paper. For the sake of clarity, we first start by describing a simpler algorithm, namely Algorithm 1 and prove in Theorems 2 and 3 that it solves the problem of finding all p-length simple cycles in time $\mathcal{O}(n^{\lfloor p/2 \rfloor} k^p)$. Then we show how to modify it to get the claimed $\mathcal{O}(n^{\lfloor p/2 \rfloor} k^{\lceil p/2 \rceil} \log k)$ and $\mathcal{O}(n^{\lfloor p/2 \rfloor} k^{\lceil p/2 \rceil})$ complexities, in Theorem 4.

Theorem 2. *At the end of Algorithm 1, all p-length simple cycles of the graph G have been outputted exactly once.*

Proof. Assume first by contradiction that there exists a p-length simple cycle $\mathcal{C}_1 : c_1, c_2, ..., c_p$ of G which has not been outputted by Algorithm 1. Without loss of generality assume that c_2 has lowest ranking in the degeneracy ordering. As defined and proved in Lemma 2, depending on the orientation of $\mathcal{C}_1$, we can compute a list of integer pairs $\mathcal{L}_1 : l_1, l_2, ..., l_{r-1}, l_r$ that corresponds to the sizes

---

Algorithm 1.

Data: A graph G and $p \in \mathbb{N}$.
Result: All simple p-length cycles of G.

1 Compute k the degeneracy of G and a degeneracy ordering σ_G. Construct an acyclic orientation of G with bounded out-degree k.

2 Compute the sorted degenerate adjacency lists of G.

3 Initialize $Cy : 1, 2, ..., p$ a simple cycle.

4 Compute all acyclic orientations of Cy.

5 **for** *each such orientation* **do**

6   Compute the ordered list $\mathcal{L} = l_1, l_2, ..., l_{r-1}, l_r$ of pairs associated to the strict t-paths of Cy.

7   **for** $j = 1$ **to** r **do**

8   Compute all possible strict t-paths associated to pair l_j in G, put them in a set $\mathcal{S}_j$.

9   **end**

10   Compute all possible lists $\mathcal{C} = c_1, ..., c_r$ with $c_i \in \mathcal{S}_i$

11   **for** *each such list* $\mathcal{C}$ **do**

12   Check if it is a simple cycle:
             – check if vertices are unique except for the end vertices of the t-paths
             – check if two consecutive t-paths have a common end vertex

         if *yes* **then**
             Let s be the string obtained by sorting the vertices of $\mathcal{C}$ by increasing identifier. Search string s in T. **if** *it does not exist in* T **then**
                 Output it.
                 Insert it in T.
             end
         end

13   **end**

14 **end**

---

of adjacent strict t-paths such that $\mathcal{C}_1$ can be decomposed into these t-paths. Up to renaming of the vertices, the orientation of $\mathcal{C}_1$ has been generated at Line 4 of Algorithm 1. This implies that there exists a list $\mathcal{L}$ computed in Line 6 which is equal to $\mathcal{L}_1$. Since in Line 8 all t-paths associated to the pairs of $\mathcal{L}$ are computed, this implies that there exists some list $\mathcal{C}$ generated in Line 10 which contains all the t-paths of $\mathcal{C}_1$, in the same order as they appear in $\mathcal{C}$. This implies in fact that cycle $\mathcal{C}_1$ is outputted at some point by Algorithm 1, which yields the contradiction.

The test done after 12 ensure that Algorithm 1 outputs simple cycles and that they are unique. □

Theorem 3. *Algorithm 1 runs in time* $\mathcal{O}(n^{\lfloor p/2 \rfloor} k^p)$.

Proof. Computing the degeneracy and the degeneracy ordering of G can be done in $\mathcal{O}(m)$ [3]. Computing the degenerate adjacency lists of G in Line 2 can be done in time $\mathcal{O}(nk \log k)$, by Lemma 1. Computing all acyclic permutations of $\mathcal{C}y$ can be done in total time $\mathcal{O}(2^p)$. Using Lemma 2, Line 6 can be done in $\mathcal{O}(p)$ for each orientation. Since there are $\mathcal{O}(2^p)$ of them this takes total time $\mathcal{O}(p2^p)$. By Corollary 2, given a pair $l_j = (a_j, b_j)$, we can compute all t-paths of size $(a_j + b_j)$ in G in time $\mathcal{O}(nk^{a_j+b_j})$. Thus for each list $\mathcal{L} : (l_1 = a_1, b_1), l_2 = (a_2, b_2), ..., l_{r-1} = (a_{r-1}, b_{r-1}), l_r = (a_r, b_r)$, Line 8 can be done in time $\mathcal{O}(n \sum_{(a_j,b_j) \in \mathcal{L}} k^{a_j+b_j}) = \mathcal{O}(nk^p)$ since $\sum_{(a_j,b_j) \in \mathcal{L}} a_j + b_j = p$ as proved in Lemma 4. By Corollary 2, each set S_j associated to a pair $l_j = (a_j, b_j)$ is of size $\mathcal{O}(nk^{a_j+b_j})$. If there are r such sets, computing all possible lists $\mathcal{C}$ in Line 10 can be done in time $\mathcal{O}(n^r k^p)$ since $\sum_{(a_j,b_j) \in \mathcal{L}} a_j + b_j = p$ as proved in Lemma 4. Now by Lemma 3, $r \leq \lfloor p/2 \rfloor$, thus Line 10 takes at most $\mathcal{O}(n^{\lfloor p/2 \rfloor} k^p)$ time, per orientation. To conclude, in Line 12, checking if a list $\mathcal{C}$ is a simple cycle can be done in time $\mathcal{O}(p)$. Since $\mathcal{C}$ is of size at most $\mathcal{O}(p + \lfloor p/2 \rfloor) = \mathcal{O}(p)$ we can check the uniqueness of the vertices and the condition on the end vertices in time $\mathcal{O}(p)$. Finally searching, sorting a string s of size $\mathcal{O}(p)$ or inserting it in a radix tree can be done in time $\mathcal{O}(p \log p)$, see [17]. In total, Algorithm 1 runs in time $\mathcal{O}(E + nk \log k + p2^p + p2^p n^{\lfloor p/2 \rfloor} k^p \log p) = \mathcal{O}(n^{\lfloor p/2 \rfloor} k^p)$.   □

Theorem 4. *Algorithm 1 can be modified to run in time $\mathcal{O}(n^{\lfloor p/2 \rfloor} k^{\lceil p/2 \rceil} \log k)$ if the graph is stored in adjacency lists or $\mathcal{O}(n^{\lfloor p/2 \rfloor} k^{\lceil p/2 \rceil})$ time if it is stored in an adjacency matrix.*

Proof. We modify Algorithm 1 in the following two ways. Assume list $\mathcal{L} : l_1 = (a_1, b_1), l_2 = (a_2, b_2), ..., l_r = (a_r, b_r)$ has been computed in Line 6.

- For each $j \in [1, r]$, transform $l_j = (a_j, b_j)$ into $(a_j, b_j - 1)$. This can be done in $\mathcal{O}(p)$.
- When checking if a list $\mathcal{C}$ is a simple cycle after line 12 we check the uniqueness of all the vertices and check if two consecutive t-paths have adjacent end vertices (before we had to check if they had common vertices). This can be done in time $\mathcal{O}(p \log k)$ using the degenerate adjacency lists computed in line 2 or $\mathcal{O}(1)$ if we have the adjacency matrix of the graph.

We show that Algorithm 1 modified in this way has the claimed complexity. When we decrease the values of the pairs in $\mathcal{L}$ as described in the first modification and assuming $|\mathcal{L}| = r$, Line 8 can be computed in total time $\mathcal{O}(nk^{p-r})$ since $\sum_{(a_j,b_j) \in \mathcal{L}} a_j + b_j = p - r$. All possible lists in Line 10 can be computed in time $\mathcal{O}(n^r k^{p-r})$ and there are $\mathcal{O}(n^r k^{p-r})$ of them. Since $r \leq \lfloor p/2 \rfloor$ as proved in Lemma 3 and since $k \leq n$ then $\mathcal{O}(n^r k^{p-r}) = \mathcal{O}(n^{\lfloor p/2 \rfloor} k^{\lceil p/2 \rceil})$. Thus the total time complexity is $\mathcal{O}(n^{\lfloor p/2 \rfloor} k^{\lceil p/2 \rceil} p2^p \log p \log k) = \mathcal{O}(n^{\lfloor p/2 \rfloor} k^{\lceil p/2 \rceil} \log k)$ with these modifications if adjacency queries can be done in $\mathcal{O}(\log k)$ or $\mathcal{O}(n^{\lfloor p/2 \rfloor} k^{\lceil p/2 \rceil})$ if they can be done in $\mathcal{O}(1)$, which completes the complexity analysis.

We prove now the correctness of Algorithm 1 when modified in this way. As shown in Lemma 2, a simple cycle $\mathcal{C}_1$, given an orientation of its vertices, can be decomposed into strict adjacent t-paths. Consider now all pairs of such adjacent t-paths. If we remove the common vertex from one t-path from each pair, cycle $\mathcal{C}_1$ can still be decomposed into these new t-paths. The previous adjacent t-paths which had a common end vertex now become new smaller t-paths that have adjacent end vertices. The two modifications described above reflect this property: by decreasing the value of the pairs in the list $\mathcal{L}$ we generate t-paths with one less vertex but we have now to check that, if they appear consecutively in list $\mathcal{C}$ at Line 10, they have adjacent end vertices.   □

Corollary 3. *With same complexities, we can output exactly all p-length induced cycles.*

Proof. Consider Algorithm 1 modified as in Theorem 4. Before outputting a cycle we can check if it is induced in time $\mathcal{O}(p^2 \log k)$ with the sorted degenerate adjacency lists: for every pair of vertices of the cycle check if they are adjacent. If we have the adjacency matrix this can be done in time $\mathcal{O}(p^2)$.   □

Theorem 5. *The maximal number of p-length simple cycles in a k-degenerate graph is $\Theta(n^{\lfloor p/2 \rfloor} k^{\lceil p/2 \rceil})$.*

Proof. Algorithm 1 modified as described in the proof of Theorem 4 generates at most $\mathcal{O}(n^{\lfloor p/2 \rfloor} k^{\lceil p/2 \rceil})$ candidate cycles, assuming a constant bound on p. Since this algorithms outputs all cycles of the graph, this yields the upper bound.

We now prove the lower bound. We construct for any k, p and $n \geq kp$ an n-order k-degenerate graph with $\Omega(n^{\lfloor p/2 \rfloor} k^{\lceil p/2 \rceil})$ simple p-length cycles.

Assume first that p and k are even. Consider $p/2$ independent sets $K_1, K_2, ..., K_{p/2}$ of size $k/2$ and $p/2$ independent sets $L_1, L_2, ... L_{p/2}$ of size $l \geq k$. Connect all the vertices of set K_i to all the vertices of set L_i for every i. Connect all the vertices of set L_i to all the vertices of set $K_{i+1 \mod p}$. (See Fig. 1). This

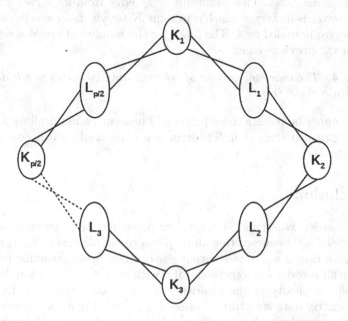

Fig. 1. A k-degenerate graph with the maximal number of p-length simple cycles.

graph is k-degenerate. To show that observe that every vertex has degree at least k, thus the graph cannot be $(k-1)$-degenerate. Every edge has one endpoint in a set K_i and the other in a set L_j. Thus orienting every edge towards its vertex which is in some K set yields an acyclic orientation with out-degree bounded by k. This graph has $n = \frac{p}{2}\frac{k}{2} + \frac{p}{2}l$ vertices which implies $l = \frac{2n}{p} - \frac{k}{2}$. The total number of simple p-length cycles is $l^{p/2}\frac{k}{2}^{p/2} = (\frac{2n}{p} - \frac{k}{2})^{p/2}\frac{k}{2}^{p/2}$. Observe that $n \geq kp$ implies that $(\frac{2n}{p} - \frac{k}{2})^{p/2}\frac{k}{2}^{p/2} = \Omega(n^{p/2}k^{p/2})$

Assume now that p is even and k odd. The construction is similar except that set K_i has $\lfloor\frac{k}{2}\rfloor$ vertices if i odd and $\lceil\frac{k}{2}\rceil$ otherwise. If $p/2$ is even the proof is exactly the same as for the case in which p and k are even. If $p/2$ is odd we only prove that the graph is k-degenerate, the proof for the number of simple p-length cycles being the same. The graph is not $(k-1)$-degenerate since the subgraph induced by the vertices of sets K_1, L_1 and K_2 has no vertex of degree less than k. Orienting the edges as in the case in which p and k are even yields an acyclic orientation with out-degree bounded by k.

Assume now that p is odd. The construction is similar. We consider $\lceil p/2 \rceil$ independent sets $K_1, K_2, ..., K_{\lceil p/2 \rceil}$ and $\lfloor p/2 \rfloor$ independent sets $L_1, L_2, ..., L_{\lfloor p/2 \rfloor}$. Connect all the vertices of K_i to all the vertices of L_i for every $i < \lceil p/2 \rceil$. Connect all the vertices of L_i to all the vertices of K_{i+1}. Finally connect all vertices of $K_{\lceil p/2 \rceil}$ to all the vertices of K_1. The proof that this graph is k-degenerate is the same as before. If k is even, take the sets K_i of size $k/2$, otherwise take set K_i of size $\lfloor k/2 \rfloor$ if i odd or of size $\lceil k/2 \rceil$ if i even. Now every vertex has degree at least k so the graph cannot be $(k-1)$-degenerate. Orient every edge between the sets $K_{\lceil p/2 \rceil}$ and K_1 arbitrarily. Every other edge has one vertex in some K set and the other in some L set. Thus orienting every edge from its vertex which is in some L set towards its vertex which is in some K set yields an acyclic orientation with out-degree bounded by k. The proof for the number of cycles is exactly the same as for the previous cases. □

Corollary 4. *The maximal number of p-length induced cycles in a k-degenerate graph is $\Theta(n^{\lfloor p/2 \rfloor}k^{\lceil p/2 \rceil})$.*

Proof. The upper bound is a consequence of Theorem 4 and Corollary 3. Observe that the cycles constructed in Theorem 5 are induced, which completes the proof.   □

5   Conclusion

Given an n-order k-degenerate graph, we presented an algorithm running in time $\mathcal{O}(n^{\lfloor p/2 \rfloor}k^{\lceil p/2 \rceil})$ enumerating all its p-length simple cycles. We then proved that this algorithm is worst-case output size optimal by constructing for any k, p and $n \geq kp$ an n-order k-degenerate graph with $\Theta(n^{\lfloor p/2 \rfloor}k^{\lceil p/2 \rceil})$ simple p-length cycles. The complexity of the algorithm is given assuming it is stored in an adjacency matrix data structure. If instead it is stored in an adjacency list data structure the complexity becomes $\mathcal{O}(n^{\lfloor p/2 \rfloor}k^{\lceil p/2 \rceil}\log k)$. Thus the first question we ask is whether or not we can achieve the optimal complexity when the graph is given through its adjacency lists. A second improvement would be to prove an output sensitive algorithm, similar to the one Kowalik presented for cycles of size less than five, see [14]. That is, we ask whether it is possible to achieve an $\mathcal{O}(n^{\mathcal{O}(1)} k^{\mathcal{O}(1)} + occ)$ complexity for this problem where occ is the number of p-length simple cycles in the graph, assuming p constant. Kowalik essentially shows that small cycles can be broken into few small special paths with at most 3 or 4 vertices and proves bounds on the number of these paths that can have common vertices. Can we extend his approach using t-paths decompositions?

References

1. Alon, N., Yuster, R., Zwick, U.: Finding and counting given length cycles. Algorithmica **17**(3), 209–223 (1997)
2. Barabási, A.-L., Albert, R.: Emergence of scaling in random networks. Science **286**(5439), 509–512 (1999)
3. Batagelj, V., Zaversnik, M.: An $\mathcal{O}(m)$ algorithm for cores decomposition of networks (2003). arXiv preprint arXiv:cs/0310049
4. Birmelé, E., Ferreira, R., Grossi, R., Marino, A., Pisanti, N., Rizzi, R., Sacomoto, G.: Optimal listing of cycles and st-paths in undirected graphs, pp. 1884–1896 (2013)
5. Björklund, A., Kaski, P., Kowalik, Ł.: Counting thin subgraphs via packings faster than meet-in-the-middle time. In: Proceedings of the Twenty-Fifth Annual ACM-SIAM Symposium on Discrete Algorithms, pp. 594–603. Society for Industrial and Applied Mathematics (2014)
6. Cai, L., Chan, S.M., Chan, S.O.: Random separation: a new method for solving fixed-cardinality optimization problems. In: Bodlaender, H.L., Langston, M.A. (eds.) IWPEC 2006. LNCS, vol. 4169, pp. 239–250. Springer, Heidelberg (2006). doi:10.1007/11847250_22
7. Chiba, N., Nishizeki, T.: Arboricity and subgraph listing algorithms. SIAM J. Comput. **14**(1), 210–223 (1985)
8. Chrobak, M., Eppstein, D.: Planar orientations with low out-degree and compaction of adjacency matrices. Theor. Comput. Sci. **86**(2), 243–266 (1991)
9. Dorn, F.: Planar subgraph isomorphism revisited. In: 27th International Symposium on Theoretical Aspects of Computer Science. Leibniz International Proceedings in Informatics (LIPIcs), Dagstuhl, Germany, vol. 5, pp. 263–274. Schloss Dagstuhl-Leibniz-Zentrum fuer Informatik (2010)
10. Eppstein, D.: Subgraph isomorphism in planar graphs and related problems. In: Proceedings of the Sixth Annual ACM-SIAM Symposium on Discrete Algorithms, SODA 1995, Philadelphia, PA, USA, pp. 632–640. Society for Industrial and Applied Mathematics (1995)
11. Giscard, P.-L., Kriege, N., Wilson, R.C.: A general purpose algorithm for counting simple cycles and simple paths of any length (2016). arXiv preprint arXiv:1612.05531
12. Goel, G., Gustedt, J.: Bounded arboricity to determine the local structure of sparse graphs. In: Fomin, F.V. (ed.) WG 2006. LNCS, vol. 4271, pp. 159–167. Springer, Heidelberg (2006). doi:10.1007/11917496_15
13. Johnson, D.B.: Finding all the elementary circuits of a directed graph. SIAM J. Comput. **4**(1), 77–84 (1975)

14. Kowalik, Ł.: Short cycles in planar graphs. In: Bodlaender, H.L. (ed.) WG 2003. LNCS, vol. 2880, pp. 284–296. Springer, Heidelberg (2003). doi:10.1007/978-3-540-39890-5_25

15. Lick, D.R., White, A.T.: d-degenerate graphs. Canad. J. Math. **22**, 1082–1096 (1970)

16. Meeks, K.: Randomised enumeration of small witnesses using a decision oracle. In: 11th International Symposium on Parameterized, Exact Computation, IPEC 2016, Aarhus, Denmark, 24–26 August 2016, pp. 22:1–22:12 (2016)

17. Morrison, D.R.: Patricia-practical algorithm to retrieve information coded in alphanumeric. J. ACM **15**(4), 514–534 (1968)

18. Papadimitriou, C.H., Yannakakis, M.: The clique problem for planar graphs. Inf. Process. Lett. **13**(4), 131–133 (1981)

19. Richards, D.: Finding short cycles in planar graphs using separators. J. Algorithms **7**(3), 382–394 (1986)

20. Tarjan, R.: Enumeration of the elementary circuits of a directed graph. SIAM J. Comput. **2**(3), 211–216 (1973)

21. Uno, T., Satoh, H.: An efficient algorithm for enumerating chordless cycles and chordless paths. In: Džeroski, S., Panov, P., Kocev, D., Todorovski, L. (eds.) DS 2014. LNCS, vol. 8777, pp. 313–324. Springer, Cham (2014). doi:10.1007/978-3-319-11812-3_27

22. Williams, V.V., Williams, R.: Finding, minimizing, and counting weighted subgraphs. SIAM J. Comput. **42**(3), 831–854 (2013)

Reliable Communication via Semilattice Properties of Partial Knowledge

Aris Pagourtzis[1], Giorgos Panagiotakos[2], and Dimitris Sakavalas[1(✉)]

[1] School of Electrical and Computer Engineering,
National Technical University of Athens, 15780 Athens, Greece
pagour@cs.ntua.gr, sakaval@corelab.ntua.gr
[2] School of Informatics, University of Edinburgh, Edinburgh EH8 9AB, Scotland, UK
giorgos.pan@ed.ac.uk

Abstract. A fundamental primitive in distributed computing is *Reliable Message Transmission* (RMT), which refers to the task of correctly sending a message from a party to another, despite the presence of Byzantine corruptions. We explicitly consider the initial knowledge possessed by the parties-players by employing the recently introduced *Partial Knowledge Model* [13], where a player has knowledge over an arbitrary subgraph of the network, and the general adversary model of Hirt and Maurer [5]. Our main contribution is a tight condition for the feasibility of RMT in the setting resulting from the combination of these two quite general models; this settles the central open question of [13].

Obtaining such a condition presents the need for knowledge exchange between players. To this end, we introduce the *joint view operation* which serves as a fundamental tool for deducing maximal useful information conforming with the exchanged local knowledge. Maximality of the obtained knowledge is proved in terms of the semilattice structure imposed by the operation on the space of partial knowledge. This in turn, allows for the definition of a novel network separator notion that yields a necessary condition for achieving RMT in this model. In order to show the sufficiency of the condition, we propose the RMT Partial Knowledge Algorithm (RMT-PKA), an algorithm which employs the joint view operation to solve RMT in every instance where the necessary condition is met. To the best of our knowledge, this is the first protocol for RMT against general adversaries in the partial knowledge model. Due to the generality of the model, our results provide, for any level of topology knowledge and any adversary structure, an exact characterization of instances where RMT is possible and an algorithm to achieve RMT on such instances.

Keywords: Reliable message transmission · Partial knowledge · Semi-lattices · General adversary · Byzantine adversary

A short version of this paper, entitled *"Brief Announcement: Reliable Message Transmission under Partial Knowledge and General Adversaries"*, appeared in PODC 2016 [12].

R. Klasing and M. Zeitoun (Eds.): FCT 2017, LNCS 10472, pp. 367–380, 2017.
DOI: 10.1007/978-3-662-55751-8_29

1 Introduction

Achieving reliable communication in unreliable networks is fundamental in distributed computing. Of course, if there is an authenticated channel between two parties then reliable communication between them is guaranteed. However, it is often the case that certain parties are only indirectly connected, and need to use intermediate parties as relays to propagate their message to the actual receiver. The *Reliable Message Transmission* problem (RMT) is the problem of achieving correct delivery of a message m from a sender S to a receiver R even if some of the intermediate nodes are corrupted and do not relay the message as agreed. Essentially, an RMT protocol simulates the functionality of a reliable communication channel between the sender and the receiver. In this work we consider the worst case corruption scenario, in which the adversary is unbounded and may control several nodes. The adversary is able to make the corrupted nodes deviate from the protocol arbitrarily by blocking, rerouting, or even altering a message that they should normally relay intact to specific nodes. An adversary with this behavior is referred to as *Byzantine adversary*.

The RMT problem has been initially considered by Dolev [2] in the context of the closely related *Reliable Broadcast* (Byzantine Generals) and *Consensus* (Byzantine Agreement) problems, introduced by Lamport et al. [9]. The two latter problems have been extensively studied in complete networks with reliable channels and studies on RMT naturally imply results for these problems since an RMT protocol simulates reliable channels between players in an incomplete network.

The problem of message transmission under Byzantine adversaries has been studied extensively in various settings: secure or reliable transmission, general or threshold adversary, perfect or unconditional security, full or local topology knowledge. Here we focus on perfectly reliable transmission under a general adversary and the partial knowledge model. In the general adversary model, introduced by Hirt and Maurer [5], the adversary may corrupt any player-set among a given family of all possible corruption sets (*adversary structure*); it subsumes both the global [9] and the local threshold adversary model [7]. For instance, the global threshold model, which assumes that the adversary can corrupt at most t players, corresponds to the family of sets with cardinality at most t. Regarding the topology knowledge, the recently introduced *Partial Knowledge Model* [13] assumes that each player only has knowledge over some arbitrary subgraph including itself and the intersection of this subgraph with the adversary structure; it encompasses both the full knowledge and the ad hoc (unknown topology) models.

The motivation for partial knowledge considerations comes from large scale networks (e.g. the Internet) where topologically local estimation of the power of the adversary may be possible, while global estimation may be hard to obtain due to geographical or jurisdiction constraints. Additionally, proximity in social networks is often correlated with an increased amount of available information, further justifying the relevance of the model.

The strength of this work lies in the combination of these two quite general models (general adversary and partial knowledge), forming the most general setting we have encountered so far within the synchronous deterministic model.

1.1   Related Work

The RMT problem under a threshold Byzantine adversary, where a fixed upper bound t is set for the number of corrupted players was addressed in [1,3], where additional secrecy restrictions were posed and in [16] where a probability of failure was allowed. Results for RMT in the general adversary model [5], where given in [8,17,18]. In general, very few studies have addressed RMT or related problems in the partial knowledge setting despite the fact that this direction was already proposed in 2002 by Kumar *et al.* [8].

The approach that we follow here stems from a line of work which addresses the Reliable Broadcast problem with an honest sender (dealer) in incomplete networks, initiated by Koo [7]. Koo studied the problem in *ad hoc* networks of specific topology under the *t-locally bounded adversary model*, in which at most a certain number t of corruptions are allowed in the neighborhood of every node. A simple, yet powerful Reliable Broadcast protocol called *Certified Propagation Algorithm* (CPA) was proposed in this work; CPA is based on the idea that if a set of $t+1$ neighbors of v provides the same information to v then the information is valid because at least one of them is honest. This work was extended in the context of generic networks by Pelc, Peleg in [14] who also pointed out how full knowledge of the topology yields better solvability results. After a series of works ([6,10,19]) tight conditions for the correctness of CPA were obtained in the *ad hoc* case. Observe that all of these aforementioned works only considered the *t*-locally bounded adversary model and did not provide tight conditions for the solvability of the problem. Finally, in [13] the Partial Knowledge Model was introduced, in which the players only have partial knowledge of the topology and the adversary structure. In [13] both the *t*-locally bounded adversary model and the general adversary model were considered and tight conditions for the solvability of the problem along with matching algorithms for the extreme cases of full topology knowledge and *ad hoc* setting were proposed. Trivially all the aforementioned results for Reliable Broadcast with an honest sender can be adapted for the RMT problem. However, it was left as an open problem in [13] to determine a necessary and sufficient condition (tight) for the most general case of the partial knowledge model. The latter issue appeared to be most challenging due to the need of sound knowledge exchange between possibly corrupted players. The deduction of maximal secure information through the combination of partial knowledge possessed by the involved players, is achieved in this work through careful analysis of the algebraic properties of partial knowledge, which in turn leads to the tight feasibility condition.

1.2 Our Results

We study the RMT problem under partial knowledge and general adversaries. Our contribution concerns the feasibility of RMT in the Partial Knowledge model. We prove a necessary and sufficient condition for achieving RMT in this setting, and present RMT-PKA, an algorithm that achieves RMT whenever this condition is met. In terminology of [14] (formally defined in [13]) this is a *unique* algorithm for the problem, in the sense that whenever any algorithm achieves RMT in a certain instance so does RMT-PKA. This settles an open question of [13] and is, to the best of our knowledge, the first algorithm with this property. It is worth mentioning that RMT-PKA can achieve RMT with the minimal amount of player's knowledge that renders the problem solvable. This new algorithm encompasses earlier algorithms such as CPA [7], PPA and $\mathcal{Z}$-CPA [13] as special cases. A remarkable property of our algorithm is its *safety*: even when RMT is not possible the receiver will never make an incorrect decision despite the increased adversary's attack capabilities, which include reporting fictitious topology and false local knowledge among others.

A key algorithmic tool that we define and use is the *joint view operation* which computes the *joint adversary structure* of (a set of) players, i.e., the worst case adversary structure that conforms to each player's initial knowledge. This operation is crucial in obtaining the tight condition mentioned above since it provides a way to safely utilize the maximal valid information from all the messages exchanged. We show that this operation actually implies a semilattice structure on the partial knowledge that players may have. In this context, we prove that the worst possible adversary structure, conforming with the initial knowledge of a set of parties, can be expressed as the supremum of the parties' knowledge under the semilattice partial order.

To obtain our result we propose a non trivial generalization of earlier separator techniques, introduced by Pelc and Peleg [14] and extended in [13] in the context of Broadcast. Analogous techniques were used in [13] to obtain characterizations of classes of graphs for which Broadcast is possible for various levels of topology knowledge and types of corruption distribution; however, an exact characterization for the partial knowledge setting was left as an open question. Here we address this question by proposing a new type of pair-cut (separator) appropriate for the partial knowledge model, coupled with a proof that RMT-PKA works exactly whenever no such pair-cut exists. This, as already mentioned, implies a tight solvability condition for RMT in the quite general model of partial knowledge with general adversaries. A useful by-product of practical interest is that the new cut notion can be used, in a network design phase, in order to determine the exact subgraph in which RMT is possible.

1.3 Model and Definitions

In this work we address the problem of Perfectly Reliable Message Transmission, hereafter simply referred as Reliable Message Transmission (RMT) under the influence of a general Byzantine adversary. In our model the players have partial knowledge of the network topology and of the adversary structure.

We assume a synchronous network represented by a graph $G = (V, E)$ consisting of the player (node) set $V(G)$ and edge set $E(G)$ which represents undirected authenticated[1] channels between players. The set of neighbors of a player v is denoted with $\mathcal{N}(v)$. In our study we will often make use of node-cuts (separators) which separate the receiver R from the sender, hence, node-cuts that do not include the sender. From here on we will simply use the term *cut* to denote such a separator. The problem definition follows.

Reliable Message Transmission. We assume the existence of a designated player $S \in V$, called the *sender*, who wants to propagate a certain value $x_S \in X$, where X is the initial message space, to a designated player R, called the receiver. We say that a distributed protocol achieves (or solves) RMT if by the end of the protocol the receiver R has *decided on* x_S, i.e. if it has been able to output the value x_S originally sent by the sender.

The Adversary Model. The *general adversary model* was introduced by Hirt and Maurer in [5]. In this work they study the security of multiparty computation protocols with respect to an *adversary structure*, that is, a family of subsets of the players; the adversary is able to corrupt one of these subsets. More formally, an adversary structure $\mathcal{Z}$ for the set of players V is a monotone family of subsets of V, i.e. $\mathcal{Z} \subseteq 2^V$, where all subsets of a set Z are in $\mathcal{Z}$ if $Z \in \mathcal{Z}$. In this work we obtain our results w.r.t. a general byzantine adversary, i.e., a general adversary which can make all the corrupted players deviate arbitrarily from the given protocol.

The Partial Knowledge Model [13]. In this setting each player v only has knowledge of the topology of a certain subgraph G_v of G which includes v. Namely if we consider the family $\mathcal{G}$ of subgraphs of G we use the *view function* $\gamma : V(G) \to \mathcal{G}$, where $\gamma(v)$ represents the subgraph of G over which player v has knowledge of the topology. We extend the domain of γ by allowing as input a set $S \subseteq V(G)$. The output will correspond to the *joint view* of nodes in S. More specifically, if $\gamma(v) = G_v = (V_v, E_v)$ then $\gamma(S) = G_S = (\bigcup_{v \in S} V_v, \bigcup_{v \in S} E_v)$. The extensively studied *ad hoc* model can be seen as a special case of the Partial Knowledge Model, where we assume that the topology knowledge of each player is limited to its own neighborhood, i.e., $\forall v \in V(G)$, $\gamma(v) = \mathcal{N}(v)$.

In order to capture partial knowledge in this setting we need to define the restriction of some structure to an a set of nodes.

Definition 1. *For an adversary structure $\mathcal{E}$ and a node set A let $\mathcal{E}^A = \{Z \cap A \mid Z \in \mathcal{E}\}$ denote the restriction of $\mathcal{E}$ to the set A.*

---

[1] As usual in the byzantine faults literature, the existence of authenticated channel (u, v), guarantees that once a message is sent from node u to node v, the message will be delivered intact to the receiver v and the receiver will be aware of the identity of the sender u, i.e. no tampering of the message or identity spoofing can be performed by the adversary.

Hence, we assume that given the actual adversary structure $\mathcal{Z}$ each player v only knows the possible corruption sets under his view $\mathcal{Z}_v$, which is equal to $\mathcal{Z}^{V(\gamma(v))}$ (the *local adversary structure*).

We denote an instance of the problem by the tuple $I = (G, \mathcal{Z}, \gamma, S, R)$. We next define some useful protocol properties.

We say that an RMT protocol is *resilient* for an instance I if it achieves RMT on instance I for any possible corruption set and any admissible behavior of the corrupted players. We say that an RMT protocol is *safe* if it never causes the receiver R to decide on an incorrect value in any instance. The importance of the safety property is pointed out in [14], where it is regarded as a basic requirement of a Broadcast algorithm; similarly, in the case of RMT it guarantees that if the receiver does not have sufficient information to decide on the sender's value, she won't eventually decide on an incorrect value or accept false data.

Definition 2 (Uniqueness of algorithm). *Let $\mathcal{A}$ be a family of algorithms. An algorithm A is unique (for RMT) among algorithms in $\mathcal{A}$ if the existence of an algorithm of family $\mathcal{A}$ which achieves RMT in an instance I implies that A also achieves RMT in I.*

A unique algorithm A among $\mathcal{A}$, naturally defines the class of instances in which the problem is solvable by $\mathcal{A}$-algorithms, namely the ones that A achieves RMT in.

2   The Algebraic Structure of Partial Knowledge

In this section we delve into the algebraic structure of the knowledge of players regarding the adversary. We do this by first defining an operation used to calculate their joint knowledge. As is proved in the following, this operation allows the combination of local knowledge in an optimal way. The operation takes into account potentially different adversarial structures, so that it is well defined even if a corrupted player provides a different structure than the real one to some honest player.

Definition 3. *Let V be a finite node set; let also $\mathbb{T} = \{(\mathcal{E}, A) \mid \mathcal{E} \subseteq 2^A, A \subseteq V, \mathcal{E} \text{ is monotone}\}$ denote the space of all pairs consisting of a monotone family of subsets of a node set along with that node set. The operation $\oplus : \mathbb{T} \times \mathbb{T} \to \mathbb{T}$, is defined as follows:*

$$(\mathcal{E}, A) \oplus (\mathcal{F}, B) = (\{Z_1 \cup Z_2 | (Z_1 \in \mathcal{E}) \wedge (Z_2 \in \mathcal{F}) \wedge (Z_1 \cap B = Z_2 \cap A)\}, A \cup B)$$

Informally, $(\mathcal{E}, A) \oplus (\mathcal{F}, B)$ unites possible corruption sets from $\mathcal{E}$ and $\mathcal{F}$ that 'agree' on $A \cap B$ (see Fig. 1). The following theorem offers further insight on the algebraic properties of this operation, by revealing a semilattice structure on the space of partial knowledge obtained by the players. The semilattice structure is shown by proving the commutativity, associativity and idempotence properties of operation $\oplus$ (see [15]). The proof is deferred to the full version.

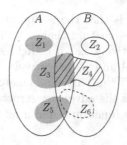

Fig. 1. Example of the $\oplus$ operation in the pairs $(\mathcal{E}, A), (\mathcal{F}, B)$, with $\mathcal{E} = \{Z_1, Z_3, Z_5\}$, $\mathcal{F} = \{Z_2, Z_4, Z_6\}$: For $(\mathcal{E}, A) \oplus (\mathcal{F}, B) = (\mathcal{H}, A \cup B)$, we observe that $Z_1 \cup Z_2$ and $Z_3 \cup Z_4$ belong to $\mathcal{H}$ while $Z_5 \cup Z_6$ and $Z_1 \cup Z_4$ do not.

Theorem 1. $\langle \mathbb{T}, \oplus \rangle$ *is a semilattice.*

From semilattice theory, it is well known that the algebraic definition of the join-semilattice [2] $\langle \mathbb{T}, \oplus \rangle$ implies a binary relation $\geq$ that partially orders $\mathbb{T}$ in the following way: for all elements $x, y \in \mathbb{T}$, $x \geq y$ if and only if $x = x \oplus y$. This binary relation provides the equivalent order theoretic definition of the same semilattice $\langle \mathbb{T}, \geq \rangle$. The following theorem reveals the binary relation implied by the $\oplus$ operation. Its proof is deferred to the full version.

Theorem 2. *The partial ordering "$\geq$" induced by the $\oplus$ operation on $\mathbb{T}$ satisfies the following: for $(\mathcal{E}, A), (\mathcal{F}, B) \in \mathbb{T}$, $(\mathcal{E}, A) \geq (\mathcal{F}, B)$ if and only if $(B \subseteq A) \wedge (\mathcal{E}^B \subseteq \mathcal{F})$.*

The semilattice structure guarantees that every non-empty finite subset of $\langle \mathbb{T}, \geq \rangle$ has a supremum with respect to the "$\geq$" relation (also called a *join*). Moreover it holds that for $(\mathcal{E}, A), (\mathcal{F}, B) \in \mathbb{T}$, $\sup\{(\mathcal{E}, A), (\mathcal{F}, B)\} = (\mathcal{E}, A) \oplus (\mathcal{F}, B)$. The latter implies a property of the $\oplus$ operation which is important in our study. Namely,

Corollary 1. *Let $\langle \mathbb{T}, \geq \rangle$ be a semilattice as defined above. For any $z \in \mathbb{T}$ it holds that if $x, y \leq z$, then $x \oplus y \leq z$.*

Proof. The join of x, y is their least upper bound. Thus, since z is an upper bound of x, y, it must also be greater or equal to their join, i.e. $x \oplus y$. The Corollary follows. □

Returning to our problem after this short detour, notice that for any adversary structure $\mathcal{Z}$ it holds that $(\mathcal{Z}^A, A), (\mathcal{Z}^B, B) \leq (\mathcal{Z}^{A \cup B}, A \cup B)$. We immediately get by Corollary 1 the following corollary.

Corollary 2. *For any adversary structure $\mathcal{Z}$ and node sets A, B:*

$$\text{if } (\mathcal{H}, A \cup B) = (\mathcal{Z}^A, A) \oplus (\mathcal{Z}^B, B) \text{ then } \mathcal{Z}^{(A \cup B)} \subseteq \mathcal{H}$$

---

[2] The notion of meet-semilattice can be used as well by inversing the ordering.

What Corollary 2 tells us is that the $\oplus$ operation gives the maximal (w.r.t inclusion) possible adversary structure that is indistinguishable by two agents that know $\mathcal{Z}^A$ and $\mathcal{Z}^B$ respectively, i.e., it coincides with their knowledge of the adversary structures on sets A and B respectively.

Now recall that $\mathcal{Z}_u = \mathcal{Z}^{V(\gamma(u))}$. This allows us to define the combined knowledge of a set of nodes B about the adversary structure $\mathcal{Z}$ as follows. For a given adversary structure $\mathcal{Z}$, a view function γ and a node set B let

$$(\mathcal{Z}_B, V(\gamma(B))) = \bigoplus_{v \in B} (\mathcal{Z}_v, V(\gamma(v))) = \bigoplus_{v \in B} (\mathcal{Z}^{V(\gamma(v))}, V(\gamma(v)))$$

Note that $\mathcal{Z}_B$ exactly captures the maximal adversary structure possible, restricted in $\gamma(B)$, relative to the initial knowledge of players in B. Also notice that using Corollary 2 we get $\mathcal{Z}^{V(\gamma(B))} \subseteq \mathcal{Z}_B$. The interpretation of this inequality in our setting, is that what nodes in B conceive as the worst case adversary structure indistinguishable to them, always contains the actual adversary structure in their scenario.

3   A Tight Condition for RMT

In RMT we want the sender S to send a message to some player R (the receiver) in the network. We assume that the sender knows the id of player R. We denote an instance of the problem by the tuple $(G, \mathcal{Z}, \gamma, S, R)$. To analyze feasibility of RMT we introduce the notion of RMT-cut.

Definition 4 (RMT-cut). *Let* $(G, \mathcal{Z}, \gamma, S, R)$ *be an RMT instance and* $C = C_1 \cup C_2$ *be a cut in* G, *partitioning* $V \setminus C$ *in two sets* $A, B' \neq \emptyset$ *where* $S \in A$ *and* $R \in B'$. *Let* $B \subseteq B'$ *be the node set of the connected component that* R *lies in. Then* C *is a RMT-cut iff* $C_1 \in \mathcal{Z}$ *and* $C_2 \cap V(\gamma(B)) \in \mathcal{Z}_B$.

The necessary condition proof adapts techniques and ideas from [13,14] to the partial knowledge with general adversary setting.

Theorem 3 (Necessity).   *Let* $(G, \mathcal{Z}, \gamma, S, R)$ *be an RMT instance. If there exists a RMT-cut in* G *then no safe and resilient RMT algorithm exists for* $(G, \mathcal{Z}, \gamma, S, R)$.

Proof. Let $C = C_1 \cup C_2$ be the RMT-cut which partitions $V \setminus C$ in sets $A, B \neq \emptyset$ s.t. $S \in A$ and $R \in B$. Without loss of generality assume that B is connected. If it is not, then by adding to A all nodes that do not belong to the connected component of R, an RMT-cut with the desired property is obtained. Consider a second instance where $\mathcal{Z}' = \mathcal{Z}_B$ and all other parameters are the same as in the original instance. Recall that $\mathcal{Z}_B$ is defined using the $\oplus$ operator and exactly captures (by Corollary 2) the worst case adversary structure possible, restricted to $V(\gamma(B))$, relative to the initial knowledge of players in B. Hence, all nodes in B have the same initial knowledge in both instances, since $\mathcal{Z}_B = \mathcal{Z}'_B$.

The proof is by contradiction. Suppose that there exists a safe algorithm $\mathcal{A}$ which is resilient for $(G, \mathcal{Z}, \gamma, S, R)$. We consider the following executions σ and σ' of $\mathcal{A}$:

- Execution σ is on instance $(G, \mathcal{Z}, \gamma, S, R)$, with sender's value $x_S = 0$, and corruption set C_1; in each round, each corrupted player in C_1 performs the actions that its corresponding player performs in the respective round of execution σ' (where C_1 consists of honest players only).
- Execution σ' is on instance $(G, \mathcal{Z}', \gamma, S, R)$, with sender's value $x_S = 1$, and corruption set C_2; in each round, each corrupted player in C_2 performs the actions that its corresponding player performs in the respective round of execution σ (where C_2 consists of honest players only).

Note that C_1, C_2 are admissible corruption sets in scenarios σ, σ' respectively since they belong to $\mathcal{Z}$ and $\mathcal{Z}'$ (resp.) It is easy to see that $C_1 \cup C_2$ is a cut which separates S from B in both instances and that actions of every node of this cut are identical in both executions σ, σ'. Consequently, the actions of any honest node $w \in B$ must be identical in both executions. Since, by assumption, algorithm $\mathcal{A}$ is resilient on $(G, \mathcal{Z}, \gamma, S, R)$, R must decide on the sender's message 0 in execution σ, and must do the same in execution σ'. However, in execution σ' the sender's message is 1. Therefore $\mathcal{A}$ makes R decide on an incorrect message in $(G, \mathcal{Z}', \gamma, S, R)$. This contradicts the assumption that $\mathcal{A}$ is safe. $\square$

3.1   The RMT Partial Knowledge Algorithm (RMT-PKA)

We next present the *RMT Partial Knowledge Algorithm* (RMT-PKA), an RMT protocol which succeeds whenever the condition of Theorem 3 (in fact, its negation) is met, rendering it a tight condition on when RMT is possible. To prove this we provide some supplementary notions.

In RMT-PKA there are two types of messages exchanged. *Type 1 messages* are used to propagate the sender's value and are of the form (x, p) where $x \in X$ and p is a path [3]. *Type 2 messages* of the form $((v, \gamma(v), \mathcal{Z}_v), p)$ are used for every node v to propagate its initial information $\gamma(v), \mathcal{Z}_v$ throughout the graph. Let M denote a subset of the messages of type 1 and 2 that the receiver node R receives at some round of the protocol on $(G, \mathcal{Z}, \gamma, S, R)$. We will say that $value(M) = x$ if and only if all the type 1 messages of M report the same sender value x, i.e., for every such message (y, p), it holds that $y = x$, for some $x \in X$. Observe that M may consist of messages which contain contradictory information. We next define the form of a message set M which contains no contradictory information in our setting (a valid set M).

Definition 5 (Valid set M). *A set M of both type 1 and type 2 messages corresponds to a valid scenario, or more simply is valid, if*

---

[3] By $p||v$ (appearing in the algorithm) we will denote the concatenation of path p with node v.

- $\exists x \in X$ s.t. $value(M) = x$. That is, all type 1 messages relay the same x as sender's value.
- $\forall m_1, m_2 \in M$ of type 2, their first component (the part of the pair which comprises of the local information of a node) is the same when they refer to the same node. That is, if $m_1 = ((v, \gamma(v), \mathcal{Z}_v), p)$ and $m_2 = (((v', \gamma'(v), \mathcal{Z}'_v), p'),$ then $v = v'$ implies that $\gamma(v) = \gamma'(v)$ and $\mathcal{Z}_v = \mathcal{Z}'_v$.

For every valid M we can define the pair (G_M, x_M) where $x_M = value(M)$; we assume that $x_M = \bot$ if no type 1 messages are included in M. To define G_M let V_M be the set of nodes u for which the information $\gamma(u), \mathcal{Z}_u$ is included in M, namely $V_M = \{v \mid ((v, \gamma(v), \mathcal{Z}_v), p) \in M$ for some path $p\}$. Then, G_M is the node induced subgraph of graph $\gamma(V_M)$ on node set V_M. Therefore, a valid message set M uniquely determines the pair (G_M, x_M). We next propose two notions that we use to check if a valid set M contains correct information.

Definition 6 (full message set). A full message set M received by R, is a valid set M, with $value(M) \neq \bot$, that contains all the $S \rightsquigarrow R$ paths which appear in G_M as part of type 1 messages.

Next we define the notion of adversary cover of a full message set M. If such a cut exists, then there is a scenario where all propagated values might be false.

Definition 7 (Adversary cover of full message set M). A set $C \subseteq V_M$ is an adversary cover of full message set M if C has the following property: C is a cut between S and R on G_M and if B is the node set of the connected component that R lies in, it holds that $(C \cap V(\gamma(B))) \in \mathcal{Z}_B$.

With the predicate $\mathtt{nocover}(M)$ we will denote the non existence of an adversary cover of M. The next theorem states the somewhat counterintuitive safety property of RMT-PKA, i.e., that the receiver will never decide on an incorrect value despite the increased adversary's attack capabilities, which includes reporting fictitious nodes and false local knowledge. The proof is deferred to the full version.

Theorem 4 (RMT-PKA Safety). *RMT-PKA is safe.*

The sufficiency proof combines techniques from [13] (correctness of the Path Propagation Agorithm) with the novel notions of full message set M, adversary cover of M and corresponding graph G_M.

Theorem 5 (Sufficiency). Let $(G, \mathcal{Z}, \gamma, S, R)$ be an RMT instance. If no RMT-cut exists, then RMT-PKA achieves reliable message transmission.

Proof. Observe that if $R \in \mathcal{N}(S)$ then R trivially decides on x_S due to the sender propagation rule, since the sender is honest. Assuming that no RMT-cut exists, we will show that if $R \notin \mathcal{N}(S)$ then R will decide on x_S due to the full message set propagation rule.

Let $T \in \mathcal{Z}$ be any admissible corruption set and consider the run e_T of RMT-PKA where T is the actual corruption set. Let P be the set of all paths

RMT Partial Knowledge Algorithm (RMT-PKA)

Input for each node v: sender's label S, $\gamma(v)$, $\mathcal{Z}_v$.
Additional input for S : value $x_S \in X$ (message space).
Type 1 message format: pair (x, p)
Type 2 message format: pair $((u, \gamma(u), \mathcal{Z}_u), p)$,
where $x \in X$, u the id of some node, $\gamma(u)$ is the view of node u, $\mathcal{Z}_u$ is the local adversary structure of node u, and p is a path of G (message's propagation trail).

Code for S: send messages $(x_S, \{S\})$ and $((S, \gamma(S), \mathcal{Z}_S), \{S\})$ to all neighbors and terminate.

Code for $v \notin \{S, R\}$: send message $((v, \gamma(v), \mathcal{Z}_v), \{v\})$ to all neighbors.

upon reception of type 1 or type 2 message (a, p) from node u **do**:

   if $(v \in p) \vee (tail(p) \neq u)$(We use $tail(p)$ to denote the last node of path p. Checking whether $tail(p) \neq u$ we ensure that at least one corrupted node will be included in a faulty propagation path.) **then discard** (a, p) **else send** $(a, p\|v)$ to all neighbours.

Code for R: Initialize $M_R \leftarrow \emptyset$
upon reception of type 1 or type 2 message (x, p) from node u **do**:

   if $(v \in p) \vee (tail(p) \neq u)$ **then discard** (x, p) **else** $M_R \leftarrow M_R \cup (a, p)$
   if (x, p) is a type 1 message **then**
      $lastmsg \leftarrow (x, p)$
      if $decision(M_R, lastmsg) = x$ **then output** x and **terminate**.

function $decision(M_R, lastmsg)$

   if $R \in \mathcal{N}(S)$ **then**
      if $lastmsg = (x_S, \{S\})$ **then return** x_S
      else return $\perp$.

   for all valid $M \subseteq M_R$ with $value(M) = value(lastmsg)$ **do**
      compute graph G_M
      $M_1 \leftarrow$ type 1 messages of M
      $\mathcal{P}_1 \leftarrow$ set of all paths p with $(x, p) \in M_1$
      $\mathcal{P}_{S,R} \leftarrow$ set of all $S \rightsquigarrow R$ paths of G_M
      if $(\mathcal{P}_{S,R} \subseteq \mathcal{P}_1) \wedge nocover(M)$ **then**   ▷ *full message set with no*
         return $value(lastmsg)$ **else return** $\perp$.   ▷ *adversary cover*

function $nocover(M)$

   $check \leftarrow$ **true**
   for all $C \subseteq V_M$ **do**
      if C is a (S, R) cut on G_M **then**
         $B \leftarrow$ connected component of R in $G_M \setminus C$
         $(\mathcal{Z}_B, V(\gamma(B))) \leftarrow \bigoplus_{v \in B} (\mathcal{Z}_v, V(\gamma(v)))$ ▷ *joint adversary structure*
         if $(C \cap V(\gamma(B))) \in \mathcal{Z}_B$ **then** $check \leftarrow$**false**
   return $check$

connecting S with R and are composed entirely by nodes in $V(G) \setminus T$ (honest nodes). Observe that $P \neq \emptyset$, otherwise T is a cut separating S from R which is trivially a RMT-cut, a contradiction.

Since paths in P are entirely composed by honest nodes, it should be clear by the protocol that by round $|V(G)|$, R will have obtained x_S through all paths in P by receiving the corresponding type 1 messages M_1. Furthermore, by round $|V(G)|$, R will have received type 2 messages set M_2 which includes information for all the nodes connected with R via paths that do not pass through nodes in T. This includes all nodes of paths in P. Consequently, R will have received the full message set $M = M_1 \cup M_2$ with $value(M) = x_S$.

We next show that there is no adversary cover for M and thus R will decide on x_S through the full message set propagation rule on M. Assume that there exists an adversary cover C for M. This, by definition means that C is a cut between S, R on G_M and if B is the node set of the the connected component that R lies in, it holds that and $(C \cap V(\gamma(B))) \in \mathcal{Z}_B$ (observe that R can compute $\mathcal{Z}_B$ using the information contained in M_2 as defined in the previous paragraph). Then obviously $T \cup C$ is a cut in G separating S from R, since every path of G that connects S with R contains at least a node in $T \cup C$. Let the cut $T \cup C$ partition $V(G) \setminus \{T \cup C\}$ in the sets A, B s.t. $S \in A$. Then clearly $T \cup C$ is an RMT cut by definition, a contradiction. Thus there is no adversary cover for M and R will decide on x_S. Moreover, since RMT-PKA is safe, the receiver will not decide on any other value different from x_S.   □

Corollary 3 (Uniqueness). *RMT-PKA is unique among safe algorithms, i.e., given an RMT instance $(G, \mathcal{Z}, \gamma, S, R)$, if there exists any safe RMT algorithm which is resilient for this instance, then RMT-PKA also achieves reliable message transmission on this instance.*

4   Conclusions and Open Questions

Regarding the partial knowledge model, the RMT-PKA protocol employs topology information exchange between players. Although topology discovery was not our motive, techniques used here (e.g. the $\oplus$ operation) may be applicable to that problem under a Byzantine adversary [4,11]. A comparison with the techniques used in this field might give further insight on how to efficiently extract information from maliciously crafted topological data.

We have shown that RMT-PKA protocol is unique for the partial knowledge model; this only addresses the feasibility issue. A natural question is whether and when we can devise a unique and also efficient algorithm for this setting. The techniques used so far to reduce the communication complexity (e.g. [8]) do not seem to be directly applicable to this model. So, exploring this direction further is particularly meaningful.

References

1. Desmedt, Y., Wang, Y.: Perfectly secure message transmission revisited. In: Knudsen, L.R. (ed.) EUROCRYPT 2002. LNCS, vol. 2332, pp. 502–517. Springer, Heidelberg (2002). doi:10.1007/3-540-46035-7_33
2. Dolev, D.: The byzantine generals strike again. J. Algorithms **3**(1), 14–30 (1982)
3. Dolev, D., Dwork, C., Waarts, O., Yung, M.: Perfectly secure message transmission. J. ACM **40**(1), 17–47 (1993). http://doi.acm.org/10.1145/138027.138036
4. Dolev, S., Liba, O., Schiller, E.M.: Self-stabilizing byzantine resilient topology discovery and message delivery. In: Gramoli, V., Guerraoui, R. (eds.) NETYS 2013. LNCS, vol. 7853, pp. 42–57. Springer, Heidelberg (2013). doi:10.1007/978-3-642-40148-0_4
5. Hirt, M., Maurer, U.M.: Complete characterization of adversaries tolerable in secure multi-party computation (extended abstract). In: Burns, J.E., Attiya, H. (eds.) PODC, pp. 25–34. ACM (1997)
6. Ichimura, A., Shigeno, M.: A new parameter for a broadcast algorithm with locally bounded byzantine faults. Inf. Process. Lett. **110**(12–13), 514–517 (2010)
7. Koo, C.Y.: Broadcast in radio networks tolerating byzantine adversarial behavior. In: Chaudhuri, S., Kutten, S. (eds.) PODC, pp. 275–282. ACM (2004)
8. Kumar, M.V.N.A., Goundan, P.R., Srinathan, K., Rangan, C.P.: On perfectly secure communication over arbitrary networks. In: Proceedings of the Twenty-first Annual Symposium on Principles of Distributed Computing, PODC 2002, pp. 193–202, ACM, New York (2002). http://doi.acm.org/10.1145/571825.571858
9. Lamport, L., Shostak, R.E., Pease, M.C.: The byzantine generals problem. ACM Trans. Program. Lang. Syst. **4**(3), 382–401 (1982)
10. Litsas, C., Pagourtzis, A., Sakavalas, D.: A graph parameter that matches the resilience of the certified propagation algorithm. In: Cichoń, J., Gębala, M., Klonowski, M. (eds.) ADHOC-NOW 2013. LNCS, vol. 7960, pp. 269–280. Springer, Heidelberg (2013). doi:10.1007/978-3-642-39247-4_23
11. Nesterenko, M., Tixeuil, S.: Discovering network topology in the presence of byzantine faults. IEEE Trans. Parallel Distrib. Syst. **20**(12), 1777–1789 (2009)
12. Pagourtzis, A., Panagiotakos, G., Sakavalas, D.: Brief announcement: reliable message transmission under partial knowledge and general adversaries. In: Giakkoupis, G. (ed.) Proceedings of the 2016 ACM Symposium on Principles of Distributed Computing, PODC 2016, Chicago, IL, USA, 25-28 July 2016. pp. 203–205. ACM (2016). http://doi.acm.org/10.1145/2933057.2933080
13. Pagourtzis, A., Panagiotakos, G., Sakavalas, D.: Reliable broadcast with respect to topology knowledge. Distrib. Comput. **30**(2), 87–102 (2017). http://dx.doi.org/10.1007/s00446-016-0279-6
14. Pelc, A., Peleg, D.: Broadcasting with locally bounded byzantine faults. Inf. Process. Lett. **93**(3), 109–115 (2005)
15. Roman, S.: Lattices and Ordered Sets. Springer Science & Business Media, Heidelberg (2008)
16. Shankar, B., Gopal, P., Srinathan, K., Rangan, C.P.: Unconditionally reliable message transmission in directed networks. In: Proceedings of the Nineteenth Annual ACM-SIAM Symposium on Discrete Algorithms, SODA 2008, pp. 1048–1055. Society for Industrial and Applied Mathematics, Philadelphia, PA, USA (2008). http://dl.acm.org/citation.cfm?id=1347082.1347197

17. Srinathan, K., Patra, A., Choudhary, A., Rangan, C.P.: Unconditionally secure message transmission in arbitrary directed synchronous networks tolerating generalized mixed adversary. In: Proceedings of the 4th International Symposium on Information, Computer, and Communications Security, ASIACCS 2009, pp. 171–182. ACM, New York (2009). http://doi.acm.org/10.1145/1533057.1533083
18. Srinathan, K., Rangan, C.P.: Possibility and complexity of probabilistic reliable communication in directed networks. In: Ruppert, E., Malkhi, D. (eds.) Proceedings of the Twenty-Fifth Annual ACM Symposium on Principles of Distributed Computing, PODC 2006, Denver, CO, USA, 23-26 July 2006, pp. 265–274. ACM (2006). http://doi.acm.org/10.1145/1146381.1146421
19. Tseng, L., Vaidya, N., Bhandari, V.: Broadcast using certified propagation algorithm in presence of byzantine faults. Inf. Process. Lett. **115**(4), 512–514 (2015). http://www.sciencedirect.com/science/article/pii/S0020019014002609

Polynomial-Time Algorithms for the Subset Feedback Vertex Set Problem on Interval Graphs and Permutation Graphs

Charis Papadopoulos[✉] and Spyridon Tzimas

Department of Mathematics, University of Ioannina, Ioannina, Greece
charis@cs.uoi.gr, roytzimas@hotmail.com

Abstract. Given a vertex-weighted graph $G = (V, E)$ and a set $S \subseteq V$, a subset feedback vertex set X is a set of the vertices of G such that the graph induced by $V \setminus X$ has no cycle containing a vertex of S. The SUBSET FEEDBACK VERTEX SET problem takes as input G and S and asks for the subset feedback vertex set of minimum total weight. In contrast to the classical FEEDBACK VERTEX SET problem which is obtained from the SUBSET FEEDBACK VERTEX SET problem for $S = V$, restricted to graph classes the SUBSET FEEDBACK VERTEX SET problem is known to be NP-complete on split graphs and, consequently, on chordal graphs. Here we give the first polynomial-time algorithms for the problem on two subclasses of AT-free graphs: interval graphs and permutation graphs. Moreover towards the unknown complexity of the problem for AT-free graphs, we give a polynomial-time algorithm for co-bipartite graphs. Thus we contribute to the first positive results of the SUBSET FEEDBACK VERTEX SET problem when restricted to graph classes for which FEEDBACK VERTEX SET is solved in polynomial time.

1 Introduction

For a given set S of vertices of a graph G, a *subset feedback vertex set* X is a set of vertices such that every cycle of $G[V \setminus X]$ does not contain a vertex from S. The SUBSET FEEDBACK VERTEX SET problem takes as input a graph $G = (V, E)$ and a set $S \subseteq V$ and asks for the subset feedback vertex set of minimum cardinality. In the weighted version every vertex of G has a weight and the objective is to compute a subset feedback vertex set with the minimum total weight. The SUBSET FEEDBACK VERTEX SET problem is a generalization of the classical FEEDBACK VERTEX SET problem in which the goal is to remove a set of vertices X such that $G[V \setminus X]$ has no cycles. Thus by setting $S = V$ the problem coincides with the NP-complete FEEDBACK VERTEX SET problem [19]. Both problems find important applications in several aspects that arise in optimization theory, constraint satisfaction, and bayesian inference [1,2,14,15]. Interestingly the SUBSET FEEDBACK VERTEX SET problem for $|S| = 1$ also coincides with the NP-complete MULTIWAY CUT problem [17] in which the task is to disconnect a predescribed set of vertices [9,20].

© Springer-Verlag GmbH Germany 2017
R. Klasing and M. Zeitoun (Eds.): FCT 2017, LNCS 10472, pp. 381–394, 2017.
DOI: 10.1007/978-3-662-55751-8_30

SUBSET FEEDBACK VERTEX SET was first introduced by Even et al. who obtained a constant factor approximation algorithm for its weighted version [14]. The unweighted version in which all vertex weights are equal has been proved to be fixed parameter tractable [13]. Moreover the fastest algorithm for the weighted version in general graphs runs in $O^*(1.87^n)$ time[1] by enumerating its minimal solutions [17], whereas for the unweighted version the fastest algorithm runs in $O^*(1.76^n)$ time [16]. As the unweighted version of the problem is shown to be NP-complete even when restricted to split graphs [17], there is a considerable effort to reduce the running time on chordal graphs, a proper superclass of split graphs, and more general on other classes of graphs. Golovach et al. considered the weighted version and gave an algorithm that runs in $O^*(1.68^n)$ time for chordal graphs [21]. Reducing the existing running time even on chordal graphs has been proved to be quite challenging and only for the unweighted version of the problem a faster algorithm was given that runs in $O^*(1.62^n)$ time [10]. In fact the $O^*(1.62^n)$-algorithm given in [10] runs for every graph class which is closed under vertex deletions and edge contractions, and on which the weighted FEED-BACK VERTEX SET problem can be solved in polynomial time. Thus there is an algorithm that runs in $O^*(1.62^n)$ time for the unweighted version of the SUB-SET FEEDBACK VERTEX SET problem when restricted to AT-free graphs [10], a graph class that properly contains permutation graphs and interval graphs. Here we show that for the classes of permutation graphs and interval graphs we design a much faster algorithm even for the weighted version of the problem.

As SUBSET FEEDBACK VERTEX SET is a generalization of the classical FEED-BACK VERTEX SET problem, let us briefly give an overview of the complexity of FEEDBACK VERTEX SET on related graph classes. Concerning the complexity of FEEDBACK VERTEX SET it is known to be NP-complete on bipartite graphs [33] and planar graphs [19], whereas it becomes polynomial-time solvable on the classes of bounded clique-width graphs [8], chordal graphs [11, 32], interval graphs [28], permutation graphs [4–6,26], cocomparability graphs [27], and, more generally, AT-free graphs [25]. Despite the many positive and negative results of the FEEDBACK VERTEX SET problem, very few similar results are known concerning the complexity of SUBSET FEEDBACK VERTEX SET. Clearly for graph classes for which the FEEDBACK VERTEX SET problem is NP-complete, so does the SUBSET FEEDBACK VERTEX SET problem. However as the SUBSET FEEDBACK VERTEX SET problem is more general that FEEDBACK VERTEX SET problem, it is natural to study its complexity for graph classes for which FEEDBACK VERTEX SET is polynomial-time solvable. In fact restricted to graph classes there is only a negative result for the SUBSET FEEDBACK VERTEX SET problem regarding its NP-completeness on split graphs [17]. Such a result, however, implies that there is an interesting algorithmic difference between the two problems, as the FEEDBACK VERTEX SET problem is known to be polynomial-time computable for split graphs [11, 32].

Both interval graphs and permutation graphs have unbounded clique-width [23] and, therefore, any algorithmic metatheorem related to MSOL formulation

---

[1] The O^* notation is used to suppress polynomial factors.

is not applicable [12]. Let us also briefly explain that extending the approach of [25] for the FEEDBACK VERTEX SET problem when restricted to AT-free graphs is not straightforward. A graph is *AT-free* if for every triple of pairwise non-adjacent vertices, the neighborhood of one of them separates the two others. The class of AT-free graphs is well-studied and it properly contains interval, permutation, and cocomparability graphs [7,22]. One of the basic tools in [25] relies on growing a small representation of an independent set into a suitable forest. Although such a representation is rather small on AT-free graphs (and, thus, on interval graphs or permutation graphs), when considering SUBSET FEEDBACK VERTEX SET it is not necessary that the fixed set induces an independent set which makes it difficult to control how the partial solution may be extended. Therefore the methodology described in [25] cannot be trivially extended towards the SUBSET FEEDBACK VERTEX SET problem.

Our Results. Here we initiate the study of SUBSET FEEDBACK VERTEX SET restricted on graph classes from the positive perspective. We consider its weighted version and give the first positive results on permutation graphs and interval graphs, both being proper subclasses of AT-free graphs. As already explained, we are interested towards subclasses of AT-free graphs since for chordal graphs the problem is already NP-complete [17]. Permutation graphs and interval graphs are unrelated to split graphs and are both characterized by a linear structure with respect to a given vertex ordering [7,22,32]. For both classes of graphs we design polynomial-time algorithms based on dynamic programming of subproblems defined by passing the vertices of the graph according to their natural linear ordering. One of our key ingredients is that during the pass of the dynamic programming we augment the considered vertex set and we allow the solutions to be chosen only from a specific subset of the vertices rather than the whole vertex set. Although for interval graphs such a strategy leads to a simple algorithm, the case of permutation graphs requires further descriptions of the considered subsolutions by augmenting the considered part of the graph with a small number of additional vertices. Moreover we consider the class of co-bipartite graphs (complements of bipartite graphs) and settle its complexity status. We show that the number of minimal solutions of a co-bipartite graph is polynomial which implies a polynomial-time algorithm of the SUBSET FEEDBACK VERTEX SET problem for the class of co-bipartite graphs. Figure 1 summarizes our overall results.

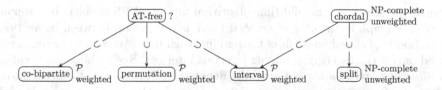

Fig. 1. The computational complexity of the SUBSET FEEDBACK VERTEX SET problem restricted to the considered graph classes. All polynomial-time results ($\mathcal{P}$) are obtained in this work, whereas the NP-completeness result of split graphs is due to [17].

2   Preliminaries

We refer to [7,22] for our standard graph terminology. A *path* is a sequence of distinct vertices $\langle v_1 v_2 \cdots v_k \rangle$ where each pair of consecutive vertices $v_i v_{i+1}$ forms an edge of G. If in addition $v_1 v_k$ is an edge then we obtain a *cycle*. In this paper, we distinguish between paths (or cycles) and *induced paths* (or *induced cycles*). By an induced path (or cycle) of G we mean a chordless path (or cycle). A chordless cycle on four vertices is referred to as *square*. A *weighted graph* $G = (V, E)$ is a graph, where each vertex $v \in V$ is assigned a *weight* that is a positive integer number. We denote by $w(v)$ the weight of each vertex $v \in V$. For a vertex set $A \subset V$, the weight of A is $\sum_{v \in A} w(v)$.

The *Subset Feedback Vertex Set* (SFVS) problem is defined as follows: given a weighted graph G and a vertex set $S \subseteq V$, find a vertex set $X \subset V$, such that all cycles containing vertices of S, also contains a vertex of X and $\sum_{v \in X} w(v)$ is minimized. In the unweighted version of the problem all weights are equal and positive. A vertex set X is defined as *minimal* subset feedback vertex set if no proper subset of X is a subset feedback vertex set for G and S. The classical FEEDBACK VERTEX SET (FVS) problem is a special case of the subset feedback vertex set problem with $S = V$. Note that a *minimum weight* subset feedback vertex set is dependent on the weights of the vertices, whereas a *minimal* subset feedback vertex set is only dependent on the vertices and not their weights. Clearly, both in the weighted and the unweighted versions, a minimum subset feedback vertex set must be minimal.

An induced cycle of G is called S-cycle if a vertex of S is contained in the cycle. We define an S-*forest* of G to be a vertex set $Y \subseteq V$ such that no cycle in $G[Y]$ is an S-cycle. An S-forest Y is *maximal* if no proper superset of Y is an S-forest. Observe that X is a minimal subset feedback vertex set if and only if $Y = V \setminus X$ is a maximal S-forest. Thus, the problem of computing a minimum weighted subset feedback vertex set is equivalent to the problem of computing a maximum weighted S-forest. Let us denote by $\mathcal{F}_S$ the class of S-forests. In such terms, given the graph G and the subset S of V, we are interested in finding a $\max_w \{Y \subseteq V \mid G[Y] \in \mathcal{F}_S\}$, where $\max_w$ selects a vertex set having the maximum weight. It is not difficult to see that for any clique C of G, an S-forest of G that contains a vertex of $S \cap C$ contains at most two vertices of C.

3   Computing SFVS on Interval Graphs

Here we present a polynomial-time algorithm for the SFVS problem on interval graphs. A graph is an *interval graph* if there is a bijection between its vertices and a family of closed intervals of the real line such that two vertices are adjacent if and only if the two corresponding intervals intersect. Such a bijection is called an *interval representation* of the graph, denoted by $\mathcal{I}$. Notice that every induced subgraph of an interval graph is an interval graph. Moreover it can be decided in linear time whether a given graph is an interval graph, and if so, an interval representation can be generated in linear time [18]. Moreover it is known that any induced cycle of an interval graph is an induced triangle [28,32].

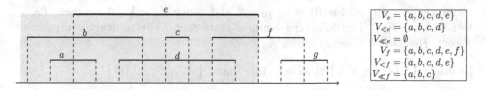

$$V_e = \{a, b, c, d, e\}$$
$$V_{<e} = \{a, b, c, d\}$$
$$V_{\ll e} = \emptyset$$
$$V_f = \{a, b, c, d, e, f\}$$
$$V_{<f} = \{a, b, c, d, e\}$$
$$V_{\ll f} = \{a, b, c\}$$

Fig. 2. An interval graph given by its interval representation and the corresponding sets of V_e and V_f. Observe that $<f = e$ whereas $\ll f = c$. Also notice that the intervals that are properly contained within the gray area form the set V_e.

As already mentioned, instead of finding a subset feedback vertex set X of minimum weight of (G, S) we concentrate on the equivalent problem of finding a maximum weighted S-forest Y of (G, S). We first define the necessary vertex sets. Let G be a weighted interval graph and let $\mathcal{I}$ be its interval representation. The left and right endpoints of an interval i, $1 \leq i \leq n$, are denoted by $\ell(i)$ and $r(i)$, respectively. The intervals are numbered from 1 to n according to their ascending $r(i)$. For technical reasons, we add an interval with label 0 that does not belong to S, has weight zero, and augment $\mathcal{I}$ to $\mathcal{I}^+$ by setting $\ell(0) = \min_{i \in \mathcal{I}} \{\ell(i)\} - 2$ and $r(0) = \min_{i \in \mathcal{I}} \{\ell(i)\} - 1$. Notice that interval 0 is non-adjacent to any vertex of G. Clearly if Y is a maximum weighted S-forest for $G[\mathcal{I}^+]$ then $Y \setminus \{0\}$ is a maximum weighted S-forest for $G[\mathcal{I}]$.

We consider the two relations on V that are defined by the endpoints of the intervals as follows: $i \leq_\ell j \Leftrightarrow \ell(i) \leq \ell(j)$ and $i \leq_r j \Leftrightarrow r(i) \leq r(j)$. Since all endpoints of the collection's intervals are distinct, $\leq_\ell$ and $\leq_r$ are total orders on V. For a set of vertices $U \subseteq V$ we write $\ell\text{-}\min U$ to denote the *minimum* vertex of U with respect to $\leq_\ell$ and we write $r\text{-}\max U$ to denote the *maximum* vertex of U with respect to $\leq_r$. For a vertex $i \in V$ we let $V_i =_{\text{def}} \{h \in V : h \leq_r i\}$. We define two types of predecessors of the interval $i \neq 0$ with respect to $\leq_r$, which correspond to the subproblems that our algorithm wants to solve:
$$<i =_{\text{def}} r\text{-}\max(V_i \setminus \{i\}) \quad \text{and} \quad \ll i =_{\text{def}} r\text{-}\max(V_i \setminus (\{i\} \cup \{h \in V : \{h, i\} \in E\})).$$
Observe that for two vertices $i, x \in V$ with $r(i) < r(x)$, $x \in V \setminus V_i$. An example of an interval representation that depicts the corresponding notation of V_i is shown in Fig. 2. By definition we get the following partitions of V-sets.

Observation 1. Let $i \in V \setminus \{0\}$ and let $j \in V \setminus V_i$ such that $\{i, j\} \in E$. Then,
(1) $V_i = V_{<i} \cup \{i\}$ and (2) $V_{<i} = V_{\ll j} \cup \{h \in V_{<i} : \{h, j\} \in E\}$.

Next we define the sets that our dynamic programming algorithm uses in order to compute the S-forest of G that has maximum weight.

A-sets: Let $i \in V$. Then, $A_i =_{\text{def}} \max_w \{X \subseteq V_i : G[X] \in \mathcal{F}_S\}$.
B-sets: Let $i \in V$, $x \in V \setminus V_i$. Then, $B_i^x =_{\text{def}} \max_w \{X \subseteq V_i : G[X \cup \{x\}] \in \mathcal{F}_S\}$.
C-sets: Let $i \in V$ and let $x, y \in V \setminus (V_i \cup S)$ such that $x <_\ell y$ and $\{x, y\} \in E$. Then, $C_i^{x,y} =_{\text{def}} \max_w \{X \subseteq V_i : G[X \cup \{x, y\}] \in \mathcal{F}_S\}$.

Since $V_0 = \{0\}$ and $w(0) = 0$, $A_0 = \emptyset$ and, since $V_n = V$, $A_n = \max_w \{X \subseteq V : G[X] \in \mathcal{F}_S\}$. The following lemmas state how to recursively compute all A-sets, B-sets and C-sets besides A_0.

Lemma 1. *Let $i \in V \setminus \{0\}$. Then $A_i = \max_w \left\{ A_{<i}, B^i_{<i} \cup \{i\} \right\}$.*

Proof. By Observation 1 (1), $V_i = V_{<i} \cup \{i\}$. If $i \notin A_i$, then we have $A_i = A_{<i}$, otherwise we have $A_i = B^i_{<i} \cup \{i\}$, since $B^i_{<i}$ is the $\max_w$ subset of $V_{<i}$ such that the graph induced by its union with $\{i\}$ contains no S-cycle by definition. $\square$

Lemma 2. *Let $i \in V$ and let $x \in V \setminus V_i$. Moreover, let $x' = \ell\text{-}\min\{i, x\}$ and let y' be the remaining vertex of $\{i, x\}$.*

(1) If $\{i, x\} \notin E$, then $B^x_i = A_i$.

(2) If $\{i, x\} \in E$, then $B^x_i = \begin{cases} \max_w \left\{ B^x_{<i}, B^{x'}_{\ll y'} \cup \{i\} \right\}, & \text{if } i \in S \text{ or } x \in S \\ \max_w \left\{ B^x_{<i}, C^{x', y'}_{<i} \cup \{i\} \right\}, & \text{if } i, x \notin S. \end{cases}$

Proof. Assume first that $\{i, x\} \notin E$. Then $r(i) < \ell(x)$, so that x has no neighbor in $G[V_i \cup \{x\}]$. Thus no subset of $V_i \cup \{x\}$ containing x induces an S-cycle of G, implying that $B^x_i = A_i$.

Next assume that $\{i, x\} \in E$. If $i \notin B^x_i$ then according to Observation 1 (1) it follows that $B^x_i = B^x_{<i}$. So let us assume in what follows that $i \in B^x_i$. Observe that $B^x_i \setminus \{i\} \subseteq V_{<i}$, by Observation 1 (1). We distinguish two cases according to whether i or x are elements of S.

- Let $i \in S$ or $x \in S$. Assume there is a vertex $h \in B^x_i \setminus \{i\}$ such that $\{h, y'\} \in E$. Then we know that $\ell(y') < r(h)$ and by definition we have $\ell(x') < \ell(y')$ and $r(h) < r(x')$. This particularly means that h is adjacent to x'. This however leads to a contradiction since $\langle h, x', y' \rangle$ is an induced S-triangle of G. Thus for any vertex $h \in B^x_i \setminus \{i\}$ we know that $\{h, y'\} \notin E$. By Observation 1 (2) notice that $B^x_i \setminus \{i\} \subseteq V_{\ll y'}$. Also observe that the neighborhood of y' in $G[V_{\ll y'} \cup \{x', y'\}]$ is $\{x'\}$. Thus no subset of $V_{\ll y'} \cup \{x', y'\}$ that contains y' induces an S-cycle of G. Therefore $B^x_i = B^{x'}_{\ll y'} \cup \{i\}$.
- Let $i, x \notin S$. Since $V_i = V_{<i} \cup \{i\}$ and $x' <_\ell y'$, we get $B^x_i = C^{x', y'}_{<i} \cup \{i\}$.

Therefore in all cases we reach the desired equations. $\square$

Lemma 3. *Let $i \in V$ and let $x, y \in V \setminus (V_i \cup S)$ such that $x <_\ell y$ and $\{x, y\} \in E$. Moreover, let $x' = \ell\text{-}\min\{i, x, y\}$ and let $y' = \ell\text{-}\min(\{i, x, y\} \setminus \{x'\})$.*

1. If $\{i, y\} \notin E$, then $C^{x, y}_i = B^x_i$.

2. If $\{i, y\} \in E$, then $C^{x, y}_i = \begin{cases} C^{x, y}_{<i} & , \text{if } i \in S \\ \max_w \left\{ C^{x, y}_{<i}, C^{x', y'}_{<i} \cup \{i\} \right\}, & \text{if } i \notin S. \end{cases}$

Proof. Assume first that $\{i, y\} \notin E$. Then $r(i), \ell(x) < \ell(y) < r(x)$, so that the neighborhood of y in $G[V_i \cup \{x, y\}]$ is $\{x\}$. Thus no subset of $V_i \cup \{x, y\}$ that contains y induces an S-cycle of G. By definitions it follows that $C_i^{x,y} = B_i^x$.

Assume next that $\{i, y\} \in E$. Then $\ell(x) < \ell(y) < r(i) < r(x), r(y)$, so that $\langle i, x, y \rangle$ is an induced triangle of G. If $i \notin C_i^{x,y}$ then by Observation 1 (1) we have $C_i^{x,y} = C_{<i}^{x,y}$. Suppose that $i \in C_i^{x,y}$. If $i \in S$ then $\langle i, x, y \rangle$ is an induced S-triangle of G, contradicting the fact that $i \in C_i^{x,y}$. By definition, $x \notin S$ and $y \notin S$. Hence $i \in C_i^{x,y}$ implies that $S \cap \{i, x, y\} = \emptyset$. We will now show that $C_i^{x,y} = C_{<i}^{x',y'} \cup \{i\}$.

Notice that $C_i^{x,y} \setminus \{i\} \subseteq V_{<i}$, so $C_i^{x,y} \subseteq C_{<i}^{x',y'} \cup \{i\}$ by definition. To complete the proof we show $C_{<i}^{x',y'} \cup \{i\} \subseteq C_i^{x,y}$ as well. Let z' be the vertex of $\{i, x, y\} \setminus \{x', y'\}$. Observe that by the leftmost ordering we have $\ell(x') < \ell(y') < \ell(z')$. By definition no subset of $C_i^{x,y} \cup \{x, y\}$ induces an S-triangle in G. Assume for contradiction that a subset of $C_{<i}^{x',y'} \cup \{x', y', z'\}$ containing z' induces an S-triangle $\langle v_1, v_2, z' \rangle$ in G. Since $S \cap \{x', y', z'\} = \emptyset$, without loss of generality, assume that $v_1 \in S$. This particularly means that $v_1 \in C_{<i}^{x',y'} \subseteq V_{<i}$. Regarding the vertex ordering notice that the S-triangle implies that $\ell(z') < r(v_1)$. By the fact that $v_1 \in V_{<i}$ we have $r(v_1) < r(x'), r(y'), r(z')$. Since $\ell(x') < \ell(y') < \ell(z')$, the previous inequalities imply that $\{v_1, x'\}, \{v_1, y'\} \in E$. Thus $\langle v_1, x', y' \rangle$ is an induced S-triangle in G, leading to a contradiction. Therefore $C_{<i}^{x',y'} \cup \{i\} \subseteq C_i^{x,y}$ as desired. $\square$

Now we are equipped with our necessary tools to obtain the main result of this section, namely a polynomial-time algorithm for SFVS on interval graphs.

Theorem 1. SUBSET FEEDBACK VERTEX SET *can be solved in* $O(n^3)$ *time on interval graphs.*

4   Computing SFVS on Permutation Graphs

Let $\pi = \pi(1), \ldots, \pi(n)$ be a permutation over $\{1, \ldots, n\}$. The position of an integer i in π is denoted by $\pi^{-1}(i)$. Given a permutation π, the *inversion graph* of π, denoted by $G(\pi)$, has vertex set $\{1, \ldots, n\}$ and two vertices i, j are adjacent if $(i - j)(\pi(i) - \pi(j)) < 0$. A graph is a *permutation graph* if it is isomorphic to the inversion graph of a permutation [7,22]. For our purposes, we assume that a permutation graph is given as a permutation π and equal to the defined inversion graph. Permutation graphs are the intersection graphs of segments between two horizontal parallel lines, that is, there is a one-to-one mapping from the segments onto the vertices of a graph such that there is an edge between two vertices of the graph if and only if their corresponding segments intersect. We refer to the two horizontal lines as *top* and *bottom* lines. This representation is called a *permutation diagram* and a graph is a permutation graph if and only if it has a permutation diagram. Given a permutation graph, its permutation diagram can be constructed in linear time [30]. It is important to note that any induced cycle of a permutation graph is either an induced triangle or an induced square [4–6,26,32].

We assume that we are given a permutation graph $G = (V, E)$ such that $G = G(\pi)$ along with $S \subseteq V$ and a weight function $w : V \to \mathbb{R}^+$ as input. We add an isolated vertex in G and augment π to π' so that $\pi'(0) = 0$. Further we assign zero value for 0's weight and assume that $0 \notin S$.

We consider the two relations on V defined as follows: $i \leq_t j$ if and only if $i \leq j$ and $i \leq_b j$ if and only if $\pi^{-1}(i) \leq \pi^{-1}(j)$ for all $i, j \in V \cup \{0\}$. It is not difficult to see that both $\leq_t$ and $\leq_b$ are total orders on V; they are exactly the orders in which the integers appear on the top and bottom line, respectively, in the permutation diagram. Moreover we write $i <_t j$ or $i <_b j$ if and only if $i \neq j$ and $i \leq_t j$ or $i \leq_b j$, respectively. We extend $\leq_t$ and $\leq_b$ to support sets of vertices as follows. For two sets of vertices L and R we write $L \leq_t R$ (resp., $L \leq_b R$) if for any two vertices $u \in L$ and $v \in R$, $u \leq_t v$ (resp., $u \leq_b v$).

Two vertices $i, j \in V$ are called *crossing pair*, denoted by ij, if $i \leq_t j$ and $j \leq_b i$. We denote by $\mathcal{X}$ the set of all crossing pairs in G. Let $\mathcal{I} = \{ii \in \mathcal{X} : i \in V\}$, so that the crossing pairs of $\mathcal{X} \setminus \mathcal{I}$ correspond exactly to the edges of G. Given two crossing pairs $gh, ij \in \mathcal{X}$ we define two partial orders $\leq_\ell$ and $\leq_r$:

$$gh \leq_\ell ij \Leftrightarrow g \leq_t i \text{ and } h \leq_b j \quad \text{and} \quad gh \leq_r ij \Leftrightarrow g \leq_b i \text{ and } h \leq_t j.$$

Given a vertex set $X \subseteq V$ we denote by $\mathcal{X}[X]$ the set of all crossing pairs of G formed exclusively from vertices of X. It is not difficult to see that the *minimum* crossing pair of $\mathcal{X}[X]$ with respect to $\leq_\ell$ and the *maximum* crossing pair contained in $\mathcal{X}[X]$ with respect to $\leq_r$ are both well defined; we write ℓ-$\min$ and r-$\max$ to denote them respectively.

We next define the predecessors of a crossing pair with respect to $\leq_r$, which correspond to the subproblems that our dynamic programming algorithm wants to solve. Let $ij \in \mathcal{X}$ be a crossing pair. We define the set of vertices that induce the part of the subproblem that we consider at each crossing pair as follows: $V_{ij} =_{\text{def}} \{h \in V : hh \leq_r ij\}$. Let x be a vertex such that $i <_b x$ or $j <_t x$. Notice that V_{ij} does not contain x by definition. The predecessors of the crossing pair $ij \neq 00$ are defined as follows:

$$\lessdot ij =_{\text{def}} r\text{-}\max \mathcal{X}[V_{ij} \setminus \{j\}], \qquad \leqslant ij =_{\text{def}} r\text{-}\max \mathcal{X}[V_{ij} \setminus \{i\}],$$
$$< ij =_{\text{def}} r\text{-}\max \mathcal{X}[V_{ij} \setminus \{i, j\}],$$
$$\ll ij =_{\text{def}} r\text{-}\max \mathcal{X}[V_{ij} \setminus (\{i, j\} \cup \{h \in V : \{h, i\} \in E \text{ or } \{h, j\} \in E\})], \text{ and}$$
$$< ij \ll xx =_{\text{def}} r\text{-}\max \mathcal{X}[V_{ij} \setminus \{h \in V : \{h, x\} \in E\}].$$

Although it seems somehow awkward to use one the symbols $\{\lessdot, \leqslant, <, \ll, <\ll\}$ for the defined predecessors, we stress that such predecessors are required only to describe the necessary subset V_{gh} of V_{ij}. Moreover it is not difficult to see that each of the symbol gravitates towards a particular meaning with respect to the top and bottom orderings as well as the non-adjacency relationship. An example of a permutation graph that illustrates the defined predecessors is given in Fig. 3. With the above defined predecessors of ij, we show how V_{ij} can be partitioned into smaller sets of vertices with respect to a suitable predecessor.

Observation 2. Let $ij \in \mathcal{X} \setminus \{00\}$ and let $x \in V \setminus V_{ij}$. Then,

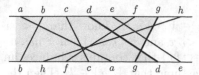

Fig. 3. A permutation graph given by its permutation diagram and the set V_{dg} of the crossing pair dg together with the corresponding predecessors of dg. Observe that the line segments that are properly contained within the gray area form the set V_{dg}.

(1) $V_{ij} = V_{\leqslant ij} \cup \{j\} = V_{\leqslant ij} \cup \{i\} = V_{<ij} \cup \{i,j\}$,

(2) $V_{<ij} = V_{\ll jj} \cup \{h \in V_{<ij} : \{h,j\} \in E\} = V_{\ll ii} \cup \{h \in V_{<ij} : \{h,i\} \in E\}$,

(3) $V_{\ll ii} = V_{\ll ij} \cup \{h \in V_{\ll ii} : \{h,j\} \in E\}$,

(4) $V_{\ll jj} = V_{\ll ij} \cup \{h \in V_{\ll jj} : \{h,i\} \in E\}$, and

(5) $V_{<ij} = V_{<ij \ll xx} \cup \{h \in V_{<ij} : \{h,x\} \in E\}$.

It is clear that for any edge $\{i,j\} \in E$ either $i <_t j$ and $j <_b i$ hold, or $j <_t i$ and $i <_b j$ hold. If further $ij \in \mathcal{X} \setminus \mathcal{I}$ then we know that $i <_t j$ and $j <_b i$.

Our dynamic programming algorithm relies on similar sets that we used for the case of interval graphs. That is, we need to describe appropriate sets that define the solutions to be chosen only from a specific part of the considered subproblem. Although for interval graphs we showed that adding two vertices into such sets is enough, for permutation graphs we need to consider at most two newly crossing pairs which corresponds to consider four newly vertices.

A-sets: Let $ij \in \mathcal{X}$. Then, $A_{ij} = \max_w \{X \subseteq V_{ij} : G[X] \in \mathcal{F}_S\}$.

B-sets: Let $ij \in \mathcal{X}$ and let $x \in V \setminus V_{ij}$. Then, $B_{ij}^{xx} =_{\text{def}} \max_w \{X \subseteq V_{ij} : G[X \cup \{x\}] \in \mathcal{F}_S\}$. Moreover, let $xy \in \mathcal{X} \setminus \mathcal{I}$ such that $j <_t y$, $i <_b x$, and $x,y \notin S$. Then, $B_{ij}^{xy} =_{\text{def}} \max_w \{X \subseteq V_{ij} : G[X \cup \{x,y\}] \in \mathcal{F}_S\}$.

C-sets: Let $ij \in \mathcal{X}$, $xy \in \mathcal{X} \setminus \mathcal{I}$, and $z \in V \setminus (V_{ij} \cup \{x,y\})$ such that $xy <_\ell zz$, at least one of x,y is adjacent to z, $j <_t y$, $i <_b x$, and $x,y,z \notin S$. Then, $C_{ij}^{xy,zz} =_{\text{def}} \max_w \{X \subseteq V_{ij} : G[X \cup \{x,y,z\}] \in \mathcal{F}_S\}$. Moreover, let $zw \in \mathcal{X} \setminus \mathcal{I}$ such that $xy <_\ell zw$, $\{x,w\}, \{y,z\} \in E$, $j <_t \{y,w\}$, $i <_b \{x,z\}$, and $x,y,z,w \notin S$. Then, $C_{ij}^{xy,zw} =_{\text{def}} \max_w \{X \subseteq V_{ij} : G[X \cup \{x,y,z,w\}] \in \mathcal{F}_S\}$.

Observe that, since $V_{00} = \{0\}$ and $w(0) = 0$, $A_{00} = \emptyset$ and, since $V_{\pi(n)n} = V$, $A_{\pi(n)n} = \max_w \{X \subseteq V : G[X] \in \mathcal{F}_S\}$. The following lemmas state how to recursively compute all A-sets, B-sets, and C-sets other than A_{00}. We first consider the crossing pairs ii for the sets A_{ii}, B_{ii}^{xx}, B_{ii}^{xy}, $C_{ii}^{xy,zz}$, and $C_{ii}^{xy,zw}$.

Lemma 4. *Let $i \in V \setminus \{0\}$. Then $A_{ii} = A_{<ii} \cup \{i\}$.*

Proof. By Observation 2 (1), $A_{<ii} \cup \{i\} \subseteq A_{ii}$. Notice that i is non-adjacent to all vertices of $V_{<ii}$. Thus no subset of V_{ii} that contains i induces an S-cycle. □

Lemma 5. *Let $i \in V$ and let $x \in V \setminus V_{ii}$.*

1. *If $\{i,x\} \notin E$ then $B_{ii}^{xx} = A_{ii}$.*
2. *If $\{i,x\} \in E$ then $B_{ii}^{xx} = B_{<ii}^{xx} \cup \{i\}$.*

Proof. Assume first that $\{i, x\} \notin E$. Since $x \in V \setminus V_{ii}$ we know that $i <_t x$ or $i <_b x$. Moreover as $\{i, x\} \notin E$ we have $i <_t x$ and $i <_b x$. Then x has no neighbor in $G[V_{ii} \cup \{x\}]$. Thus no subset of $V_{ii} \cup \{x\}$ that contains x induces an S-cycle in G. Hence $B_{ii}^{xx} = A_{ii}$ follows. Next assume that $\{i, x\} \in E$. Then the neighborhood of i in $G[V_{ii} \cup \{x\}]$ is $\{x\}$. This means that no subset of $V_{ii} \cup \{x\}$ that contains i induces an S-cycle in G, so that $i \in B_{ii}^{xx}$. By Observation 2 (1) it follows that $B_{ii}^{xx} = B_{<ii}^{xx} \cup \{i\}$. $\qquad\square$

Lemma 6. *Let $i \in V$ and let $xy \in \mathcal{X} \setminus \mathcal{I}$ such that $i <_t y$, $i <_b x$, and $x, y \notin S$.*

1. *If $\{i, y\} \notin E$ then $B_{ii}^{xy} = B_{ii}^{xx}$.*
2. *If $\{i, x\} \notin E$ then $B_{ii}^{xy} = B_{ii}^{yy}$.*
3. *If $\{i, x\}, \{i, y\} \in E$ then $B_{ii}^{xy} = \begin{cases} B_{<ii}^{xy} & , if\, i \in S \\ B_{<ii}^{xy} \cup \{i\}, if\, i \notin S. \end{cases}$*

Lemma 7. *Let $i \in V$, $xy \in \mathcal{X} \setminus \mathcal{I}$, and let $z \in V \setminus (V_{ii} \cup \{x, y\})$ such that $xy <_\ell zz$, at least one of x, y is adjacent to z, $i <_t y$, $i <_b x$, and $x, y, z \notin S$.*

1. *If $\{i, z\} \notin E$ then $C_{ii}^{xy, zz} = B_{ii}^{xy}$.*
2. *If $\{i, z\} \in E$ then $C_{ii}^{xy, zz} = \begin{cases} C_{<ii}^{xy, zz}, & if\, i \in S \\ C_{<ii}^{xy, zz} \cup \{i\}, if\, i \notin S. \end{cases}$*

Lemma 8. *Let $i \in V$ and let $xy, zw \in \mathcal{X} \setminus \mathcal{I}$ such that $xy <_\ell zw$, $\{x, w\}, \{y, z\} \in E$, $i <_t \{y, w\}$, $i <_b \{x, z\}$, and $x, y, z, w \notin S$.*

1. *If $\{i, w\} \notin E$ then $C_{ii}^{xy, zw} = C_{ii}^{xy, zz}$.*
2. *If $\{i, z\} \notin E$ then $C_{ii}^{xy, zw} = C_{ii}^{xy, ww}$.*
3. *If $\{i, z\}, \{i, w\} \in E$ then $C_{ii}^{xy, zw} = \begin{cases} C_{<ii}^{xy, zw}, & if\, i \in S \\ C_{<ii}^{xy, zw} \cup \{i\}, if\, i \notin S. \end{cases}$*

Lemmas 4–8 describe the subsolutions for all crossing pairs $ii \in \mathcal{I} \setminus \{00\}$. Next we give the recursive formulations for A_{ij}, B_{ij}^{xx}, B_{ij}^{xy}, $C_{ij}^{xy, zz}$, and $C_{ij}^{xy, zw}$ whenever $ij \in \mathcal{X} \setminus \mathcal{I}$ which particularly means that i and j are distinct vertices in G.

Lemma 9. *Let $ij \in \mathcal{X} \setminus \mathcal{I}$. Then,*

$$A_{ij} = \begin{cases} \max_w \left\{ A_{\leqslant ij}, A_{\leqslant ij}, B_{\ll jj}^{ii} \cup \{i, j\}, B_{\ll ii}^{jj} \cup \{i, j\} \right\}, if\, i \in S \, or \, j \in S \\ \max_w \left\{ A_{\leqslant ij}, A_{\leqslant ij}, B_{<ij}^{ij} \cup \{i, j\} \right\}, \qquad\qquad if\, i, j \notin S. \end{cases}$$

With the next two lemmas we describe recursively the sets B_{ij}^{xx} and B_{ij}^{xy}.

Lemma 10. *Let $ij \in \mathcal{X} \setminus \mathcal{I}$ and let $x \in V \setminus V_{ij}$. Moreover let $x'y' = \ell\text{-}\min \mathcal{X}[\{i, j, x\}]$ and let z' be the remaining vertex of $\{i, j, x\}$.*

1. *If $\{i, x\}, \{j, x\} \notin E$ then $B_{ij}^{xx} = A_{ij}$.*

2. If $\{i,x\} \in E$ and $\{j,x\} \notin E$ then

$$B_{ij}^{xx} = \begin{cases} \max_w \left\{ B_{<ij}^{xx}, B_{\leq ij}^{xx}, B_{\ll jj}^{ii} \cup \{i,j\}, B_{\ll ix}^{jj} \cup \{i,j\} \right\}, & if\ i \in S\ or\ j \in S \\ \max_w \left\{ B_{<ij}^{xx}, B_{\leq ij}^{xx}, B_{<ij \ll xx}^{ij} \cup \{i,j\} \right\}, & if\ i,j \notin S, x \in S \\ \max_w \left\{ B_{<ij}^{xx}, B_{\leq ij}^{xx}, C_{<ij}^{x'y',z'z'} \cup \{i,j\} \right\}, & if\ i,j,x \notin S. \end{cases}$$

3. If $\{i,x\} \notin E$ and $\{j,x\} \in E$ then

$$B_{ij}^{xx} = \begin{cases} \max_w \left\{ B_{<ij}^{xx}, B_{\leq ij}^{xx}, B_{\ll xj}^{ii} \cup \{i,j\}, B_{\ll ii}^{jj} \cup \{i,j\} \right\}, & if\ i \in S\ or\ j \in S \\ \max_w \left\{ B_{<ij}^{xx}, B_{\leq ij}^{xx}, B_{<ij \ll xx}^{ij} \cup \{i,j\} \right\}, & if\ i,j \notin S, x \in S \\ \max_w \left\{ B_{<ij}^{xx}, B_{\leq ij}^{xx}, C_{<ij}^{x'y',z'z'} \cup \{i,j\} \right\}, & if\ i,j,x \notin S. \end{cases}$$

4. If $\{i,x\}, \{j,x\} \in E$ then

$$B_{ij}^{xx} = \begin{cases} \max_w \left\{ B_{<ij}^{xx}, B_{\leq ij}^{xx} \right\}, & if\ i \in S\ or\ j \in S\ or\ x \in S \\ \max_w \left\{ B_{<ij}^{xx}, B_{\leq ij}^{xx}, C_{<ij}^{x'y',z'z'} \cup \{i,j\} \right\}, & if\ i,j,x \notin S. \end{cases}$$

Let $ij, xy \in \mathcal{X} \setminus \mathcal{I}$ such that $\{i,y\}, \{j,x\} \in E$. It is not difficult to see that if we remove the vertices of a crossing pair $uv \in \mathcal{X}[\{i,j,x,y\}]$ from $\{i,j,x,y\}$ then the remaining two vertices are adjacent.

Lemma 11. *Let $ij, xy \in \mathcal{X} \setminus \mathcal{I}$ such that $j <_t y$, $i <_b x$ and $x,y \notin S$. Moreover, if $\{i,y\}, \{j,x\} \in E$ then let $x'y' = \ell\text{-}\min \mathcal{X}[\{i,j,x,y\}]$ and let $z'w' = \ell\text{-}\min \mathcal{X}[\{i,j,x,y\} \setminus \{x',y'\}]$.*

1. *If $\{i,y\} \notin E$ then $B_{ij}^{xy} = B_{ij}^{xx}$.*
2. *If $\{j,x\} \notin E$ then $B_{ij}^{xy} = B_{ij}^{yy}$.*
3. *If $\{i,y\}, \{j,x\} \in E$ then*

$$B_{ij}^{xy} = \begin{cases} \max_w \left\{ B_{<ij}^{xy}, B_{\leq ij}^{xy} \right\}, & if\ i \in S\ or\ j \in S \\ \max_w \left\{ B_{<ij}^{xy}, B_{\leq ij}^{xy}, C_{<ij}^{x'y',z'w'} \cup \{i,j\} \right\}, & if\ i,j \notin S. \end{cases}$$

Lemma 12. *Let $ij, xy \in \mathcal{X} \setminus \mathcal{I}$ and let $z \in V \setminus V_{ij}$ such that $xy <_\ell zz$, at least one of x,y is adjacent to z, $j <_t y$, $i <_b x$, and $x,y,z \notin S$. Moreover, if $\{i,z\} \in E$ or $\{j,z\} \in E$ then let $x'y' = \ell\text{-}\min \mathcal{X}[\{i,j,x,y,z\}]$ and let $z'w' = \ell\text{-}\min \mathcal{X}[\{i,j,x,y,z\} \setminus \{x',y'\}]$.*

1. *If $\{i,z\}, \{j,z\} \notin E$ then $C_{ij}^{xy,zz} = B_{ij}^{xy}$.*
2. *If $\{i,z\} \in E$ or $\{j,z\} \in E$ then*

$$C_{ij}^{xy,zz} = \begin{cases} \max_w \left\{ C_{<ij}^{xy,zz}, C_{\leq ij}^{xy,zz} \right\}, & if\ i \in S\ or\ j \in S \\ \max_w \left\{ C_{<ij}^{xy,zz}, C_{\leq ij}^{xy,zz}, C_{<ij}^{x'y',z'w'} \cup \{i,j\} \right\}, & if\ i,j \notin S. \end{cases}$$

The next lemma shows how to recursively compute $C_{ij}^{xy,zw}$. Note that in each case we describe $C_{ij}^{xy,zw}$ as a predefined smaller set of a subsolution that is either in the same form or has already been described in one of the previous lemmas.

Lemma 13. *Let* $ij, xy, zw \in \mathcal{X} \setminus \mathcal{I}$ *such that* $xy <_\ell zw$, $\{x,w\}, \{y,z\} \in E$, $j <_t \{y,w\}$, $i <_b \{x,z\}$, *and* $x,y,z,w \notin S$. *Moreover, if* $\{i,w\}, \{j,z\} \in E$, *let* $x'y' = \ell\text{-}\min \mathcal{X}[\{i,j,x,y,z,w\}]$ *and let* $z'w' = \ell\text{-}\min \mathcal{X}[\{i,j,x,y,z,w\} \setminus \{x',y'\}]$.

1. *If* $\{i,w\} \notin E$ *then* $C_{ij}^{xy,zw} = C_{ij}^{xy,zz}$.
2. *If* $\{j,z\} \notin E$ *then* $C_{ij}^{xy,zw} = C_{ij}^{xy,ww}$.
3. *If* $\{i,w\}, \{j,z\} \in E$ *then*

$$C_{ij}^{xy,zw} = \begin{cases} \max_w \left\{ C_{\leqslant ij}^{xy,zw}, C_{\leqslant ij}^{xy,zw} \right\}, & if\ i \in S\ or\ j \in S \\ \max_w \left\{ C_{\leqslant ij}^{xy,zw}, C_{\leqslant ij}^{xy,zw}, C_{<ij}^{x'y',z'w'} \cup \{i,j\} \right\}, & if\ i,j \notin S. \end{cases}$$

It is important to notice that all described formulations are given recursively based on Lemmas 4–13. Now we are in position to state our claimed polynomial-time algorithm for the SFVS problem on permutation graphs.

Theorem 2. SUBSET FEEDBACK VERTEX SET *can be solved in* $O(n+m^3)$ *time on permutation graphs.*

5   Concluding Remarks

From the complexity point of view, since FVS is polynomial-time solvable on the class of AT-free graphs [25], a natural problem is to settle the complexity of SFVS on AT-free graphs. Interestingly most problems that are hard on AT-free graphs are already hard on co-bipartite graphs (see for e.g., [29]). Also notice that SFVS remains NP-complete on bipartite graphs, as FVS is NP-complete on bipartite graphs [33]. Co-bipartite graphs are the complements of bipartite graphs and are unrelated to permutation graphs or interval graphs. Here we show that SFVS admits a simple solution on co-bipartite graphs, therefore excluding such an approach through a hardness result on co-bipartite graphs.

Theorem 3. *The number of maximal S-forests of a co-bipartite graph is at most* $22n^4$ *and these can be enumerated in time* $O(n^4)$.

Moreover it is interesting to settle the complexity of SFVS on other related graph classes such as strongly chordal graphs or subclasses of AT-free graphs like trapezoid graphs or complements of triangle-free graphs. Regarding graphs of bounded structural parameter and due to the nature of the dynamic programming used for SFVS on interval and permutation graphs, it is interesting to consider graphs of bounded maximum induced matching width [3].

Another interesting open question is concerned with problems related to *terminal-sets* such as the MULTIWAY CUT problem in which we want to disconnect a given set of terminals by removing vertices of minimum total weight.

As already mentioned in the Introduction, the MULTIWAY CUT problem reduces to the SFVS problem by adding a vertex s with $S = \{s\}$ that is adjacent to all terminals and whose weight is larger than the sum of the weights of all vertices in the original graph [17]. Notice that such a reduction does not directly work on interval or permutation graphs, since the augmented graph might not belong to the same graph class. Despite the polynomial-time algorithms for the MULTIWAY CUT problem on permutation graphs [31] and interval graphs [24], it is still interesting whether we can apply our algorithms for the SFVS problem with respect to the MULTIWAY CUT problem.

References

1. Bar-Yehuda, R., Geiger, D., Naor, J., Roth, R.M.: Approximation algorithms for the feedback vertex set problem with applications to constraint satisfaction and bayesian inference. SIAM J. Comput. **27**(4), 942–959 (1998)
2. Becker, A., Geiger, D.: Optimization of Pearl's method of conditioning and greedy-like approximation algorithms for the vertex feedback set problem. Artif. Intell. **83**(1), 167–188 (1996)
3. Belmonte, R., Vatshelle, M.: Graph classes with structured neighborhoods and algorithmic applications. Theor. Comput. Sci. **511**, 54–65 (2013)
4. Brandstädt, A.: On improved time bounds for permutation graph problems. In: Mayr, E.W. (ed.) WG 1992. LNCS, vol. 657, pp. 1–10. Springer, Heidelberg (1993). doi:10.1007/3-540-56402-0_30
5. Brandstädt, A., Kratsch, D.: On the restriction of some NP-complete graph problems to permutation graphs. In: Budach, L. (ed.) FCT 1985. LNCS, vol. 199, pp. 53–62. Springer, Heidelberg (1985). doi:10.1007/BFb0028791
6. Brandstädt, A., Kratsch, D.: On domination problems for permutation and other graphs. Theor. Comput. Sci. **54**(2), 181–198 (1987)
7. Brandstädt, A., Le, V.B., Spinrad, J.P.: Graph Classes: A Survey. Society for Industrial and Applied Mathematics, Philadelphia (1999). http://dx.doi.org/10.1137/1.9780898719796
8. Bui-Xuan, B.-M., Suchý, O., Telle, J.A., Vatshelle, M.: Feedback vertex set on graphs of low clique-width. Eur. J. Comb. **34**(3), 666–679 (2013)
9. Calinescu, G.: Multiway cut. In: Kao, M.-Y. (ed.) Encyclopedia of Algorithms. Springer, New York (2008). doi:10.1007/978-0-387-30162-4_253
10. Chitnis, R.H., Fomin, F.V., Lokshtanov, D., Misra, P., Ramanujan, M.S., Saurabh, S.: Faster exact algorithms for some terminal set problems. J. Comput. Syst. Sci. **88**, 195–207 (2017)
11. Corneil, D.G., Fonlupt, J.: The complexity of generalized clique covering. Discret. Appl. Math. **22**(2), 109–118 (1988)
12. Courcelle, B., Makowsky, J.A., Rotics, U.: Linear time solvable optimization problems on graphs of bounded clique-width. Theory Comput. Syst. **33**, 125–150 (2000)
13. Cygan, M., Pilipczuk, M., Pilipczuk, M., Wojtaszczyk, J.O.: Subset feedback vertex set is fixed-parameter tractable. SIAM J. Discret. Math. **27**(1), 290–309 (2013)
14. Even, G., Naor, J., Zosin, L.: An 8-approximation algorithm for the subset feedback vertex set problem. SIAM J. Comput. **30**(4), 1231–1252 (2000)
15. Festa, P., Pardalos, P.M., Resende, M.G.C.: Feedback set problems. In: Floudas, C.A., Pardalos, P.M. (eds.) Encyclopedia of Optimization, pp. 1005–1016. Springer, New York (2009). doi:10.1007/978-0-387-74759-0_178

16. Fomin, F.V., Gaspers, S., Lokshtanov, D., Saurabh, S.: Exact algorithms via monotone local search. Proc. STOC **2016**, 764–775 (2016)
17. Fomin, F.V., Heggernes, P., Kratsch, D., Papadopoulos, C., Villanger, Y.: Enumerating minimal subset feedback vertex sets. Algorithmica **69**(1), 216–231 (2014)
18. Fulkerson, D.R., Gross, O.A.: Incidence matrices and interval graphs. Pac. J. Math. **15**, 835–855 (1965)
19. Garey, M.R., Johnson, D.S.: Computers and Intractability. W. H. Freeman and Co., San Francisco (1978)
20. Garg, N., Vazirani, V.V., Yannakakis, M.: Multiway cuts in node weighted graphs. J. Algorithms **50**(1), 49–61 (2004)
21. Golovach, P.A., Heggernes, P., Kratsch, D., Saei, R.: Subset feedback vertex sets in chordal graphs. J. Discret. Algorithms **26**, 7–15 (2014)
22. Golumbic, M.C.: Algorithmic Graph Theory and Perfect Graphs. Annals of Discrete Mathematics, vol. 57. Elsevier, Amsterdam (2004)
23. Golumbic, M.C., Rotics, U.: On the clique-width of some perfect graph classes. Int. J. Found. Comput. Sci. **11**(3), 423–443 (2000)
24. Guo, J., Hüffner, F., Kenar, E., Niedermeier, R., Uhlmann, J.: Complexity and exact algorithms for vertex multicut in interval and bounded treewidth graphs. Eur. J. Oper. Res. **186**, 542–553 (2008)
25. Kratsch, D., Müller, H., Todinca, I.: Feedback vertex set on AT-free graphs. Discret. Appl. Math. **156**(10), 1936–1947 (2008)
26. Liang, Y.D.: On the feedback vertex set problem in permutation graphs. Inf. Process. Lett. **52**(3), 123–129 (1994)
27. Liang, Y.D., Chang, M.-S.: Minimum feedback vertex sets in cocomparability graphs and convex bipartite graphs. Acta Inform. **34**(5), 337–346 (1997)
28. Lu, C.L., Tang, C.Y.: A linear-time algorithm for the weighted feedback vertex problem on interval graphs. Inf. Process. Lett. **61**(2), 107–111 (1997)
29. Maw-Shang, C.: Weighted domination of cocomparability graphs. Discret. Appl. Math. **80**(2), 135–148 (1997)
30. McConnell, R.M., Spinrad, J.P.: Modular decomposition and transitive orientation. Discret. Math. **201**, 189–241 (1999)
31. Papadopoulos, C.: Restricted vertex multicut on permutation graphs. Discret. Appl. Math. **160**(12), 1791–1797 (2012)
32. Spinrad, J.P.: Efficient Graph Representations. Fields Institute Monograph Series, vol. 19. American Mathematical Society, Providence (2003)
33. Yannakakis, M.: Node-deletion problems on bipartite graphs. SIAM J. Comput. **10**(2), 310–327 (1981)

Determinism and Computational Power of Real Measurement-Based Quantum Computation

Simon Perdrix[1]([✉]) and Luc Sanselme[2]

[1] CNRS, LORIA, Université de Lorraine, Inria-Carte, Nancy, France
simon.perdrix@loria.fr
[2] LORIA, CNRS, Université de Lorraine, Inria-Caramba, Lycée Poincaré,
Nancy, France

Abstract. Measurement-based quantum computing (MBQC) is a universal model for quantum computation. The combinatorial characterisation of determinism in this model, powered by measurements, and hence, fundamentally probabilistic, is the cornerstone of most of the breakthrough results in this field. The most general known sufficient condition for a deterministic MBQC to be driven is that the underlying graph of the computation has a particular kind of flow called Pauli flow. The necessity of the Pauli flow was an open question. We show that Pauli flow is not necessary, providing several counter examples. We prove however that Pauli flow is necessary for determinism in the *real* MBQC model, an interesting and useful fragment of MBQC.

We explore the consequences of this result for real MBQC and its applications. Real MBQC and more generally real quantum computing is known to be universal for quantum computing. Real MBQC has been used for interactive proofs by McKague. The two-prover case corresponds to real-MBQC on bipartite graphs. While (complex) MBQC on bipartite graphs are universal, the universality of real MBQC on bipartite graphs was an open question. We show that real bipartite MBQC is not universal proving that all measurements of real bipartite MBQC can be parallelised leading to constant depth computations. As a consequence, McKague's techniques cannot lead to two-prover interactive proofs.

1 Introduction

Measurement-based quantum computing [19,20] (MBQC for short) is a universal model for quantum computation. This model is not only very promising in terms of the physical realisations of the quantum computer [17,22], MBQC has also several theoretical advantages, e.g. parallelisation of quantum operations [3,5] (logarithmic separation with the traditional model of quantum circuits), blind quantum computing [2] (a protocol for delegated quantum computing), fault tolerant quantum computing [21], simulation [9], contextuality [18], interactive proofs [2,12].

In MBQC, a computation consists of performing local quantum measurements over a large entangled resource state. The resource state is described by a

© Springer-Verlag GmbH Germany 2017
R. Klasing and M. Zeitoun (Eds.): FCT 2017, LNCS 10472, pp. 395–408, 2017.
DOI: 10.1007/978-3-662-55751-8_31

graph – using the so-called graph state formalism [11]. The *tour de force* of this model is to tame the fundamental non-determinism of the quantum measurements: the number of possible outputs of a measurement-based computation on a given input is exponential in the number of measurements, and each of these branches of the computation is produced with an exponentially small probability. The only known technique to make such a fundamentally probabilistic computation exploitable is to implement a correction strategy which makes the overall computation deterministic: it does not affect the probability for each branch of the computation to occur, but it guarantees that all the branches produce the same output.

The existence of a correction strategy relies on the structures of the entanglement of the quantum state on which the measurements are performed. Deciding whether a given resource state allows determinism is a central question in MBQC. Several sufficient conditions for determinism have been introduced. First in [6] the notion of *causal flow* has been introduced: if the graph describing the entangled resource state has a causal flow then a deterministic MBQC can be driven on this resource. Causal flow has been generalized to a weaker condition called Generalized flow (Gflow) which is also sufficient for determinism. Gflow has been proved to be necessary for a robust variant of determinism and when roughly speaking there is no Pauli measurement, a special class of quantum measurements (see Sect. 2 for details) [4]. In the same paper, the authors have introduced a weaker notion of flow called Pauli Flow, allowing some measurements to be Pauli measurements. Pauli flow is the weakest known sufficient condition for determinism and its necessity was a crucial open question as the characterisation of determinism in MBQC is the cornerstone of most of the applications of MBQC.

In Sect. 2, we present the MBQC model, and the tools that come with it. Our first contribution is to provide a simpler characterisation of the Pauli flow (Proposition 1), with three instead of nine conditions to satisfy for the existence of a Pauli flow. Our main contribution is to prove in Sect. 3 that the Pauli flow is not necessary in general – by pointing out several counter examples – but is actually necessary for *real* MBQC (Theorem 3). Real MBQC is a restriction of MBQC where only real observables are used, i.e. observables which eigenstates are quantum states that can be described using real numbers. Quantum mechanics, and hence models of quantum computation, are traditionally based on complex numbers. Real quantum computing is universal for quantum computation [1] and has been crucially used recently in the study of contextuality and simulation by means of quantum computing by state injection [9]. Real MBQC [14] may lead to several other applications. One of them is an interactive proof protocol built by McKague [12]. McKague introduced a protocol where a verifier using a polynomial number of quantum provers can perform a computation, with the guaranty that, if a prover has cheated, it will be able to detect it. An open question left in [12] by McKague is to know whether this model can bring to an interactive proof protocol with only two quantum provers. We answer negatively to this question in Sect. 4.2. Our third contribution is to point out the existence of a kind of supernormal-form for Pauli flow in real MBQC on bipartite graphs (Lemma 1). This result enables us to prove in Theorem 4 that real MBQC on bipartite graphs is not very powerful: all measurements of a real bipartite MBQC can be parallelised. As a consequence, only problems that can be solved in constant depth can be solved using real bipartite MBQC.

2 Measurement-Based Quantum Computation, Generalized Flow and Pauli Flow

Notations. We assume the reader familiar with quantum computing notations, otherwise one can refer to the Appendix A of the pre-print version of the present article [16] or to [15]. We will use the following set/graph notations: First of all, the *symmetric difference* of two sets A and B will be denoted $A \Delta B :=$ $(A \cup B) \setminus (A \cap B)$. We will use intensively the *open* and *closed neighbourhood*. Given a simple undirected graph $G = (V, E)$, for any $u \in V$, $N(u) := \{v \in V \mid (u, v) \in E\}$ is the (open) neighbourhood of u, and $N[u] := N(u) \cup \{u\}$ is the closed neighbourhood of u. For any subset A of V, $\mathsf{Odd}(A) := \Delta_{v \in A} N(v)$ (resp. $\mathsf{Odd}[A] := \Delta_{v \in A} N[u]$) is the odd (resp. odd closed) neighbourhood of A. Also, we will use the notion of *extensive maps*. A map $f : A \to 2^B$, with $A \subseteq B$ is extensive if the transitive closure of $\{(u, v) : v \in f(u)\}$ is a strict partial order. We say that f is extensive with respect to a strict partial order $\prec$ if $(v \in f(u) \Rightarrow u \prec v)$.

2.1 MBQC, Concretely, Abstractly

In this section, a brief description of the measurement-based quantum computation is given, a more detailed introduction can be found in [7,8]. Starting from a low-level description of measurement-based quantum computation using the so-called patterns of the Measurement-Calculus – an assembly language composed of 4 kinds of commands: creation of ancillary qubits, entangling operation, measurement and correction – we end up with a graph theoretical description of the computation and in particular of the underlying entangled resource of the computation.

2.2 Measurement-Calculus Patterns: An Assembly Language

An assembly language for MBQC is the Measurement-Calculus [7,8]: a pattern is a sequence of commands, each command is either:

- N_u: initialisation of a fresh qubit u in the state $|+\rangle = \frac{|0\rangle + |1\rangle}{\sqrt{2}}$;
- $E_{u,v}$ entangling two qubits u and v by applying Control-Z operation $\Lambda Z :$ $|x, y\rangle \mapsto (-1)^{xy} |x, y\rangle$ to the qubits u and v;
- $M_u^{\lambda_u, \alpha_u}$ measurement of qubit u according to the observable $\mathcal{O}_{\lambda_u, \alpha_u}$ described below;

– $X_u^{s_v}$ (resp. $Z_u^{s_v}$), a correction which consists of applying Pauli $X : |x\rangle \mapsto |1-x\rangle$ (resp. $Z : |x\rangle \mapsto (-1)^x |x\rangle$) to qubit u iff s_v (the classical outcome of the measurement of qubit v) is 1.

A pattern is subject to some basic well-formedness conditions like: no operation can be applied on a qubit u after u being measured; a correction cannot depend on a signal s_u if qubit u is not yet measured.

The qubits which are not initialised using the N command are the input qubits, and those which are not measured are the output qubits. The measurement of a qubit u is characterized by $\lambda_u \subset \{X, Y, Z\}$ a subset of one or two Pauli operators, and an angle $\alpha_u \in [0, 2\pi)$:

– when $\lambda_u = \{M\}$ is a singleton, u is measured according to $\mathcal{O}_{\lambda_u, \alpha_u} := M$ if $\alpha_u = 0$ or $\mathcal{O}_{\lambda_u, \alpha_u} := -M$ if $\alpha_u = \pi$.
– when $|\lambda_u| = 2$, u is measured in the λ_u-plane of the Bloch sphere with an angle α_u, i.e. according to the observable:

$$\mathcal{O}_{\lambda_u, \alpha_u} := \begin{cases} \cos(\alpha_u)X_u + \sin(\alpha_u)Y_u & \text{if } \lambda_u = \{X, Y\} \\ \cos(\alpha_u)Y_u + \sin(\alpha_u)Z_u & \text{if } \lambda_u = \{Y, Z\} \\ \cos(\alpha_u)Z_u + \sin(\alpha_u)X_u & \text{if } \lambda_u = \{Z, X\} \end{cases}$$

Measurement of qubit u produces a classical outcome $(-1)^{s_u}$ where $s_u \in \{0, 1\}$ is called *signal*, or simply *classical outcome* with a slight abuse of notation.

2.3   A Graph-Based Representation

In the Measurement-Calculus, the patterns are equipped with an equational theory which captures some basic invariant properties, e.g. two operations acting on distinct qubits commute, or $E_{u,v}$ is equivalent to $E_{v,u}$. It is easy to show using the equations of the Measurement-Calculus that any pattern can be transformed into an equivalent pattern of the form:

$$\left(\prod_{u \in O^c}^{\prec} Z_{\mathbf{z}(u)}^{s_u} X_{\mathbf{x}(u)}^{s_u} M_u^{\lambda_u, \alpha_u} \right) \left(\prod_{(u,v) \in G} E_{u,v} \right) \left(\prod_{u \in I^c} N_u \right)$$

where $G = (V, E)$ is a simple undirected graph, $I, O \subseteq V$ are respectively the input and output qubits, and $\mathbf{x}, \mathbf{z} : O^c \to 2^V$ are two extensive maps, i.e. the relation $\prec$ defined as the transitive closure of $\{(u, v) : v \in \mathbf{x}(u) \cup \mathbf{z}(u)\}$ is a strict partial order. Notice that $O^c := V \setminus O$ and $X_{\mathbf{x}(u)}^{s_u} := \prod_{v \in \mathbf{x}(u)} X_v^{s_u}$. Moreover the product $\prod_{(u,v) \in G}$ means that the indices are the edges of the G, in particular each edge is taken once.

The septuple $(G, I, O, \lambda, \alpha, \mathbf{x}, \mathbf{z})$ is a graph-based representation which captures entirely the semantics of the corresponding pattern. We simply call an MBQC such a septuple.

2.4 Semantics and Determinism

An MBQC $(G, I, O, \lambda, \alpha, \mathbf{x}, \mathbf{z})$ has a fundamentally probabilistic evolution with potentially $2^{|O^c|}$ possible branches as the computation consists of $|O^c|$ measurements. For any $s \in \{0,1\}^{|O^c|}$, let $A_s : \mathbb{C}^{\{0,1\}^I} \to \mathbb{C}^{\{0,1\}^O}$ be

$$A_s(|\phi\rangle) = \left(\prod_{u \in O^c}^{\prec} Z_{\mathbf{z}(u)}^{s_u} X_{\mathbf{x}(u)}^{s_u} \langle \phi_{s_u}^{\lambda_u, \alpha_u} |_u \right) \left(\prod_{(u,v) \in G} \Lambda Z_{u,v} \right) \left(|\phi\rangle \otimes \frac{\sum_{x \in \{0,1\}^{I^c}} |x\rangle}{\sqrt{2^{|I^c|}}} \right)$$

where $|\phi_{s_u}^{\lambda_u, \alpha_u}\rangle$ is the eigenvalue of $\mathcal{O}^{\lambda_u, \alpha_u}$ associated with the eigenvalue $(-1)^{s_u}$.

Given an initial state $|\phi\rangle \in \mathbb{C}^{\{0,1\}^I}$ and $s \in \{0,1\}^{O^c}$, the outcome of the computation is the state $A_s |\Psi\rangle$ (up to a normalisation factor), with probability $\langle \phi | A_s^\dagger A_s |\phi\rangle$. In other words the MBQC implements the cptp-map[1] $\rho \mapsto \sum_{s \in \{0,1\}^{O^c}} A_s \rho A_s^\dagger$.

Among all the possible measurement-based quantum computations, those which are deterministic are of peculiar importance. In particular, deterministic MBQC are those which are used to simulate quantum circuits (cornerstone of the proof that MBQC is a universal model of quantum computation), or to implement a quantum algorithm. An MBQC $(G, I, O, \lambda, \alpha, \mathbf{x}, \mathbf{z})$ is **deterministic** if the output of the computation does not depend on the classical outcomes obtained during the computation: for any input state $|\phi\rangle \in \mathbb{C}^{\{0,1\}^I}$ and branches $s, s' \in \{0,1\}^{O^c}$, $A_s |\phi\rangle$ and $A_{s'} |\phi\rangle$ are proportional.

Notice that the semantics of a deterministic MBQC $(G, I, O, \lambda, \alpha, \mathbf{x}, \mathbf{z})$ is entirely defined by a single branch, e.g. the branch $A_{0^{|O^c|}}$. Moreover, this particular branch $A_{0^{|O^c|}}$ is correction-free by construction (indeed all corrections are controlled by a signal, which is 0 in this particular branch). As a consequence, intuitively, when the evolution is deterministic, the corrections are only used to make the overall evolution deterministic but have no effect on the actual semantics of the evolution. Thus the correction can be abstracted away leading to the notion of **abstract MBQC** $(G, I, O, \lambda, \alpha)$. There is however a caveat when the branch $A_{0^{|O^c|}}$ is 0: for instance $M_1^{X,\pi} N_1 N_2$ and $Z_2^{s_1} M_1^{X,\pi} N_1 N_2$ are both deterministic[2] and share the same abstract open graph, however they do not have the same semantics: the outcome of the former pattern is $\frac{|0\rangle + |1\rangle}{\sqrt{2}}$, whereas the outcome of the latter is $\frac{|0\rangle - |1\rangle}{\sqrt{2}}$.

To avoid these pathological cases and guarantee that the corrections can be abstracted away, a stronger notion of determinism has been introduced in [4]: an MBQC is **strongly deterministic** when all the branches are not only proportional but equal up to a global phase. The strongness assumption guarantees that for any input state $|\phi\rangle$, $A_{0^{|O^c|}} |\phi\rangle$ is non zero, and thus guarantees that the

---

[1] A completely positive trace-preserving map describes the evolution of a quantum system which state is represented by a density matrix. See for instance [15] for details.

[2] In both cases the unique measurement consists of measuring a qubit in state $|+\rangle$ according to the observable $-X$ which produces the signal $s_1 = 1$ with probability 1.

overall evolution is entirely described by the correction-free branch, or in other words by the knowledge of the abstract MBQC $(G, I, O, \lambda, \alpha)$.

Whereas deterministic MBQC are not necessarily invertible (e.g. $M_1^{(X,0)} N_2$ which maps any state $|\phi\rangle$ to the state $|+\rangle$), strongly deterministic MBQC correspond to the invertible deterministic quantum evolutions: they implement isometries ($\exists U : \mathbb{C}^{\{0,1\}^I} \to \mathbb{C}^{\{0,1\}^O}$ s.t. $U^\dagger U = I$ and $\forall s \in \{0,1\}^{|O^c|}$, $\exists \theta$ s.t. $A_s = 2^{-|O^c|} e^{i\theta} U$).

We consider a variant of strong determinism which is robust to variation of the angles of measurements (which is a continuous parameter, so a priori subject to small variations in an experimental setting for instance), and to partial computation i.e., roughly speaking if one aborts the computation, the partial outcome does not depend on the branch of the computation.

Definition 1 (Robust Determinism). $(G, I, O, \lambda, \alpha, \mathbf{x}, \mathbf{z})$ *is robustly deterministic if for any lowerset $S \subseteq O^c$ and for any $\beta : S \to [0, 2\pi)$, $(G, I, O \cup S^c, \lambda|_S, \beta, \mathbf{x}|_S, \mathbf{z}|_S)$ is strongly deterministic, where S is a lowerset for the partial order induced by $\mathbf{x}$ and $\mathbf{z}$: $\forall v \in S, \forall u \in O^c, v \in \mathbf{x}(u) \cup \mathbf{z}(u) \Rightarrow u \in S$.*

The notion of *robust determinism* we introduce is actually a short cut for *uniformly strong and stepwise determinism* which has been already extensively studied in the context of measurement-based quantum computing [4, 8, 13].

A central question in measurement-based quantum computation is to decide whether an abstract MBQC can be implemented deterministically: given $(G, I, O, \lambda, \alpha)$, does there exist correction strategies $\mathbf{x}, \mathbf{z}$ such that $(G, I, O, \lambda, \alpha, \mathbf{x}, \mathbf{z})$ is (robustly) deterministic? This question is related to the power of postselection in quantum computing: allowing postselection one can select the correction-free branch and thus implement any abstract MBQC $(G, I, O, \lambda, \alpha)$. Post-selection is a priori a non physical evolution, but in the presence of a correction strategy, postselection can be simulated using measurements and corrections.

The robustness assumption allows one to abstract away the angles and focus on the so-called **open graph** (G, I, O, λ) i.e. essentially the initial entanglement. For which initial entanglement – or in other words for which resource state – a deterministic evolution can be performed? This is a fundamental question about the structures and the computational power of entanglement.

Several graphical conditions for determinism have been introduced: causal flow, Generalized flow (Gflow) and Pauli Flow [4, 6, 8]. These are graphical conditions on open graphs which are sufficient to guarantee the existence of a robust deterministic evolution. Gflow has been proved to be a necessary condition for Pauli-free MBQC (i.e. for any open graph (G, I, O, λ) s.t. $\forall u \in O^c, |\lambda_u| = 2$). The necessity of Pauli flow was an open question[3]. In this paper we show that

---

[3] In [4], an example of deterministic MBQC with no Pauli flow is given. This is however not a counter example to the necessity of the Pauli flow as the example is not robustly deterministic. More precisely not all the branches of computation occur with the same probability: with the notation of Fig. 8 in [4] if measurements of qubits 4, 6, 8 produce the outcome 0, then the measurement of qubit 10 produces the outcome 0 with probability 1.

Pauli flow fails to be necessary in general, but is however necessary for real MBQC, i.e. when $\forall u \in O^c$, $\lambda_u \subseteq \{X, Z\}$. In the next section, we review the graphical sufficient conditions for determinism.

2.5   Graphical Conditions for Determinism

Several flow conditions for determinism have been introduced to guarantee robust determinism. Causal flow has been the first sufficient condition for determinism [6]. This condition has been extended to Generalized flow (Gflow) and Pauli flow [4]. Our first contribution is to provide a simpler description of the Pauli flow, equivalent to the original one.

Property 1. (G, I, O, λ) has a Pauli flow iff there exist a strict partial order $<$ over O^c and $p : O^c \to 2^{I^c}$ s.t. $\forall u \in O^c$,

$$(c_X) \quad X \in \lambda_u \Rightarrow u \in \mathsf{Odd}(p(u)) \setminus \left(\bigcup_{\substack{v \geq u \\ v \notin O \cup \{u\}}} \mathsf{Odd}(p(v)) \right)$$

$$(c_Y) \quad Y \in \lambda_u \Rightarrow u \in \mathsf{Odd}[p(u)] \setminus \left(\bigcup_{\substack{v \geq u \\ v \notin O \cup \{u\}}} \mathsf{Odd}[p(v)] \right)$$

$$(c_Z) \quad Z \in \lambda_u \Rightarrow u \in p(u) \setminus \left(\bigcup_{\substack{v \geq u \\ v \notin O \cup \{u\}}} p(v) \right)$$

where $v \geq u$ iff $\neg(v < u)$

Remark 1. Notice that the existence of a Pauli flow forces the input qubits to be measured in the $\{X, Y\}$-plane: If (G, I, O, λ) has a Pauli flow then for any $u \in I \cap O^c$, $u \notin p(u)$ since $p(u) \subseteq I^c$. It implies, according to condition (c_Z), that $Z \notin \lambda_u$.

Gflow and Causal flows are special instances of Pauli flow: A Pauli flow is a Gflow when all measurements are performed in a plane (i.e. $\forall u, |\lambda_u| = 2$); a Causal flow [6] is nothing but a Gflow $(p, <)$ such that $\forall u, |p(u)| = 1$. GFlow has been proved to be a necessary and sufficient condition for robust determinism:

Theorem 1 [4]. *Given an abstract MBQC $(G, I, O, \lambda, \alpha)$ such that $\forall u \in O^c, |\lambda_u| = 2$, (G, I, O, λ) has a GFlow $(p, <)$ if and only if there exists $\mathbf{x}, \mathbf{z}$ extensive with respect to $<$ s.t. $(G, I, O, \lambda, \alpha, \mathbf{x}, \mathbf{z})$ is robustly deterministic.*

Pauli flow is the most general known sufficient condition for determinism for robust determinism:

Theorem 2 [4]. *If (G, I, O, λ) has a Pauli flow $(p, <)$, then for any $\alpha : O^c \to [0, 2\pi)$, $(G, I, O, \lambda, \alpha, \mathbf{x}, \mathbf{z})$ is robustly deterministic where $\forall u \in O^c$,*

$$\mathbf{x}(u) = \{v \in p(u) \mid u < v\}$$
$$\mathbf{z}(u) = \{v \in \mathsf{Odd}(p(u)) \mid u < v\}$$

Is there a converse? This is the purpose of next section.

3   Characterising Robust Determinism

In this section, we show the main result of the paper: Pauli flow is necessary for robust determinism in the real case, i.e. when all the measurements are in the $\{X, Z\}$-plane ($\forall u, \lambda_u \subseteq \{X, Z\}$).

We investigate in the subsequent sections the consequences of this result for real MBQC which is a universal model of quantum computation with several crucial applications.

A **real open graph** (G, I, O, λ) is an open graph such that $\forall u \in O^c, \lambda_u \subseteq \{X, Z\}$. We define similarly **real abstract MBQC** and **real MBQC**. Pauli flow conditions on real open graphs can be simplified as follows:

Property 2. A real open graph (G, I, O, λ) has a Pauli flow iff there exist a strict partial order $<$ over O^c and $p : O^c \to 2^{I^c}$ s.t. $\forall u \in O^c$,

$$(i) \quad X \in \lambda_u \Rightarrow u \in \mathsf{Odd}(p(u)) \setminus \left(\bigcup_{\substack{v \geq u \\ v \notin O \cup \{u\}}} \mathsf{Odd}(p(v)) \right)$$

$$(ii) \quad Z \in \lambda_u \Rightarrow u \in p(u) \setminus \left(\bigcup_{\substack{v \geq u \\ v \notin O \cup \{u\}}} p(v) \right)$$

Theorem 3. *Given a real abstract MBQC $(G, I, O, \lambda, \alpha)$, (G, I, O, λ) has a Pauli flow $(p, \prec)$ if and only if there exist $\mathbf{x}, \mathbf{z}$ extensive with respect to $\prec$ s.t. $(G, I, O, \lambda, \alpha, \mathbf{x}, \mathbf{z})$ is robustly deterministic.*

The proof[4] is fundamentally different from the proof that Gflow is necessary for Pauli-free robust determinism (Theorem 1 in [4]). Roughly speaking, the proof that Pauli flow is necessary goes as follows: first we fix the inputs to be either $|0\rangle$ or $|+\rangle$ and all the measurements to be Pauli measurements (i.e. if $\lambda_u = \{X, Z\}$

----

[4] The proof of Theorem 3 is available in the pre-print version of the present article [16].

we fix the measurement of u to be either X or Z). For each of these choices the computation can be described in the so-called stabilizer formalism which allows one to point out the constraints the corrections should satisfy for each of these particular choices of inputs and measurements. Then, as the corrections of a robust deterministic MBQC should not depend on the choice of the inputs and the angles of measurements, one can combine the constraints the corrections should satisfy and show that they coincide with the Pauli flow conditions.

Remark 2. We consider in this paper a notion of real MBQC which corresponds to a constraint on the measurements ($\forall u \in O^c, \lambda_u \in \{X, Z\}$), it can also be understood as an additional constraint on the inputs: the input of the computation is in $\mathbb{R}^I$ instead of $\mathbb{C}^I$. This distinction might be important, for instance the pattern $M_1^Y N_2$ is strongly deterministic on real inputs but not on arbitrary complex inputs. It turns out that the proof of Theorem 3 only consider real inputs, and as a consequence is valid in both cases (i.e. when both inputs and measurements are real; or when inputs are complex and measurements are in the $\{X, Z\}$-plane).

Pauli flow is necessary for real robust determinism. This property is specific to real measurements: Pauli flow is not necessary in general even when the measurements are restricted to one of the other two planes of measurements. In the following $\{X, Y\}$-MBQC (resp. $\{Y, Z\}$-MBQC) refers to MBQC where all measurements are performed in the $\{X, Y\}$-plane (resp. $\{Y, Z\}$-plane).

Property 3. There exists robustly deterministic $\{X, Y\}$-MBQC (resp. $\{Y, Z\}$-MBQC) $(G, I, O, \lambda, \alpha, \mathbf{x}, \mathbf{z})$ such that (G, I, O, λ) has no Pauli flow $(p, \prec)$ where $\mathbf{x}$ and $\mathbf{z}$ are extensive with respect to $\prec$.

Proof. We consider the pattern $\mathcal{P} = Z_3^{s_2} M_2^{X,0} X_2^{s_1} M_1^{\{X,Y\},\alpha} E_{1,2} E_{1,3} N_1 N_2 N_3$ which is an implementation of the $\{X, Y\}$-MBQC given in Fig. 1 (the other example is similar). Notice that the correction $X_2^{s_1}$ is useless as qubit 2 is going to be measured according to M^X. Thus $\mathcal{P}$ has the same semantics as $\mathcal{P}' = Z_3^{s_2} M_2^{X,0} M_1^{\{X,Y\},\alpha} E_{1,2} E_{1,3} N_1 N_2 N_3$. Notice in $\mathcal{P}'$ that the two measurements commute since there is no dependency between them, leading to the pattern $\mathcal{P}'' = M_1^{\{X,Y\},\alpha} Z_3^{s_2} M_2^{X,0} E_{1,2} E_{1,3} N_1 N_2 N_3$. It is easy to check that $\mathcal{P}''$ has a Pauli flow so is robustly deterministic. All but the stepwise property are transported by the transformations from $\mathcal{P}''$ to $\mathcal{P}$. Notice that $\mathcal{P}'$ is not stepwise deterministic as $M_1^{\{X,Y\},\alpha} E_{1,2} E_{1,3} N_1 N_2 N_3$ is not deterministic. However, $\mathcal{P}$ enjoys the stepwise property since $X_2^{s_1} M_1^{\{X,Y\},\alpha} E_{1,2} E_{1,3} N_1 N_2 N_3$ has a Pauli flow so is robustly deterministic. Finally, it is easy to show that the open graph has no Pauli flow $(p, \prec)$ such that $1 \prec 2$, which is necessary to guarantee that $\mathbf{x}$ is extensive with respect to $\prec$.

Remark 3. This is the last step of the proof of Theorem 3 which fails with the examples of Fig. 1. For instance in the $\{X, Y\}$-MBQC example, in both cases of Pauli measurements of qubit 1 (according to X or according to Y), a Pauli

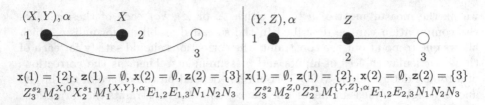

Fig. 1. Robustly deterministic $\{X,Y\}$-MBQC and $\{Y,Z\}$-MBQC with no compatible Pauli flow. The two MBQC are described by means of there abstract MBQC (G, I, O, α) and the corrective maps $\mathbf{x}$ and $\mathbf{z}$. In both cases there is no input and the output is located on qubit 3. A description using the measurement-pattern formalism is also provided (commands should be read from right to left). Notice that the only order that makes $\mathbf{x}$ and $\mathbf{z}$ extensive has to verify $1 \prec 2$, and there is no Pauli flow for this order.

flow exists, sharing the same partial order $1 \prec 2$. However the two Pauli flows are distinct and none of them is a Pauli flow when qubit 1 is measured in the $\{X, Y\}$-plane.

Remark 4. The examples given in Fig. 1 do have a Pauli flow but with a partial order not compatible with the order of measurements. It is important that the orders of the flow and the measurements coincide for guaranteeing that the depth of the flow (longest increasing sequence) corresponds to the depth of the MBQC. Because of the logarithmic separation between the quantum circuit model and MBQC in terms of depth (e.g. PARITY can be computed with a constant quantum depth MBQC but requires a logarithmic depth quantum circuit) [5], it is also important that Pauli flow characterises not only the ability to perform a robust deterministic evolution, but characterizes also the depth of such evolution. There exists an efficient polynomial time which, given an open graph, compute a Gflow of optimal depth (when it exists) [13], the existence of such an algorithm in the Pauli case is an open question.

4   Applications: Computational Power of Real Bipartite MBQC

In this section we focus on the real MBQC which underlying graph are bipartite (real bipartite MBQC for short). Bipartite graphs (or equivalently 2-colorable graphs) play an important role in MBQC, the square grid is universal for quantum computing: any quantum circuit can be simulated by an MBQC whose underlying graph is a square grid. The brickwork graph [2] is bipartite and universal for $\{X, Y\}$-MBQC. Regarding real MBQC, the (non bipartite) triangular grid is universal for real MBQC [14] but there is no known universal family of bipartite graphs. We show in this section that there is no universal family of bipartite graphs for real MBQC, by showing that any real bipartite MBQC can be done in constant depth.

4.1   Real Bipartite MBQC in Constant Depth

In this section we show that real bipartite MBQC can always be parallelized:

Theorem 4. *All measurements of a robustly deterministic real bipartite MBQC can be performed in parallel.*

The rest of the section is dedicated to the proof of Theorem 4. According to Theorem 3, a real MBQC is robustly deterministic if and only if the underlying open graph has a Pauli flow. To prove that all the measurements can be performed in parallel in the bipartite case we point out the existence of a particular correction strategy which ensures that each measurement is corrected using output qubits only.

Lemma 1. *Given a bipartite graph G, $I, O \subseteq V(G)$ and $\lambda : O^c \to \{\{X\},$ $\{Z\}, \{X, Z\}\}$, if (G, I, O, λ) has a Pauli flow then there exists $p : O^c \to 2^{I^c}$ s.t.:*

$$\mathrm{Odd}(p(u)) \setminus (O \cup \lambda^{-1}(\{Z\})) = \{u\} \setminus \lambda^{-1}(\{Z\})$$
$$p(u) \setminus (O \cup \lambda^{-1}(\{X\})) = \{u\} \setminus \lambda^{-1}(\{X\})$$

This particular correction strategy corresponds to a king of *super-normal form*. Indeed it is known that Gflow can be put into the so called *Z-* or *X-normal form* but not both at the same time (see [10] for details). Lemma 1 shows, roughly speaking, that the Pauli flow in the real bipartite case can be put in both normal forms at the same time.

Proof (Proof of Theorem 4). Given a robustly deterministic real bipartite MBQC $(G, I, O, \lambda, \alpha, \mathbf{x}, \mathbf{z})$, according to Theorem 3, (G, I, O, λ) has a Pauli flow, so according to Lemma 1 there exists p s.t. $\mathrm{Odd}(p(u)) \setminus (O \cup \lambda^{-1}(\{Z\})) = \{u\} \setminus \lambda^{-1}(\{Z\})$ and $p(u) \setminus (O \cup \lambda^{-1}(\{X\})) = \{u\} \setminus \lambda^{-1}(\{X\})$. Notice that $(p, \emptyset)$ is a Pauli flow for (G, I, O, λ), thus according to Theorem 2, $(G, I, O, \lambda, \alpha, \mathbf{x}', \mathbf{z}')$ is robustly deterministic where $\mathbf{x}' = u \mapsto p(u) \setminus (\lambda^{-1}(\{X\}) \cup \{u\})$ and $\mathbf{z}' = u \mapsto \mathrm{Odd}(p(u)) \setminus (\lambda^{-1}(\{Z\}) \cup \{u\})$. Both $(G, I, O, \lambda, \alpha, \mathbf{x}, \mathbf{z})$ and $(G, I, O, \lambda, \alpha, \mathbf{x}', \mathbf{z}')$ implement the same computation, and $\forall u \in O^c$ $\mathbf{x}'(u) \subseteq O$ and $\mathbf{z}'(u) \subseteq O$ which implies that all measurements of the latter MBQC can be performed in parallel.

4.2   Interactive Proofs

The starting point of our work has been a sentence of McKague in [12]. In the future work section, McKague wonders how his work could be used to build an interactive prover with only two provers. The problem that McKaque wants to solve is the following. We imagine a classical verifier, which is a computer with classical resources, who wants to perform a computation using some non-communicating quantum provers. The quantum provers are computers with quantum resources. In fact, the classical verifier wants to achieve his computation using the quantum power of quantum provers. In this model, the hard point to breakthrough is that we want the verifier to detect cheating behavior of some provers. The model should guarantee the verifier that the result of the computation made by the provers is correct: if a prover has cheated and not computed what he was asked, the verifier should be able to detect it. We specify that the provers, in this model, cannot communicate one with the others: each prover can try to cheat on his own but he does not have the power to do it by exchanging information with the others. McKague, in [12], proves that it is possible to imagine a protocol in which the computation can be performed by the classical verifier using a polynomial number of quantum provers. To achieve this goal, McKague uses two main tools, one of them being Measurement Based Quantum Computation in the (X, Z) plane. Mhalla and Perdrix, in [14], prove that there exists a grid that enables to perform a universal computing in the (X, Z) plane. Usually, the (X, Y) plane, first known to allow universal computation is preferred. In his work, McKague needs the (X, Z) plane: to be able to detect cheating behavior, McKague needs to compute in the reals. The conjugation operation that can be performed in other planes is a problem to detect some cheatings.

In his future work section, McKague argues that most his work could be used to improve his result to the use of only two provers. The main difficulty he points out is to build a bipartite graph to compute with. His self-testing skill, which is the second important tool of his work, can be applied only if the graph does not have any odd cycle. Therefore, the question we wanted to answer was whether one could build a universal bipartite grid for the (X, Z)-plane. Our Theorem 4 shows that in the real case a bipartite graph is not very powerful to compute: it is far from being universal. Therefore, at best, new skills will be needed to adapt McKague's method to interactive proofs with two provers.

5   Conclusion and Future Work

In this paper, we made substantial steps in understanding MBQC world. The first important one is this equivalence between being robustly deterministic and having a Pauli flow for a real-MBQC. Since it does not hold for $\{X, Y\}$- and $\{Y, Z\}$-planes, a natural question is how one can modify the Pauli flow definition to obtain a characterisation of determinism in these cases? A bi-product of the characterisation of robust determinism for real MBQC is the low computational power of real bipartite MBQC. It would be interesting to compare the computational power of real bipartite MBQC and of commuting quantum circuits. There are some good reasons to think that the power of real bipartite MBQC is exactly the same as those commuting quantum circuits. Taking a global view of the MBQC domain, some advances we make in this paper, and a good direction for further research should be to better understand the specificity of each plane in the power of the MBQC model and how the ability to perform a deterministic computation is linked to this power. Finally, another open question is the existence of an efficient algorithm for deciding whether a given open graph has a Pauli flow, and which produces a Pauli flow of optimal depth when it exists. Such an algorithm exists for Gflow [13].

References

1. Bernstein, E., Vazirani, U.: Quantum complexity theory. SIAM J. Comput. **26**, 1411–1478 (1997)
2. Broadbent, A., Fitzsimons, J., Kashefi, E.: Universal blind quantum computation. In: 50th Annual IEEE Symposium on Foundations of Computer Science, FOCS 2009 (2009). http://www.citebase.org/abstract?id=oai:arXiv.org:0807.4154
3. Broadbent, A., Kashefi, E.: Parallelizing quantum circuits. Theor. Comput. Sci. **410**(26), 2489–2510 (2009)
4. Browne, D.E., Kashefi, E., Mhalla, M., Perdrix, S.: Generalized flow and determinism in measurement-based quantum computation. New J. Phys. (NJP) **9**(8) (2007). http://iopscience.iop.org/1367-2630/9/8/250/fulltext/
5. Browne, D., Kashefi, E., Perdrix, S.: Computational depth complexity of measurement-based quantum computation. In: van Dam, W., Kendon, V.M., Severini, S. (eds.) TQC 2010. LNCS, vol. 6519, pp. 35–46. Springer, Heidelberg (2011). doi:10.1007/978-3-642-18073-6_4
6. Danos, V., Kashefi, E.: Determinism in the one-way model. Phys. Rev. A **74**(052310) (2006)
7. Danos, V., Kashefi, E., Panangaden, P.: The measurement calculus. J. ACM **54**(2) (2007)
8. Danos, V., Kashefi, E., Panangaden, P., Perdrix, S.: Extended Measurement Calculus. Cambridge University Press, Cambridge (2010)
9. Delfosse, N., Guerin, P.A., Bian, J., Raussendorf, R.: Wigner function negativity and contextuality in quantum computation on rebits. Phys. Rev. X **5**(2), 021003 (2015)
10. Hamrit, N., Perdrix, S.: Reversibility in extended measurement-based quantum computation. In: Krivine, J., Stefani, J.-B. (eds.) RC 2015. LNCS, vol. 9138, pp. 129–138. Springer, Cham (2015). doi:10.1007/978-3-319-20860-2_8
11. Hein, M., Eisert, J., Briegel, H.J.: Multi-party entanglement in graph states. Phys. Rev. A **69**, 062311 (2004). doi:10.1103/PhysRevA.69.062311
12. McKague, M.: Interactive proofs for BQP via self-tested graph states. Theory Comput. **12**(3), 1–42 (2016)
13. Mhalla, M., Perdrix, S.: Finding optimal flows efficiently. In: Aceto, L., Damgård, I., Goldberg, L.A., Halldórsson, M.M., Ingólfsdóttir, A., Walukiewicz, I. (eds.) ICALP 2008. LNCS, vol. 5125, pp. 857–868. Springer, Heidelberg (2008). doi:10.1007/978-3-540-70575-8_70
14. Mhalla, M., Perdrix, S.: Graph states, pivot minor, and universality of (X, Z)-measurements. Int. J. Unconv. Comput. **9**(1–2), 153–171 (2013)
15. Nielsen, M.A., Chuang, I.L.: Quantum Computation and Quantum Information. Cambridge University Press, New York (2000)
16. Perdrix, S., Sanselme, L.: Determinism and computational power of real measurement-based quantum computation (2016). arXiv preprint arXiv:1610.02824
17. Prevedel, R., Walther, P., Tiefenbacher, F., Bohi, P., Kaltenbaek, R., Jennewein, T., Zeilinger, A.: High-speed linear optics quantum computing using active feedforward. Nature **445**(7123), 65–69 (2007). doi:10.1038/nature05346
18. Raussendorf, R.: Contextuality in measurement-based quantum computation. Phys. Rev. A **88**(2), 022322 (2013)
19. Raussendorf, R., Briegel, H.J.: A one-way quantum computer. Phys. Rev. Lett. **86**, 5188–5191 (2001)

20. Raussendorf, R., Browne, D.E., Briegel, H.J.: Measurement-based quantum computation with cluster states. Phys. Rev. A **68**, 022312 (2003). http://arxiv.org/abs/quant-ph/0301052
21. Raussendorf, R., Harrington, J., Goyal, K.: A fault-tolerant one-way quantum computer. Ann. Phys. **321**(9), 2242–2270 (2006)
22. Walther, P., Resch, K.J., Rudolph, T., Schenck, E., Weinfurter, H., Vedral, V., Aspelmeyer, M., Zeilinger, A.: Experimental one-way quantum computing. Nature **434**(7030), 169–176 (2005). doi:10.1038/nature03347

Busy Beaver Scores and Alphabet Size

Holger Petersen[✉]

Reinsburgstr. 75, 70197 Stuttgart, Germany
dr.holger.petersen@googlemail.com

Abstract. We investigate the Busy Beaver Game introduced by Rado [13] generalized to non-binary alphabets. Harland [5] conjectured that activity (number of steps) and productivity (number of non-blank symbols) of candidate machines grow as the alphabet size increases. We prove this conjecture for any alphabet size under the condition that the number of states is sufficiently large. For the measure activity we show that increasing the alphabet size from two to three allows an increase. By a classical construction it is even possible to obtain a two-state machine increasing activity and productivity of any machine if we allow an alphabet size depending on the number of states of the original machine. We also show that an increase of the alphabet by a factor of three admits an increase of activity.

1 Introduction

The Busy Beaver Game, as originally defined by Rado [13], is to determine for a given number n of states of deterministic Turing machines over the alphabet $\{0,1\}$ (0 is the blank symbol) the maximum number of ones produced on an initially blank two-way infinite tape. In each step such a machine reads a tape symbol and—depending on the current state–writes a symbol, shifts its head one square to the left or to the right, and enters a new state. There is a single halt state (which is traditionally not counted), and on the transition to this state the machine also writes a symbol. What we have just described is sometimes called the *quintuple variant* of Turing machines in view of the five pieces of information that define a transition. In contrast, the *quadruple variant* can either move the tape head or write a symbol but not both.

Rado introduced the function $\Sigma(n)$ as the maximum number of ones produced by machines with n states. The function $S(n)$ denotes the maximum number of steps performed (shift-number) by such machines. He proved that these functions are non-computable and even grow faster than any computable function. Rado also pointed out that these are very simple examples of non-computable functions and that no (explicit) enumeration of computable functions is used in their definition.

The functions are of metamathematical interest as well, since open problems like Goldbach's conjecture, which can be refuted in a constructive way by a counterexample, would be settled if $S(n)$ would be computable for an n large enough to determine a counterexample by running a Turing machine [3,4]. Recently

R. Klasing and M. Zeitoun (Eds.): FCT 2017, LNCS 10472, pp. 409–417, 2017.
DOI: 10.1007/978-3-662-55751-8_32

explicit bounds on such an n have been determined for Goldbach's conjecture and the Riemann hypothesis along with a Turing machine that cannot be proved to run forever in ZFC [15].

Here we consider the generalization of the Busy Beaver Game to alphabets with more than two symbols. As in [12] we denote by $\Sigma(n, m)$ the maximum number of non-blanks produced by any halting deterministic Turing machine with n states and m symbols (called *productivity*) working on an initially blank two-way infinite tape. Similarly, we denote by $S(n, m)$ the maximum number of steps performed (called *activity*). Thus the functions defined by Rado are now special cases with $m = 2$. For a specific Turing machine M we denote the two measures by productivity(M) and activity(M).

A Turing machine M participating in the generalized Busy Beaver competition can be represented by a table of the form

input symbol

current state	0	1	$\cdots$	$m - 1$
1	$w_1^0 \delta_1^0 s_1^0$	$w_1^1 \delta_1^1 s_1^1$	$\cdots$	$w_1^{m-1} \delta_1^{m-1} s_1^{m-1}$
2	$w_2^0 \delta_2^0 s_2^0$	$w_2^1 \delta_2^1 s_2^1$	$\cdots$	$w_2^{m-1} \delta_2^{m-1} s_2^{m-1}$
$\vdots$	$\vdots$	$\vdots$	$\cdots$	$\vdots$
n	$w_n^0 \delta_n^0 s_n^0$	$w_n^1 \delta_n^1 s_n^1$	$\cdots$	$w_n^{m-1} \delta_n^{m-1} s_n^{m-1}$

where $w_k^i \in \{0, 1, \ldots, m - 1\}$ indicates the symbol written by M after reading i in state k, $\delta_k^i \in \{L, R\}$ is the direction of the head movement, and $s_k^i \in \{1, \ldots, n+1\}$ is the new state M enters. State 1 is the initial state and state $n+1$ is the halting state.

As early as 1966, the lower bounds $\Sigma(3, 3) \geq 12$ and $S(3, 3) \geq 57$ were reported in [7] for a non-binary alphabet[1]. Over the following decades, investigations concentrated on computing $\Sigma(4, 2)$ and on improving lower bounds for larger numbers of states in the classical setting of a binary tape alphabet. The progress in the chase of Busy Beavers is reflected in the table below.

With the exception of lower bounds due to Brady ($S(2, 3) \geq 38$, $S(2, 4) \geq 7,195$ [3]), the search for high scoring machines with more than two symbols did not continue before 2004. As outlined in the survey [12], Michel and Brady improved the lower bounds on $\Sigma(3, 3)$ and $S(3, 3)$ during that year. Between 2005 and 2008 many new machines for non-binary alphabets were found mainly by two teams: Grégory Lafitte and Christophe Papazian, and Terry and Shawn Ligocki (father and son). Lafitte and Papazian could also establish that $\Sigma(2, 3) = 9$ and $S(2, 3) = 38$, confirming Michel's conjecture from [11] that Brady's lower bound dating back almost two decades was tight.

---

[1] The origin of these bounds communicated to Korfhage by C. Y. Lee of Bell Telephone Lab. is not clear. In [7] Lee, Tibor Rado, Shen Lin, Patrick Fischer, Milton Green, and David Jefferson are mentioned in connection with these lower bounds and other early results.

n	$\Sigma(n,2)$	$S(n,2)$	References
1	1	1	Rado [13]
2	4	6	Rado [13]
3	6	21	Lin, Rado [8]
4	13	107	Brady [2]
5	$\geq$4098	$\geq$47,176,870	Marxen, Buntrock [9]
6	$\geq 3.514 \cdot 10^{18,267}$	$\geq 7.412 \cdot 10^{36,534}$	Kropitz, see [10]

Given the known values and lower bounds for non-binary alphabets, it is natural to expect that $\Sigma(n,m)$ and $S(n,m)$ are increasing in both parameters (it is easily shown that they are increasing in their first parameter, see Lemma 1 below). An even stronger conjecture was stated by Harland [5].

Before presenting our results we cite Harland's conjecture:

Conjecture 28 (in [5]). *Let M be a k-halting Turing machine with n states and m symbols for some $k \geq 1$ with finite activity. Then there is a k-halting n-state $(m+1)$-symbol Turing machine M' with finite activity such that*

$$\text{activity}(M') > \text{activity}(M) \text{ and productivity}(M') > \text{productivity}(M).$$

Here k-halting means that there are k transitions to the halting state.

For $n = 1$, an n-state Turing machine has to halt after the first step on a blank in order to have finite activity. As this holds independently of the size of the alphabet, no increase of activity and productivity is possible. We therefore exclude the trivial case $n = 1$.

Notice that the conjecture is stronger than just stating that Σ and S are increasing as m grows and n is kept fixed (which it implies by taking highest scoring machines as M). The conjecture considers for any *specific* machine *both* activity and productivity at the same time. A machine maximizing one of the measures may in fact not maximize the other, as is the case for $n = 3$ where machines with activity 21 produce at most $5 < \Sigma(3)$ ones.

In addition, Harland's conjecture imposes a restriction on the structure of a machine increasing these measures, namely that the number of halting transitions is kept constant for machine M'.

Highest scores for small machines still provide evidence in support of the conjecture. We have

$$\Sigma(2,2) = 4 < \Sigma(2,3) = 9 < 2,050 \leq \Sigma(2,4),$$

$$S(2,2) = 6 < S(2,3) = 38 < 3,932,964 \leq S(2,4),$$

$$\Sigma(3,2) = 6 < 374,676,383 \leq \Sigma(3,3),$$

and

$$S(3,2) = 21 < 119,112,334,170,342,540 \leq S(3,3)$$

(results of Rado, Lin, Lafitte, Papazian, T. Ligocki and S. Ligocki, see [12] for references).

2   Results

It is well-known that activity and productivity grow with the number of states, see the figure on p. 77 of [6] or Proposition 27 of [5].

Lemma 1. *Let M be a Turing machine with n states and m symbols with finite activity. Then there is an $(n + 1)$-state m-symbol Turing machine M' with finite activity such that* activity$(M') >$ activity(M) *and* productivity$(M') >$ productivity(M).

The lemma can be proved for any alphabet by redirecting the (unique) halting transition to the new state and having it skip symbols different from the blank while moving the head in one direction. The first blank encountered is replaced with a non-blank and then the machine halts.

An encoding scheme originally developed by Ben-Amram and Petersen [1] and called *introspective computing* by Luke Schaeffer [15] will be essential in proving Harland's conjecture for sufficiently large numbers of states.

Theorem 1. *For every $m \geq 2$ and $k \geq 1$ there is an $N_{m,k}$ such that for every k-halting Turing machine M with $n \geq N_{m,k}$ states and m symbols with finite activity there is an n-state, $(m + 1)$-symbol k-halting Turing machine M' with finite activity such that* activity$(M') >$ activity(M) *and* productivity$(M') >$ productivity(M).

Proof. We first outline the idea of the construction. While for the proof of Lemma 1 a microscopic approach suffices, we will apply introspective computing here on a more general level. A fraction of the available n states is used for encoding the m-symbol machine, where the larger alphabet compensates the missing states. Thus we can efficiently convert from alphabet symbols to states. This encoding is embedded into a simulator M' having the properties required by the theorem.

Let M be a Turing machine as described in the theorem with $n > m$ states. We first notice that w.l.o.g. all n states appear in the unique halting computation of M on the blank tape. For otherwise we omit an unused state s (reducing the number of halting transitions by at most m) and redirect all transitions with target s to some remaining state. The resulting Turing machine $\hat{M}$ with $n - 1$ states is equivalent to M on a blank tape, since none of the modified transitions is ever reached in the course of the computation. We apply Lemma 1 to $\hat{M}$ resulting in a machine M' with a new state s' having activity$(M') >$ activity(M) and productivity$(M') >$ productivity(M). Since the construction for Lemma 1 preserves the number of halting transitions, it suffices to add at most one halting transition on the new symbol m for each of the $n - 1$ states different from s' in order to transform M' into a k-halting machine. The remaining non-halting transitions on m can be arbitrary. These additional transitions will not influence the computation, because symbol m is never written onto the tape. In the following we let $N_{m,k} \geq m$.

The next normalization of M is the observation from [1] that in its computation on a blank tape "new" states (states not previously visited) appear in increasing order, i.e., the first state visited and not in the set $\{1, \ldots, s\}$ is $s+1$. This can be achieved by renaming the states appearing in the unique computation of M on a blank tape. A transition followed when a state s is first arrived at is called *special*, all other transitions are *ordinary*. Targets of special transitions can be omitted from a description of M, as long as there is a flag indicating whether a transition is special. We further note that the number of special transitions is exactly n, since by the normalization above all states (including the halt state) are reached.

Finally halting transitions (except the one appearing in the halting computation) are modified, such that they target another state. Obviously this does not influence the computation.

After these transformations, M can be described by the following information:

1. The number $n-1$ in a self-delimiting binary notation, using at most $2\lceil \log_2 n \rceil$ bits.
2. An array containing $m(\lceil \log_2 m \rceil + 2)$ bits for every state $i \in \{1, \ldots, n\}$. These bits correspond to the components (symbol written, head movement, and next state) of a row of the transition table encoding all transitions from a state. The next state is replaced by a flag that is 1 if and only if the transition is special.
3. A list of $n(m-1)$ destinations of ordinary transitions. The list is sorted according to their first appearance in the computation on a blank tape. A destination can be encoded in $\lceil \log_2 n \rceil$ bits, since the halting transition is always special and by the transformation above the halting state does not appear in another transition.

In summary, the description of M requires $nm\lceil \log_2 n \rceil - n\lceil \log_2 n \rceil + cn$ bits for some constant c if m is fixed.

Next we consider the information content of n' states acting as a ROM in the finite control of a Turing machine with $m+1$ symbols. By the technique of introspective computing [1] generalized to $m+1$ tape symbols, $n'm\lfloor \log_2 n' \rfloor$ bits can be extracted from these states by a fixed extractor machine E with n_E states. We will briefly outline how this can be done. To each pair of state q among the n' states of the ROM and symbol α (except for the blank) a binary string $R_{q,\alpha}$ of length $\lfloor \log_2 n' \rfloor$ is assigned and the numerical value of this string is denoted by $r_{q,\alpha}$. A base state b for the ROM is chosen and we assume that $b \leq q < b + n'$. Next we include into the finite control of the simulator M' the following transitions for the n' states acting as a ROM:

- On the blank state q goes to $q+1$, writes a blank, and moves the head to the right.
- On a non-blank symbol α state q goes to $b + n' - r_{q,\alpha}$, writes α, and moves the head to the right.

In order to extract $R_{q,\alpha}$, simulator M' sets up a tape segment $0^{q-1}\alpha 0^{n'}$ and starts a computation in state b on the first symbol. Notice that the transitions defined above move the head to the right until the α in this segment is encountered, which happens in state q. Then state $b + n' - r_{q,\alpha}$ is entered and another $r_{q,\alpha}$ steps are carried out before state $b + n'$ is entered, which is not part of the portion of M' acting as a ROM. In state $b + n'$ the distance between the head and symbol α to the left determines $r_{q,\alpha}$ and thus the string $R_{q,\alpha}$. All these strings are in turn extracted and recorded by M' on its tape.

The extracted bits can be processed by a universal Turing machine U having n_U states and simulating machines with m symbols step by step. As opposed to usual simulators, we let U write an extra non-blank symbol after it has reached the halting transition of the machine being simulated (notice that this will make sure that activity as well as productivity increase in comparison to M). A further specific requirement is that U keeps track of the first appearance of a state and finalizes the transition table according to the flags while simulating a machine. Finally an ordinary universal Turing machine would have exactly one halting transition. In order to meet the requirements of Harland's conjecture we add a sufficient number of (unreachable) states to accommodate $k - 1$ additional halting transitions.

We let $d = n_E + n_U$, $n' = n - d$ and observe that $n'm\lfloor \log_2 n' \rfloor = (n - d)m\lfloor \log_2(n-d) \rfloor \geq (n-d)m(\lfloor \log_2 n \rfloor - 1) \geq nm\lfloor \log_2 n \rfloor - dm\lfloor \log_2 n \rfloor - nm + dm \geq nm\lceil \log_2 n \rceil - dm\lfloor \log_2 n \rfloor - 2nm + dm \geq nm\lceil \log_2 n \rceil - n\lceil \log_2 n \rceil + cn$ for $n \geq N_{m,k}$ with a sufficiently large $N_{m,k}$. Therefore n' states suffice to encode M.

Finally we compose the Turing machine over $m + 1$ symbols with n' states encoding machine M, the extractor E, and the universal Turing machine U to obtain machine M' with n states simulating M and satisfying the theorem.   □

Next we consider weaker versions of Harland's conjecture. But first we show some technical lemmas.

Lemma 2. *For all $n, m \geq 2$ we have $S(n, m) > n$*

Proof. We have $S(2, 2) = 6 > 2$. Suppose $S(n, 2) > n$ for some $n \geq 2$. By Lemma 1 we get $S(n + 1, 2) > n + 1$ and $S(n, m) \geq S(n, 2) > n$ for $m \geq 3$ by adding transitions on $m - 2$ symbols for a two-symbol champion.   □

Lemma 3. *If all transitions of Turing machine M with n states on the blank move the head in the same direction and M has finite activity, then we have activity$(M) \leq n$.*

Proof. If M makes more than n steps in one direction, then a state repeats and M does not stop.   □

The next result is inspired by the construction in Fig. 14 of [5]. In contrast to Theorem 1 it does not preserve the number of halting transitions.

Theorem 2. *For every Turing machine M with $n \geq 2$ states and two symbols having finite activity there is an n-state, three-symbol Turing machine M' with finite activity such that activity$(M') >$ activity(M).*

Proof. Without loss of generality M has maximum activity among all n state, two-symbol Turing machines and the first transition of M moves the head to the right.

We let M' have the basic structure of M and add transitions on the new (third) symbol to every state. For a state s to be determined below this transition is halting, while the other transitions are non-halting and can otherwise be arbitrary, since they will never be used.

Consider the tape cell i at the final position of the head in the computation of M on a blank tape. We modify the halting transition taken by M to write the new symbol and move the head depending on the symbols in neighboring cells $i - 1$ (left) and $i + 1$ (right) of i.

If cell $i - 1$ contains a blank, we modify the halting transition to move left and go to the initial state. By the normalization of the first transition, M' will move right on the blank (it cannot halt due to Lemma 2) to a state which is chosen as s. Then M halts on the new symbol increasing activity by two.

If cell $i - 1$ contains 1 and there is a state with a transition moving right on 1, we modify the last transition to move left and go to such a state. This will increase activity by one if the transition moving right on 1 is halting, in which case we choose the current state as s. Otherwise activity increases by two as in the previous case if M' returns to cell i in a state chosen as s.

If all transitions move left on 1, we consider tape cell $i + 1$. If it contains 1, we modify the halting transition to move right and go to an arbitrary state. Machine M' will either halt immediately or return to cell i in a state chosen as s and halt.

Finally consider a blank in cell $i + 1$. Since for all $n \geq 2$ there is a machine with activity exceeding n by Lemma 2, we conclude from Lemma 3 that at least one transition moves the head left on a blank. Go to a state with such a transition and move the head to the right. The resulting Turing machine will halt either when reading cell $i + 1$ or when it returns to cell i in a state chosen as s.

In each case we have activity$(M') \in \{\text{activity}(M) + 1, \text{activity}(M) + 2\}$ and activity$(M') > \text{activity}(M)$. $\qquad\qquad\qquad\qquad\qquad\qquad\qquad\qquad\qquad\qquad\qquad\qquad$ $\square$

Next we turn to constructions that increase the alphabet by more than one symbol.

Theorem 3. *For every Turing machine M with $n \geq 2$ states and $m \geq 2$ symbols having finite activity there is a 2-state, $(4nm + 5m)$-symbol Turing machine M' with finite activity such that* activity$(M') > $ activity(M) *and* productivity$(M') > $ productivity(M).

Proof. Let M be a Turing machine with n states and m symbols. By Lemma 1 there is a machine M' with $n + 1$ states and m symbols increasing activity and productivity. The classical construction from [14] transforms it into an equivalent 2-state machine with $4m(n + 1) + m = 4nm + 5m$ symbols. $\qquad\qquad\qquad$ $\square$

Theorem 4. *For every Turing machine M with $n \geq 2$ states and $m \geq 3$ symbols having finite activity there is an n-state, $3m$-symbol Turing machine M' with finite activity such that* activity$(M') > $ activity(M).

Proof. If among the Turing machines with n states and m symbols M does not have maximum activity, we choose as M' such a machine and no increase of the tape alphabet is necessary.

Otherwise for every symbol a of M we add new symbols a_L and a_R to the transition table of M'. A transition of M on an old symbol writing a is modified to write a_R if it moves the head to the left (indicating that a_R is to the right of the tape head) and similarly a_L if it moves the head to the right. On new symbols a_R and a_L machine M' replaces the new symbol with a and "bounces" back to the right if the symbol was a_L and to the left on a_R. Observe that all symbols with subscript L are to the left of the tape head or under it and all symbols with subscript R are to the right of the tape head or under it in the course of the computation of M'.

Consider the homomorphism h defined by $h(a) = h(a_L) = h(a_R) = a$ for every symbol a of M. We claim that for every instantaneous description of M at step k with a tape inscription w of cells visited by M and its head on cell i there is an instantaneous description of M' at step $k' \geq k$ with a tape inscription w' satisfying $h(w') = w$ with its head on cell i. This clearly holds for step 0 when there are no modified cells. If M' reads an old symbol a it writes some b_R or b_L while M writes b and both move their heads in the same direction. This clearly maintains the property $h(w') = w$ and that the head positions correspond. If M' reads a_L it has just moved its head left and the neighboring cell contains some symbol b_R. Now M' writes a, moves right, replaces b_R with b, and returns to a. In comparison to M, two additional steps have been performed while $h(a_L b_R) = ab$. In a similar way M' behaves on a_R. We conclude that M' halts if and only M does and activity$(M') \geq$ activity(M).

To see that activity$(M') >$ activity(M) we make use of the assumption that M has maximum activity among the Turing machines with n states and m symbols. By Lemmas 2 and 3 machine M' has to make at least one turn, which adds at least two steps to the computation of M' in comparison to M.   □

3   Discussion

We have partially proved Harland's conjecture. It holds for n sufficiently large and (restricted to the measure activity and without maintaining the number of halting transitions) for $m = 2$. An increase of the alphabet size exceeding one admits similar results for all n. In the former construction we have used the technique of interpretation instead of instrumentation (in terms of [1]).

If Harland's conjecture is true in general, it provides further evidence for the symmetry of symbols and states discussed by Shannon in the concluding remarks of [14], since an increase in one of the parameters adds power to the machines.

Acknowledgements. Thanks are due to the reviewers for suggesting several improvements of the presentation.

References

1. Ben-Amram, A.M., Petersen, H.: Improved bounds for functions related to Busy Beavers. Theory Comput. Syst. **35**(1), 1–11 (2002)
2. Brady, A.H.: The determination of the value of Rado's noncomputable function $\Sigma(k)$ for four-state Turing machines. Math. Comput. **40**(162), 647–665 (1983)
3. Brady, A.H.: The Busy Beaver game and the meaning of life. In: Herken, R. (ed.) The Universal Turing Machine: A Half-Century Survey, 2nd Edition, pp. 237–254. Springer, Heidelberg (1995)
4. Chaitin, G.: Computing the Busy Beaver function. In: Cover, T.M., Gopinath, B. (eds.) Open Problems in Communication and Computation, pp. 108–112. Springer, Heidelberg (1987). doi:10.1007/978-1-4612-4808-8_28
5. Harland, J.: Generating Candidate Busy Beaver Machines (Or How to Build the Zany Zoo). https://arxiv.org/abs/1610.03184v1 (2016)
6. Hopcroft, J.E.: Turing machines. Sci. Am. **250**(5), 70–80 (1984)
7. Korfhage, R.R.: Logic and Algorithms: With Applications to the Computer and Information Sciences. Wiley, New York (1966)
8. Lin, S., Rado, T.: Computer studies of Turing machine problems. J. Assoc. Comput. Mach. **12**(2), 196–212 (1965)
9. Marxen, H., Buntrock, J.: Attacking the Busy Beaver 5. Bull. Eur. Assoc. Theoret. Comput. Sci. (EATCS) **40**, 247–251 (1990)
10. Marxen, H.: Currently Known Results. (Download 25 April 2017). http://www.drb.insel.de/~heiner/BB
11. Michel, P.: Small Turing machines and generalized Busy Beaver competition. Theoret. Comput. Sci. **326**(1–3), 45–56 (2004)
12. Michel, P.: The Busy Beaver Competition: A Historical Survey. https://arxiv.org/abs/0906.3749v4 (2016)
13. Rado, T.: On non-computable functions. Bell Syst. Tech. J. **41**, 877–884 (1962)
14. Shannon, C.E.: A universal Turing machine with two internal states. In: Shannon, C.E., McCarthy, J. (eds.) Automata Studies (AM-34), pp. 157–166. Princeton University Press (1956)
15. Yedidia, A., Aaronson, S.: A relatively small Turing machine whose behavior is independent of set theory. Complex Syst. **25**(4) (2016). http://www.complex-systems.com/issues/25-4.html

Automatic Kolmogorov Complexity and Normality Revisited

Alexander Shen[⊠]

LIRMM CNRS, University of Montpellier, Montpellier, France
alexander.shen@lirmm.fr

Abstract. It is well known that normality (all factors of a given length appear in an infinite sequence with the same frequency) can be described as incompressibility via finite automata. Still the statement and the proof of this result as given by Becher and Heiber (2013) in terms of "lossless finite-state compressors" do not follow the standard scheme of Kolmogorov complexity definition (an automaton is used for compression, not decompression). We modify this approach to make it more similar to the traditional Kolmogorov complexity theory (and simpler) by explicitly defining the notion of automatic Kolmogorov complexity and using its simple properties. Other known notions (Shallit and Wang [15], Calude et al. [8]) of description complexity related to finite automata are discussed (see the last section).

As a byproduct, this approach provides simple proofs of classical results about normality (equivalence of definitions with aligned occurrences and all occurrences, Wall's theorem saying that a normal number remains normal when multiplied by a rational number, and Agafonov's result saying that normality is preserved by automatic selection rules).

1 Introduction

What is an individual random object? When could we believe, looking at an infinite sequence α of zeros and ones, that α was obtained by tossing a fair coin? The minimal requirement is that zeros and ones appear "equally often" in α: both have limit frequency $1/2$. Moreover, it is natural to require that all 2^k bit blocks of length k appear equally often. Sequences that have this property are called *normal* (see the exact definition in Sect. 3; a historic account can be found in [4,6]).

Intuitively, a reasonable definition of an individual random sequence should require much more than just normality; the corresponding notions are studied in the algorithmic randomness theory (see [9,13] for the detailed exposition, [17] for a textbook and [16] for a short survey). The most popular definition is called *Martin-Löf randomness*; the classical Schnorr–Levin theorem says that this notion is equivalent to *incompressibility*: a sequence α is Martin-Löf random

---

A. Shen—Supported by ANR-15-CE40-0016-01 RaCAF grant.

On leave from IITP RAS, Moscow, Russia.

R. Klasing and M. Zeitoun (Eds.): FCT 2017, LNCS 10472, pp. 418–430, 2017.

DOI: 10.1007/978-3-662-55751-8_33

if an only if prefixes of α are incompressible (do not have short descriptions). See again [9, 13, 16, 17] for exact definitions and proofs.

It is natural to expect that normality, being a weak randomness property, corresponds to some weak incompressibility property. The connection between normality and finite-state computations was noticed long ago, as the title of [1] shows. This connection led to a characterization of normality as "finite-state incompressibility" (see [4]). However, the notion of incompressibility that was used in [4] does not fit well the general framework of Kolmogorov complexity (finite automata are considered as *compressors*, while in the usual definition of Kolmogorov complexity we restrict the class of allowed *decompressors*).

In this paper we give a definition of automatic Kolmogorov complexity that restricts the class of allowed decompressors and is suitable for the characterization of normal sequences as incompressible ones. This definition and its properties are considered in Sect. 2. In Sect. 3 we recall the notion of normal sequence. Then in Sect. 4 we provide a characterization of normal sequences in terms of automatic Kolmogorov complexity. In Sect. 5 we show how this characterization can be used to give simple proofs for classical results about normality, including Wall's theorem (normal numbers remain normal when multiplied by a rational factor). In a similar way one can prove Agafonov's result [1], but we omit the proof due to space restrictions (see the arxiv version of this paper). Finally, in Sect. 7 we compare our definition of automatic complexity with other similar notions.

2   Automatic Kolmogorov Complexity

Let us recall the definition of algorithmic (Kolmogorov) complexity. It is usually defined in the following way: $C(x)$, the complexity of an object x, is the minimal length of its "description". (We assume that both objects and descriptions are binary strings; the set of binary strings is denoted by $\mathbb{B}^*$, where $\mathbb{B} = \{0, 1\}$.) Of course, this definition makes sense only after we explain which type of "descriptions" we consider, but most versions of Kolmogorov complexity can be described according to this scheme [19]:

Definition 1. *Let $D \subset \mathbb{B}^* \times \mathbb{B}^*$ be a binary relation; we read $(p, x) \in D$ as "p is a D-description of x". Then complexity function C_D is defined as*

$$C_D(x) = \min\{|p| : (p, x) \in D\},$$

i.e., as the minimal length of a D-description of x.

Here $|p|$ stands for the length of a binary string p and $\min(\varnothing) = +\infty$, as usual. We say that D is a *description mode* and $C_D(x)$ is the *complexity of x with respect to the description mode D*.

We get the original version of Kolmogorov complexity ("plain complexity") if we consider all computable functions as description modes, i.e., if we consider relations $D_f = \{(p, f(p))\}$ for arbitrary computable partial functions f

as description modes. Equivalently, we may say that we consider (computably) enumerable relations D that are graphs of functions (for every p there exists at most one x such that $(p, x) \in D$; each description describes at most one object). Then the Kolmogorov–Solomonoff optimality theorem says that there exists an optimal D in this class that makes C_D minimal (up to an $O(1)$ additive term). (We assume that the reader is familiar with basic properties of Kolmogorov complexity, see, e.g., [11, 17]; for a short introduction see also [16].)

Note that we could get a trivial C_D if we take, e.g., the set of all pairs as a description mode D (in this case all strings have complexity zero, since the empty string describes all of them). So we should be careful and do not consider description modes where the same string describes too many objects.

To define our class of descriptions, let us first recall some basic notions related to finite automata. Let A and B be two finite alphabets. Consider a directed graph G whose edges are labeled by pairs (a, b) of letters (from A and B respectively). We also allow pairs of the form (a, ε), (ε, b), and $(\varepsilon, \varepsilon)$ where ε is a special symbol (not in A or B) that informally means "no letter". For such a graph, consider all directed paths in it (no restriction on starting or final points), and for each path concatenate all the first and all the second components of the pairs; ε is replaced by an empty word. For each path we get some pair (u, v) where $u \in A^*$ and $v \in B^*$ (i.e., u and v are words over alphabets A and B). Consider all pairs that can be read in this way along all paths in G. For each labeled graph G we obtain a relation (set of pairs) R_G that is a subset of $A^* \times B^*$. For the purposes of this paper, we call the relations obtained in this way "automatic". This notion is similar to *rational relations* defined by transducers [5, Sect. 3.6]. The difference is that we do not fix initial/finite states (so every subpath of a valid path is also valid) and that we do not allow arbitrary words as labels, only letters and ε. (This will be important, e.g., for the statement (j) of Theorem 1.)

Definition 2. *A relation $R \subset A^* \times B^*$ is* automatic *if there exists a labeled graph (automaton) G such that $R = R_G$.*

Now we define automatic description modes as automatic relations where each string describes at most $O(1)$ objects:

Definition 3. *A relation $D \subset \mathbb{B}^* \times \mathbb{B}^*$ is an* automatic description mode *if*

– *D is automatic in the sense of Definition 2;*
– *D is a graph of an $O(1)$-valued function: there exists some constant c such that for each p there are at most c values of x such that $(p, x) \in D$.*

For every automatic description mode D we consider the corresponding complexity function C_D. There is no optimal mode D that makes C_D minimal (see Theorem 1 below). So, stating some properties of complexity, we need to mention D explicitly and say something like "for every automatic description mode D there exists another automatic description mode D' such that...", and then make a statement that involves both C_D and $C_{D'}$. (A similar approach is needed when we try to adapt inequalities for Kolmogorov complexity to the case of resource-bounded complexities.)

Let us first mention some basic properties of automatic description modes.

Proposition 1.

(a) *The union of two automatic description modes is an automatic description mode.*

(b) *The composition of two automatic description modes is an automatic description mode.*

(c) *If D is a description mode, then $\{(p, x0)\colon (p, x) \in D\}$ is a description mode (here $x0$ is the binary string x with 0 appended); the same is true for $x1$ instead of $x0$.*

Proof. There are two requirements for an automatic description mode: (1) the relation is automatic and (2) the number of images is bounded. The second one is obvious in all three cases. The first one can be proven by a standard argument (see, e.g., [5, Theorem 4.4]) that we reproduce for completeness.

(a)   The union of two relations R_G and R'_G for two automata G and G' corresponds to an automaton that is a disjoint union of G and G'.

(b)   Let S and T be automatic relations that correspond to automata K and L. Consider a new graph that has set of vertices $K \times L$. (Here we denote an automaton and the set of vertices of its underlying graph by the same letter.)

– If an edge $k \to k'$ with a label (a, ε) exists in K, then the new graph has edges $(k, l) \to (k', l)$ for all $l \in L$; all these edges have the same label (a, ε).

– In the same way an edge $l \to l'$ with a label (ε, c) in L causes edges $(k, l) \to (k, l')$ in the new graph for all k; all these edges have the same label (ε, c).

– Finally, if K has an edge $k \to k'$ labeled (a, b) and at the same time L has an edge $l \to l'$ labeled (b, c), where b is the same letter, then we add an edge $(k, l) \to (k', l')$ labeled (a, c) in the new graph.

Any path in the new graph is projected into two paths in K and L. Let (p, q) and (u, v) be the pairs of words that can be read along these projected paths in K and L respectively, so $(p, q) \in S$ and $(u, v) \in T$. The construction of the graph $K \times L$ guarantees that $q = u$ and that we read (p, v) in the new graph along the path. So every pair (p, v) of strings that can be read in the new graph belongs to the composition of S and T.

On the other hand, assume that (p, v) belong to the composition, i.e., there exists q such that (p, q) can be read along some path in K and (q, v) can be read along some path in L. Then the same word q appears in the second components in the first path and in the first components in the second path. If we align the two paths in such a way that the letters of q appear at the same time, we get a valid transition of the third type for each letter of q. Then we complete the path by adding transitions inbetween the synchronized ones (interleaving them in arbitrary way); all these transitions exist in the new graph by construction.

(c)   We add an additional outgoing edge labeled $(\varepsilon, 0)$ for each vertex of the graph; all these edges go to a special vertex that has no outgoing edges.

Remark 1. Given a graph, one can check in polynomial time whether the corresponding relation is $O(1)$-valued [21, Theorem 5.3, p. 777].

Now we are ready to prove the following simple result about the properties of *automatic Kolmogorov complexity* functions, i.e., of functions C_R where R is some automatic description mode.

Theorem 1 (Basic properties of automatic Kolmogorov complexity).

(a) *There exists an automatic description mode R such that $C_R(x) \leq |x|$ for all strings x.*

(b) *For every automatic description mode R there exists some automatic description mode R' such that $C_{R'}(x0) \leq C_R(x)$ and $C_{R'}(x1) \leq C_R(x)$ for all x.*

(c) *For every automatic description mode R there exists some automatic description mode R' such that $C_{R'}(\bar{x}) \leq C_R(x)$, where $\bar{x}$ stands for the reversed x.*

(d) *For every automatic description mode R there exists some constant c such that $C(x) \leq C_R(x)+c$. (Here C stands for the plain Kolmogorov complexity.)*

(e) *For every $c > 0$ there exists an automatic description mode R such that $C_R(1^n) \leq n/c$ for all n.*

(f) *For every automatic description mode R there exists some $c > 0$ such that $C_R(1^n) \geq n/c - 1$ for all n.*

(g) *For every two automatic description modes R_1 and R_2 there exists an automatic description mode R such that $C_R(x) \leq C_{R_1}(x)$ and $C_R(x) \leq C_{R_2}(x)$ for all x.*

(h) *There is no optimal automatic description mode. (A mode R is called optimal in some class if for every mode R' in this class there exists some c such that $C_R(x) \leq C_{R'}(x) + c$ for all strings x.)*

(i) *For every automatic description mode R, if x' is a substring of x, then $C_R(x') \leq C_R(x)$.*

(j) *Moreover, $C_R(xy) \geq C_R(x) + C_R(y)$ for every two strings x and y.*

(k) *For every automatic description mode R and for every constant $\varepsilon > 0$ there exists an automatic description mode R' such that $C_{R'}(xy) \leq (1+\varepsilon)C_R(x) + C_R(y)$ for all strings x and y.*

(l) *Let S be an automatic description mode. Then for every automatic description mode R there exists an automatic description mode R' such that $C_{R'}(y) \leq C_R(x)$ for every $(x,y) \in S$.*

(m) *If we allow a bigger alphabet B instead of $\mathbb{B}$ as an alphabet for descriptions, we divide the complexity by $\log |B|$, up to a constant factor that can be chosen arbitrarily close to 1.*

Proof.

(a) Consider an identity relation as a description mode; it corresponds to an automaton with one state.

(b) This is a direct corollary of Proposition 1(c).

(c) The definition of an automaton is symmetric (all edges can be reversed), and the $O(1)$-condition still holds.

(d) Let R be an automatic description mode. An automaton defines a decidable (computable) relation, so R is decidable. Since R defines a $O(1)$-valued function, a Kolmogorov description of some y that consists of its R-description x and the ordinal number of y among all strings that are in R-relation to x, is only $O(1)$ bits longer than x.

(e) Consider an automaton that consists of a cycle where it reads one input symbol 1 and then produces c output symbols 1. (Since we consider the relation as an $O(1)$-multivalued function, we sometimes consider the first components of pairs as "input symbols" and the second components as "output symbols".) Recall that there is no restrictions on initial and finite states, so this automaton produces all pairs $(1^k, 1^l)$ where $(k-1)c \le l \le (k+1)c$.

(f) Consider an arbitrary description mode, i.e., an automaton that defines some $O(1)$-valued relation. Then every cycle in the automaton that produces some output letter should also produce some input letter, otherwise an empty input string corresponds to infinitely many output strings. For any sufficiently long path in the graph we can cut away a minimal cycle, removing at least one input letter and at most c output letters, where c is the number of states, until we get a path of length less than c.

(g) This follows from Proposition 1(a).

(h) This statement is a direct consequence of (e) and (f). Note that for finitely many automatic description modes there is a mode that is better than all of them, as (g) shows, but we cannot do the same for all description modes (as was the case for Kolmogorov complexity).

(i) If R is a description mode, (p, x) belongs to R and x' is a substring of x, then there exists some substring p' of p such that $(p', x') \in R$. Indeed, we may consider the input symbols used while producing x'.

(j) Note that in the previous argument we can choose disjoint p' for disjoint x'.

(k) Informally, we modify the description mode as follows: a fixed fraction of input symbols is used to indicate when a description of x ends and a description of y begins. More formally, let R be an automatic description mode; we use the same notation R for the corresponding automaton. Consider $N+1$ copies of R (called 0-, 1-,..., Nth layers). The outgoing edges from the vertices of ith layer that contain an input symbol are redirected to $(i+1)$th layer (the new state remains the same, only the layer changes, so the layer number counts the input length). The edges with no input symbol are left unchanged (and go to ith layer as before). The edges from the Nth layer are of two types: for each vertex x there is an edge with label $(0, \varepsilon)$ that goes to the same vertex in 0th layer, and edges with labels $(1, \varepsilon)$ that connect each vertex of Nth layer to all vertices of an additional copy of R (so we have $N+2$ copies in total). If both x and y can be read (as outputs) along the edges of R, then xy can be read, too (additional zeros should be added to the input string after groups of N input symbols). We switch from x to y using the edge that goes from Nth layer to the additional copy of R (using additional symbol 1 in the input string). The overhead in the description is one symbol per every N input symbols used to describe x. We get the required bound, since N can be arbitrarily large.

The only thing to check is that the new automaton is $O(1)$-valued. Indeed, the possible switch position (when we move to the states of the additional copy of R) is determined by the positions of the auxiliary bits modulo $N + 1$: when this position modulo $N + 1$ is fixed, we look for the first 1 among the auxiliary bits. This gives only a bounded factor $(N + 1)$ for the number of possible outputs that correspond to a given input.

(l) The composition $S \circ R$ is an automatic description mode due to Proposition 1.

(m) Take the composition of a given description mode R with a mode that provides block encoding of inputs. Note that block encoding can be implemented by an automaton. There is some overhead when $|B|$ is not a power of 2, but the corresponding factor becomes arbitrarily close to 1 if we use block code with large block size.

Remark 2. Not all these results are used in the sequel; we provide them for comparison with the properties of the standard Kolmogorov complexity function.

3 Normal Sequences and Numbers

Consider an infinite bit sequence $\alpha = a_0 a_1 a_2 \ldots$ and some integer $k \geq 1$. Split the sequence α into k-bit blocks: $\alpha = A_0 A_1 \ldots$. For every k-bit string r consider the limit frequence of r among the A_i, i.e. the limit of $\#\{i: i < N \text{ and } A_i = r\}/N$ as $N \to \infty$. This limit may exist or not; if it exists for some k and for all r, we get a probability distribution on k-bit strings.

Definition 4. *A sequence α is* normal *if for every number k and every string r of length k this limit exists and is equal to 2^{-k}.*

Sometimes sequences with these properties are called *strongly normal* while the name "normal" is reserved for sequences that have this property for $k = 1$.

There is a version of this definition that considers all occurences of some string r in α, not only aligned ones (whose starting point is a multiple of k). In this version we require that the limit of $\#\{i < N : \alpha_i \alpha_{i+1} \ldots \alpha_{i+k-1} = r\}/N$ equals 2^{-k} for all k and for all strings r of length k. A classical result (see, e.g., [12, Chap. 1, Sect. 8]) says that this is an equivalent notion, and we give below a simple proof of this equivalence using automatic complexity. Before this proof is given, we will distinguish the two definitions by using the name "non-aligned-normal" for the second version.

A real number is called *normal* if its binary expansion is normal (we ignore the integer part). If a number has two binary expansions, like $0.0111\ldots = 0.1000\ldots$, both expansions are not normal, so this is not a problem.

A classical example of a normal number is the *Champernowne number* [7]

$$0.0\,1\,10\,11\,100\,101\,110\,111\,1000\,1001\ldots$$

(the concatenation of all positive integers in binary). Let us sketch the proof of its normality (not used in the sequel) using the non-aligned version of normality definition. All N-bit numbers in the Champernowne sequence form a block that starts with 10^{N-1} and ends with 1^N. Note that every string of length $k \ll N$ appears in this block with probability close to 2^{-k}, since each of 2^{N-1} strings (after the leading 1 for the N-bit numbers in the Champernowne sequence) appears exactly once. The deviation is caused by the leading 1's and also by the boundaries between the consecutive N-bit numbers where the k-bit substrings are out of control. Still the deviation is small since $k \ll N$.

This is not enough to conclude that C is (non-aligned) normal, since the definition speaks about frequencies in all prefixes; the prefixes that end on a boundary between two blocks are not enough. The problem appears because the size of a block is comparable to the length of the prefix before it. To deal with arbitrary prefixes, let us note that if we ignore *two* leading digits in each number (first 10 and then 11) instead of one, the rest is periodic in the block (the block consists of two periods). If we ignore three leading digits, the block consists of four periods, etc. An arbitrary prefix is then close to the boundary between these sub-blocks, and the distance can be made small compared to the total length of the prefix. (End of the proof sketch.)

The definition of normality can be given for an arbitrary alphabet (instead of the binary one), and we get the notion of b-*normality* of a real number for every base $b \geq 2$. It is known that for different bases we get non-equivalent notions (a rather difficult result). The numbers in $[0, 1]$ that are normal for every base are called *absolutely normal*. Their existence can be proved by a probabilistic argument. For every base b, almost all reals are b-normal (the non-normal numbers have Lebesgue measure 0); this is guaranteed by the Strong Law of Large Numbers. Therefore the numbers that are not absolutely normal form a null set (a countable union of the null sets for each b). The constructive version of this argument shows that there exist computable absolutely normal numbers. This result goes back to an unpublished note of Turing (1938, see [2]).

In the next section we prove the connection between normality and automatic complexity: a sequence α is normal if for every automatic description mode D the corresponding complexity C_D of its prefix never becomes much smaller than the length of this prefix.

4   Normality and Incompressibility

Theorem 2. *A sequence $\alpha = a_0 a_1 a_2 \ldots$ is normal if and only if*

$$\liminf_{n \to \infty} \frac{C_R(a_0 a_1 \ldots a_{n-1})}{n} \geq 1$$

for every automatic description mode R.

Proof. First, let us show that a sequence that is not normal is compressible. Assume that for some bit sequence α and for some k the requirement for aligned

k-bit blocks is not satisfied. Using a compactness argument, we can find a sequence of lengths N_i such that for the prefixes of these lengths the frequencies of k-bit blocks do converge to some probability distribution A on $\mathbb{B}^k$, but this distribution is not uniform. Then its Shannon entropy $H(A)$ is less than k.

The Shannon theorem can then be used to construct a block code of average length close to $H(A)$, namely, of length at most $H(A) + 1$ (this "+1" overhead is due to rounding if the frequencies are not powers of 2). Since this code can be easily converted into an automatic description mode, it will give the desired result if $H(A) < k - 1$. It remains to show that it is the case for long enough blocks.

Selecting a subsequence, we may assume without loss of generality that the limit frequencies exist also for (aligned) $2k$-bit blocks, so we get a random variable $A_1 A_2$ whose values are $2k$-bit blocks (and A_1 and A_2 are their first and second halves of length k). The variables A_1 and A_2 may be dependent, and their distributions may differ from the initial distribution A for k-bit blocks. Still we know that A is the average of A_1 and A_2 (since A is computed for all blocks, and A_1 [resp. A_2] corresponds to odd [resp. even] blocks). A convexity argument (the function $p \mapsto -p \log p$ used in the definition of entropy has negative second derivative) shows that $H(A) \geq [H(A_1) + H(A_2)]/2$. Then

$$H(A_1 A_2) \leq H(A_1) + H(A_2) \leq 2H(A),$$

so $A_1 A_2$ has twice bigger difference between entropy and length (at least). Repeating this argument, we can find k such that the difference between length and entropy is greater than 1. This finishes the proof in one direction.

Now we need to prove that every normal sequence α is incompressible. Let R be an arbitrary automatic description mode. Consider some k and split the sequence into k-bit blocks: $\alpha = A_0 A_1 A_2 \ldots$. (Now A_i are just the blocks in α, not random variables.) We will show that

$$\liminf_{n \to \infty} C_R(A_0 A_1 \ldots A_{n-1})/nk$$

cannot be much smaller than 1. More precisely, we will show that

$$\liminf_{n \to \infty} \frac{C_R(A_0 A_1 \ldots A_{n-1})}{nk} \geq 1 - \frac{O(1)}{k},$$

where the constant in $O(1)$ does not depend on k. This will be enough: note that (i) we may consider only prefixes whose length is a multiple of k, because adding the last incomplete block can only increase the complexity and the change in length is negligible, and (ii) the value of k may be arbitrarily large.

Now let us prove this bound for some fixed k. Recall that

$$C_R(A_0 A_1 \ldots A_{n-1}) \geq C_R(A_0) + C_R(A_1) + \ldots + C_R(A_{n-1})$$

and that $C(x) \leq C_R(x) + O(1)$ for all x and some $O(1)$-constant that depends only on R (Theorem 1). By assumption, all k-bit strings appear with the same limit frequency among $A_0, A_1, \ldots, A_{n-1}$. It remains to note that the average

Kolmogorov complexity $C(x)$ of all k-bit strings is $k - O(1)$; indeed, the fraction of k-bit strings that can be compressed by more than d bits ($C(x) < k - d$) is at most 2^{-d}, and the series $\sum d2^{-d}$ (the upper bound for the average number of bits saved by compression) has finite sum.

A small modification of this proof adapts it to the non-aligned definition of normality. Let α be a sequence that is not normal in the non-aligned version. This means that for some k the k-bit blocks do not have a correct limit distribution (non-aligned). These blocks can be split into k groups according to their starting positions modulo k. In one of the groups blocks do not have a correct limit distribution (otherwise the average distribution would be correct, too). So we can delete some prefix (less than k symbols) of our sequence and get a sequence that is not normal in the aligned sense. Its prefixes are compressible (as we have seen). The same is true for the original sequence since adding a fixed finite prefix (or suffix) changes complexity at most by $O(1)$.

In the other direction: let us assume that the sequence is normal in the non-aligned sense. The aligned frequency of some compressible-by-d-bits block (as well as any other block) can be only k times bigger than its non-aligned frequency, which is exponentially small in d (the number of saved bits), so we can choose the parameters to get the required bound.

Indeed, let us consider blocks of length k whose C_R-complexity is smaller than $k - d$. Their Kolmogorov complexity is then smaller than $k - d + O(1)$, and the fraction of these blocks (among all k-bit strings) is at most $2^{-d+O(1)}$. So their frequency among aligned blocks is at most $2^{-d+O(1)} \cdot k$. For all other blocks R-compression saves at most d bits, and for compressible blocks it saves at most k bits, so the average number of saved bits (per k-bit block) is bounded by

$$d + k2^{-d+O(1)} \cdot k = d + O(k^2 2^{-d}).$$

We need this bound to be $o(k)$, i.e., we need that

$$\frac{d}{k} + O(k2^{-d}) = o(1)$$

as $k \to \infty$. This can be achieved, for example, if $d = 2 \log k$.

In this way we get the following corollary:

Corollary 1. *The aligned and non-aligned definitions of normality are equivalent.*

Note also that adding/deleting a finite prefix does not change the compressibility, and, therefore, normality. (For the non-aligned version of the normality definition it is obvious anyway, but for the aligned version it is not so easy to see directly.)

Another corollary is a result proven by Piatetski-Shapiro in [18]: if for some c and for all k every k-bits block appears in a sequence with $\limsup$-frequency at most $c2^{-k}$, then the sequence is normal. Indeed, in the argument above we had a constant factor in $O(k2^{-d})$ anyway. (We can even allow the constant c to depend on k if its growth as a function of k is not too fast.)

5   Wall's Theorem

Now we obtain a known result about normal numbers (Wall's theorem) as an easy corollary. Recall that a real number is normal if its binary expansion is normal. We agreed to ignore the integer part (since it has only finitely many digits, adding it as a prefix would not matter anyway).

Theorem 3 (Wall [20]). *If p and q are rational numbers and α is normal, then $\alpha p + q$ is normal.*

Proof. It is enough to show that multiplication and division by an integer c preserve normality (note that adding an integer preserves it by definition, since the integer part is ignored). This fact follows from the incompressibility characterization (Theorem 2), the non-increase of complexity under automatic $O(1)$ mappings (Theorem 1(l)) and the following lemma:

Lemma 1. *Let c be an integer. Consider the relation R_c that consists of pairs of strings x and y such that x and y have the same length and can be prefixes of the binary expansions of the fractional parts of γ and $c\gamma$ for some real γ. This relation, as well as its inverse, is contained in an automatic description mode.*

Assuming Lemma 1, we conclude that the prefixes of γ and $c\gamma$ have the same automatic complexity. More precisely, for every automatic description mode R there exists another automatic description mode R' such that $C_{R'}(y) \leq C_R(x)$ if x and y are prefixes of γ and $c\gamma$ respectively. So if γ is compressible, then $c\gamma$ is also compressible. The same is true if we consider the inverse relation; if $c\gamma$ is compressible, then γ is also compressible.

It remains to prove Lemma 1. Indeed, the school division algorithm can be represented by an automaton; integer parts can be different, but this creates $O(1)$ different possible remainders. We have to take care of two representations of the same number (note that while dividing $0.29999\ldots$ by 3, we obtain only $0.09999\ldots$, not $0.10000\ldots$), but at most two representations are possible and the relation between them is automatic, so we still get an automatic description mode.

6   Pairs as Descriptions and Agafonov's Theorem

The incompressibility criterion for normality can also be used for an easy proof of Agafonov's theorem from [1]. This result says that an automatic selection rule (a term a_n of a sequence is selected or not depending on whether $a_0 \ldots a_{n-1}$ is accepted by a finite automaton), being applied to a normal sequence, selects either finite or normal sequence.

The idea of the proof: a sequence can be split into two: the selected subsequence and the rest. The selection process guarantees that the original sequence can be reconstructed from these two subsequences. If one of them (the selected one) is compressible, then this compression can be used to compress the prefixes of the original sequence (the unselected part remains unchanged, but the selected part is compressed). There are two technical points needed to implement this plan: first, one should prove that the selected subsequence has positive density (using the normality of the original sequence); second, one should generalize the notion of automatic complexity by using pairs as descriptions. Due to space restrictions, the details of this argument are omitted (see the arxiv version).

7   Discussion

The connection between normality and finite-state computations was noticed long ago, as the title of [1] shows; see also [14] where normality was related to martingales arising from finite automata. This connection led to a characterization of normality as incompressibility (see [4] for a direct proof). On the other hand, it was also clear that the notion of Kolmogorov complexity is not directly practical since it considers arbitrary algorithms as decompressors, and this makes it non-computable. So restricted classes of decompressors are of interest, and finite-state computations are a natural candidate for such a class.

Shallit and Wang [15] suggested to consider, for a given string x, the minimal number of states in an automaton that accepts x but not other strings of the same length. Later Hyde and Kjos-Hanssen [10] considered a similar notion using nondeterministic automata. The intrinsic problem of this approach is that it is not naturally "calibrated" in the following sense: measuring the information in bits, we would like to have about 2^n objects of complexity at most n.

Another (and "calibrated") approach was suggested by Calude et al. [8]: in their definition a deterministic transducer maps a description string to a string to be described, and the complexity of y is measured as the combination of the sizes of a transducer and an input string needed to produce y (the minimum over all transducers and all input strings producing y is taken). The size of the transducer is measured via some encoding, so the complexity function depends on the choice of this encoding. The open question posed in [8, Sect. 6] asks whether this notion of complexity can be used to characterize normality.

The incompressibility notion used in [4] provides such a characterization for a different definition. It uses deterministic transducers and requires additionally that for every output string y and every final state s there exists at most one input string that produces y and brings the automaton into the state s. Our approach is a refinement of this one: we consider non-deterministic automata without initial/final states and require only that decompressor is an $O(1)$-valued function. The proofs then become simpler, mainly for two reasons: (1) we use the comparison of the automatic complexity and the plain Kolmogorov complexity and apply standard results about Kolmogorov complexity; (2) we explicitly state and prove the property $C_R(xy) \geq C_R(x) + C_R(y)$ that is crucial for the proofs.

Acknowledgements. I am grateful to Veronica Becher, Olivier Carton and Paul Heiber for many discussions of their paper [3] and the relations between incompressibility and normality, and for the permission to use observations made during these discussions in the current paper. I am also grateful to my colleagues in LIRMM (ESCAPE team) and Moscow (Kolmogorov seminar, Computer Science Department of the HSE). I am thankful to the anonymous referees of an earlier version of this paper submitted to ICALP (and rejected) and to the anonymous referees of the final version.

References

1. Agafonov, V.N.: Normal sequences and finite automata. Doklady AN SSSR **179**, 255–256 (1968). See also the paper of V.N. Agafonov with the same name: Problemy Kibernetiki. vol. 20, pp. 123–129. Nauka, Moscow (1968)
2. Becher, V.: Turing's normal numbers: towards randomness. In: Cooper, S.B., Dawar, A., Löwe, B. (eds.) CiE 2012. LNCS, vol. 7318, pp. 35–45. Springer, Heidelberg (2012). doi:10.1007/978-3-642-30870-3_5
3. Becher, V., Carton, O., Heiber, P.: Finite-state independence, 12 November 2016. https://arxiv.org/pdf/1611.03921.pdf
4. Becher, V., Heiber, P.: Normal number and finite automata. Theoret. Comput. Sci. **477**, 109–116 (2013)
5. Berstel, J.: Transductions and Context-Free Languages. Vieweg+Teubner Verlag, Wiesbaden (1969). For a revised 2006–2009 version see the author's homepage. ISBN 978-3-519-02340-1. http://www-igm.univ-mlv.fr/~berstel
6. Bugeaud, Y.: Distribution Modulo One and Diophantine Approximation. Cambridge Tracts in Mathematics, vol. 193. Cambridge University Press, Cambridge (2012)
7. Champernowne, D.: The construction of decimals normal in the scale of ten. J. London Math. Soc. **8**(4), 254–260 (1933). (Received 19 April, read 27 April 1933)
8. Calude, C.S., Salomaa, K., Roblot, T.K.: Finite state complexity. Theoret. Comput. Sci. **412**, 5668–5677 (2011)
9. Downey, R.G., Hirschfeldt, D.R.: Algorithimic Randomness and Complexity, xxviii+855 p. Springer, New York (2010). ISBN 978-0-387-68441-3
10. Hyde, K.K., Kjos-Hanssen, B.: Nondeterministic complexity of overlap-free and almost square-free words. Electron. J. Comb. **22**, 3 (2015)
11. Li, M., Vitányi, P.: An Introduction to Kolmogorov Complexity and Its Applications, 3rd edn, pp. 1–792. Springer, New York (2008). ISBN 978-0-387-49820-1
12. Kuipers, L., Niederreiter, H.: Uniform Distribution of Sequences. Wiley, Hoboken (1974)
13. Nies, A.: Computability and Randomness. Oxford Logic Guides. Oxford University Press, Oxford (2009). ISBN 978-0199652600
14. Schnorr, C., Stimm, H.: Endliche Automaten und Zufallsfolgen. Acta Informatica **1**(4), 345–359 (1972)
15. Shallit, J., Wang, M.-W.: Automatic complexity of strings. J. Automata Lang. Comb. **6**(4), 537–554 (2001)
16. Shen, A.: Around Kolmogorov complexity: basic notions and results. In: Vovk, V., Papadopoulos, H., Gammerman, A. (eds.) Measures of Complexity, pp. 75–115. Springer, Cham (2015). doi:10.1007/978-3-319-21852-6_7. see also http://arxiv.org/abs/1504.04955
17. Shen, A., Uspensky, V.A., Vereshchagin, N.: Kolmogorov Complexity and Algorithmic Randomness. MCCME, Moscow (2013). (in Russian), English version accepted for publication by AMS. www.lirmm.fr/ashen/kolmbook-eng.pdf
18. Piatetski-Shapiro, I.I.: On the laws of distribution of the fractional parts of an exponential function. Izvestia Akademii Nauk SSSR Seriya Matematicheskaya **15**(1), 47–52 (1951). (in Russian)
19. Uspensky, V.A., Shen, A.: Relations between varieties of Kolmogorov complexities. Math. Syst. Theory **29**, 271–292 (1996)
20. Wall, D.D.: Normal numbers. Thesis, University of California (1949)
21. Weber, A.: On the valuedness of finite transducers. Acta Informatica **27**(8), 749–780 (1990)

Author Index

Anshu, Anurag 41
Atserias, Albert 56

Banerjee, Indranil 69
Bärtschi, Andreas 82
Bentert, Matthias 96
Bläser, Markus 111
Boros, Endre 123
Bousquet, Nicolas 136

Cabessa, Jérémie 150
Čepek, Ondřej 123
Colcombet, Thomas 3
Crochemore, Maxime 164

Daviaud, Laure 3
De Marco, Gianluca 177
Dereniowski, Dariusz 190

Eickmeyer, Kord 204
Engelfriet, Joost 217
Engels, Christian 230

Filiot, Emmanuel 243
Finkel, Olivier 150
Fluschnik, Till 96, 257

Ganty, Pierre 271
Grigoriev, Dima 284
Gutiérrez, Elena 271

Héliou, Alice 164
Høyer, Peter 41
Hromkovič, Juraj 11

Joglekar, Pushkar S. 298
Jurdziński, Tomasz 177, 312

Kanovich, Max 326
Kawarabayashi, Ken-ichi 204
Kolaitis, Phokion G. 56
Kucherov, Gregory 164
Kuznetsov, Stepan 326

Lingas, Andrzej 190
Löding, Christof 341

Makino, Kazuhisa 123
Maletti, Andreas 217
Manoussakis, George 355
Mary, Arnaud 136
Mazzocchi, Nicolas 243
Mhalla, Mehdi 41
Morik, Marco 257
Mouchard, Laurent 164
Muscholl, Anca 29

Nichterlein, André 96
Niedermeier, Rolf 96

Pagourtzis, Aris 367
Panagiotakos, Giorgos 367
Papadopoulos, Charis 381
Parreau, Aline 136
Perdrix, Simon 41, 395
Persson, Mia 190
Petersen, Holger 409
Pin, Jean-Éric 34
Pissis, Solon P. 164
Podolskii, Vladimir V. 284

Ramusat, Yann 164
Rao, B.V. Raghavendra 111, 230, 298
Raskin, Jean-François 243
Richards, Dana 69
Rossmanith, Peter 11
Różański, Michał 177, 312

Sakavalas, Dimitris 367
Sanselme, Luc 395
Sarma, Jayalal 111
Scedrov, Andre 326
Severini, Simone 56
Shen, Alexander 418
Sivakumar, Siddhartha 298
Sorge, Manuel 257
Spinrath, Christopher 341

Sreenivasaiah, Karteek   230
Stachowiak, Grzegorz   177

Tschager, Thomas   82
Tzimas, Spyridon   381

Urbańska, Dorota   190

Zuleger, Florian   3
Żyliński, Paweł   190